Biodiversity

Contents at a Glance

1. Living Resources
2. Importance of Biodiversity
3. Biodiversity and Conservation
4. Species Extinction
5. Population Ecology
6. Biodiversity in India
7. Species Diversity
8. Impact of Biotechnology on Biodiversity
9. Biotechnology and Agrobiodiversity
10. Impact of Genetically Engineered and Genetically Modified Crops on Biodiversity
11. Agrobiodiversity Informatics with Special Reference to Spices
12. Dry Land Biodiversity
13. Biodiversity in Grasslands
14. Effects of Pesticides on Biodiversity
15. Salt Marsh
16. Biodiversity in Mangrove Ecosystems
17. Marine Biodiversity: Threats and Conservation Needs
18. Microbial Diversity and Soil Functions
19. Microbial Biodiversity: Strategies for its Recovery
20. Tropical Forest Conservation
21. Forest Fire and Biological Diversity
22. Forecasting the Effects of Global Warming on Biodiversity
23. Effect of Global Warming on Biodiversity in India
24. Biodiversity in Asia and the Pacific
25. Thailand's Biodiversity
26. Biodiversity in China

Biodiversity

MN William

CBS

CBS Publishers & Distributors Pvt Ltd

New Delhi • Bengaluru • Chennai • Kochi • Kolkata • Mumbai

Bhopal • Bhubaneswar • Hyderabad • Jharkhand • Nagpur • Patna • Pune • Uttarakhand • Dhaka (Bangladesh)

Biodiversity

ISBN: 978-93-88902-92-2

Copyright © Author and Publisher

First Edition: 2019

Published by Satish Kumar Jain and produced by Varun Jain for

CBS Publishers & Distributors Pvt Ltd

4819/XI Prahlad Street, 24 Ansari Road, Daryaganj, New Delhi 110 002, India.
Ph: 23289259, 23266861, 23266867   Fax: 011-23243014   Website: www.cbspd.com
e-mail: delhi@cbspd.com; cbspubs@airtelmail.in.
Corporate Office; 204 FIE, Industrial Area, Patparganj, Delhi 110 092
Ph: 4934 4934   Fax: 4934 4935   e-mail: publishing@cbspd.com; publicity@cbspd.com

Branches

- **Bengaluru:** Seema House 2975, 17th Cross, K.R. Road,
  Banasankari 2nd Stage, Bengaluru 560 070, Karnataka
  Ph: +91-80-26771678/79   Fax: +91-80-26771680   e-mail: bangalore@cbspd.com
- **Chennai:** 7, Subbaraya Street, Shenoy Nagar, Chennai 600 030, Tamil Nadu
  Ph: +91-44-26680620, 26681266   Fax: +91-44-42032115   e-mail: chennai@cbspd.com
- **Kochi:** 42/1325, 1326, Power House Road, Opp KSEB Power House,
  Ernakulam 682 018, Kochi, Kerala
  Ph: +91-484-4059061-65   Fax: +91-484-4059065   e-mail: kochi@cbspd.com
- **Kolkata:** 6/B, Ground Floor, Rameswar Shaw Road, Kolkata-700 014, West Bengal
  Ph: +91-33-22891126, 22891127, 22891128   e-mail: kolkata@cbspd.com
- **Mumbai:** 83-C, Dr E Moses Road, Worli, Mumbai-400018, Maharashtra
  Ph: +91-22-24902340/41   Fax: +91-22-24902342   e-mail: mumbai@cbspd.com

Representatives

• Bhopal	0-8319310552	• Bhubaneswar	0-9911037372	• Hyderabad	0-9885175004
• Jharkhand	0-9811541605	• Nagpur	0-9421945513	• Patna	0-9334159340
• Pune	0-9623451994	• Uttarakhand	0-9716462459	• Dhaka (Bangladesh)	01912-003485

Printed at: Mudrak, Noida, UP, India

Preface

The variety of life on earth, its biological diversity is commonly referred to as biodiversity. The number of species of plants, animals, and micro-organisms, the enormous diversity of genes in these species, the different ecosystems on the planet, such as deserts, rainforests and coral reefs are all part of a biologically diverse earth. Appropriate conservation and sustainable development strategies attempt to recognise this as being integral to any approach. Almost all cultures have in some way or form recognised the importance that nature and its biological diversity has had upon them and the need to maintain it. Yet, power, greed and politics have affected the precarious balance.

Biodiversity boosts ecosystem productivity where each species, no matter how small, all have an important role to play. For example, a larger number of plant species means a greater variety of crops; greater species diversity ensures natural sustainability for all life forms; and healthy ecosystems can better withstand and recover from a variety of disasters.

It has long been feared that human activity is causing massive extinctions. Despite increased efforts at conservation, it has not been enough and biodiversity losses continue. The costs associated with deteriorating or vanishing ecosystems will be high. However, sustainable development and consumption would help avert ecological problems. Preserving species and their habitats is important for ecosystems to self-sustain themselves. Yet, the pressures to destroy habitat for logging, illegal hunting and other challenges are making conservation a struggle. Rapid global warming can affect an ecosystems chances to adapt naturally. The subject matter of this book is accommodated in 26 chapters.

Chapter 1 is devoted to living resources. Chapter 2 deals with importance of biodiversity. Biodiversity is defined as 'the variability among living organisms from all sources, including terrestrial, marine and other aquatic ecosystems and the ecological complexes of which they are a part; this includes diversity within species, between species and of ecosystems'. Chapter 3 acquaints the readers with biodiversity and conservation. Chapter 4 concentrates on species extinction. Species are groups of interbreeding natural population that are reproductively isolated from other such groups. In biology and ecology, extinction is the end of an organism or of a group of organisms (taxon), normally a species. The moment of extinction is generally considered to be the death of the last individual of the species, although the capacity to breed and recover may have been lost before this point.

Chapter 5 focuses on population ecology which is the study of the processes that affect the distribution and abundance of animal and plant populations. Chapter 6 is devoted to biodiversity in India. Chapter 7 deals with species diversity which is the effective number of different species that are represented in a collection of individuals. Chapter 8 concentrates on impact of biotechnology on biodiversity. Modern biotechnology offers new means of improving rather than threatening biodiversity. If properly tested for both risks and benefits to humans and the environment, transgenic crops are more likely to increase agricultural biodiversity and help maintain native biodiversity rather than to endanger it. Chapter 9 acquaints the readers with biotechnology and agrobiodiversity. Agrobiodiversity has become an increasingly useful source for crop and animal improvement since biotechnology has widened and facilitated the potential use of genetic resources. Chapter 10 is devoted to impact of genetically engineered and genetically modified crops on biodiversity. Crop genetic diversity is considered a source of continuing advances in yield, pest resistance and quality improvement. It is widely accepted that greater varietal and species diversity would enable agricultural systems to maintain productivity over a wide range of conditions. Chapter 11 concentrates on agrobiodiversity with special reference to spices.

Dry lands have an immense scientific, economic and social value. They are the habitat and source of livelihood for about one quarter of the earth's population. It is estimated that these ecosystems cover one-third of the earth's total land surface and about half of this area is in economically productive use as range or agricultural land. Keeping this in view, chapter 12 focuses on dry land biodiversity. Chapter 13 deals with biodiversity in grasslands. Grasslands, in common with other major biomes, are experiencing the effect of major global changes.

Chapter 14 acquaints the effect of pesticides on biodiversity. It is often difficult to separate the individual aspects and their effects on biological diversity from one another. The choice of crop, the type of crop rotation and the use of fertilisers and pesticides codetermine which plants and animals survive and establish themselves on agricultural land, which are displaced or harmed, and whether natural regulatory systems are disrupted or supported. Pesticides allow the type of agriculture that contributes to the loss of biological diversity. Special attention must therefore be paid to coherency between the protection of biodiversity and pesticide legislation.

Chapter 15 concentrates on salt marsh. A salt marsh is an environment in the upper coastal intertidal zone between land and salt water or brackish water, it is dominated by dense stands of halophytic (salt-tolerant) plants such as herbs, grasses or low shrubs. Chapter 16 is devoted to biodiversity in mangrove ecosystems which are various kinds of tres up to medium height and shrubs that grow in saline coastal sediment habitats in the tropics and subtropics. Mangrove dominate three quarters of tropical coastlines. Chapter 17 acquaints the readers with marine biodiversity: threats and conservation needs. Chapter 18 focuses on microbial diversity and soil functions. Chapter 19 concentrates on microbial biodiversity: strategies for its recovery. This chapter discusses bacteria with some comment on the fungi, the two groups of greatest interest in industrial microbiology, and on recovering diversity through retrieving DNA from nature. Chapter 20 is devoted to tropical forest conservation. Tropical forests have a special role in the conservation of biodiversity and influence the local and global climates. Chapter 21 acquaints the readers with forest fire and biological diversity. Fire is a vital and natural part of the functioning of numerous forest ecosystems. Fire ecology is concerned with the processes linking the natural incidence of fire in an ecosystem and the ecological effects of this fire.

Chapter 22 concentrates on forecasting the effects of global warming on biodiversity. The purpose of an environmental forecast is either to support a decision process or to test a scientific hypothesis. To support a decision process, it must be clear which decisions the forecast expects to improve. To mitigate the effects of global warming on biodiversity, two distinct kinds of actions are needed: long-term actions, such as reducing emissions of greenhouse gases, and short-term ones, such as designing an appropriate nature reserve. Chapter 23 deals with effects of global warming on biodiversity in India. Chapter 24 focuses on biodiversity in Asia and the Pacific. Asia and Pacific encompasses some of the world's greatest biological, cultural and economic diversity. Chapter 25 acquaints the readers with Thailand's biodiversity. Chapter 26 is devoted to biodiversity in China, which has a vast territory with complex climate, varied geomorphic types, a large river network, many lakes and long coastline. Such complicated conditions inevitably form diversified habitants and ecosystems.

This book is intended for students of BTech and MTech (biotechnology, chemical, environmental engineering). Besides students, the book will prove useful also for consultants, industrialist and decision-makers.

References, glossary and index have been provided at the end for quick reference. Diagrams, figures, and tables supplement the text. All the topics have been covered in a cogent and lucid style to help the reader grasp the information quickly and easily.

It will not be wrong to hold that the book *Biodiversity* is an essential reading for all students and teachers involved in all branches of engineering. It has been prepared with meticulous care, aiming at making the book error-free. Constructive suggestions are always welcome from the users of this book.

MN William

Contents

Preface   *v*

1 Living Resources   1–15

Introduction 1
Biodiversity 1
Species 2
    Present Problems 2
Conservation of Biodiversity 3
    In Situ and *Ex Situ* Conservation 3
    Which Areas to Conserve 3
Integrating Conservation and Development 4
Earth Summit 5
Forest Management and Policies 5
Developing an Information System to Prioritise Biodiversity Conservation Areas and
    Management Zones 6
Assessment of Biodiversity 6
    The Habitat/Ecosystem Oriented Approach 7
    The Species Oriented Approach 7
    GIS to Integrate Both Approaches 7
Monitoring the Dynamics of Biodiversity 7
    Land Use and Land Cover Changes 7
    Ecosystem Uses and Forest Products 8
    Forest Dynamics 8
Perspectives 8
Threats to Biodiversity 9
    Over-hunting 9
Protecting and Restoring Biodiversity 10
    Extent of Protected Areas 10
Placement and Design of Nature Reserves 10
Value of Biodiversity 11
    Consumptive Use Value 12
    Productive Use Value 12
    Social Values 12
    Aesthetic Value 13
    Option Value 13
Biodiversity and Ethics: Do We Have a Responsibility to Preserve 13
Conclusion 15

2 Importance of Biodiversity 16–34

Introduction 16
Where it Came From 17
Why it Matters 17
Definition of Biodiversity 18
Species Diversity and Biodiversity 19
Ecological Effects of Biodiversity 20
    Productivity and Stability as Indicators of Ecosystem Health 21
    Effects on Community Productivity 21
    Effects on Community Stability 23
    Theory and Preliminary Effects from Examining Food Webs 24
Genetic Diversity 24
    Importance of Genetic Diversity 24
    Survival and Adaptation 25
    Agricultural Relevance 25
    Coping with Poor Genetic Diversity 25
    Measures of Genetic Diversity 26
    Other Measures of Diversity 26
Distribution 26
Evolution 27
    Evolutionary Diversification 27
Human Benefits 28
    Agriculture 28
    Human Health 28
    Business and Industry 29
    Leisure, Cultural and Aesthetic Value 29
    Other Services 29
Species Loss Rates 30
Threats 30
    Habitat Destruction 30
    Introduced and Invasive Species 30
    Overexploitation 31
    Hybridisation, Genetic Pollution/Erosion and Food Security 31
    Climate Change 32
    Overpopulation 32
Holocene Extinction 32
Conservation 32
Protection and Restoration Techniques 33
    Resource Allocation 33
Legal Status 33
Analytical Limits 34
    Taxonomic and Size Relationships 34

3 Biodiversity and Conservation   35–70

Introduction 35
Losses of Biodiversity 36
    Why Conserve Biodiversity 37
Ex Situ Conservation (Out of the Natural Habitat) 39
    Zoos 39
    Aquaria 40
    Plant Collections 41
In Situ Conservation—Within the Natural Habitat 41
    Protected Sites or Reserves 41
    Restoration 43
    Recovery of Threatened Species 43
    Introduced Species 43
    Genetic Modification 44
    Legal Protection 44
    Advantages, Risks and Opportunities 44
    Integrated Regional Development Planning and Implementation 45
Biodiversity Conservation and Traditional Agroecosystems 48
    Agriculture and the Conservation of Biodiversity 49
    Threats to Traditional Agroecosystems 50
    Tenure: The Key Issue 53
Need and Potential for Private Biodiversity Conservation 54
    A Basic Assumption 54
    Multiple Faces of Biodiversity 55
    Why We are Losing Biodiversity 56
    An Evolving Strategy 56
    Biodiversity in a Life-zone Context 57
    Land-capability Assessment 57
    Hunting and Fishing Reserves 58
    Private Conservation 59
    Coral-reef Parks, Tourists and Fishermen 60
    Alternatives to Timber Management for Biodiversity Conservation 61
Integrating Park and Regional Planning Through An Ecosystem Approach 64
    Ecosystem Science 65
    Ecosystem Approaches 66
Conclusions and Recommendations 67
    Social Concerns and Biodiversity Conservation 67
    Political Will and Biodiversity Conservation 69
    Information and Biodiversity Conservation 70
    Development Planning and Biodiversity Conservation 70

4 Species Extinction   **71–98**

Introduction 71
Darwinian Species Criteria 72
    Darwin's Morphological Species Criterion 72
    Polytypic Species 72
Philosophisation of Species, The 'Interbreeding' Concept 72
Alternative Species Concepts 74
    Ecological Species Concept 74
    Recognition Concept of Species 74
Species Concepts Based on History 75
    Monophyly 75
    Genealogy 75
    Diagnostic Species Concept 76
Combined Species Concepts 77
    Evolutionary and Lineage Concepts 77
    Cohesion Concept of Species 77
Dissent: Maybe Species are Not Real 78
    Taxonomic Practice 78
    Populations are Evolutionary Units, Not Species 79
    Phenetic Species Concept 79
    Genotypic Cluster or Genomic Cluster Criterion 80
    Unreality of Species in Space and Time 82
Importance of Species Concepts for Biodiversity and Conservation 83
    Traditional: Species as Real Entities 83
    Alternatives: Genetic Differences More Valuable than Species Status 84
    Species Differences as Ecologically Important Markers 84
    Biodiversity in Space and Time 84
Species Influences Upon Ecosystem Function 85
    Species Traits 85
    Dominance 87
    Keystones 88
    Ecosystem Engineering 88
Rare Species and Ecosystem Functioning 89
    Less Common Species as Keystone Species 91
Extinction 93
    Pseudoextinction 94
    Causes 95
    Mass Extinctions 97

5 Population Ecology   **99–141**

Introduction 99
Population Density and Growth 99
    Life histories and the Structure of Populations 99
    Life Tables and the Rate of Population Growth 100

Natality in Population Ecology 103
    Definitions 103
    Animal Natality 103
    Plant Natality 103
    Natality in Humans 104
    Calculations Where Natality is a Factor 104
    Fluctuation in Population Size 104
    Population Bioenergetics 107
    Biological Dispersal 107
Origin of Species 110
    Isolating Mechanisms 111
    Reproductive Isolation 111
    Speciation 120
    Isolation of Population 124
    Genetic Divergence 124
    Operation of Reproductive Isolation 128
    Ring Species 129
    Darwinism 131
Modern Evolutionary Synthesis 136
    Summary of the Modern Synthesis 137
Continental Drift 137
Gaia Hypothesis 138
    Processing of CO_2 140
Geologic Time Scale 140
Molecular Clock 141

6 Biodiversity in India   **142–157**

Introduction 142
Pressure 143
    Major Problems with Biodiversity Conservation 144
State impact 146
    Status of Biodiversity in India 146
    Biogeographic Classification of India 146
    Biodiversity Hotspots 149
    Biodiversity Contribution to Indian Economy 150
Response 151
    Existing Policy Response 152
    Policy Gaps 156
    Knowledge/Information/Data 156
    Policy Recommendations 156

7 Species Diversity   **158–189**

Introduction 158
Calculation of Diversity 158
Diversity Indices 159

Sampling Considerations 159
Trends in Species Diversity 160
Species Richness 160
    Sampling Considerations 160
    Trends in Species Richness 160
    Applications 161
Species Evenness 161
Types of Diversity 161
    Alpha Diversity 161
    Beta Diversity 163
    Gamma Diversity 164
Relative Species Abundance 165
    Distribution Plots 165
    Understanding Relative Species Abundance Patterns 167
Species-area Curve 171
Latitudinal Gradients in Species Diversity 173
    Patterns in the Past 173
    Hypotheses for Pattern 174
    Synthesis and Conclusions 177
Diversity Index 177
    Terms 177
    Indices that Measure Diversity 178
Species Distribution 180
    Clumped Distribution 181
    Regular or Uniform Distribution 182
    Random Distribution 183
    Species Distribution Model 183
    Abiotic and Biotic Factors 183
    Statistical Determination of Distribution Patterns 184
Keystone Species 185
    Predators 185
    Mutualists 185
    Engineers 186
Biodiversity Hotspot 186
    Hotspot Conservation Initiatives 186
    Biodiversity Hotspots by Region 187
    Critiques of Hotspots 188
Megadiverse Countries 189
    Megadiverse Countries 189
    Cancún Initiative and Declaration of Like-minded Megadiverse Countries 189

8 Impact of Biotechnology on Biodiversity   **190–202**

Introduction 190
Essence of Biodiversity 190
    Native Biodiversity 190

Agricultural Biodiversity 191
Human Population Expansion 192
International Agreements 192
Convention on Biological Diversity (CBD) 193
Cartagena Protocol on Biosafety 193
Loss of Biodiversity and Conservation 194
Reduction of Biodiversity 194
Conservation Strategies 196
Applications of Biotechnology and its Effect on Biodiversity 196
Biotechnology for the Acquisition of Knowledge 196
Direct Gene Transfer to Crops and Farm Animals 197
Native Biodiversity and Biotechnology 198
Agricultural Biodiversity and Biotechnology 200
Social Consequences 201
Economic Considerations 201
Ethical Considerations 201

9 Biotechnology and Agrobiodiversity 203-229

Introduction 203
Technical Aspects of Agricultural Biotechnology at the Interface with
Agrobiodiversity 205
In Vitro Technologies 205
Genetic Modification and the Sourcing of Genes 206
Loss of Agrobiodiversity: Plants and Animals for Food and Agriculture 208
The International Framework 209
International Treaty for Plant Genetic Resources for Food and Agriculture 210
Global System on Plant Genetic Resources 210
Global Strategy for the Management of Farm Animal Genetic Resources 211
Sustainable Livelihoods Approach 211
People-centred 212
Holistic 212
Dynamic 212
Building on Strength and Assets 212
Macro-micro Linkages 213
Sustainability 213
Balance 215
Vulnerability Context 215
Policies, Institutions and Processes (PIPs) 216
Livelihood Strategies 217
Livelihood Outcomes 217
Key Points 218
Linkages between Agrobiodiversity, Local Knowledge and Gender from a Livelioods
Perspective 218
Relationships between Assets 218
Relationships with other Framework Components 219

Linkages between Policies, Institutions and Processes within the Framework 220
Key Points 220
Farmers and the Future of Agrobiodiversity 221
    Adapting to Climate Change 222
    Agriculture and Mitigation 223
    Challenges 223
    Agricultural Innovations and their Impact on Agrobiodiversity 224
    Biofuels and Agrobiodiversity 226
    Governance and Agriculture 227

10 Impact of Genetically Engineered and Genetically Modified Crops on Biodiversity   **230–255**
Introduction 230
Genetically Engineered (GE) Crops 230
    Crop Diversity 230
    Farm-scale Diversity 230
    Landscape-scale Diversity 231
    Indirect Indicators 232
Genetically Modified (GM) Crops 233
    Crop Diversity 234
    Farm-scale Diversity 235
    Landscape-Scale Diversity 239
    Indirect Indicators 250

11 Agrobiodiversity Informatics with Special Reference to Spices   **256–265**
Introduction 256
Importance of Biodiversity 256
Biodiversity Informatics 257
    Current Biodiversity Informatics Issues 258
    Mobilising Primary Biodiversity Information 258
    Biodiversity Informatics Standards and Protocols 259
    Current Biodiversity Informatics Activities 259
World Biodiversity Database CD-ROM Series 260
Biodiversity Database Initiatives in India 260
    Black Pepper 263
    Curcuma Species 263
    Nutmeg 264
    Phytophthora spp 264
Plasbid 265

12 Dry Land Biodiversity   **266–277**
Introduction 266
Definition and Extent of Drylands 266
Dryland Biodiversity Status and Trends 267
    Driving Forces of Biodiversification in Drylands 267
    Status of Dryland Biodiversity 268

Special Features of Dryland Biodiversity 268
Threats to Dryland Ecosystems and Species Diversity 272
Biodiversity, Desertification and Climate Interactions in Drylands 273
Conservation and Sustainable Use of Dryland Biodiversity: A Call for Action 274
Lessons Learnt From Past and Ongoing Efforts 275
Framework for Action 276
Conclusion 277

13 Biodiversity in Grasslands   **278–293**

Introduction 278
Biodiversity as a Hierarchy 279
Global Change and Biodiversity Scenarios in Grasslands 280
Land Use Effects on Biodiversity 281
Grazing by Domestic Herbivores 281
Replacement of Grassland by Other Land Cover Types 283
Nitrogen Inputs and Atmospheric CO_2 Increases 285
Biotic Exchange 286
After Extinction 287
Management of Biocomplexity—The Next Frontier 289
Conclusion 292

14 Effects of Pesticides on Biodiversity   **294–310**

Introduction 294
Water 295
Soil 295
Air 295
Effects on Biota 296
Plants 296
Animals 296
Human 296
Aquatic Life 297
Birds 297
Threatening Reports on Hazardous Effects of Pesticides 298
Alternative Methods for Eliminating Pesticides 299
Diversified Planting 299
Low Toxicity Pesticides 299
Case Study: Pesticides and Loss of Biodiversity in Europe 299
Biodiversity Loss and the Use of Pesticides 299
Bird Species Decline Owing to Pesticides 301
Risk to Mammals of Hazardous Pesticides 303
Impact of Pesticides on Butterflies, Bees and Natural Enemies 304
Pesticides Affecting Amphibians and Aquatic Species 305
Effect of Pesticides on Plant Communities 306
Are Pesticides Diminishing Soil Fertility 307
Policies and Methods for Biodiversity Conservation 308
Need for a Biodiversity Rescue Plan 309

15 Salt Marsh 311–321

Introduction 311
Basic Information on Salt Marsh 311
Tidal Flooding and Vegetation Zonation 312
Human Impacts 313
    Land Reclamation 313
    Nitrogen Loading 313
    Mosquito Control 314
Restoration and Management 314
Restoration of the Marsh 315
    Fish and Shellfish Habitat 316
    Wildlife Habitat 318
    Flood Flow Alteration 319
    Educational/Scientific Value 319
    Production and Export 320
    Uniqueness/Heritage 320
    Visual Quality/Aesthetics 321

16 Biodiversity in Mangrove Ecosystems 322–332

Introduction 322
Floral Diversity: Bacteria and Fungi 322
Marine Microbial Diversity 323
Marine Bacterial Diversity 324
Archaebacteria 325
Eubacteria 325
Chemoautotrophic Eubacteria 326
Chemoheterotrophic Eubacteria 326
    Gram-positive Bacteria 326
    Gram-negative Bacteria 327
Bacterial Life in Extreme Environments 327
Marine Fungal Diversity 328
Bacterial Flora in Mangrove Ecosystems 329
Sampling Methods 329
    Isolation Method 329
    Identification Method 329
Identification Using Diagnostic Kits 332
Identification Using Microbial Identification System (MIS) 332
Fungal Flora in Mangrove Ecosystems 332
Collection Techniques 332

17 Marine Biodiversity: Threats and Conservation Needs 333–346

Introduction 333
Biodiversity 333
    Genetic Diversity 334
    Species Diversity 334

Phyletic Diversity 335
Functional Diversity 336
Community and Ecosystem Diversity 336
Habitat Diversity 337
Characteristic Patterns of Marine Biodiversity 338
Latitudinal Pattern of Diversity 338
Longitudinal Pattern of Tropical Diversity 338
Other Marine Biodiversity Patterns 339
Threats to Marine Biodiversity 339
Habitat Degradation, Fragmentation and Loss 340
Global Climate Change 341
UV-B Radiation 342
Effects of Fishing and Other Forms of Over Exploitation 342
Pollution and Marine Litter 343
Species Introductions/Invasions 344
Watershed Alteration and Physical Alterations of Coasts 344
Tourism 344
Human Perceptions of the Oceans 345
Summary of Threats 345
Legal Framework of Biodiversity Conservation 345
How Can Marine Biodiversity Best be Conserved? 346

18 Microbial Diversity and Soil Functions

347–362

Introduction 347
Soil as a Microhabitat 348
Definition and Measurement of Microbial Diversity 350
Plate and Direct Counts 350
Molecular Techniques 351
Phospholipid Fatty Acid (PLFA) Analysis 353
Microbial Diversity in Soil 353
Microbial and Biochemical Functions in Soil 354
Microbial Diversity and Soil Functions 357
Holistic Approach and Soil Functioning 359
Conclusion 361

19 Microbial Biodiversity: Strategies for Its Recovery

363–381

Introduction 363
Microbial Diversity on Earth 363
Extent of Microbial Diversity on Earth 363
Importance of Microbial Diversity 364
The Problem 365
Where is New Diversity to be Found? 365
Biodiversity of Culturable Bacteria 367
What Level of Bacterial Diversity Matters? 367
Isolation Strategies 369

Has the Isolated Strain been Seen Before 372
Fungal Biodiversity: Isolation and Identification 373
Recovering Biodiversity Using Environment DNA 373
Accessing Uncultivated Microbes 373
Environmental Genomics 373
Screening Environmental Libraries 375
Barriers and Challenges 376
Microbial Biodiversity Investigation Techniques 376
Culture Based Molecular Techniques 378
Nonculturable Technique 379
Construction of Phylogenetic Trees 380
Examples of 16S rRNA Sequencing 380
Conclusion 381

20 Tropical Forest Conservation   382–398
Introduction 382
Importance of Tropical Forests 382
Environmental Importance 382
Socioeconomic Importance 383
Production of Forest Products in Developing Countries 383
Types of Tropical Forests 384
Rainforests 384
Dry Forests 384
Value of Tropical Forests 385
Wood and Other Products 386
Other Economic Values 387
Environmental Benefits 388
Deforestation 388
Reasons for Deforestation 388
Endangered Wildlife 389
Practice of Forestry 390
Ancient Forestry Practices 390
Sustainable Forestry 391
Harvest-Regeneration Methods 391
Forestry Research 392
Tropical Forestry 393
Plantation Forestry 393
Forest Reserves 394
Agroforestry 394
Taungya System 395
Shade Cropping 395
Support Crops 395
Alley Cropping 395
Living Fences 395
Windbreaks 395

New Directions in Tropical Forestry 395
    Global Warming 395
    Role of Forests 396
    Forestry Issues 396
Tropical Rainforests—Four Ways to Stop Deforestation 397
    Step 1: Education 397
    Step 2: Conservation Policies 397
    Step 3: Restore and Regrow 398
    Step 4: Support Ecotourism 398

21 Forest Fire and Biological Diversity   **399–417**

Introduction 399
Ecosystem Effects of Fire 400
Effects of Fire on Plants and Animals: Individual Level 400
    Plants 400
Fire Ecology 401
    Physical and Chemical Nature of Fire 401
    Fire Behaviour 401
    Effects of Fire on Ecosystems 402
Impacts of Human-Induced or Severe Natural Wildfire on Plant Diversity 404
    Wildfire 404
Effects of Fire on Forest Fauna 409
    Loss of Habitat, Territories and Shelter 409
    Loss of Food 410
    Fire-adapted Fauna 410
Effects of Suppression of the Natural Fire Regime 410
Economics of Fire Management Policy 411
    Detection 412
    Wildfire Modelling 412
Volcanoes 413
    Plate Tectonics and Hotspots 413
    Effects of Volcanoes 416

22 Forecasting the Effects of Global Warming on Biodiversity   **418–427**

Introduction 418
Eight Ways to Improve Biodiversity Forecasting 419
Recommendations 427

23 Effect of Global Warming on Biodiversity in India   **428–437**

Introduction 428
Biodiversity: What is it, Where is it and Why is it Important 429
Global Warming Effects on Biodiversity and Animals 429
Global Warming Effects on Various Seasonal Processes of Plants and Animals 430
Impacts on Forests and Wildlife 430
Shifts in Climatic Envelopes 430

Coral Bleaching 431
Increases in Extreme Events 431
Rises in Concentrations of Carbon Dioxide 431
    Sea-level Rise 431
    The First Victims 432
    Critically Endangered 432
    Endangered 432
    Vulnerable 433
    Threatened 435
What Would Rapid Species Extinction Mean for India 435
Adapting to Change 435
National Mission for Sustaining the Himalayan Ecosystem 436
National Mission for a Green India 436
Increase in Forest Cover and Density 437
Conserving Biodiversity 437

24 Biodiversity in Asia and the Pacific   **438-453**

Introduction 438
Introduction to Asia and the Pacific 438
Key Biodiversity Challenges in Asia and the Pacific 439
Regional Efforts to Achieve the Biodiversity Targets set for 2010 and Beyond 440
Recommendations on Development of the Updated Strategic Plan of the
        Convention for the Post-2010 Period 442
    Overall Suggestions for the New Strategic Plan 442
    Specific Suggestions for the Strategic Plan 443
    Suggestions for the Implementation of the New Strategic Plan 443
Biodiversity: West Asia 444
    Resources 444
    Habitat Degradation and Loss 444
    Loss of Terrestrial Species 446
    Addressing Biodiversity Loss 446
Latin America and the Caribbean 446
    Biodiversity in 2010 447
    Terrestrial Ecosystems 447
    Inland Waters Ecosystems 448
    Marine and Coastal Ecosystems 448
    Species Populations and Extinction Risks 449
    Genetic Diversity 449
    Current Pressures on Biodiversity 449
Regional Biodiversity Trends for the Twenty-first Century 450
    Terrestrial Ecosystems 451
    Inland Water Ecosystems 451
    Coastal and Marine Ecosystems 451
    Strategy and Vision for Reducing Biodiversity Loss 452

25 Thailand's Biodiversity   **454–459**

Introduction 454
Biogeography 454
Plant Diversity 455
Animal Diversity 456
Lost of Biodiversity 457
Causes of the Biodiversity Loss 458
Future Prospect 459

26 Biodiversity in China   **460–480**

Introduction 460
Habitat and Ecosystem Diversity 461
    Forest 463
    Meadow 464
    Steppe 464
    Savanna 464
    Desert 464
    Marsh 464
    Freshwater and Marine Ecosystems 465
Species Diversity 465
    Megadiversity 465
    Endemism 467
    Status 468
    Threats 473
Genetic Diversity 474
Recent Progress 475
    Survey and Bioinventory 475
    Species Conservation 475
    In Situ Conservation-Reserves 476
    Ex Situ Conservation 477
Recommendations 478
    General Recommendations for Strategy and Actions 479
    Scientific Research 479

Glossary   481–487
References   489
Index   491–496

Chapter 1

Living Resources

INTRODUCTION

Life occurs in a limited space on earth. The thin layer of the earth extending from the tropics to the poles and centred on sea level, supports all the life we know. The conditions are limited in general by temperature, light, nutrients, availability of water and other factors. Within the biosphere, there are several million different kinds of organisms, sufficiently different from one another to be called species.

The species are not independent units of life. Each are the evolutionary products of an evolving environment. All the species do not exist alone. There are interactions which are fundamental to their proper functioning. For instance, the bacterial or microbial populations of the human intestine play a significant role in defence mechanisms of the body.

BIODIVERSITY

Biological diversity in its narrowest sense this term refers to the number of species on the planet, and it also is used more broadly as an umbrella term. Biological diversity refers to the variety and variability among living organisms and the ecological complexes in which they occur. Diversity can be defined as the number of different items and their relative frequency. For biological diversity, these items are organised at many levels, ranging from complete ecosystems to the chemical structures that are the molecular basis of heredity. Thus, the term encompasses different ecosystems, species, genes and their relative abundance. Or to paraphrase: biodiversity is the number and variety of species, ecological systems, and the genetic variability they contain.

Our estimates of the number of unknown species greatly exceed our count of the number of known species. Most experts estimate the world's species diversity at 10 to 30 million, but that is very approximate. Fewer than 2 million species are 'known to science'—meaning that they have been classified by a specialist. Most published accounts put the number of known species at 1.4 million, but it may be approaching 1.8 million as on 2010. The estimates of 10 to 30 million species are based on expert opinion of how many species are yet to be formally identified. One study of insects in the forest canopy found five out of six to be new species. Even vertebrates are not completely known—it is estimated that nearly half of the freshwater fishes of South America are undescribed. New finds are made continuously in the tropics, and exploration of deep-sea hydrothermal vents recently led to the discovery not just of new species, but of new life forms at the family level (20 families or subfamilies). When you consider that virtually every species has its own parasite, and how many groups such as nematodes and bacteria have yet to be well-studied, it is apparent that the estimates of 10 to 30 million are not out of line. The global distribution of biodiversity—its geography—is interesting in its own right, and relevant to conservation. Biological diversity is greatest near the equator and declines toward higher latitudes. Tropical rain forests are especially known for their exceptional diversity. Some locations, known as 'hotspots', harbour an unusually rich local diversity, perhaps because conditions there favoured evolutionary diversification.

SPECIES

In biology, a species is one of the basic units of biological classification and a taxonomic rank. A species is often defined as a group of organisms capable of interbreeding and producing fertile offspring. While in many cases this definition is adequate, more precise or differing measures are often used, such as similarity of DNA, morphology or ecological niche. Presence of specific locally adapted traits may further subdivide species into subspecies.

Species that are believed to have the same ancestors are grouped together, and this group is called a genus. A species can only belong to one genus that it was grouped into. The belief is best checked by a similarity of their DNA, but for practical reasons, other similar properties are used. For plants similarities of flowers are used. All species are given a two part name (a 'binomial name'). The first part of a binomial name is the generic name, the genus of the species. The second part is either the specific name (a term used only in zoology, never in botany, for the second part of a binomial) or the specific epithet (the term always used in botany, which can also be used in zoology).

For example, *Boa constrictor*, which is commonly called by its binomial name, and is one of four species of the *Boa* genus. The first part of the name is capitalised, and the second part has a lower case. The two part name is written in italics. A usable definition of the word 'species' and reliable methods of identifying particular species are essential for stating and testing biological theories and for measuring biodiversity, though other taxonomic levels such as families may be considered in broad scale studies. Extinct species known only from fossils are generally difficult to assign precise taxonomic rankings, which is why higher taxonomic levels such as families are often used for fossil based studies. The total number of non-bacterial species in the world has been estimated at 8.7 million, with previous estimates ranging from two million to 100 million.

A usable definition of the word 'species' and reliable methods of identifying particular species is essential for stating and testing biological theories and for measuring biodiversity. Traditionally, multiple examples of a proposed species must be studied for unifying characters before it can be regarded as a species. It is generally difficult to give precise taxonomic rankings to extinct species known only from fossils.

Some biologists may view species as statistical phenomena, as opposed to the traditional idea, with a species seen as a class of organisms. In that case, a species is defined as a separately evolving lineage that forms a single gene pool. Although properties such as DNA-sequences and morphology are used to help separate closely related lineages, this definition has fuzzy boundaries.

However, the exact definition of the term 'species' is still controversial, particularly in prokaryotes, and this is called the species problem. Biologists have proposed a range of more precise definitions, but the definition used is a pragmatic choice that depends on the particularities of the species of concern.

Present Problems

1. Our country faces of overpopulation, large number of cattle heads, growing demand for land, energy and water supply.

2. Unplanned developmental works and over-exploitation of resources have made its living resources most vulnerable. Over-exploitation has not only resulted in scarcity of various materials but also left our biodiversity exposed to various ecological threats and even extinction.

3. Already the forest has been reduced to approximately 55 per cent of its original cover and the rate of deforestation is in excess of 1,00,000 sq. km; this equals more than the area of Switzerland and Netherlands combined.

4. In India, the rate of destruction of forests is 13,000 sq. km annually. Fifty years ago Kerala had a forest cover of almost 38 per cent of its total land area and that has now dwindled to 10 per cent.

If this rate of deforestation continues, one can imagine the ultimate fate of our forest wealth and biological richness. In addition to this, ozone depletion, acid rain and global warming are posing further threats to biodiversity.

CONSERVATION OF BIODIVERSITY

Conservation is the protection, preservation, management or restoration of wildlife and natural resources such as forests and water. Through the conservation of biodiversity the survival of many species and habitats which are threatened due to human activities can be ensured. Other reasons for conserving biodiversity include securing valuable natural resources for future generations and protecting the well being of ecosystem functions.

In situ and *Ex situ* Conservation

Conservation can broadly be divided into two types:

1. *In situ*: Conservation of habitats, species and ecosystems where they naturally occur. This is *in situ* conservation and the natural processes and interaction are conserved as well as the elements of biodiversity.

2. *Ex situ*: The conservation of elements of biodiversity out of the context of their natural habitats is referred to as *ex situ* conservation. Zoos, botanical gardens and seed banks are all example of *ex situ* conservation.

In situ conservation is not always possible as habitats may have been degraded and there may be competition for land which means species need to be removed from the area to save them.

Which Areas to Conserve

Hotspots of biodiversity

A popular approach for selecting priority areas has been to select hotspots of diversity. Since it is not possible to conserve all biodiversity due to lack of resources and the need to use land for human activities, areas are prioritised to those which are most in need of conservation. 'Hotspot' a term used to define regions of high conservation priority combining high richness, high endemism and high threat.

Threatened species

Over the last 200 years many species have become extinct and the extinction rate is on the increase due to the influence of human activity. The status of species has been assessed on a global scale by the World Conservation Union. Taxa that are facing a high risk of global extinction are catalogued and highlighted in the IUCN Red List of Threatened Species.

Threatened habitats

Habitat destruction comes in many forms from clear felling of forests to simple changes in farming practices that change the overall surrounding habitat. If a habitat is degraded or disappears a species may also become threatened. The UK is in danger of losing diverse habitats ranging from lowland calcareous grassland to mudflats and wet woodland. The UK BAP has specific Habitat Action Plans in place in order to try and manage and conserve these precious places. Many of these areas lie within SSSIs which are designated prioritised areas of conservation.

Flagship and keystone species

Conservation efforts are often focused on a single species. This is usually for two reasons:

1. Some species are key to the functioning of a habitat and their loss would lead to greater than average change in other species populations or ecosystem processes. These are known as keystone species.
2. Humans will find the idea of conserving one species more appealing than conserving others. For example it would be easier to persuade people that it is necessary to conserve tigers that it is to persuade people to conserve the Zayante band-winged grasshopper. Using a flagship species such as a tiger will attract more resources for conservation which can be used to conserve areas of habitat.

Complementarity

Complementarity is a method used to select areas for conservation. These methods are used to find areas that in sum total have the highest representation of diversity. For example using complementarity methods, areas could be selected that would contain the most species between them but not necessarily be the most species rich areas individually and take into account pressures of development.

Distinguishing higher from lower priority areas for urgent conservation is the purpose of such area-selection methods. However, an acceptance of priorities must recognise that this idea also implies that some areas will be given lower priority. This is not to say that they have no conservation values rather that in relation to agreed goals the actions are not as urgent.

Where identities of species or other biodiversity indicators are known, complementarity methods can be applied.

INTEGRATING CONSERVATION AND DEVELOPMENT

Conservation cannot be conducted in isolation from humans and for conservation to be successful and sustainable there needs to be local community involvement. In the UK most biodiversity is found in countryside which is farmed. It is, therefore, necessary to integrate conservation into farming practices. In other areas of the world livelihood and development priorities of local communities must be taken into account if the conservation measures are to be sustainable.

Community-based natural resource management (CBNRM) is a process through which grass-roots institutions are involved in the decision-making and have rights to manage and control their environment. CBNRM Net (Community-Based Natural Resource Management Network) is a website that provides useful networking tools so that people can exchange experiences, manage relevant knowledge, and support learning across countries and cultures and in this way achieve better results. IIED have set up a Biodiversity and Livelihoods Group which aims through sustainable management of biodiversity to improve the livelihoods of the poor. BLG researches, analyses and implements new projects and strategies around the world.

A good example of this type of integration is being shown by an earthwatch project in Africa. The project helped create a wildlife reserve that was initiated by the Wechiau and Tokali local communities. The Wechiau Community Hippo Sanctuary now generates income for the local communities via eco-tourism which helps conserve and protect hippos and other sanctuary wildlife at the same time.

EARTH SUMMIT

1. At the Earth Summit held at Rio de Janeiro in June 1992, the majority of the world's nations signed a Convention on Biological Diversity.
2. Ninety per cent of the world's total biodiversity lies in the southern nations. India is one of the richest areas in the world in terms of biological diversity as it is a home to 6.5 per cent of the world's wild species of plants and animals.
3. Of the 1.5 million species known to inhabit the earth, an estimated one-third is likely to become extinct within the next few decades.
4. The species destruction, it is estimated, is taking place at the rate of almost one a day. This is the result of extensive habitat changes brought about by mankind.

FOREST MANAGEMENT AND POLICIES

The concept of the management of forests was introduced in India 150 years ago. Colonial legacy and princely states carved out Reserved Forests and State Forests under different forest laws for the 'scientific' management of forests. While doing so, nearly an equal per cent of forests were kept outside the purview of the Reserved Forest for community use, that is, for obtaining their bonafide requirements of small timber, fuel, fodder, green manure and a host of other non-timber products. At that time, the population was less and consequently the pressure on the forest was minimal. Most of the requirements of the local population were obtained from the Revenue Forests.

Currently, the State Forest Departments, which are the custodians of forests, control large areas of forest as state property. However, substantial areas of natural vegetation still remain either in private control or under the Revenue Department's authority. The management of the Reserved Forests (RF) under the State Forest Departments has traditionally revolved around protection, silviculture and plantation. On the other hand, Revenue Forests are under the control of the Revenue Departments without any kind of management. Locals have had free access to revenue forests, and due to heavy pressures, vast areas of such lands are being converted into private croplands by granting title deeds to local communities under different schemes and programmes. With the disappearance of Revenue Forests, pressure is being brought on the Reserved Forests to provide the bonafide requirements of the local people. This has resulted in the degradation of Reserved Forests, too.

In northeast India, the forests are still worse from the management point of view. Only 5–10 per cent of forests in the region is under the control of the forest department and the rest is vested with autonomous district forest councils. These councils do not follow any management guidelines and most of the decisions on forest utilisation are made following other than scientific considerations. The result is a continuous onslaught on the forests.

Until recently, forest management policies have been largely a legacy of the colonial era with its industrial/commercial bias. Biodiversity, preservation of the environment and people's dependence on forests were given secondary importance. With the increasing range of stakeholders in forests—indigenous/local communities, public sectors, private sectors—the existing forest management system and policies are incapable of limiting the onslaught on forests. The rapid changes in the biological and socioeconomic environments make it relevant to adopt management policies in tune with sustainable development, equitable access to forest resources by local communities and environmental conservation.

Moreover, the implementation of the management plans faces numerous constraints, both through poor correspondence with national policy and the lack of an effective integrative approach in the field. India is a signatory to the International Convention on Biological Diversity, which emphasises the identification of biodiversity components and its conservation and sustainable use. In the spirit of this convention, the current Indian Forest Policy has given importance to Joint Forest Management (JFM) and the formation of Village Forest Committees (VFC) in order to use and manage forest resources without jeopardising the biodiversity.

The change in the National Forest Policy has necessitated a new strategy and action plans for sustainable conservation and management of biodiversity through an integrative approach which takes into account ecological, social, economic and institutional aspects. There are several strategies and action plans being worked out at global and national levels. However, these strategies become irrelevant, unless they are site-specific and can be implemented through existing mechanisms like working and management plans, which are the operational tools of State Forest Departments.

Moreover, developing strategies requires a good information system. Still, there are major gaps in information resources pertaining to forest biodiversity, causative factors of degradation and threats. The available data are often inadequate to provide a lucid picture of the current status and the ongoing losses in biodiversity.

DEVELOPING AN INFORMATION SYSTEM TO PRIORITISE BIODIVERSITY CONSERVATION AREAS AND MANAGEMENT ZONES

Developing a good strategy requires a highly reliable and meaningful information system at different levels. In the wide field of biodiversity, the French Institute of Pondicherry (FIP) research programs have been focusing for about four decades on species and ecosystem diversity at the local (i.e. stand and community), landscape and regional levels. The Institute has been concentrating on plant ecology with a strong emphasis on trees and forests, from open woodland to dense moist evergreen forests, considering their present status as well as their long-term history. Geographically speaking, most of the studies are being carried out in the western ghats and some projects in the eastern ghats and mangroves.

The biodiversity-related programs of the FIP could be listed under two main headings: 'assessment of biodiversity' and 'monitoring the dynamics of biodiversity'. These programs are being carried out in collaboration with Forest Departments in Karnataka, Kerala, Tamil Nadu and Andhra Pradesh, the School of Environmental Sciences (JNU), the Kerala Forest Research Institute, the Centre for Ecological Sciences (IISc), the Salim Ali School of Ecology (Pondicherry University) and the National Remote Sensing Agency (Department of Space).

ASSESSMENT OF BIODIVERSITY

A key feature of biodiversity assessment is the very duality of its nature. If we consider the species and ecosystems in a given region, a first viewpoint is to examine the geographical distribution and the ecological niche of each species; a second is to study the various ecosystems and characterise their floristic composition and structure. In order to address this duality, the following approaches are carried out at the FIP and are being integrated under Geographic Information Systems (GIS).

The Habitat/Ecosystem Oriented Approach

This approach is derived from biogeography and phytoecology and was the cornerstone of the ecological mapping program initiated in the late 50s by the FIP. It consists in studying and classifying vegetation in relation to ecological conditions (climate and soil), in characterising the species composition, structure and physiognomy of the vegetation units, in analysing their dynamics and succession under 'natural' and 'disturbed' regimes. The ultimate outputs of this approach are the following vegetation and land use maps, along with floristic lists attached to each vegetation type.

The Species Oriented Approach

This approach is in direct lineage of taxonomic and botanical studies. It is best illustrated by the 'Atlas of Endemic Plants of the Western Ghats' published by the FIP. The species oriented approach consists in collecting information on the location of the species from various sources: herbaria, literature and field surveys. This information may be extended to include the ecological conditions (bioclimate, soil, altitude, topography) and the type of ecosystems in which the plant is encountered, the role it plays in these ecosystems, as well as its biological traits (morphology, architecture, growth and reproductive strategy). The ultimate goal is to have a sort of 'identity card' for each species. This information is most crucial for rare and endangered species in the perspective of their *in situ* conservation.

GIS to Integrate Both Approaches

Both of the above approaches end with large sets of spatial information and especially with maps. A major issue is to ensure the consistency of this information and to recombine it according to various viewpoints. In order to perform this, different 'layers' of GIS data have been created using Arc/Info to generate the following information:

1. Vegetation physiognomy and human pressures (deduced from density of population or road network) to assess disturbance levels.
2. The spatial distribution of several species to determine biodiversity 'hotspots'.
3. Past and present maps for monitoring land cover and land use changes.
4. Conservation value maps using biodiversity indicators (richness, diversity, endemicity, uniqueness, etc.) to prioritise the area for conservation and management.

MONITORING THE DYNAMICS OF BIODIVERSITY

Biodiversity assessment has the following outputs: lists of species, sets of values for several diversity indices and land use and vegetation maps. More often these statistics bear no meaning by themselves. Their significance depends more on their absolute and relative variations over space and time. Thus, it is crucial not only to study the biological diversity but also: (i) to monitor it in relation to factors (ecological, human and social), which influence its dynamics and (ii) to study the processes (biological, ecological, human and social) that govern its evolution.

Land Use and Land Cover Changes

The first step in monitoring changes in biodiversity consists in comparing successive observations. At the local level, this can be done by observing the appearance and disappearance of species: it requires that the same sites be sampled on several occasions. In order to observe this, two permanent plots have been set up in the Biligirirangan hills (3.5 ha) and the Kadamakal RF (28 ha), both in Karnataka. In addition to these, initial data have also been collected from one hundred 1-ha permanent plots, established by the Karnataka Forest Department in the Karnataka, part of the western ghats.

At the regional and landscape levels, this can be done using past and present land cover and land use maps. The joint development of satellite imagery, image analysis techniques and GIS has opened avenues for such studies. At the regional level, such studies are being carried out for the entire western ghats of Karnataka and at landscape level it was done for the Agastyamalai area, which is one of the 'super hot-spots' in the southern western ghats.

Ecosystem Uses and Forest Products

Understanding changes in biodiversity requires the analysis of the processes that are at play. A first major set of processes is constituted by those related to human activities, especially the direct exploitation of the ecosystems and species. This is where the social sciences play a key role: the land tenure system, the representation of ecosystem and species and the sacred and economic values of the resources are important factors to explain the changes.

In order to understand the processes and to assess the impact of anthropogenic activities on biodiversity, the following work is being carried out in the Kodagu district of Karnataka:

1. Impact of extension of coffee and cardamom plantations.
2. Assessment of biodiversity and disturbance gradient.
3. Modelling of the sacred grove system using the Multi-Agent System.

Forest Dynamics

Biological processes and ecological factors temporally govern plant demography and constitute a major set of processes, which have a strong influence on changes in biodiversity. It is thus important to analyse, in 'natural' and 'disturbed' conditions, how the plants regenerate, grow and die when they interact with each other. Such studies are best carried out at the local level in large permanent plots where the environmental conditions can be described.

Since the mid-80s, the FIP has been monitoring such plots in the low-elevation wet evergreen forests of the Kadamakal Reserved Forests in Karnataka, comparing an unlogged compartment to a once-selectively-logged compartment, analysing the spatial variation of diversity according to topographical heterogeneity, studying silvigenesis, tree regeneration and growth strategy in relation to environmental factors and monitoring phenology.

PERSPECTIVES

These long-term efforts have already been able to put together a sum of knowledge that can help to better define conservation strategies. Further, using these data, it may be possible to construct models that simulate disturbance regimes and their impacts on the forest physiognomy and species composition. Modelling the effect of various types of activity, particularly on sensitive areas, would allow an informed assessment of the potential environmental impact and a comparison of costs and benefits, which also takes into account the losses of biological diversity.

There is a need to carry out or incorporate data from studies on a finer scale of forest and landscape change linked to social and economic studies of forest use and management. These would shed light on some of the proximate and underlying causes of deforestation and loss of biodiversity. New approaches that are holistic, integrative and involve multiple agents would require a degree of coordination and cooperation between institutions. It would be useful for the State Forest Department to assume a leading role in this and to invite participating institutions to share their findings and to propose resource management alternatives based on empirical studies. There also needs to be better coordination among government agencies and research institutions, for example, between the State Forest Departments and the Revenue Department, which between them administer vast tracts of land in India.

The links between research and development and extension services need to be better defined and better implemented. The Joint Forest Planning and Management (JFPM) experiment represents a good overall attempt by the Forest Department to enter into a participatory mode. These projects will need to be continued, extended to cover a broader geographical area and closely monitored for their impact on forests and on the livelihood security of the participants.

Most critically perhaps, we reiterate here that biodiversity conservation must be integrated with national planning and economic/infrastructural priorities. Article 6 of the Convention on Biological Diversity calls upon signatory countries to approach biodiversity planning in a comprehensive manner. Biological diversity must be assigned a high priority based upon long-term planning and studies of human-ecosystem relationships.

THREATS TO BIODIVERSITY

Extinction is a natural event and, from a geological perspective, routine. We now know that most species that have ever lived have gone extinct. The average rate over the past 200 million years is 1–2 species per million species present per year. The average duration of a species is 1–10 million years (based on the last 200 million years). There have also been several episodes of mass extinction, when many taxa representing a wide array of life forms have gone extinct in the same blink of geological time. In the modern era, due to human actions, species and ecosystems are threatened with destruction to an extent rarely seen in earth history. Probably only during the handful of mass extinction events have so many species been threatened, in so short a time. What are these human actions that threaten biodiversity? There are many ways to conceive of these; let's consider two. First, we can attribute the loss of species and ecosystems to the accelerating transformation of the earth by a growing human population. As the human population passes the 6 billion mark, we have transformed, degraded or destroyed roughly half of the world's forests. We appropriate roughly half of the world's net primary productivity for human use. We appropriate most available freshwater, and we harvest virtually all of the available productivity of the oceans. It is little wonder that species are disappearing and ecosystems are being destroyed. Second, we can examine six specific types of human actions that threaten species and ecosystems—the 'sinister sextet'.

Over-hunting

Over-hunting has been a significant cause of the extinction of hundreds of species and the endangerment of many more, such as whales and many African large mammals. Most extinctions over the past several hundred years are mainly due to over-harvesting for food, fashion and profit. Commercial hunting, both legal and illegal (poaching), is the principal threat. The snowy egret, passenger pigeon and heath hen are US examples. The recent expansion of road networks into previously remote tropical forests enables the bushmeat trade, resulting in what some conservationist describe as 'empty forests' as more and more wild animals are shot for food. The pet and decorative plant trade falls within this commercial hunting category, and includes a mix of legal and illegal activities. Sport or recreational hunting causes no endangerment of species where it is well regulated, and may help to bring back a species from the edge of extinction. Many wildlife managers view sport hunting as the principal basis for protection of wildlife. While over-hunting, particularly illegal poaching, remains a serious threat to certain species, for the future, it is globally less important than other factors mentioned next.

PROTECTING AND RESTORING BIODIVERSITY

Many species and ecosystems will disappear over the next century. However, starting with recognising the problem, and then identifying management objectives, much can be done to alleviate this trend. A sound strategy would emphasise improving our management of existing protected land, and strategically adding new protected areas. Ecological systems have considerable potential to recover if appropriate restoration measures are taken. Ultimately we wish to manage populations and ecosystems sustainably, so that they may be utilised and enjoyed by future generations. These are the goals of science-based management.

Extent of Protected Areas

The World Resources Institute estimates that there are 8163 protected areas worldwide, managed for various objectives ranging from strict nature protection to controlled harvesting. They cover 750 million hectares (1 hectare = 2.5 acres) of marine and terrestrial ecosystems, which is about 1.5 per cent of the earth's surface, and 5.1 per cent of national land area. In many developing countries, the existence of protected areas creates conflicts for local people, who may depend upon that area for their subsistence. Often, enforcement of laws protecting parks is minimal.

PLACEMENT AND DESIGN OF NATURE RESERVES

Where to concentrate one's efforts is a critical issue facing conservation groups and government agencies. The uniqueness of an ecosystem, the number of species, especially endemic species (species found only in that area) it supports, and the imminence of the threats to its survival all play a role in the targeting of conservation activities. The World Wildlife Fund has identified 25 ecosystems around the world for highest priority.

The size and placement of existing protected areas around the world is determined by many factors, and not necessarily primarily by conservation needs. Many are located in remote and unproductive lands, or in areas of great scenic beauty or as a result of a conservation-minded national leader, philanthropist or member of royalty. Some are in logged-over areas, and are only beginning to recover to their original splendour. A map of US wilderness areas is a useful reminder that only a few per cent of rivers and total land area, at most, are protected. In the US, there is more protected land in the West, partly due to its scenic beauty, but partly because much of this area is less productive.

At least until recently, the system of protected areas has been almost entirely haphazard. Gap analysis is a new approach based on mapping of vegetation, animals (usually terrestrial vertebrates) and land ownership in order to identify gaps in the network of parks, reserves and public lands that hopefully protect the biodiversity they contain.

Gap analysis relies on three primary data layers. These are:
1. The distribution of actual vegetation types, delineated from satellite imagery.
2. The distribution of terrestrial vertebrates, predicted from the vegetation distribution by associating individual species with the vegetation that characterises their habitat.
3. The distribution of land ownership.

The process can be as simple as placing layers of transparent mylar over a base map, such as a topographic map, and tracing the information onto separate layers of transparencies. The first gap analysis looked at the distribution of three species of endangered Hawaiian honeycreepers (forest birds) on the island of Hawaii. It was possible to see the distribution of each bird species, and also locations where all three species coincided. Logically, these areas would be conservation priorities. Although the island had a number of nature reserves, none overlapped the birds' distributions. These maps determined the site for a new reserve.

Extending this approach to a large scale depends on computer mapping of satellite images. Landsat TM (thematic mapper) or other remotely sensed imagery may be used to construct vegetation maps. Landsat TM receives seven spectral bands of reflected infrared light, in individual cells or pixels of 30 m × 30 m. Image classification uses those spectral data to develop a map of vegetation classes, which is compared to ground measurements to improve accuracy. The resulting vegetation map is geo-referenced, meaning that every location has a latitude and longitude or some other X-Y grid location.

Specialised computer software, along with these spatial data and other information, make-up a Geographic Information System (GIS).

Using these data layers and a GIS, one can ask:

1. What fraction of threatened species occurs within existing reserves?
2. What fraction of each major vegetation type falls within existing reserves?
3. Are areas of highest species richness found within existing reserves?
4. Are areas of high endemism found within existing reserves?

VALUE OF BIODIVERSITY

Environmental services from species and ecosystems are essential at global, regional and local levels. Production of oxygen, reducing carbon dioxide, maintaining the water cycle, protecting soil are important services. The world now acknowledges that the loss of biodiversity contributes to global climatic changes. Forests are the main mechanism for the conversion of carbon dioxide into carbon and oxygen. The loss of forest cover, coupled with the increasing release of carbon dioxide and other gases through industrialisation contributes to the 'greenhouse effect'. Global warming is melting ice caps, resulting in a rise in the sea level which will submerge the low lying areas in the world. It is causing major atmospheric changes, leading to increased temperatures, serious droughts in some areas and unexpected floods in other areas.

Biological diversity is also essential for preserving ecological processes, such as fixing and recycling of nutrients, soil formation, circulation and cleansing of air and water, global life support (plants absorb CO_2, give out O_2), maintaining the water balance within ecosystems, watershed protection, maintaining stream and river flows throughout the year, erosion control and local flood reduction.

Food, clothing, housing, energy, medicines, are all resources that are directly or indirectly linked to the biological variety present in the biosphere. This is most obvious in the tribal communities who gather resources from the forest or fisherfolk who catch fish in marine or freshwater ecosystems. For others, such as agricultural communities, biodiversity is used to grow their crops to suit the environment. Urban communities generally use the greatest amount of goods and services, which are all indirectly drawn from natural ecosystems.

It has become obvious that the preservation of biological resources is essential for the well-being and the long-term survival of mankind. This diversity of living organisms which is present in the wilderness, as well as in our crops and livestock, plays a major role in human 'development'. The preservation of 'biodiversity' is therefore integral to any strategy that aims at improving the quality of human life.

Consumptive Use Value

The direct utilisation of timber, food, fuelwood, fodder by local communities. The biodiversity held in the ecosystem provides forest dwellers with all their daily needs, food, building material, fodder, medicines and a variety of other products. They know the qualities and different uses of wood from different species of trees, and collect a large number of local fruits, roots and plant material that they use as food, construction material or medicines.

Fisherfolk are highly dependent on fish and know where and how to catch fish and other edible aquatic animals and plants.

Productive Use Value

The biotechnologist uses biorich areas to 'prospect' and search for potential genetic properties in plants or animals that can be used to develop better varieties of crops that are used in farming and plantation programs or to develop better livestock. To the pharmacist, biological diversity is the raw material from which new drugs can be identified from plant or animal products. To industrialists, biodiversity is a rich storehouse from which to develop new products. For the agricultural scientist the biodiversity in the wild relatives of crop plants is the basis for developing better crops.

Genetic diversity enables scientists and farmers to develop better crops and domestic animals through careful breeding. Originally this was done by selecting or pollinating crops artificially to get a more productive or disease resistant strain. Today this is increasingly being done by genetic engineering, selecting genes from one plant and introducing them into another. New crop varieties (cultivars) are being developed using the genetic material found in wild relatives of crop plants through biotechnology.

Even today, species of plants and animals are being constantly discovered in the wild. Thus these wild species are the building blocks for the betterment of human life and their loss is a great economic loss to mankind. Among the known species, only a tiny fraction have been investigated for their value in terms of food or their medicinal or industrial potential.

Preservation of biodiversity has now become essential for industrial growth and economic development. A variety of industries such as pharmaceuticals are highly dependent on identifying compounds of great economic value from the wide variety of wild species of plants located in undisturbed natural forests. This is called biological prospecting.

Social Values

While traditional societies which had a small population and required less resources had preserved their biodiversity as a life supporting resource, modern man has rapidly depleted it even to the extent of leading to the irrecoverable loss due to extinction of several species. Thus apart from the local use or sale of products of biodiversity there is the social aspect in which more and more resources are used by affluent societies. The biodiversity has to a great extent been preserved by traditional societies that valued it as a resource and appreciated that its depletion would be a great loss to their society.

The consumptive and productive value of biodiversity is closely linked to social concerns in traditional communities. 'Ecosystem people' value biodiversity as a part of their livelihood as well as through cultural and religious sentiments. A great variety of crops have been cultivated in traditional agricultural systems and this permitted a wide range of produce to be grown and marketed throughout the year and acted as an insurance against the failure of one crop. In recent years farmers have begun to receive economic incentives to grow cash crops for national or international markets, rather than to supply local needs. This has resulted in local food shortages, unemployment (cash crops are usually mechanised), landlessness and increased vulnerability to drought and floods.

Aesthetic Value

Knowledge and an appreciation of the presence of biodiversity for its own sake is another reason to preserve it. Quite apart from killing wildlife for food, it is important as a tourist attraction. Biodiversity is a beautiful and wonderful aspect of nature. Sit in a forest and listen to the birds. Watch a spider weave its complex web. Observe a fish feeding. It is magnificent and fascinating.

Symbols from wild species such as the lion of Hinduism, the elephant of Buddhism and deities such as Lord Ganesh, and the vehicles of several deities that are animals, have been venerated for thousands of years. Valmiki begins his epic story with a couplet on the unfortunate killing of a crane by a hunter. The 'Tulsi' has been placed at our doorsteps for centuries.

Option Value

Keeping future possibilities open for their use is called option value. It is impossible to predict which of our species or traditional varieties of crops and domestic animals will be of great use in the future. To continue to improve cultivars and domestic livestock, we need to return to wild relatives of crop plants and animals. Thus the preservation of biodiversity must also include traditionally used strains already in existence in crops and domestic animals.

BIODIVERSITY AND ETHICS: DO WE HAVE A RESPONSIBILITY TO PRESERVE

Biodiversity is the dynamic variability among extant organisms, including that within and between species and environments. An estimated 20 million species exist in the global ecosystem, the evolutionary progeny of a lineage stretching back 3.5 billion years. However, current levels of biodiversity are threatened by what scientists describe as a modern, anthropogenic mass extinction—a 'human meteor'— with researchers estimating the accelerated extinction rate at 100 to 1000 times that of natural background rates. The mounting urgency to protect both individual species and whole ecosystems has thrust the matter of global biodiversity loss to the fore of international politics, precipitating the ratification of treaties such as the United Nations Convention on Biological Diversity. As politicians and scientists alike call for strengthened efforts to preserve threatened habitats and endangered species, one must wonder, are we morally obligated to do so? What responsibility, if any, do humans owe to other species, or to their environment? In this essay, we will address two reigning arguments in the conservation debate, one advocating stewardship as a moral imperative, the other attributing the current spate of extinctions to wholly natural, ecological processes. Here we contend that both stances are inherently flawed, and, ultimately, preservation as a solution to the current ecological crisis is neither in conflict with nature nor a normative duty, but rather a practical necessity.

Stewardship, as a defense of ecological health, appoints humans as custodians of biodiversity. The vanguards of the natural world, we have a moral obligation to defend it from harm, specifically harm instigated by our own actions. We are no longer mere cogs in the ecological wheel, we are the mechanics as well. The motivations behind the stewardship argument stem primarily from either value claims or rights claims, whether our own or those of potential persons.

Many environmentalists take it as uncontroversial that there is an inherent value in biodiversity, in both ecosystems and individual species, and, perhaps, individual animals. By ascribing an inborn value to all things great and small, exponents of stewardship provide grounds for 'the principles that we ought to try to preserve those species that have been driven to the brink of extinction by human activities and that we should make sure that no additional species are endangered or threatened by our behaviour'. Thus, in order to ensure the survival and propagation of endangered species, and the recovery of fragmented habitats, we must preserve those ecosystems in which they are found.

However, the nature of the value attributed to components of biodiversity is a nebulous concept. Are tigers and salamanders and Pacific Yew trees 'good' in and of themselves? If yes, how so? An object with an intrinsic value has an ostensibly infinite value; it is, for all intents and purposes, priceless. While we may enjoy telling ourselves that plants, animals and ecosystems have an essential value independent of their utilisation by man, it is a very difficult philosophical position to justify. That is not to say that ecological components do not have any value; in fact, biodiversity has a very real instrumental value to humans, as plants and animals can contribute to human welfare through economic, medical and agricultural exploitation. Instrumental value of a plant or animal notwithstanding, the absence of any intrinsic value or worth nullifies both an object's moral standing and the obligatory consideration of its well-being.

Defenders of rights-based stewardship cite the concept of intergenerational justice, that is, the duty of one generation to consider and ensure the well-being of later generations, as the foundation for current preservation efforts. As it relates to the conservation argument, they contend, older generations have a moral duty to maintain the earth's natural capital for future generations; we are merely borrowing the earth and its resources from our as-yet unborn descendents, and in the interest of intergenerational equity, we are obligated to pass it down to them in as good, if not better, shape as when we received it. While this line of reasoning may seem compelling at first, it is based on a rather specious assumption: that future generations can have rights at all. Nonexistence bars possession of anything, whether tangible or metaphysical. Similarly, potential persons lack the right to an unadulterated planet replete with healthy populations of the flora and fauna contained therein, and therefore cannot make any rights-claim upon ancestral generations, by virtue of their non-actuality. As remarked by noted Oxford economist Wilfred Beckerman, 'It makes little sense to say that our right to see a live Dodo has been violated by the inhabitants of the Mauritius islands three centuries ago.'

Whereas the stewardship argument suffers from an egoism that divorces man from nature and grants him dominion over other living things, the argument from naturalism places man squarely in competition with other organisms, and maintains that as such, we have no moral obligation to preserve species. This contention emphasises the interdependency of all forms of life, affording humans no special privilege or designation, unlike stewardship.

Proponents of man's active ecological role recognise that humans are subject to the same density-dependent regulations, primarily competition for food, resources and habitat, that affect all other living things. Humans, despite our language, culture and higher-order reasoning skills, are the 'result of the same processes that produced all other species' and are governed by the same natural laws.

It is important to note here that in order to advocate such a position, we must accept that the perceived loss of biodiversity resulting from human economic and social development is a purely natural phenomenon, despite scientists' warnings that we are ushering in a sixth major extinction event. If extinction is a normal by-product of unbalanced interspecific interactions and the ultimate fate of all species, then the anthropogenic extinctions we are seeing today are a testament to man's efficiency as a competitor. To wit, we are perhaps the earth's most successful cosmopolitan predator, generalists who consistently emerge victorious from interspecific showdowns. We are so adept at out-competing other species that we have even gone so far as to outsource competition, recruiting third parties to take care of the dirty work (e.g. biological pest control).

So, if we are doing nothing more than what comes naturally to all of earth's estimated 20 million species, are we doing anything wrong? While it is true that we have no moral obligations to preserve current levels of biodiversity (as demonstrated earlier), that is not to imply that (i) the argument from naturalism is immune to criticism, and (ii) there are not pragmatic reasons to maintain species and habitat diversity. Addressing (i), the very competitive strength that is the selling point of the naturalism argument is also its flaw. We are, without a doubt, excellent competitors; however, we are also unapologetic overachievers in this arena, and continuing to drive other species to extinction will ultimately spell trouble for us. As each subsequent lineage is lost to time, the species richness of a given ecosystem is reduced, thereby weakening the community as a whole. Ecosystems that suffer considerable loss of diversity are less likely to rebound successfully from disturbance events, natural or man-made. Further, by destroying the habitats that we exploit for capital and medical gain, we are acting in a manner contrary to our own self-interests, specifically that of improving the human condition. While we may not be duty-bound to our grandchildren's children to maintain biodiversity, perhaps we owe it to ourselves.

CONCLUSION

An integrated approach is necessary for conserving the global biodiversity. Establishment of nature reserves or biospheres with lot of biophysical variability, maintenance of corridors with different nature reserves for the possible migration of species in response to climatic change, etc. are the immediate steps to be taken for conserving the very precious biological diversity on our earth.

An international consensus on establishing global network of gene banks, microbiological resource centres, and marine parks is also important. At the same time conservation must be closely coupled with socioeconomic development especially in countries where mounting human pressure threatens the very existence of mankind.

Chapter 2
Importance of Biodiversity

INTRODUCTION

Biodiversity is the variability among living organisms on the earth, including the variability within and between species and within and between ecosystems. Biological diversity, often shortened to biodiversity, is the variation of life at all levels of biological organisation, referring not only to the sum total of life forms across an area, but also to the range of differences between those forms. Biodiversity runs the gamut from the genetic diversity in a single population to the variety of ecosystems across the globe.

Greater biodiversity in ecosystems, species and individuals leads to greater stability. For example, species with high genetic diversity and many populations that are adapted to a wide variety of conditions are more likely to be able to weather disturbances, disease and climate change. Greater biodiversity also enriches us with more varieties of foods and medicines.

The measurement of biodiversity is complex and has a qualitative as well as a quantitative aspect. If a species is genetically unique—if, for instance, it's far out on a remote arm of the evolutionary tree, like the distinctive, peculiar platypus—its biodiversity value is greater than that of a species clustered with many similar species because it preserves a unique part of the evolutionary history of the planet. This means that biodiversity can't be defined merely as the aggregate total of genes, species or habitats, but must also be understood as a measure of the variety of their differences.

That said, the easiest shorthand way to describe biodiversity is often through species counts. Current estimates of global species diversity vary between 2 million and 100 million species, with a popular estimate of somewhere near 13 to 14 million. The majority of them are arthropods. But very little is known about most species. Only roughly 1.5 million species have been described, and only 40,000 to 50,000 species have had their conservation status assessed. Roughly a third of these are believed to be at some risk of extinction.

Diversity is concentrated in certain areas and is highest in the tropics, in a band around the equator, declining progressively toward the polar regions. Specific places with high overall diversity or high levels of endemism—arrays of species found nowhere else—are often called hotspots and include parts of the southwestern United States and Mexico, Brazil, California and South Africa, as well as Hawaii, Madagascar, New Zealand, and other islands across the world. Of course, every intact ecosystem—hotspot or not—is important to preserve, not only because they all provide services, such as clean water and climate moderation, but also because each contains a unique ecological composition and priceless evolutionary information.

WHERE IT CAME FROM

The earth's biodiversity is the result of 4 billion years of evolution—change in the inherited traits of a population of organisms from one generation to the next. Up until about 600 million years ago, life consisted of single-celled organisms.

The history of biodiversity during the Phanerozoic era (the past 540 million years) begins with the rapid growth of the Cambrian explosion—the period in which most phyla of multicellular organisms appeared. Over the next 400 million years, global diversity showed little overall trend and was marked by periodic, massive losses of diversity classified as mass extinction events. The largest of these occurred about 250 million years ago and is often called the P-Tr or Permian-Triassic extinction event; various mechanisms, ranging from increased volcanic eruptions to a drastic decrease in the air's oxygen, are thought to have contributed to the P-Tr, which killed about 96 per cent of all marine species and an estimated 70 per cent of land species. Recovery from this 'Great Dying' didn't even begin for 4 to 6 million years, during which only a small number of resilient species roamed the earth. The most recent mass extinction, the K-T event, happened 65 million years ago and ended the reign of the dinosaurs.

The causes of previous mass extinctions aren't fully understood, but extinctions tend to occur when long-term stresses like climate change are compounded by sudden shocks. Four of the previous five mass extinction events were probably caused by greenhouse gas emissions and associated global warming; only the K-T event appears to have been a non-warming impact.

Most biologists agree that our era comprises a new, sixth mass extinction, the Holocene extinction event, caused by ongoing human impacts on the biosphere—primarily numerous forms of habitat destruction and fragmentation. Global warming will soon trump other threats as the leading cause of extinction. Scientists such as EO Wilson argue that the current extinction rate, which is between 100 and 10,000 times the 'background' rate (and projected to rise), will eliminate most species on earth within the next century.

WHY IT MATTERS

Beyond its intrinsic value, biodiversity is necessary to human survival. Ecosystem diversity is crucial to ecosystem integrity, which in turn enables our life support, giving us a livable climate, breathable air, and drinkable water. Food-crop diversity and pollinating insects and bats allow agriculture to support our populations; when disease strikes a food crop, only diversity can save the system from collapse. Plant and animal diversity provide building blocks for medicine, both current and potential; almost half of the pharmaceuticals used in the United States today are manufactured using natural compounds, many of which cannot be synthesised. They also provide critical industrial products used to build our homes and businesses, from wood and rubber to the fuels that underpin our economies—even coal and oil are the products of ancient plant matter and preserved zooplankton remains.

Biodiversity plays a central mythic and symbolic role in our language, religion, literature, art and music, making it a key component of human culture with benefits to society that have not been quantified but are clearly vast. From our earliest prehistory, people have never lived in a world with low biodiversity. We've always been dependent on a varied and rich natural environment for both our physical survival and our psychological and spiritual health. As extinctions multiply, and cannot be undone, we tread further and further into unexplored terrain—a journey from which there is no return.

The huge variety of life on earth is one of the great puzzles of modern science. Why do so many species coexist? Is such diversity inevitable given the laws of evolution? What is the history of life? How did species diverge and succeed one another over the course of evolutionary history? What impact have dominant species had on the formation of today's environment? What role do they currently play in the modification of environmental processes? These questions have fascinated biologists for many decades and are at the heart of one of the greatest intellectual adventures of our time. Since the Rio de Janeiro conference in 1992 biodiversity has also become a social issue and its maintenance one of the major challenges for sustainable development. Biodiversity needs to be protected and managed, but why? The simple answer to this question is that life on our planet depends on biodiversity. Humans draw upon it for the food and the raw materials necessary for their survival.

Biodiversity is a source both of concern and of hope. It is a source of concern due to the incredibly fast rate at which species are vanishing today, which leads the general public, understandably, to ask about the seriousness of the situation. In scientific terms, this issue is fuelling a huge debate about the functional value of biodiversity. For instance, we need to define the role that species play in the biophysicochemical organisations of which they are a part or put another away, their position in ecosystem structure and function. We also need to determine whether a minimum number of species is required for ecosystem survival and whether or not genetic diversity plays the same role as species diversity with respect to ecosystem performance.

These problems are not only of theoretical interest to ecologists. They also have a direct bearing on the quality of our environment. One only has to consider the key role played by biodiversity in what are called ecosystem services, i.e. the ecological functions listed in the Millennium Ecosystem Assessment, which affect the chemical composition of water and the atmosphere, the spread of disease, etc. Or the response of ecosystems to climate change, which will depend above all on the number of species present in a given ecosystem, the nature of the interactions and relationships among the species, their ability to disperse and the impact of such changes on genetic variability. For instance, the amazing plasticity of genomes such as microbial genomes represents an evolutionary response to the spatially and temporally variable environment in which humans live. The future of our planet will thus depend on our capacity to manage this biodiversity.

Biodiversity is also a source of hope. Firstly, there is nothing inevitable about the loss of species. It can be slowed down or even halted by using innovative methods of land management and species reintroduction based on the most recent research findings. Secondly, living organisms are an almost inexhaustible source of molecules of interest to the pharmaceutical and chemical industries, which every day enable us to fight disease or produce a number of substances which are essential to industry. Knowledge of natural substances, their variability and how they change in space and time is also an investment for the future. Finally, manipulating populations of plants, animals or micro-organisms *in situ* makes it possible to rehabilitate degraded environments or maximise certain of their characteristics or functions, depending on the environmental problems, as well as issues related to the exploitation of natural resources, which arise.

DEFINITION OF BIODIVERSITY

Biodiversity is the degree of variation of life forms within a given ecosystem, biome or an entire planet. Biodiversity is a measure of the health of ecosystems. Greater biodiversity implies greater health. Biodiversity is in part a function of climate. In terrestrial habitats, tropical regions are typically rich whereas polar regions support fewer species. Rapid environmental changes typically cause extinctions. One estimate is that less than one per cent of the species that have existed on earth are extant.

Since life began on earth, five major mass extinctions and several minor events have led to large and sudden drops in biodiversity. The Phanerozoic eon (the last 540 million years) marked a rapid growth in biodiversity via the Cambrian explosion — a period during which nearly every phylum of multicellular organisms first appeared. The next 400 million years included repeated, massive biodiversity losses classified as mass extinction events. In the Carboniferous, rainforest collapse led to a great loss of plant and animal life. The Permian–Triassic extinction event, 251 million years ago, was the worst; vertebrate recovery took 30 million years. The most recent, the Cretaceous–Tertiary extinction event, occurred 65 million years ago, and has often attracted more attention than others because it resulted in the extinction of the dinosaurs.

The period since the emergence of humans has displayed an ongoing biodiversity reduction and an accompanying loss of genetic diversity. Named the Holocene extinction, the reduction is caused primarily by human impacts, particularly habitat destruction. Biodiversity's impact on human health is a major international issue.

'Biological diversity' or 'biodiversity' can have many interpretations. It is most commonly used to replace the more clearly defined and long established terms, species diversity and species richness. Biologists most often define biodiversity as the 'totality of genes, species, and ecosystems of a region'. An advantage of this definition is that it seems to describe most circumstances and presents a unified view of the traditional three levels at which biological variety has been identified:

1. Species diversity.
2. Ecosystem diversity.
3. Genetic diversity.

Measuring diversity at one level in a group of organisms may not precisely correspond to diversity at other levels. However, tetrapod (terrestrial vertebrates) taxonomic and ecological diversity shows a very close correlation.

SPECIES DIVERSITY AND BIODIVERSITY

Biologists are not completely sure how many different species live on the earth. Estimates of how many species exist on the earth range from low of 2 million to high of about 100 million. To date, about 2.1 million species have been classified, primarily in the habitats of the middle latitudes. Most of the unclassified species on this planet are invertebrates. This group of organisms includes insects, spiders, mollusks, sponges, flatworms, starfish, urchins, earthworms and crustaceans. These species are often difficult to find and identify because of their small size and the fact that they live in habitats that are difficult to explore. In the tropical rainforest, the cataloguing of species has been quite limited because of this later reason. Scientists estimate that this single biome may contain 50 to 90 per cent of the earth's biodiversity.

Many species have gone extinct over the earth's geologic history. The primary reason for these extinctions is environmental change or biological competition. Since the beginning of the Industrial Revolution, a large number of biologically classified species have gone extinct due to the actions of humans. This includes 83 species of mammals, 113 species of birds, 23 species of amphibians and reptiles, 23 species of fish, about 100 species of invertebrates and over 350 species of plants. Scientists can only estimate the number of unclassified species that have gone extinct. Using various methods of extrapolation, biologists estimate that in 1991 between 4000 to 50,000 unclassified species became extinct, mainly in the tropics, due to our activities. This rate of extinction is some 1000 to 10,000 times greater than the natural rate of species extinction (2–10 species per year) prior to the appearance of human beings. The continued extinction of species on this planet by human activities is one of the greatest environmental problems facing humankind. Several times during the earth's history there have been periods of mass extinctions, when many species became extinct in a relatively short time period (a few million years is a relatively short time when compared to the age of the earth). Scientists are unsure of the causes of both background extinction and mass extinction. Possible explanations for mass extinctions include climate changes or catastrophes such as the earth being hit by a meteor. Since the beginning of time, five or six mass extinctions have occurred that eliminated between 35 and 96 per cent of all species on earth (Table 2.1). Further, it is believed that of all species that ever inhabited the earth over 99 per cent of them are now extinct.

Table 2.1. Major extinction events during the Phanerozoic.

Date of the extinction event	Per cent species lost	Species affected
65 million years ago (Cretaceous)	85%	Dinosaurs, plants (except ferns and seed bearing plants), marine vertebrates and invertebrates. Most mammals, birds, turtles, crocodiles, lizards, snakes, and amphibians were unaffected.
213 million years ago (Triassic)	44%	Marine vertebrates and invertebrates.
248 million years ago (Permian)	75–95%	Marine vertebrates and invertebrates.
380 million years ago (Devonian)	70%	Marine invertebrates.
450–440 million years ago (Ordovician)	50%	Marine invertebrates.

Assessment of the number of different organisms that live on this planet is plagued with difficulties. First and foremost, biologists lack a precise definition of what exactly defines a species. The concept of a species often refers to a population of physically similar individuals that can successfully mate between each other, but cannot produce fertile offspring with other organisms. However, many species are composed of a number of distinct populations that can interbreed even though they display physiological and anatomical differences. Scientists developed the notion of biodiversity to overcome some of the difficulties of species concept. To accomplish this task, biodiversity describes the diversity of life at the following three biological levels:

1. Genetic level or genetic diversity—Genetic diversity refers to the total number of genetic characteristics expressed and recessed in all of the individuals that comprise a particular species.
2. Species level or species diversity—Species diversity is the number of different species of living things living in an area. As mentioned above, a species is a group of plants or animals that are similar and able to breed and produce viable offspring under natural conditions.
3. Ecosystem level or ecosystem diversity—Ecosystem diversity is the variation of habitats, community types and abiotic environments present in a given area. An ecosystem consists of all living and non-living things in a given area that interact with one another.

The biodiversity found on earth today is the product of 3.5 billion years of evolution. In fact, the earth supports more biodiversity today than in any other period in history. However, much of this biodiversity is now facing the threat of extinction because of the actions of humans.

ECOLOGICAL EFFECTS OF BIODIVERSITY

The diversity of species and genes in ecological communities affects the functioning of these communities. These ecological effects of biodiversity in turn affect both climate change through enhanced greenhouse gases, aerosols and loss of land cover, and biological diversity, causing a rapid loss of ecosystems and extinctions of species and local populatior.s. The current rate of extinction is sometimes considered a mass extinction, with current species extinction rates on the order of 100 to 1000 times as high as in the past. The two main areas where the effect of biodiversity on ecosystem function have been studied are the relationship between diversity and productivity, and the relationship between diversity and community stability. More biologically diverse communities appear to be more productive (in terms of biomass production) than are less diverse communities, and they appear to be more stable in the face of perturbations. Also animals that inhabit an area may alter the surviving conditions by factors assimilated by climate.

In order to understand the effects that changes in biodiversity will have on ecosystem functioning, it is important to define some terms. Biodiversity is not easily defined, but may be thought of as the number and/or evenness of genes, species and ecosystems in a region. This definition includes genetic diversity, or the diversity of genes within a species, species diversity, or the diversity of species within a habitat or region, and ecosystem diversity or the diversity of habitats within a region.

Two things commonly measured in relation to changes in diversity are productivity and stability. Productivity is a measure of ecosystem function. It is generally measured by taking the total above ground biomass of all plants in an area. Many assume that it can be used as a general indicator of ecosystem function and that total resource use and other indicators of ecosystem function are correlated with productivity.

Stability is much more difficult to define, but can be generally thought of in two ways. General stability of a population is a measure that assumes stability is higher if there is less of a chance of extinction. This kind of stability is generally measured by measuring the variability of aggregate community properties, like total biomass, over time. The other definition of stability is a measure of resilience and resistance, where an ecosystem that returns quickly to an equilibrium after a perturbation or resists invasion is thought of as more stable than one that doesn't.

Productivity and Stability as Indicators of Ecosystem Health

The importance of stability in community ecology is clear. An unstable ecosystem will be more likely to lose species. Thus, if there is indeed a link between diversity and stability, it is likely that losses of diversity could feedback on themselves, causing even more losses of species. Productivity, on the other hand, has a less clear importance in community ecology. In managed areas like cropland, and in areas where animals are grown or caught, increasing productivity increases the economic success of the area and implies that the area has become more efficient, leading to possible long-term resource sustainability. It is more difficult to find the importance of productivity in natural ecosystems.

Beyond the value biodiversity has in regulating and stabilising ecosystem processes, there are direct economic consequences of losing diversity in certain ecosystems and in the world as a whole. Losing species means losing potential foods, medicines, industrial products, and tourism, all of which have a direct economic effect on peoples, lives.

Effects on Community Productivity

1. Complementarity: Plant species coexistence is thought to be the result of niche partitioning, or differences in resource requirements among species. By complementarity, a more diverse plant community should be able to use resources more completely, and thus be more productive. Also called niche differentiation, this mechanism is a central principle in the functional group approach, which breaks species diversity down into functional components.

2. Facilitation: Facilitation is a mechanism whereby certain species help or allow other species to grow by modifying the environment in a way that is favourable to a co-occurring species. Plants can interact through an intermediary like nitrogen, water, temperature, space, or interactions with weeds or herbivores among others. Some examples of facilitation include large desert perennials acting as nurse plants, aiding the establishment of young neighbours of other species by alleviating water and temperature stress, and nutrient enrichment by nitrogen-fixers such as legumes.

3. The sampling effect: The sampling effect of diversity can be thought of as having a greater chance of including a species of greatest inherent productivity in a plot that is more diverse. This provides for a composition effect on productivity, rather than diversity being a direct cause. However, the sampling effect may in fact be a compilation of different effects. The sampling effect can be separated into the greater likelihood of selecting a species that is: (i) adapted well to particular site conditions, (ii) of a greater inherent productivity. Additionally, one can add to the sampling effect a greater likelihood of including (iii) a pair of species that highly complement each other and (iv) a certain species with a large facilitative effect on other members of the community.

Field experiments

Field experiments to test the degree to which diversity affects community productivity have found many things, but many long-term studies in grassland ecosystems have found that diversity does indeed enhance the productivity of ecosystems. Evidence of the relationship has also been found in grassland microcosms. However, these different studies have come to different conclusions as to whether the cause was due more to diversity or to species composition. Recent mathematical models have highlighted the importance of ecological context in unravelling this problem. Some models have indicated the importance of disturbance rates and spatial heterogeneity of the environment, others have indicated that the time since disturbance and the habitat's carrying capacity can cause differing relationships. Each ecological context should yield not only a different relationship, but a different contribution to the relationship due to diversity and to composition.

Future research

In order to correctly identify the consequences of diversity on productivity and other ecosystem processes, many things must happen. First, it is imperative that scientists stop looking for a single relationship. It is obvious now from the models, the data, and the theory that there is no one overarching effect of diversity on productivity. Scientists must try to quantify the differences between composition effect and diversity effects, as many experiments never quantify the final realised species diversity (instead only counting numbers of species of seeds planted) and confound a sampling effect for facilitators (a compositional factor) with diversity effects.

Relative amounts of overyielding (or how much more a species grows when grown with other species than it does in monoculture) should be used rather than absolute amounts as relative overyielding can give clues as to the mechanism by which diversity is influencing productivity, however if experimental protocols are incomplete, one may be able to indicate the existence of a complementary or facilitative effect in the experiment, but not be able to recognise its cause. Experimenters should know what the goal of their experiment is, that is, whether it is meant to inform natural or managed ecosystems, as the sampling effect may only be a real effect of diversity in natural ecosystems (managed ecosystems are composed to maximise complementarity and facilitation regardless of species number). By knowing this, they should be able to choose spatial and temporal scales that are appropriate for their experiment. Lastly, to resolve the diversity-function debate, it is advisable that experiments be done with large amounts of spatial and resource heterogeneity and environmental fluctuation over time, as these types of experiments should be able to demonstrate the diversity-function relationship more easily.

Effects on Community Stability

1. Averaging effect: If all species have differential responses to changes in the ecosystem over time, then the averaging of these responses will cause a more temporally stable ecosystem if more species are in the ecosystem. This effect is a statistical effect due to summing random variables.
2. Negative covariance effect: If some species do better when other species are not doing well, then when there are more species in the ecosystem, their overall variance will be lower than if there were fewer species in the system. This lower variance indicates higher stability. This effect is a consequence of competition as highly competitive species will negatively covary.
3. Insurance effect: If an ecosystem contains more species then it will have a greater likelihood of having redundant stabilising species, and it will have a greater number of species that respond differently to perturbations. This will enhance an ecosystem's ability to buffer perturbations.
4. Resistance to invasion: Diverse communities may use resources more completely than simple communities because of a diversity effect for complementarity. Thus invaders may have reduced success in diverse ecosystems or there may be a reduced likelihood that an invading species will introduce a new property or process to a diverse ecosystem.
5. Resistance to disease: A decreased number of competing plant species may allow the abundances of other species to increase, facilitating the spread of diseases of those species.

Review of temporal stability data

Models have predicted that empirical relationships between temporal variation of community productivity and species diversity are indeed real, and that they almost have to be. Some temporal stability data can be almost completely explained by the averaging effect by constructing null models to test the data against. Competition, which causes negative covariances, only serves to strengthen these relationships.

Review of resistance and resilience stability data

This area is more contentious than the area of temporal stability, mostly because some have tried generalising the findings of the temporal stability models and theory to stability in general. While the relationship between temporal variations in productivity and diversity has a mathematical cause, which will allow the relationship to be seen much more often than not, it is not the case with resistance/resilience stability. Some experimenters have seen a correlation between diversity and reduced invasibility, though many have also seen the opposite. The correlation between diversity and disease is also tenuous, though theory and data do seem to support it.

Future research

In order to more fully understand the effects of diversity on the temporal stability of ecosystems it is necessary to recognise that they are bound to occur. By constructing null models to test the data against it becomes possible to find situations and ecological contexts where ecosystems become more or less stable than they should be. Finding these contexts would allow for mechanistic studies into why these ecosystems are more stable, which may allow for applications in conservation management.

More importantly more complete experiments into whether diverse ecosystems actually resist invasion and disease better than their less diverse equivalents as invasion and disease are two important factors that lead to species extinctions in the present day.

Theory and Preliminary Effects from Examining Food Webs

One major problem with both the diversity-productivity and diversity-stability debates discussed up to this point is that both focus on interactions at just a single trophic level. That is, they are concerned with only one level of the food web, namely plants. Other research, unconcerned with the effects of diversity, has demonstrated strong top-down forcing of ecosystems. There is very little actual data available regarding the effects of different food webs, but theory helps us in this area. First, if a food web in an ecosystem has a lot of weak interactions between different species, then it should have more stable populations and the community as a whole should be more stable. If upper levels of the web are more diverse, then there will be less biomass in the lower levels and if lower levels are more diverse they will better be able to resist consumption and be more stable in the face of consumption. Also, top-down forcing should be reduced in less diverse ecosystems because of the bias for species in higher trophic levels to go extinct first.

Lastly, it has recently been shown that consumers can dramatically change the biodiversity-productivity-stability relationships that are implied by plants alone. Thus, it will be important in the future to incorporate food web theory into the future study of the effects of biodiversity. In addition this complexity will need to be addressed when designing biodiversity management plans.

GENETIC DIVERSITY

Genetic diversity, the level of biodiversity, refers to the total number of genetic characteristics in the genetic make-up of a species. It is distinguished from genetic variability, which describes the tendency of genetic characteristics to vary. Genetic diversity serves as a way for populations to adapt to changing environments. With more variation, it is more likely that some individuals in a population will possess variations of alleles that are suited for the environment. Those individuals are more likely to survive to produce offspring bearing that allele. The population will continue for more generations because of the success of these individuals.

The academic field of population genetics includes several hypotheses and theories regarding genetic diversity. The neutral theory of evolution proposes that diversity is the result of the accumulation of neutral substitutions. Diversifying selection is the hypothesis that two subpopulations of a species live in different environments that select for different alleles at a particular locus. This may occur, for instance, if a species has a large range relative to the mobility of individuals within it. Frequency-dependent selection is the hypothesis that as alleles become more common, they become more vulnerable. This is often invoked in host-pathogen interactions, where a high frequency of a defensive allele among the host means that it is more likely that a pathogen will spread if it is able to overcome that allele.

Importance of Genetic Diversity

There are many different ways to measure genetic diversity. The modern causes for the loss of animal genetic diversity have also been studied and identified. A 2009 study conducted by the National Science Foundation found that genetic diversity and biodiversity are dependent upon each other—that diversity within a species is necessary to maintain diversity among species, and *vice versa*. According to the lead researcher in the study, Dr. Richard Lankau, 'If any one type is removed from the system, the cycle can break down, and the community becomes dominated by a single species.'

The interdependence between genetic and biological diversity is delicate. Changes in biological diversity lead to changes in the environment, leading to adaptation of the remaining species. Changes in genetic diversity, such as in loss of species, leads to a loss of biological diversity.

Survival and Adaptation

Genetic diversity plays a very important role in survival and adaptability of a species because when a species's environment changes, slight gene variations are necessary to produce changes in the organisms' anatomy that enables it to adapt and survive. A species that has a large degree of genetic diversity among its population will have more variations from which to choose the most fit alleles. Increase in genetic diversity is also essential for a species to evolve. Species that have very little genetic variation are at a great risk. With very little gene variation within the species, healthy reproduction becomes increasingly difficult, and offspring often deal with similar problems to those of inbreeding. The vulnerability of a population to certain types of diseases can also increase with reduction in genetic diversity.

Agricultural Relevance

When humans initially started farming, they used selective breeding to pass on desirable traits of the crops while omitting the undesirable ones. Selective breeding leads to monocultures: entire farms of nearly genetically identical plants. Little to no genetic diversity makes crops extremely susceptible to widespread disease. Bacteria morph and change constantly. When a disease causing bacterium changes to attack a specific genetic variation, it can easily wipe out vast quantities of the species. If the genetic variation that the bacterium is best at attacking happens to be that which humans have selectively bred to use for harvest, the entire crop will be wiped out.

A very similar occurrence is the cause of the infamous Potato Famine in Ireland. Since new potato plants do not come as a result of reproduction but rather from pieces of the parent plant, no genetic diversity is developed, and the entire crop is essentially a clone of one potato, it is especially susceptible to an epidemic. In the 1840s, much of Ireland's population depended on potatoes for food. They planted namely the 'lumper' variety of potato, which was susceptible to a rot-causing plasmodiophorid called *Phytophthora infestans*. This plasmodiophorid destroyed the vast majority of the potato crop, and left one million people to starve to death.

Coping with Poor Genetic Diversity

The natural world has several ways of preserving or increasing genetic diversity. Among oceanic plankton, viruses aid in the genetic shifting process. Ocean viruses, which infect the plankton, carry genes of other organisms in addition to their own. When a virus containing the genes of one cell infects another, the genetic make-up of the latter changes. This constant shift of genetic make-up helps to maintain a healthy population of plankton despite complex and unpredictable environmental changes.

Cheetahs are a threatened species. Extremely low genetic diversity and resulting poor sperm quality has made breeding and survivorship difficult for cheetahs—only about 5 per cent of cheetahs survive to adulthood. About 10,000 years ago, all but the jubatus species of cheetahs died out. The species encountered a population bottleneck and close family relatives were forced to mate with each other or inbreed. However, it has been recently discovered that female cheetahs can mate with more than one male per litter of cubs. They undergo induced ovulation, which means that a new egg is produced every time a female mates. By mating with multiple males, the mother increases the genetic diversity within a single litter of cubs.

Measures of Genetic Diversity

Genetic diversity of a population can be assessed by some simple measures:
1. Gene diversity is the proportion of polymorphic loci across the genome.
2. Heterozygosity is the mean number of individuals with polymorphic loci.
3. Alleles per locus is also used to demonstrate variability.

Other Measures of Diversity

Alternatively, other types of diversity may be assessed for organisms:
1. Taxonomic diversity.
2. Ecological diversity.
3. Morphological diversity.
4. Degeneracy.

There are broad correlations between different types of diversity. For example, there is a close link between vertebrate taxonomic and ecological diversity.

DISTRIBUTION

Selection bias amongst researchers may contribute to biased empirical research for modern estimates of biodiversity. Biodiversity is not evenly distributed. Flora and fauna diversity depends on climate, altitude, soils and the presence of other species. Diversity consistently measures higher in the tropics and in other localised regions such as Cape Floristic Province and lower in polar regions generally. In 2008 many species were formally classified as rare or endangered or threatened; moreover, scientists have estimated that millions more species are at risk which have not been formally recognised. About 40 per cent of the 40,177 species assessed using the IUCN Red List criteria are now listed as threatened with extinction—a total of 16,119.

Even though terrestrial biodiversity declines from the equator to the poles, this characteristic is unverified in aquatic ecosystems, especially in marine ecosystems. In addition, several assessments reveal tremendous diversity in higher latitudes. Generally terrestrial biodiversity is up to 25 times greater than ocean biodiversity.

A biodiversity hotspot is a region with a high level of endemic species. Hotspots were first named in 1988 by Dr. Norman Myers. Many hotspots have large nearby human populations. Most hotspots are located in the tropics and most of them are forests.

Brazil's Atlantic Forest is considered one such hotspot, containing roughly 20,000 plant species, 1350 vertebrates, and millions of insects, about half of which occur nowhere else. The island of Madagascar, particularly the unique Madagascar dry deciduous forests and lowland rainforests, possess a high ratio of endemism. Since the island separated from mainland Africa 65 million years ago, many species and ecosystems have evolved independently. Indonesia's 17,000 islands cover 7,35,355 square miles (19,04,560 km^2) contain 10 per cent of the world's flowering plants, 12 per cent of mammals and 17 per cent of reptiles, amphibians and birds—along with nearly 240 million people. Many regions of high biodiversity and/or endemism arise from specialised habitats which require unusual adaptations, for example alpine environments in high mountains or Northern European peat bogs.

EVOLUTION

Biodiversity is the result of 3.5 billion years of evolution. The origin of life has not been definitely established by science, however some evidence suggests that life may already have been well-established only a few hundred million years after the formation of the earth. Until approximately 600 million years ago, all life consisted of archaea, bacteria, protozoans and similar single-celled organisms.

The fossil record suggests that the last few million years featured the greatest biodiversity in history. However, not all scientists support this view, since there is uncertainty as to how strongly the fossil record is biased by the greater availability and preservation of recent geologic sections. Some scientists believe that corrected for sampling artifacts, modern biodiversity may not be much different from biodiversity 300 million years ago, whereas others consider the fossil record reasonably reflective of the diversification of life.

Estimates of the present global macroscopic species diversity vary from 2 million to 100 million, with a best estimate of somewhere near 13–14 million, the vast majority arthropods. Diversity appears to increase continually in the absence of natural selection.

Evolutionary Diversification

The existence of a 'global carrying capacity', limiting the amount of life that can live at once, is debated, as is the question of whether such a limit would also cap the number of species. While records of life in the sea shows a logistic pattern of growth, life on land (insects, plants and tetrapods) shows an exponential rise in diversity. As one author states, 'Tetrapods have not yet invaded 64 per cent of potentially habitable modes, and it could be that without human influence the ecological and taxonomic diversity of tetrapods would continue to increase in an exponential fashion until most or all of the available ecospace is filled.'

On the other hand, changes through the Phanerozoic correlate much better with the hyperbolic model (widely used in population biology, demography and macrosociology, as well as fossil biodiversity) than with exponential and logistic models. The latter models imply that changes in diversity are guided by a first-order positive feedback (more ancestors, more descendants) and/or a negative feedback arising from resource limitation.

Hyperbolic model implies a second-order positive feedback. The hyperbolic pattern of the world population growth arises from a second-order positive feedback between the population size and the rate of technological growth. The hyperbolic character of biodiversity growth can be similarly accounted for by a feedback between diversity and community structure complexity. The similarity between the curves of biodiversity and human population probably comes from the fact that both are derived from the interference of the hyperbolic trend with cyclical and stochastic dynamics.

Most biologists agree however that the period since human emergence is part of a new mass extinction, named the Holocene extinction event, caused primarily by the impact humans are having on the environment. It has been argued that the present rate of extinction is sufficient to eliminate most species on the planet earth within 100 years.

New species are regularly discovered (on average between 5–10,000 new species each year, most of them insects) and many, though discovered, are not yet classified (estimates are that nearly 90 per cent of all arthropods are not yet classified). Most of the terrestrial diversity is found in tropical forests.

HUMAN BENEFITS

Biodiversity supports ecosystem services including air quality, climate (e.g. CO_2 sequestration), water purification, pollination and prevention of erosion. Since the stone age, species loss has accelerated above the prior rate, driven by human activity. Estimates of species loss are at a rate 100–10,000 times as fast as is typical in the fossil record. Non-material benefits include spiritual and aesthetic values, knowledge systems and the value of education.

Agriculture

The reservoir of genetic traits present in wild varieties and traditionally grown landraces is extremely important in improving crop performance. Important crops, such as potato, banana and coffee, are often derived from only a few genetic strains. Improvements in crop species over the last 250 years have been largely due to incorporating genes from wild varieties and species into cultivars. Crop breeding for beneficial traits has helped to more than double crop production in the last 50 years as a result of the Green Revolution. A biodiverse environment preserves the genome from which such productive genes are drawn.

Crop diversity aids recovery when the dominant cultivar is attacked by a disease or predator:

1. The Irish potato blight of 1846 was a major factor in the deaths of one million people and the emigration of another million. It was the result of planting only two potato varieties, both vulnerable to the blight.
2. When rice grassy stunt virus struck rice fields from Indonesia to India in the 1970s, 6,273 varieties were tested for resistance. Only one was resistant, an Indian variety, and known to science only since 1966. This variety formed a hybrid with other varieties and is now widely grown.
3. Coffee rust attacked coffee plantations in Sri Lanka, Brazil and Central America in 1970. A resistant variety was found in Ethiopia. Although the diseases are themselves a form of biodiversity.

Monoculture was a contributing factor to several agricultural disasters, including the European wine industry collapse in the late 19th century, and the US Southern Corn Leaf Blight epidemic of 1970.

Although about 80 per cent of humans' food supply comes from just 20 kinds of plants, humans use at least 40,000 species. Many people depend on these species for food, shelter, and clothing. Earth's surviving biodiversity provides resources for increasing the range of food and other products suitable for human use, although the present extinction rate shrinks that potential.

Human Health

Biodiversity's relevance to human health is becoming an international political issue, as scientific evidence builds on the global health implications of biodiversity loss. This issue is closely linked with the issue of climate change, as many of the anticipated health risks of climate change are associated with changes in biodiversity (e.g. changes in populations and distribution of disease vectors, scarcity of freshwater, impacts on agricultural biodiversity and food resources, etc.). Some of the health issues influenced by biodiversity include dietary health and nutrition security, infectious disease, medical science and medicinal resources, social and psychological health. Biodiversity is also known to have an important role in reducing disaster risk, and in post-disaster relief and recovery efforts.

Biodiversity provides critical support for drug discovery and the availability of medicinal resources. A significant proportion of drugs are derived, directly or indirectly, from biological sources: at least 50 per cent of the pharmaceutical compounds on the US market are derived from plants, animals and micro-organisms, while about 80 per cent of the world population depends on medicines from nature (used in either modern or traditional medical practice) for primary healthcare. Only a tiny fraction of wild species has been investigated for medical potential. Biodiversity has been critical to advances throughout the field of bionics. Evidence from market analysis and biodiversity science indicates that the decline in output from the pharmaceutical sector since the mid-1980s can be attributed to a move away from natural product exploration (bioprospecting) in favour of genomics and synthetic chemistry; meanwhile, natural products have a long history of supporting significant economic and health innovation. Marine ecosystems are particularly important, although inappropriate bioprospecting can increase biodiversity loss, as well as violating the laws of the communities and states from which the resources are taken. Higher biodiversity also limits the spread of infectious diseases as many different species act as buffers to them.

Business and Industry

Many industrial materials derive directly from biological sources. These include building materials, fibres, dyes, rubber and oil. Biodiversity is also important to the security of resources such as water, timber, paper, fibre and food. As a result, biodiversity loss is a significant risk factor in business development and a threat to long-term economic sustainability.

Leisure, Cultural and Aesthetic Value

Biodiversity enriches leisure activities such as hiking, birdwatching or natural history study. Biodiversity inspires musicians, painters, sculptors, writers and other artists. Many cultures view themselves as an integral part of the natural world which requires them to respect other living organisms.

Popular activities such as gardening, fishkeeping and specimen collecting strongly depend on biodiversity. The number of species involved in such pursuits is in the tens of thousands, though the majority do not enter commerce. The relationships between the original natural areas of these often exotic animals and plants and commercial collectors, suppliers, breeders, propagators and those who promote their understanding and enjoyment are complex and poorly understood. The general public responds well to exposure to rare and unusual organisms, reflecting their inherent value.

Philosophically it could be argued that biodiversity has intrinsic aesthetic and spiritual value to mankind in and of itself. This idea can be used as a counterweight to the notion that tropical forests and other ecological realms are only worthy of conservation because of the services they provide.

Other Services

Biodiversity supports many ecosystem services that are often not readily visible. It plays a part in regulating the chemistry of our atmosphere and water supply. Biodiversity is directly involved in water purification, recycling nutrients and providing fertile soils. Experiments with controlled environments have shown that humans cannot easily build ecosystems to support human needs; for example insect pollination cannot be mimicked, and that activity alone represents tens of billions of dollars in ecosystem services per year to humankind. Ecosystem stability is also positively related to biodiversity, protecting against disruption by extreme weather or human exploitation.

According to the Global Taxonomy Initiative and the European Distributed Institute of Taxonomy, the total number of species for some phyla may be much higher than what was known in 2010:

1. 10–30 million insects; (of some 0.9 million we know today).
2. 5–10 million bacteria.

3. 1.5 million fungi; (of some 0.075 million we know today).
4. 1 million mites.
5. The number of microbial species is not reliably known, but the Global Ocean Sampling Expedition dramatically increased the estimates of genetic diversity by identifying an enormous number of new genes from near-surface plankton samples at various marine locations, initially over the 2004–2006 period. The findings may eventually cause a significant change in the way science defines species and other taxonomic categories.

Since the rate of extinction has increased, many extant species may become extinct before they are described.

SPECIES LOSS RATES

During the last century, decreases in biodiversity have been increasingly observed. In 2007, German Federal Environment Minister Sigmar Gabriel cited estimates that up to 30 per cent of all species will be extinct by 2050. Of these, about one-eighth of known plant species are threatened with extinction. Estimates reach as high as 1,40,000 species per year (based on Species-area theory). This figure indicates unsustainable ecological practices, because few species emerge each year. Almost all scientists acknowledge that the rate of species loss is greater now than at any time in human history, with extinctions occurring at rates hundreds of times higher than background extinction rates.

THREATS

Habitat Destruction

Habitat destruction has played a key role in extinctions, especially related to tropical forest destruction. Factors contributing to habitat loss are: overpopulation, deforestation, pollution (air pollution, water pollution, soil contamination) and global warming or climate change.

Habitat size and numbers of species are systematically related. Physically larger species and those living at lower latitudes or in forests or oceans are more sensitive to reduction in habitat area. Conversion to 'trivial' standardised ecosystems (e.g. monoculture following deforestation) effectively destroys habitat for the more diverse species that preceded the conversion. In some countries lack of property rights or lax law/regulatory enforcement necessarily leads to biodiversity loss (degradation costs having to be supported by the community).

A 2007 study conducted by the National Science Foundation found that biodiversity and genetic diversity are codependent — that diversity among species requires diversity within a species, and *vice versa*. 'If any one type is removed from the system, the cycle can break down, and the community becomes dominated by a single species.' At present, the most threatened ecosystems are found in freshwater, according to the Millennium Ecosystem Assessment 2005, which was confirmed by the 'Freshwater Animal Diversity Assessment', organised by the biodiversity platform, and the French Institute of Research Development.

Co-extinctions are a form of habitat destruction. Co-extinction occurs when the extinction or decline in one accompanies the other, such as in plants and beetles.

Introduced and Invasive Species

Barriers such as large rivers, seas, oceans, mountains and deserts encourage diversity by enabling independent evolution on either side of the barrier. Invasive species occur when those barriers are blurred. Without barriers such species occupy new niches, substantially reducing diversity. Repeatedly humans have helped these species circumvent these barriers, introducing them for food and other purposes. This has occurred on a time scale much shorter than the eons that historically have been required for a species to extend its range.

Not all introduced species are invasive, nor all invasive species deliberately introduced. In cases such as the zebra mussel, invasion of US waterways was unintentional. In other cases, such as mongooses in Hawaii, the introduction is deliberate but ineffective (nocturnal rats were not vulnerable to the diurnal mongoose). In other cases, such as oil palms in Indonesia and Malaysia, the introduction produces substantial economic benefits, but the benefits are accompanied by costly unintended consequences.

Finally, an introduced species may unintentionally injure a species that depends on the species it replaces. In Belgium, *Prunus spinosa* from Eastern Europe leafs much sooner than its West European counterparts, disrupting the feeding habits of the *Thecla betulae* butterfly (which feeds on the leaves). Introducing new species often leaves endemic and other local species unable to compete with the exotic species and unable to survive. The exotic organisms may be predators, parasites or may simply outcompete indigenous species for nutrients, water and light.

At present, several countries have already imported so many exotic species, particularly agricultural and ornamental plants, that the own indigenous fauna/flora may be outnumbered.

Genetic pollution

Endemic species can be threatened with extinction through the process of genetic pollution, i.e. uncontrolled hybridisation, introgression and genetic swamping. Genetic pollution leads to homogenisation or replacement of local genomes as a result of either a numerical and/or fitness advantage of an introduced species. Hybridisation and introgression are side-effects of introduction and invasion. These phenomena can be especially detrimental to rare species that come into contact with more abundant ones. The abundant species can interbreed with the rare species, swamping its gene pool. This problem is not always apparent from morphological (outward appearance) observations alone.

Some degree of gene *flow* is normal adaptation, and not all gene and genotype constellations can be preserved. However, hybridisation with or without introgression may, nevertheless, threaten a rare species existence.

Overexploitation

Overexploitation occurs when a resource is consumed at an unsustainable rate. This occurs on land in the form of overhunting, excessive logging, poor soil conservation in agriculture and the illegal wildlife trade. Joe Walston, director of the Wildlife Conservation Society's Asian programs, called the latter the 'single largest threat' to biodiversity in Asia. The international trade of endangered species is second in size only to drug trafficking.

About 25 per cent of world fisheries are now overfished to the point where their current biomass is less than the level that maximises their sustainable yield. The overkill hypothesis explains why earlier megafaunal extinctions occurred within a relatively short period of time. This can be connected with human migration.

Hybridisation, Genetic Pollution/Erosion and Food Security

In agriculture and animal husbandry, the green revolution popularised the use of conventional hybridisation to increase yield. Often hybridised breeds originated in developed countries and were further hybridised with local varieties in the developing world to create high yield strains resistant to local climate and diseases. Local governments and industry have been pushing hybridisation. Formerly huge gene pools of various wild and indigenous breeds have collapsed causing widespread genetic erosion and genetic pollution. This has resulted in loss of genetic diversity and biodiversity as a whole.

GM organisms have genetic material altered by genetic engineering procedures such as recombinant DNA technology. GM crops have become a common source for genetic pollution, not only of wild varieties but also of domesticated varieties derived from classical hybridisation.

Genetic erosion coupled with genetic pollution may be destroying unique genotypes, thereby creating a hidden crisis which could result in a severe threat to our food security. Diverse genetic material could cease to exist which would impact our ability to further hybridise food crops and livestock against more resistant diseases and climatic changes.

Climate Change

Global warming is also considered to be a major threat to global biodiversity. For example coral reefs—which are biodiversity hotspots will be lost in 20 to 40 years if global warming continues at the current trend. In 2007, an international collaborative study on four continents estimated that 10 per cent of species would become extinct by 2050 because of global warming. 'We need to limit climate change or we wind up with a lot of species in trouble, possibly extinct', said Dr. Lee Hannah, a co-author of the paper and chief climate change biologist at the Center for Applied Biodiversity Science at Conservation International.

Overpopulation

From 1950 to 2005, world population increased from 2.5 billion to 6.5 billion and is forecast to reach a plateau of more than 9 billion during the 21st century. Sir David King, former chief scientific adviser to the UK government, told a parliamentary inquiry: 'It is self-evident that the massive growth in the human population through the 20th century has had more impact on biodiversity than any other single factor.'

HOLOCENE EXTINCTION

Rates of decline in biodiversity in this sixth mass extinction match or exceed rates of loss in the five previous mass extinction events in the fossil record. Loss of biodiversity results in the loss of natural capital that supplies ecosystem goods and services. The economic value of 17 ecosystem services for earth's biosphere (calculated in 2007) has an estimated value of US$ 37 trillion (3.7×10^{13}) per year.

CONSERVATION

Conservation biology matured in the mid-20th century as ecologists, naturalists and other scientists began to research and address issues pertaining to global biodiversity declines. The conservation ethic advocates management of natural resources for the purpose of sustaining biodiversity in species, ecosystems, the evolutionary process, and human culture and society.

Conservation biology is reforming around strategic plans to protect biodiversity. Preserving global biodiversity is a priority in strategic conservation plans that are designed to engage public policy and concerns affecting local, regional and global scales of communities, ecosystems and cultures. Action plans identify ways of sustaining human well-being, employing natural capital, market capital and ecosystem services.

PROTECTION AND RESTORATION TECHNIQUES

The most powerful technique is to preserve habitat. Exotic species removal allows less competitive species to recover their ecological niches. Exotic species that have become a pest can be identified taxonomically [e.g. with digital automated identification system (DAISY)], using the barcode of life. Removal is practical only given large groups of individuals due to the economic cost.

Once the preservation of the remaining native species in an area is assured. Missing species can be identified and reintroduced using databases such as the Encyclopedia of Life and the Global Biodiversity Information Facility.

Other techniques include:

1. Biodiversity banking places a monetary value on biodiversity. One example is the Australian Native Vegetation Management Framework.
2. Gene banks are collections of specimens and genetic material. Some banks intend to reintroduce banked species to the ecosystem (e.g. via tree nurseries).
3. Reducing and better targeting of pesticides allows more species to survive in agricultural and urbanised areas.
4. Location-specific approaches are less useful for protecting migratory species. One approach is to create wildlife corridors that correspond to the animals' movements. National and other boundaries can complicate corridor creation.

Resource Allocation

Focusing on limited areas of higher potential biodiversity promises greater immediate return on investment than spreading resources evenly or focusing on areas of little diversity but greater interest in biodiversity.

A second strategy focuses on areas that retain most of their original diversity, which typically require little or no restoration. These are typically non-urbanised, non-agricultural areas. Tropical areas often fit both criteria, given their natively high diversity and relative lack of development.

LEGAL STATUS

Biodiversity is taken into account in some political and judicial decisions:

1. The relationship between law and ecosystems is very ancient and has consequences for biodiversity. It is related to private and public property rights. It can define protection for threatened ecosystems, but also some rights and duties (for example, fishing and hunting rights).
2. Law regarding species is more recent. It defines species that must be protected because they may be threatened by extinction. The US Endangered Species Act is an example of an attempt to address the 'law and species' issue.
3. Laws regarding gene pools are only about a century old. Domestication and plant breeding methods are not new, but advances in genetic engineering has led to tighter laws covering distribution of genetically modified organisms, gene patents and process patents. Governments struggle to decide whether to focus on for example, genes, genomes or organisms and species.

Global agreements such as the 'Convention on Biological Diversity', give 'sovereign national rights over biological resources' (not property). The agreements commit countries to 'conserve biodiversity', 'develop resources for sustainability' and 'share the benefits' resulting from their use. Biodiverse countries that allow bioprospecting or collection of natural products, expect a share of the benefits rather than allowing the individual or institution that discovers/exploits the resource to capture them privately. Bioprospecting can become a type of biopiracy when such principles are not respected.

Sovereignty principles can rely upon what is better known as Access and Benefit Sharing Agreements (ABAs). The Convention on Biodiversity implies informed consent between the source country and the collector, to establish which resource will be used and for what, and to settle on a fair agreement on benefit sharing.

Uniform approval for use of biodiversity as a legal standard has not been achieved, however. Bosselman argues that biodiversity should not be used as a legal standard, claiming that the remaining areas of scientific uncertainty cause unacceptable administrative waste and increase litigation without promoting preservation goals.

ANALYTICAL LIMITS

Taxonomic and Size Relationships

Less than one per cent of all species that have been described have been studied beyond simply noting their existence. The vast majority of earth's species are microbial. Contemporary biodiversity physics is 'firmly fixated on the visible [macroscopic] world'. For example, microbial life is metabolically and environmentally more diverse than multicellular life . 'On the tree of life, based on analyses of small-subunit ribosomal RNA, visible life consists of barely noticeable twigs. The inverse relationship of size and population recurs higher on the evolutionary ladder — to a first approximation, all multicellular species on earth are insects.' Insect extinction rates are high — supporting the Holocene extinction hypothesis.

Chapter 3

Biodiversity and Conservation

INTRODUCTION

Biodiversity is a modern term which simply means 'the variety of life on earth'. This variety can be measured on several different levels.

Genetic: Variation between individuals of the same species. This includes genetic variation between individuals in a single population, as well as variations between different populations of the same species. Genetic differences can now be measured using increasingly sophisticated techniques. These differences are the raw material of evolution.

Species: Species diversity is the variety of species in a given region or area. This can either be determined by counting the number of different species present, or by determining taxonomic diversity. Taxonomic diversity is more precise and considers the relationship of species to each other. It can be measured by counting the number of different taxa (the main categories of classification) present. For example, a pond containing three species of snails and two fish, is more diverse than a pond containing five species of snails, even though they both contain the same number of species. High species biodiversity is not always necessarily a good thing. For example, a habitat may have high species biodiversity because many common and widespread species are invading it at the expense of species restricted to that habitat.

Ecosystem: Communities of plants and animals, together with the physical characteristics of their environment (e.g. geology, soil and climate) interlink together as an ecological system or 'ecosystem'. Ecosystem diversity is more difficult to measure because there are rarely clear boundaries between different ecosystems and they grade into one another. However, if consistent criteria are chosen to define the limits of an ecosystem, then their number and distribution can also be measured.

Estimates of global species diversity vary enormously because it is so difficult to guess how many species there may be in less well explored habitats such as untouched rainforest. Rainforest areas which have been sampled have shown such amazing biodiversity (nineteen trees sampled in Panama were found to contain 1200 different beetle species alone!) that the mind boggles over how many species there might remain to be discovered in unexplored rain forest areas and microhabitats.

Global species estimates range from 2 million to 100 million species. Ten million is probably nearer the mark. Only 1.4 million species have been named. Of these, approximately 2,50,000 are plants and 7,50,000 are insects.

New species are continually being discovered every year. The number of species present in little-known ecosystems such as the soil beneath our feet and the deep-sea can only be guessed at. It has been estimated that the deep-sea floor may contain as many as a million undescribed new species. To put it simply, we really have absolutely no idea how many species there are!

LOSSES OF BIODIVERSITY

Extinction is a fact of life. Species have been evolving and dying out ever since the origin of life. One only has to look at the fossil record to appreciate this. (It has been estimated that surviving species constitute about one per cent of the species that have ever lived.)

However, species are now becoming extinct at an alarming rate, almost entirely as a direct result of human activities. Previous mass extinctions evident in the geological record are thought to have been brought about mainly by massive climatic or environmental shifts. Mass extinctions as a direct consequence of the activities of a single species are unprecedented in geological history.

The loss of species in tropical ecosystems such as the rainforests, is extremely well-publicised and of great concern. However, equally worrying is the loss of habitat and species closer to home in Britain. This is arguably on a comparable scale, given the much smaller area involved.

Predictions and estimates of future species losses abound. One such estimate calculates that a quarter of all species on earth are likely to be extinct, or on the way to extinction within 30 years. Another predicts that within 100 years, three quarters of all species will either be extinct, or in populations so small that they can be described as 'the living dead'.

It must be emphasised that these are only predictions. Most predictions are based on computer models and as such, need to be taken with a very generous pinch of salt. For a start, we really have no idea how many species there are on which to base our initial premise. There are also so many variables involved that it is almost impossible to predict what will happen with any degree of accuracy. Some species actually benefit from human activities, while many others are adversely affected. Nevertheless, it is indisputable that if the human population continues to soar, then the ever increasing competition with wildlife for space and resources will ensure that habitats and their constituent species will lose out.

It is difficult to appreciate the scale of human population increases over the last two centuries. Despite the horrendous combined mortality rates of two World Wars, Hitler, Stalin, major flu pandemics and Aids, there has been no dampening effect on rising population levels. In 1950, the world population was 2.4 billion. Just over 50 years later, the world population has almost tripled, reaching 6.5 billion.

In the UK alone, the population increases by the equivalent of a new city every year. Corresponding demands for a higher standard of living for all, further exacerbates the problem. It has been estimated that if everyone in the world lived at the UK standard of living (and why should people elsewhere be denied this right) then we would either need another three worlds to supply the necessary resources or alternatively, would need to reduce the world population to 2 billion.

The only possible conclusion is that unless human populations are substantially reduced, it is inevitable that biodiversity will suffer further major losses.

Some species are more vulnerable to extinction than others. These include:

1. Species at the top of food chains, such as large carnivores: Large carnivores usually require fairly extensive territories in order to provide them with sufficient prey. As human populations increasingly encroach on wild areas and as habitats shrink in extent, the number of carnivores which can be accommodated in the area also decreases. These animals may also pose a threat to people, as populations expand into wilder areas inhabited by large carnivores. Protective measures, including elimination of offending animals in the area, further reduces numbers.

2. Endemic local species (species found only in one geographical area) with a very limited distribution: These are very vulnerable to local habitat disturbance or human development.
3. Species with chronically small populations: If populations become too small, then simply finding a mate or interbreeding, can become serious problems.
4. Migratory species: Species which need suitable habitats to feed and rest in widely spaced locations (which are often traditional and 'wired' into behaviour patterns) are very vulnerable to loss of these 'way stations'.
5. Species with exceptionally complex life-cycles: If completion of a particular life-cycle requires several different elements to be in place at very specific times, then the species is vulnerable if there is disruption of any single element in the cycle.
6. Specialist species with very narrow requirements such as a single specific food source, e.g. a particular plant species.

Loss of an individual species can have various different effects on the remaining species in an ecosystem. These effects depend upon the how important the species is in the ecosystem. Some species can be removed without apparent effect, while removal of others may have enormous effects on the remaining species. Species such as these are termed 'keystone' species.

Why Conserve Biodiversity

Ecological reasons

Individual species and ecosystems have evolved over millions of years into a complex interdependence. This can be viewed as being akin to a vast jigsaw puzzle of interlocking pieces. If you remove enough of the key pieces on which the framework is based then the whole picture may be in danger of collapsing. We have no idea how many key 'pieces' we can afford to lose before this might happen, nor even in many cases, which are the key pieces. The ecological arguments for conserving biodiversity are therefore based on the premise that we need to preserve biodiversity in order to maintain our own life support systems. Forests not only harbour untold numbers of different species, but also play a critical role in regulating climate. The destruction of forest, particularly by burning, results in great increases in the amount of carbon in the atmosphere. This happens for two reasons. Firstly, there is a great reduction in the amount of carbon dioxide taken in by plants for photosynthesis and secondly, burning releases huge quantities of carbon dioxide into the atmosphere. (The 1997 fires in Indonesia's rainforests are said to have added as much carbon to the atmosphere as all the coal, oil and gasoline burned that year in western Europe.) This is significant because carbon dioxide is one of the main greenhouse gases implicated in the current global warming trend.

Average global temperatures have been showing a steadily increasing trend. Snow and ice cover have decreased, deep ocean temperatures have increased and global sea levels have risen by 100–200 mm over the last century. If current trends continue, scientists predict that the earth could be on average 1°C warmer by 2025 and 3°C warmer by 2100. These changes, while small, could have drastic effects. As an example, average temperatures in the last Ice Age were only 5°C colder than current temperatures.

Rising sea levels which could drown many of our major cities, extreme weather conditions resulting in drought, flooding and hurricanes, together with changes in the distribution of disease-bearing organisms are all predicted effects of climate change.

Forests also affect rainfall patterns through transpiration losses and protect the watershed of vast areas. Deforestation therefore results in local changes in the amount and distribution of rainfall. It often also results in erosion and loss of soil and often to flooding. Devastating flooding in many regions of China over the past few years has been largely attributed to deforestation. These are only some of the ecological effects of deforestation. The effects described translate directly into economic effects on human populations.

Economic reasons

Environmental disasters such as floods, forest fires and hurricanes indirectly or directly caused by human activities, all have dire economic consequences for the regions afflicted. Clean-up bills can run into the billions, not to mention the toll of human misery involved. Susceptible regions are often also in the less-developed and poorer nations to begin with. Erosion and desertification, often as a result of deforestation, reduce the ability of people to grow crops and to feed themselves. This leads to economic dependence on other nations.

Non-sustainable extraction of resources (e.g. hardwood timber) will eventually lead to the collapse of the industry involved, with all the attendant economic losses. It should be noted that even if 'sustainable' methods are used, for example when harvested forest areas are replanted, these areas are in no way an ecological substitute for the established habitats which they have replaced.

Large-scale habitat and biodiversity losses mean that species with potentially great economic importance may become extinct before they are even discovered. The vast, largely untapped resource of medicines and useful chemicals contained in wild species may disappear forever. The wealth of species contained in tropical rainforests may harbour untold numbers of chemically or medically useful species. Many marine species defend themselves chemically and this also represents a rich potential source of new economically important medicines. Additionally, the wild relatives of our cultivated crop plants provide an invaluable reservoir of genetic material to aid in the production of new varieties of crops. If all these are lost, then our crop plants also become more vulnerable to extinction.

There is an ecological caveat here of course. Whenever a wild species is proved to be economically or socially useful, this automatically translates into further loss of natural habitat. This arises either through large-scale cultivation of the species concerned or its industrial production/harvesting. Both require space, inevitably provided at the expense of natural habitats. Perhaps the rainforests and the seas should be allowed to keep their secrets.

Ethical reasons

Do we have the right to decide which species should survive and which should die out? Do we have the right to cause a mass extinction?

Most people would instinctively answer 'No!'. However, we have to realise that most biodiversity losses are now arising as a result of natural competition between humans and all other species for limited space and resources.

If we want the luxury of ethics, we need to reduce our populations.

Aesthetic reasons

Most people would agree that areas of vegetation, with all their attendant life forms, are inherently more attractive than burnt, scarred landscapes, or acres of concrete and buildings. Who would not prefer to see butterflies dancing above coloured flowers, rather than an industrial complex belching smoke?

Human well-being is inextricably linked to the natural world. In the western world, huge numbers of people confined to large urban areas derive great pleasure from visiting the countryside. The ability to do so is regarded not so much as a need, but as a right. National governments must therefore juggle the conflicting requirements for more housing, industry and higher standards of living with demands for countryside for recreational purposes.

EX SITU CONSERVATION—OUT OF THE NATURAL HABITAT

Zoos

In the past, zoos were mainly display facilities for the purpose of public enjoyment and education. As large numbers of the species traditionally on display have become rarer in the wild, many zoos have taken on the additional role of building up numbers through captive breeding programs.

Although comparatively far more invertebrates than vertebrates face extinction, most captive breeding programs in zoos focus on vertebrates. Threats to vertebrate extinction tend to be well publicised (e.g. Dormouse, Panda). People find it easier to relate to and have sympathy with animals which are more similar to ourselves, particularly if they are cute and cuddly (at least in appearance, if not in fact!). Not many visitors to zoos are likely to get excited over the prospect of the zoo 'saving' a tiny beetle, which they can barely see, let alone spiders or other invertebrates which often invite horror rather than wonder. Vertebrates therefore serve as a focus for public interest. This can help to generate financial support for conservation and extend public education to other issues. This is a very important consideration, as conservation costs money and needs to be funded from somewhere.

The focus on vertebrates is not solely pragmatic. Many of the most threatened vertebrates are large top carnivores, which the world stands to lose in disproportionate numbers. Such species require extensive ranges to provide sufficient prey to sustain them. In many cases, whole habitats for these predators have all disappeared. Some biased expenditure on their survival may therefore be justified.

Several species are now solely represented by animals in captivity. Captive breeding programs are in place for numerous species. At least 18 species have been reintroduced into the wild following such programs. In many cases the species was actually extinct in the wild at the time of reintroduction (Arabian Oryx, Pere David Deer, American Bison). In some cases, all remaining individuals of a species, whose numbers are too low for survival in the wild, have been captured and the species has then been reintroduced after captive breeding.

The role of zoos in conservation is limited both by space and by expense. At population sizes of roughly 100–150 individuals per species, it has been estimated that world zoos could sustain roughly 900 species. Populations of this size are just large enough to avoid inbreeding effects. However, zoos are now shifting their emphasis from long-term holding of species, to returning animals to the wild after only a few generations. This frees up space for the conservation of other species.

Genetic management of captive populations via stud records is essential to ensure genetic diversity is preserved as far as possible. There are now a variety of international computerised stud record systems which catalogue genealogical data on individual animals in zoos around the world. Mating can therefore be arranged by computer, to ensure that genetic diversity is preserved and inbreeding minimised (always assuming the animals involved are prepared to co-operate).

Research has led to great advances in technologies for captive breeding. This includes techniques such as artificial insemination, embryo transfer and long-term cryogenic (frozen) storage of embryos. These techniques are all valuable because they allow new genetic lines to be introduced without having to transport the adults to new locations. Therefore the animals are not even required to co-operate any longer. However, further research is vital. The success of zoos in maintaining populations of endangered species is limited. Only 26 of 274 species of rare mammals in captivity are maintaining self-sustaining populations.

Reintroduction of species to the wild poses several different problems

1. Diseases: The introduction of new diseases to the habitat, which can decimate existing wild populations. Alternatively, the loss of resistance to local diseases in captive-bred populations.
2. Behaviour: Behaviour of captive-bred species is also a problem. Some behaviour is genetically determined and innate, but much has to be learned from other adults of the species or by experience. Captive-bred populations lack the *in situ* learning of their wild relatives and are therefore at a huge disadvantage in the wild. In one case of reintroduction, a number of monkeys starved because they had no concept of having to search for food to eat—it had always been supplied to them in captivity. In the next attempt, the captive monkeys were taught that they had to look for food, by hiding it in their cages, rather than just supplying it.
3. Genetic races: Reintroduced populations may be of an entirely different genetic make-up to original populations. This may mean that there are significant differences in reproduction habits and timing, as well as differences in general ecology. Reintroduction of individuals of a species into an area where the species has previously become extinct, is in many cases just like introducing a foreigner. The Large Copper Butterfly is a good example of this. Although extinct in Britain, it persists in continental Europe. There have been over a dozen attempts to re-establish it in Britain over the last century, but none have been successful. This is probably due to the differing ecology of the introduced races. Replacement of extinct populations by reintroduction from other areas may not therefore be an option.
4. Habitat: The habitat must be there for reintroduction to take place. In many cases, so much habitat has been destroyed, that areas must first be restored to allow captive populations to be reintroduced. Suitable existing habitats will also (unless the species is extinct in the wild) usually already contain wild members of the species. In this case, it is likely that within the habitat, there are already as many individuals as the habitat can support. The introduction of new individuals will only lead to stress and tension as individuals fight for limited territory and resources such as food. In this case, nothing positive has been accomplished by reintroduction, it has merely increased the stress on the species. It may even in some cases result in a decrease in numbers. In contrast, the provision of additional restored habitat nearby can allow wild populations to expand into it without the need for reintroduction.

Aquaria

The role of aquaria has largely been as display and educational facilities. However, they are assuming new importance in captive breeding programs. Growing threats to freshwater species in particular, are leading to the development of *ex situ* breeding programs. The World Conservation Union (IUCN) is currently developing captive breeding programs for endangered fish. Initially this will cover those from Lake Victoria in Africa, the desert fishes of N. America and Appalachian stream fishes. Natural habitats will be restored as part of the program.

Marine, as well as freshwater species are also the subject of captive breeding programs. For example, The National Marine Aquarium, in South West England, is playing an important role in the conservation of sea horse species through their captive breeding program.

Plant Collections

Populations of plant species are much easier than animals to maintain artificially. They need less care and their requirements for particular habitat conditions can be provided more readily. It is also much easier to breed and propagate plant species in captivity.

There are roughly 1500 botanic gardens worldwide, holding 35,000 plant species (more than 15 per cent of the world's flora). The Royal Botanic Gardens of England (Kew Gardens) contains an estimated 25,000 species. IUCN classifies 2700 of these as rare, threatened or endangered. Many botanic gardens house collections of particular taxa which are of major conservation value. There is however, a general geographic imbalance. Only 230 of the world's 1500 gardens are in the tropics. Considering the greater species richness of the tropics, this is an imbalance that needs to be addressed.

A more serious problem with *ex situ* collections involves gaps in coverage of important species, particularly those of significant value in tropical countries. One of the most serious gaps is in the area of crops of regional importance, which are not widely traded on world markets. These often have recalcitrant seeds (unsuited to long-term storage) and are poorly represented in botanic collections. Wild crop relatives are also under-represented. These are a potential source of genes conferring resistance to diseases, pests and parasites and as such are a vital gene bank for commercial crops. Plant genetic diversity can also be preserved *ex situ* through the use of seed banks. Seeds are small but tough and have evolved to survive all manner of adverse conditions and a host of attackers. Seeds can be divided into two main types, orthodox and recalcitrant. Orthodox seeds can be dried and stored at temperatures of $-20°C$. Almost all species in a temperate flora can be stored in this way. Surprisingly, many tropical seeds are also orthodox. Recalcitrant seeds, in contrast, die when dried and frozen in this manner. Acorns of oaks are recalcitrant and it is believed that so are the seeds of most tropical rainforest trees.

The result of storing seeds under frozen conditions is to slow down the rate at which they lose their ability to germinate. Seeds of crop plants such as maize and barley could probably survive thousands of years in such conditions, but for most plants, centuries is probably the norm. This makes seed banking an attractive conservation option, particularly when all others have failed. It offers an insurance technique for other methods of conservation.

All of the *ex situ* conservation methods discussed have their role to play in modern conservation. Generally, they are more expensive to maintain and should be regarded as complementary to *in situ* conservation methods. For example they may be the only option where *in situ* conservation is no longer possible.

IN SITU CONSERVATION (WITHIN THE NATURAL HABITAT)

Protected Sites or Reserves

In situ conservation maintains not only the genetic diversity of species, but also the evolutionary adaptations that enable them to adapt continually to shifting environmental conditions, such as changes in pest populations or climate. *In situ* conservation also ensures that along with target species, a host of other interlinked species are also preserved as a by-product. It is generally cheaper than *ex situ* methods (although not cheap). It may often be the only conservation option, for example for species with recalcitrant seeds.

In situ conservation measures involve designating specific areas as protected sites. Protection may be offered at various levels, from complete protection and restriction of access, through various levels of permitted human use. In practice, complete protection is rarely necessary or advisable in a terrestrial context. Human beings have been a major part of the landscape for many thousands of years. Over the course of that time, human cultures have emerged and adapted to the local environment, discovering, using and altering biotic resources. Many areas that now appear 'natural' bear the hallmarks of millennia of human influence. Other species have evolved along with that influence and in many cases require the disturbance provided by humans to provide the necessary conditions for their survival. In other words, it is rarely advisable to relegate the countryside to the status of a museum piece.

This applies particularly in the less economically developed areas of the world, where in many cases, the livelihood of the local people depends on using the natural resources available to them. Prohibiting the use of such resources in protected areas means that expensive enforcement measures usually have to be put in place. It is far better to involve local people in conservation and to find creative ways for them to make a sustainable living while still protecting valuable habitats or species.

The biosphere reserve concept has been developed through the Man and Biosphere (MAB) Program of the United Nations Educational, Scientific and Cultural Organisation (UNESCO). Biosphere reserves are an attempt to reconcile the problems of conserving biodiversity and biological resources, with sustainable use of natural resources for people. They form an international network of sites, nominated by national governments, but designated by UNESCO. The first reserves were nominated as long ago as 1976.

Marine conservation areas lag behind terrestrial ones. Protected areas have existed on land for over a century, but there is no tradition of managing marine areas for conservation. The only current statutory marine reserve in England is Lundy Island. This harbours a huge variety of marine life due to the diversity of underwater habitats present there.

Marine reserves may be vital tools in preserving species-rich areas such as tropical coral reefs, which are being devastated by non-sustainable fishing methods in many areas. The rationale of such reserves is not to lock away fish from fishermen, but rather to create refuges inside which populations can build up and spill over to repopulate adjacent areas. Marine Reserves need to be carefully designed to take into account, movement patterns, dispersal rates and population dynamics of particular target species. For example, it would be pointless having a reserve where the resident species regularly travelled to non-protected areas. It would also be pointless protecting the habitat of an adult, but neglecting the geographically different breeding grounds and habitats of juveniles, or *vice versa*. Such factors should also be taken into consideration in the design of terrestrial reserves.

Management of nature reserves

Nature reserves are usually designated to protect a particular species, assemblage of species or specific habitats. As such they can rarely be left in isolation to manage themselves. Management is necessary in order to prevent natural processes such as succession from taking place. Such resulting changes in habitat may mean the loss of particular species for which the reserve was originally designated.

Succession is a natural process which will tend to replace particular species with different ones. This is often as a result of ecological change induced by the organisms themselves. In former times, while an area of a particular habitat might be lost through succession (such as a wetland drying out and eventually becoming a woodland), there would always be other wetland habitats at an earlier stage of the succession process elsewhere. These would act as species reservoirs. With the drastic loss of area of natural habitat occurring worldwide, this is often no longer the case. The decision therefore has to be made to halt succession at a particular stage in order to preserve the species associated with it. Examples of this include wild flower meadows, wetlands and heathlands.

Restoration

Restoration attempts to bring land modified by human use back to its original state. Because determining the original natural state is often difficult and because ecosystems continually change, this is rarely a realistic goal. Changes may also in some cases be irreversible, so that restoration to an original condition is not an option. For these reasons, restoration is often limited to the approximate recreation of habitat.

Restoration does not necessarily require intervention. Left to natural processes, many ecosystems will return to their original condition provided populations of the original species still exist nearby. This may however be a lengthy process. In Brazil's caatinga forest, natural recovery of slash and burn agricultural sites takes more than a century. Sites cleared by bulldozer may take a thousand or more years to recover. Intervention may be used to speed up the process. It becomes vital where an ecosystem will never recover naturally, either because it has been physically transformed, or because species cannot migrate to repopulate the area. Restoration may rely heavily upon species maintained by *ex situ* methods and is an example of the complementary nature of *in situ* and *ex situ* techniques.

Recovery of Threatened Species

The recovery of threatened species generally hinges on providing suitable habitat and conditions in which they can thrive. Most biodiversity losses can be directly attributed to habitat loss, so provision of habitat is often the key requirement for recovery of a species. The dormouse (*Muscardinus avellanarius*) is a good example of this.

Dormice are the subject of a species recovery program sponsored by English Nature. They thrive in deciduous woodland and overgrown hedgerows. They are arboreal and require networks of interlinking low branches to provide aerial highways to food sources (e.g. hazel nuts and honeysuckle). In former times, the practice of coppicing woodland (cutting trees near to ground level and then allowing them to regenerate numerous shoots) provided ideal conditions for dormice. However, coppicing is little practised these days, with the result that large areas of former habitat have become unsuitable for dormice. Recovery therefore hinges on providing suitable areas for the species. Dormice also suffer from competition with an introduced species, the grey squirrel. Both species compete for the food resource of hazel nuts.

Introduced Species

The vast majority of exotic introduced species die out because they are unsuited to local conditions. A few however, appear to be superbly adapted to particular local conditions and will tend to out-compete native species.

Regulatory policies are necessary to curb the introduction of exotic species and genetic resources, as the consequences can be disastrous. Introduced species can wipe out innumerable other local species. Between 1967 and 1972, an African cichlid fish introduced into a lake in Panama wiped out 6 of the 8 previously common fish species, drastically reduced populations of a seventh and affected aquatic invertebrates, algae and fish-eating birds up and down the food chain.

In Britain, particularly in the west, rhododendron has taken over large areas, virtually eliminating native plants and their associated faunas. This one species alone has decimated tens of thousands of acres. Grey squirrels introduced from America have also all but replaced native red squirrel populations.

Even introducing the same species can present hazards through the mixing of genetic stocks. Fish populations can be contaminated by interbreeding with introduced varieties. The genetic integrity of native British Red Deer is now also threatened through hybridisation with introduced Sika Deer. Pure Red Deer may in the end be confined to isolated islands in Scotland.

Genetic Modification

Humans now have the technology to alter life on earth in a totally unique way. Genetic engineering can involve the transfer of genetic material between widely separated taxonomic groups, e.g. genes from a fish have been introduced into tomatoes. Entirely new species can intentionally or unintentionally be produced. There is no guarantee that all the results will be beneficial or can even be controlled.

Legal Protection

Nature reserves need to be properly protected from the adverse results of human activity. This might be indirect, as in for example, habitat degradation through pollution brought about by activities elsewhere. Alternatively, it may involve direct damage, as in unauthorised extraction of resources from the reserve. Such protection requires the presence of a legal framework which can be effectively enforced.

Advantages, Risks and Opportunities

In situ maintenance of biodiversity through the establishment of conservation and multiple-use areas offers distinct advantages over off-site methods in terms of coverage, viability of the resource, and the economic sustainability of the methods:

1. Coverage: A worldwide system of protected and multiple-use areas would allow a significant number of indigenous species and systems to be protected, thus taking care of the unknowns until such time as methods are found for their investigation and utilisation.
2. Viability: Natural selection and community evolution continue and new communities, systems and genetic material are produced.
3. Economic sustainability: A country that maintains specific examples of biodiversity stores up future economic benefits. When the need develops and this diversity is thoroughly examined, commercially valuable genetic and biochemical material may be found.

It is not sufficient to establish a conservation area and then assume its biodiversity is automatically protected and without risk. Many risks, both natural and man-created, remain. An extreme example was the near-obliteration of the entire remaining habitat of the golden lion tamarin (*Leontopithecus r. rosalia*) in 1992 by fire. Shaffer cites four broad categories of natural risk:

1. Demographic uncertainty resulting from random events in the survival and reproduction of individuals.
2. Environmental uncertainty due to random or at least unpredictable, changes in weather, food supply, and the populations of competitors, predators, parasites, etc.
3. Natural catastrophes such as floods, fires or droughts, which may occur at random intervals.
4. Genetic uncertainty or random changes in genetic make-up due to genetic drift or inbreeding that alter the survival and reproductive probabilities of individuals.

The greatest uncertainties, however, are often anthropogenic. The elimination of habitat to make way for human settlement and associated development activities is the most important factor contributing to the diminishing mosaic of biodiversity. These uncertainties can only be met with a full array of conservation programs, including those that use *ex situ* methods.

Despite the long list of uncertainties and risk, there is hope for progress. In the last decade not only have pressures from the scientific community and the efforts of non-governmental organisations led to stronger language in international agreements, but segments of the development community have accepted the idea that a large degree of compatibility exists between the need to develop and the need to maintain biodiversity. Further acceptance depends, however, on a number of attitudinal adjustments on the part of many who call for *in situ* conservation, as well as on a clearer understanding of the rationale behind it by those whose activities conflict with it. The success of conservation also requires a modification of how we cost economic goods and services in the short, medium and long-term.

Globally, the possibilities for undertaking *in situ* programs such as national parks, biological reserves, and other conservation areas appear to be somewhat favourable. However, the status of these protected areas is often not healthy and unforeseen problems repeatedly arise. The establishment of the Gurupi Biological Reserve in the eastern Brazilian Amazon, for example, significantly increased the level of threat by causing a rush of illegal extraction of forest resources. This site is probably the most endangered conservation unit in the Amazon basin. Worldwide, the list of endangered protected areas is growing in number, and additional human-dominated activities such as water development, mining, road construction and resulting development, livestock grazing, poaching, logging, and other removal of vegetation continue to threaten their integrity.

Integrated Regional Development Planning and Implementation

As is the case with all human activities, alleviating threats to conservation areas requires the involvement of those most affected by the various land-use alternatives in the decision-making process. Integrated regional development planning (IRDP) is a response to the need for better integration of the numerous interests holding conflicting views of how the resources of a region are to be used. Support from the conservation community has also been noted and is given below:

> 'The idea of basing conservation of particular species on the maintenance of the natural diversity of species will become even less tenable as the number of threatened species increases and their refuges disappear. Natural areas will have to be designed in conjunction with the goals of regional development and justified on the basis of ecological processes operating within the entire developed region and not just within natural areas.'

Not only does IRDP address the intersections between terrestrial, aquatic and marine systems, it considers the demands of those who use and who would use these systems. However, instead of confronting environmental complexity by subdividing issues into sectoral components, it divides the region into smaller spatial units and looks at the sectoral interactions in each. Interactions of this kind are often conflictive for two reasons: (i) competition is established for the same goods or services by two or more interest groups and (ii) the mix of available goods and services is changed, and one sector is harmed as a result of the activities of another sector to make use of goods and services selected from the whole. IRDP also analyses the interactions of the region with neighbouring areas. Once the regional and sub-regional systems are defined, connections between neighbouring units are better understood.

Whereas sectoral planning (including the planning of resource conservation activities) designs programs and projects to meet the needs of specific target populations, IRDP uses methods of systems analysis and conflict management to attempt an appropriate distribution of the costs and benefits of development activities throughout the affected populations or sectors. Thus, conflict identification and its management are fundamental requirements for a development plan to be 'integrated'. Sectoral integration is necessary because individual sectoral activities often hinder the activities of other sectors in their efforts to appropriate goods and services from the same and allied systems. This can be illustrated by agricultural development that affects water quality, resulting in impacts on the goods and services provided by clean water. The decision as to which activities are the correct ones or how each can be adjusted to reduce conflict can only be made through negotiation by the parties involved and not by one of the individual sectors trying to dictate to others—be that sector forestry, agriculture, livestock production or biodiversity conservation.

In this context, IRDP has a number of advantages over sectoral planning. Conflicts are always easier to manage and even resolve before time, funds, and political prestige have been invested in a specific project. Participants in the regional planning exercise, though representing individual interests, have a shared commitment to rules and procedures of the process that can be controlled. Under this model, the various parties (sectoral interests) operate with a similar rationale and can be easily encouraged to focus on criteria rather than on positions. Further, each can insist that evaluation criteria be objective. This provides an opportunity to invest in options that offer mutual gain and minimal conflict. The result is a strategy for development that demonstrates concern for both the target and the affected populations of development projects and programs, including those that have a conservation orientation. Where integrated analysis and planning show them to be necessary, projects of biodiversity conservation are in all ways development activities, as defensible as any other.

To be successful, the recommendations, strategies and policies that come out of the integrated regional planning process must consider the needs of those affected by linkages between development sectors— which are little more than a reflection of the integrated nature of the ecosystems and landscapes where development takes place. Economic interactions are an important part of these systems and landscapes. 'The more we overlook the linkages, the more we shall find the sectors fail to function efficiently and productively, with all that implies for sustainable development.'

The principles of IRDP, if adapted to the problems of *in situ* conservation of biodiversity, can provide conservationists with welcome tools. Although much of the effort is based on the agreement of parties, this planning has a scientific foundation, and once it has been clarified and accepted by the planning team, strategic alliances will have been developed for the implementation of the programs. Demonstration projects can then be initiated to help 'sell' the strategy or plan. The process then goes forward in a series of iterations of implementation, experimentation, evaluation, and, as needed, modification (Fig. 3.1).

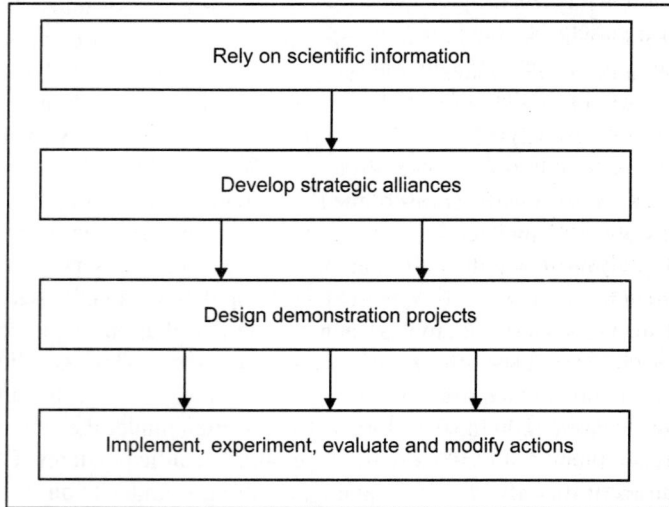

Fig. 3.1. Principles of integrated regional development planning and implementation.

Scientific information: Scientific information helps the credibility of the planning process by providing a sound basis for logical decision-making. In more traditional planning efforts, linkages tend to be ignored because our view of the world is traditionally grounded in splitting it up into manageable units. Although science is often responsible for this fragmentation, we also know from science that everything is related, in some way and to some degree, to everything else. That interrelatedness requires that information be managed in an integrated way if we are to make fact-based decisions. 'Specialised knowledge by itself produces nothing. It can become productive only when it is integrated into a task.'

Information from the social as well as the natural sciences can help develop a framework for decision-making only if it is both valid and accessible. The collection, storage, and use of relevant information should be designed to provide data rather than assumptions. Thus, one objective of any regional planning effort is to establish a permanent and dynamic database that looks toward addressing information needs on the physical, biotic and socioeconomic characteristics of the region.

These collections of information should be linked to a larger system of information management to provide complete transparency for all the collaborators. Likewise, if the principal collaborators on the database can help in its design and modification, they will have a sense of ownership in the outcome. Other considerations for designing an information system include the establishment of goals, the determination of the methods and scales to be used, the resolution of the management structure, and the design of evaluation programs.

Develop strategic alliances: We are often faced with a broad array of development alternatives the selection and implementation of which require the formation of strategic alliances.

To select the best among a number of development alternatives requires the ability to identify costly and undesirable effects of the possible alternative projects and the opportunity to modify these projects during their design stage. Review and discussion in a variety of forums, including those that are cross-sectoral and those where public participation is encouraged, can greatly strengthen planning because of the alliances that are built up in the process.

By making planning a democratic process through IRDP, people can define their problems and design their own solutions. By the same token, IRDP is strengthened through the formation of networks which make use of a variety of communication techniques to assemble knowledgeable and informed constituencies. Four types of networks are valuable to a communication strategy:

1. Advice network: Both planning and implementation require an 'advice network' of prominent collaborators on whom others depend (or will depend) to solve problems and provide technical information concerning the strategy, program, or project being developed.
2. Trust network: A 'trust network' consists of the inner circle of participants who share sensitive political information and back each other in a crisis. These are often the project initiators, who may act as catalysts to help in conflict management or to ensure project implementation through their leadership.
3. Communication network: Participants in the effort should discuss project-related matters regularly. Such a network can be specifically designed to inform politically important individuals on a timely basis, thus ensuring their feedback and support.
4. Constituent network: The need to form an alliance with a number of diverse constituencies often takes the planner into the realm of the unknown. Part of communication is recognising that we can learn from others. An exchange of ideas is important and at least four approaches can be used to foster this exchange, including information feedback, consultation, joint planning

and delegated authority. Information feedback consists of formal and informal one-way presentations, newspaper articles, notices, etc. Consultations are made through such mechanisms as public hearings, ombudspersons or representatives of the interest groups concerned. Joint planning can take place through the use of advisory committees, structured workshops, etc. And delegated authority is amenable to the use of citizens' review boards and planning commissions.

Design demonstration projects: The value of an interdisciplinary and multifunctional planning team cannot be overstated; alliances forged in the process are indispensable to the success of the decisions that are taken. Innovative demonstration projects must include cross-sectoral representation from the initial stages of their formulation. The design of both the planning process and its products should have clear goals and objectives, review existing legislation and public policy, assess the social and economic conditions of the region, describe important ecosystem structure and function, and analyse direct and indirect threats to both the process and its products.

All of this is more easily done if the region is divided into smaller homogeneous units for analysis and if partnerships are created from the variety of constituencies. A series of strategies, with timetables and benchmarks, should be established with detailed financial goals and budget projections. Visual and graphic tools can be used successfully to communicate the process and products to those who require timely information for decision-making. The successful plan also contains criteria and methods for evaluating progress towards meeting the established goals.

Implement, experiment, evaluate and modify actions: The goal of any planning process is its implementation. Countless development plans that have never been implemented sit on shelves because of problems with funding, politics, myopic vision, or the lack of qualified personnel to take the planning recommendations and make them a reality. In many government settings, the fear of being associated with a project or program that is connected to a current or previous administration can be risky. Developing communication strategies to transfer knowledge to decision-makers can ameliorate this fear.

The implementation of a plan requires that strategies and specific actions be prioritised and that public policy be formulated to do so. Accountability must be maintained, but those who are charged with the implementation of a program must recognise that there are productive failures and that implementation can be modified in positive ways either because of such failures or because of experimentation.

Thus, there are no easy solutions to the complex challenges of integrated regional development planning. A process which provides a thoughtful structure for gathering and utilising scientific information, which involves stakeholders in a genuine analysis of development alternatives, and which establishes clear and measurable objectives can provide for rational development activities, including those for the conservation of biodiversity. Using IRDP as a vehicle to communicate with decision-makers provides a framework for action. Planners often look for what is politically expedient; however, the participation of non-governmental organisations can play an extremely important role in tempering politically motivated development. This nexus between government and non-governmental organisations is one of the more important areas where 'integrated' regional development takes place. Both should represent the interests of both present and future generations.

BIODIVERSITY CONSERVATION AND TRADITIONAL AGROECOSYSTEMS

For rural people, wild plants and animals provide food, medicine, building materials, income and a source of inspiration. Rivers and lakes give them transportation, water and fish; and the coastal zone offers them a permanent source of sustenance and building materials. But instead of a sustainable flow of renewable resources, mostly furnished by nature, recent patterns of agricultural development are depleting soils and genetic and species diversity both in the cropped areas and in the surrounding ecosystems.

Agricultural lands, livestock grazing areas, manipulated forests, and other human-managed ecosystems cover at least two-thirds of the terrestrial surface of the planet, whereas protected areas cover only about 8 per cent. The remaining percentage is wilderness, urban lands, etc. These human-managed ecosystems contain an important segment of global biodiversity, and if they are managed with this in mind— especially if they are managed in conjunction with a system of protected areas—they can significantly contribute to the maintenance of global biodiversity.

Since the beginning of this century alone, about 75 per cent of the genetic diversity of the most important crops has disappeared from farmers' fields. This has increased agricultural vulnerability and reduced the essential variety of the diets of rural people. Other traditional and local species and races of domesticated plants and animals, vital for the nutrition of the poorest people, are underutilised or neglected. In fisheries and aquaculture, the introduction and transfer of exotic organisms has helped local economies but sometimes at the expense of natural systems, cultural stability and social equity. The symmetry between the development of rural areas and the conservation of many forms of established land-use is a critical issue that regional planning needs to address if biodiversity is to be maintained in the long-term.

Agriculture and the Conservation of Biodiversity

Each agricultural village is part of an ecosystem. These agricultural ecosystems vary widely—from broad expanses of river deltas with the possibility of year-round irrigation, to areas of seasonally irrigated fields interspersed among forests, to areas dominated by rain-fed crops. Legumes, cereals, tubers, herbs, fruits, trees, livestock, wild animals and fish all play important roles in most agricultural villages and must therefore be considered in planning agricultural development projects. Further, the relationships of each agricultural community extend far beyond the village itself, and these must be considered also. For example, in the hills of Nepal, each hectare of farmland needs the support of 3.48 hectares of forest and these forests require expert management if they are to continue to provide benefits in terms of food, fodder, firewood, construction materials, medicines, water, clothing, etc.

Agricultural systems will change dramatically over the coming decades because of climate change, new technologies based on genetic engineering and agroecology, and shifts in international markets. Governments and farmers will need to adapt to these changes through planning. Indicative planning, for example, is a system of dynamic planning informed by, and constantly adjusting to, changes in leading indicators. These indicators could be modified to include those related to agriculture and biodiversity in order to help ensure that agricultural systems will robustly resist mere transitory changes. In conjunction with ongoing efforts to develop environmental accounting systems, research should be initiated to find the most effective indicators and monitoring systems.

Traditional agriculture can contribute to easing the stress of changing conditions in rural areas, help conserve biodiversity, and maintain healthy relationships between rural people and the land. For example, in traditional systems of shifting cultivation or swidden agriculture, a wide range of crops—often over 100 at one time—can be grown, essentially transforming a natural forest to one that is cultured.

The species and varieties grown in the swiddens are in a state of continuous adaptation, and in many places the crops are enriched by gene exchange with wild or weedy relatives. Altieri and Merrick contend that 'maintenance of traditional agroecosystems is the only sensible strategy to preserve *in situ* repositories of crop germplasm'.

Traditional agriculture has adapted to a wide variety of local conditions, produced a diverse and reliable food supply, reduced the incidence of disease and insect problems, used labour efficiently, intensified production with limited resources, and earned maximum returns with low levels of technology. It makes use of a wide range of species and land races that vary in their reaction to diseases and insect pests, as well as to different conditions of soil, rainfall and sunlight. Traditional agriculture provides sustainable yields by drawing on centuries of accumulated experience by farmers who have not depended on scientific information, external inputs, capital, credit or markets.

But with growing populations, steps need to be taken to enhance the productivity of lands under traditional agriculture. In the forested uplands, modern agricultural development should take existing traditional systems as starting points and use modern agricultural science to improve on their productivity. The essential element is to design self-sustaining agroecosystems that assure the maintenance of the local genetic diversity available to farmers, thereby enabling rural communities to maintain control over their production systems. In addition, the maintenance of a stable, permanent link with forested land, such as that contained in some categories of protected areas, enables farmers to invest time and effort in other assets like fruit trees, fenced gardens, terraces and irrigation canals. Such mixed systems will often make possible a marriage of modern and traditional agricultural techniques leading to the establishment of more permanent villages.

Agricultural ecologists and modern land-use planners have learned to respect the wisdom inherent in much traditional practice. If it is seen as part of an overall system of conservation-oriented management, traditional farming can continue to be a meaningful part of the total agricultural productivity of a region and to contribute to the conservation of its biodiversity.

Planners should also be aware that strict protection does not always lead to more biodiversity. Nabhan, studying two oases in the Sonoran Desert on either side of the Mexico-United States border, found that the customary land-use practices of Papago farmers on the Mexican side of the border contributed to the biodiversity of the oasis but that the protection of an oasis 54 km to the northwest, within the US Organpipe Cactus National Monument, resulted in a decline in species diversity over a 25-year period.

On the other hand, some conservation measures can help preserve traditional agroecosystems. The 2000 ha Rock Coral Canyon Reserve, for example, which is owned by the US Department of Agriculture and is one of just a handful of places in North America where wild varieties of chili peppers grow naturally, is the focus of the first proposed government-sponsored *in situ* conservation plan for wild native crops. The project is run by Native Seeds/SEARCH, an NGO that aims to preserve and exploit some of the wild edible species in the region. It is estimated that colonisers since Columbus have wiped out two-thirds of North America's native crop varieties. Apart from peppers, the reserve is home to four other important wild varieties of native crops: tepary beans, cotton, squashes and *Agave* sp., from which tequila is made. These species have traditionally been gathered by the local Tohono O'Odham people. As recently as 70 years ago, the Tohono O'Odham cultivated 4000 ha of farms in Arizona without having to pump groundwater—an impossible dream for most farms in the state today. Eating tepary and lima beans, pods from mesquite trees, acorns and corn, they had an extremely healthful diet.

Threats to Traditional Agroecosystems

Modern farming technology is now removing innovation from the farm and placing it instead in the laboratory. The uniform varieties produced at the research center, with their dependence on chemical fertilisers and pesticides, are displacing farm-bred varieties. Once these traditional varieties are gone, the knowledge of their cultivation and use is also lost.

But neither is the 'museum' approach to conservation sufficient. Fencing off ecosystems valued for their diversity as protected areas, keeping plants in botanic gardens, and storing germ plasm in seed banks is hardly an adaptive long-term solution. It seems apparent that preserving genetic variety is pointless unless the farming system that produced it is also preserved, along with its climate and soil and the accumulated knowledge of its cultivation and use.

But traditional agroecosystems are under threat in virtually all parts of the world. Above all, these threats come from agricultural policies that favour centralised control and the subsidies required to achieve them. While these policies have undoubtedly increased total agricultural productivity, they have also led to considerable economic inefficiencies and vulnerabilities. The solutions are to be found in correcting inappropriate agricultural policies, including those that guide land-use planning.

Despite impressive increases in agricultural productivity in recent decades, many current agricultural policies are economically inefficient and environmentally unsound. They benefit farmers with large landholdings growing few crops and penalise farmers with smaller holdings that often cultivate many crops. Food price controls and subsidies for agricultural inputs help meet short-term consumer demands but remove incentives for increased agricultural production, and they often tend to undermine food security. Such policies have also decreased the diversity of species used by farmers, increased the uniformity of crops and livestock breeds, and made farmers dependent on expensive and often unreliable sources of agricultural inputs. Although many agronomists argue that uniformity in agricultural practices can improve productivity, the Global Biodiversity Strategy points out five current policies that are likely to be contrary to the interests of long-term agricultural productivity:

1. Agricultural input subsidies: Reducing the cost of inputs such as water, pesticides and chemical fertilisers leads to the promotion of 'industrial' agriculture based on a small number of highly uniform crops at the expense of farming systems based on a wider variety of crops. Subsidised inputs sometimes also replace natural processes based on biodiversity that are equally effective at lower cost to people and have less impact on the supporting ecosystems. The growing use of pesticides, for example, has displaced natural enemies of agricultural pests such as micro-organisms and invertebrates.

2. Food price subsidies: Policies to reduce food prices for urban consumers can cut into farm profits. Combined with subsidies for inputs, such price controls can greatly reduce agricultural diversity. The use of modern crop varieties, which require irrigation and heavy inputs of agrochemicals, can enable some farmers to neutralise the impact of food price controls. But farmers using low-input systems and traditional varieties receive no such offsetting benefit, which discourages them from developing new varieties of their own; this leads indirectly to the erosion of knowledge of traditional varieties.

3. Overvalued exchange rates: Many governments of developing countries have overvalued their currencies as a means of subsidising imported capital goods for industry, reducing the costs of imported food, and lowering the price of food for export. Basically, such policies 'tax' all agriculture, but farmers who use fewer manufactured inputs are taxed relatively more than those who use more of these inputs. Like the combination of subsidies and food price controls discussed above, this combination favours industrial agriculture with its attendant reduction in biodiversity.

4. Research biased toward high-input agriculture: Much national agricultural research has been directed toward increasing the production of a few major crops through technology change. This research model has been exported from the industrialised to the developing world through the Consultative Group on International Agricultural Research (CGIAR), and may have provided much-needed breathing room in the race between production and population. But to meet future production needs, national governments must support agricultural systems that meet food needs while maintaining important components of diversity.

5. Credit policies that discriminate against 'minor' crops and traditional varieties. All too often, governments fail to extend agricultural credit to farmers planting traditional crop varieties or growing crops consumed locally. Particularly in developing countries, where the benefits of 'improved' varieties may be negligible in marginal agriculture, reduced productivity and accelerated loss of crop diversity may result.

Traditional agriculture is now also threatened by the new global consumer culture, which is spreading through television, trade and other means. Management systems that were effective for thousands of years have become obsolete in a few decades, replaced by systems of exploitation that bring short-term profits for a few and long-term costs for many. A few examples will indicate the range of factors driving this process.

Land-use management throughout much of sub-Saharan Africa is evolving from a precolonial communal system to systems that are more formal and individualistic. Most traditional communities do not have effective title to or control over their lands, nor do they have an effective way to make their views felt at the national policy level. As a consequence, the colonial period was marked by a taking of many of the most desirable lands from long-term resident communities, and the post-colonial period of nationhood has further served to provide legal vehicles for a taking of land and resources from local communities in the national interest. Added to this are the population pressures on the land that contribute to a breakdown of traditional methods of control. For the Shona of Zimbabwe this scenario of land divestiture has been all too evident. Traditionally, the Shona managed their lands communally on the basis of ancestral relationships. Sacred sites and sites of historical importance were preserved throughout the Shona domain, though outsiders were generally unaware of these areas or of the values attached to them. Consequently, the break-up of Shona lands into small parcels under individual ownership schemes failed to maintain traditional land protection and management systems, and resulted in a loss of cultural heritage and its associated sustainable farming practices.

Robinson describes how colonists have been moving into the territory of the Yuqui Indians in Peru, primarily for the purpose of producing coca. These colonists tend to remain on their farms only during coca planting and harvesting, and return to their highland settlements at other times of the year. Their activities appear to have had a major impact on the fish and game available to the Yuqui because they use technology (such as fishing with dynamite) that leads to considerable overexploitation of resources. This is just one of many examples that could be provided of how new colonists have moved into traditional lands and disrupted the traditional systems that had worked over a period of many generations. Crosby describes the impacts Europeans have had on both cultures and ecosystems during the thousand years of their 'ecological imperialism'.

In the Moluccas of eastern Indonesia, rapidly rising consumer desires, stimulated by television images and the objects of a growing Indonesian middle class, are pushing local governments and officials to shorten the interval between traditionally controlled fish harvests. The rationale is that the increased population densities on isolated islands lead to further needs for alternative sources of income. Despite evidence that shortened intervals result in drastically decreased stocks of marine resources, local government officials claim that the needs of villagers for income—to conduct religious rituals, pay school fees, or acquire consumer goods—are forcing them to extend the period of harvest.

Hunting has long been an important part of the economy on the island of Sumbawa, in the eastern part of the Indonesian archipelago. Because most of the villagers are Muslims, pigs are not a particularly desirable game animal, but feral buffalo and cattle, as well as the local species of deer, *Cervus timorensis*, are commonly hunted. As grazers, these species do far better in grasslands than in the forest—the normal vegetative cover of the island. Today, however, grassland covers 17 per cent of Sumbawa's land area. These grasslands are several hundred years old and have always been used for grazing and hunting. The grasslands are maintained by annual fires which, while preventing reforestation, replace older and less palatable grasses with younger and more edible ones, eliminate dead plant material, and actually increase overall herbaceous productivity. The creation of grasslands by these villagers is sensible habitat management that creates conditions favouring the grazing animals at the expense of pigs (which prefer the forest). Furthermore, the replacement of forest by grassland has been of net benefit to the wild herbivores that are hunted, with populations kept at such a high level that they could be harvested virtually at will. The hunters accept communal control that proscribes hunting during the period from November to May when the deer give birth and rear their young. Government conservation programs prohibit both the burning of the savannas and the hunting of the main game animals—excluding wild pigs, the only species the Islamic islanders avoided. Because of this insensitivity to the local reality, a genuine symbiosis that had proved sustainable over long periods of time was broken, the acceptance of the program by peasant hunters was lost, and their traditional conservation measures were undermined. The process has led to the loss of both biological and cultural diversity.

Tenure: The Key Issue

A key concern for planners is the traditional links between indigenous cultures and the natural world; it deals with the responsibility over resources. Tenure systems upon which responsibility is built are based on legitimacy drawn from the community in which they operate rather than from the nation-state in which they are located. Indigenous systems of resource tenure are extremely variable, complex mixtures of individual and community rights, enforced by the local culture. These systems are flexible and constantly evolving, often in response to changing environmental conditions. Such systems invariably are being disrupted by nation-states claiming ownership of the most important areas. The institutional control of resources by local peoples tends to be strongest when the groups are the most independent. Once they become integrated into larger systems, the social and economic center of gravity shifts away from the community and rural institutions become increasingly marginalised politically.

Local people need the rights to self-determination, and to set their own development agenda. Although this does not guarantee success, it does put responsibility firmly in the hands of those who will earn the benefits and pay the costs. We might reasonably expect that communities will behave in their enlightened self-interest, if empowered to do so.

Security of tenure offers opportunities for communities to gain benefits from their resources, but at least some market forces typically exist exclusively outside local communities. Therefore, resources are perceived differently at national and community levels, and the benefits are derived differently. As a result, governments should consider returning at least some nationalised resource systems, such as forests and wildlife, to community-based tenure systems, which are often more cost-effective. Putting resource management back in the hands of local communities also helps governments divest themselves of responsibility for functions they cannot adequately fulfil. The legitimacy of community-based tenure systems can be recognised through cadastral surveys, assessments of wildlife populations, demarcation, registration, and community infrastructure that can defend against outside pressure.

The full implications of such an 'indigenous privatisation scheme' need to be considered. Transferring the control of access rights from a national to a local authority puts power into the hands of those making the local decisions. As Murphree points out, the way that natural resources are used in any particular place and time is the result of conflicting interests between groups of people having different objectives. Seldom does any one group dominate, and resources can be used in a number of different ways at the same time and place. So the variation in resource management is part of an ongoing process in which the different interests and struggles of the various actors are located. Some local actors are likely to benefit more than others, thereby creating new tensions in the community.

It is clear to all farmers living in such systems, says Rappaport, 'that their survival is contingent upon the maintenance, rather than the mere exploitation, of the larger community of which they know themselves to be only parts'. Regional plans that incorporate means of protecting the larger ecosystem within which agricultural communities survive and flourish are far more likely to succeed than those that are too narrowly based. Such considerations will often involve ensuring that the relevant communities are given management responsibility for the natural areas upon which their continued prosperity depends. Governments should therefore use regional planning as a means to promote closer collaboration between the supporters of agriculture and the supporters of protected areas, building on the common interest in maintaining the diversity and productivity of biotic resources.

NEED AND POTENTIAL FOR PRIVATE BIODIVERSITY CONSERVATION

This section discusses here some thoughts on terrestrial and aquatic biodiversity conservation outside of parks, in places where biodiversity is not a primary or often even a conscious concern. The pre-eminent threat to biodiversity is seen to be the conversion of natural ecosystems to crops or to grazing land for domestic livestock, or changes initiated with the building of human-made infrastructure. Emphasis is given to private initiatives that result incidentally or purposely in the maintenance of a measure of biodiversity, and to the role of government in guiding and encouraging land use that results in biodiversity conservation. This emphasis is not meant to denigrate the role of and need for parks and reserves located strategically to preserve key ecosystems and important species assemblages. However, because of the limited capacity of governments to achieve a desirable level and extent of protection, a broader approach is needed. It appears prudent to focus more efforts on the 90 per cent of the earth's land area that is either in private hands or in public ownership but exploited by private interests.

Important representatives of terrestrial biota exist in all ecosystems, but more attention is given here to forested tropical ecosystems where human pressure appears to be highest. The Holdridge life-zone system is offered as an organising framework for establishing biodiversity conservation strategies, particularly at the national and regional levels. Land-capability assessment is introduced as a means of justifying the protection of areas with severe biophysical limitations. Marine biodiversity concerns are focused on the diverse estuarine and near-shore marine ecosystems most threatened by human exploitation and settlements.

A Basic Assumption

It is a critical assumption that maintaining a diverse array of species and their habitats has value to society. It is also assumed that publicly and privately held natural areas outside of parks must produce goods and services of value to society competitively with alternative uses that usually result in major loses of biodiversity. Where natural-area use is not fully competitive with conversion to other uses, we can assume that society is willing to promote resource use that results in the maintenance of biodiversity by: (i) restricting concessional use of public lands, (ii) subsidising noncompetitive private uses and (iii) becoming more clever in attributing economic value to ecosystem services, such as flood control, that incidentally result in biodiversity maintenance. Anything short of subsidised preservation (*de facto* parks) involves some sacrifice of biodiversity.

While life exists everywhere on the planet, the diversity of life happens to be greatest in the developing countries and it is also there that threats to diversity are most intractable. This section is focused on the logic that peoples who threaten biodiversity through over-consumption require less immediate attention than those who deforest, burn and overgraze in order to survive.

Multiple Faces of Biodiversity

At the interface of natural science and society's concern for nature, ambiguities and differing perceptions are inevitable. Even among scientists biodiversity conservation has generated controversy, over time and today. Aldo Leopold believed that the habitat needs of large carnivores should govern efforts to maintain biodiversity: carnivores must be present 'to preserve the integrity, stability and beauty of the biotic community'. Although serious efforts continue to maintain megafauna populations, a broader multispecies perspective has evolved. Four of the comparisons that both scientists and planners of resource use address are successional (including weeds) versus climax associations, various small areas versus a few large areas, species-rich versus depauperate systems, and umbrella species (large carnivores) versus overall system diversity. With severe budget limitations choices must be made. The distribution of species that are actually or potentially useful, endangered, scientifically interesting, or beautiful is difficult to predict. The preservation of as large a variety of ecosystems within a particular country as possible appears to be a viable target.

Biodiversity is the total number of species and the distribution of a particular species calculated using formulae to index different attributes of diversity in a specific ecosystem. Conservation can be defined as managed resource use that maintains the capacity of natural habitats and agricultural land to produce crops, livestock, timber, fish or wildlife. Biodiversity could be added to the list of 'products' of managed resource use, though a bit awkwardly. Maintaining biodiversity may be either a management objective in itself or a condition on other resource uses, such as timber production. Biodiversity can be an important yardstick of the relative success of our conservation efforts if the formal methods of measuring it are employed consistently over time in accordance with an appropriate sampling regime.

The conservation community must look carefully at what is meant by 'biodiversity conservation'. Public interest in the subject is sparked by the educational and fund-raising activities of conservation groups that focus on preserving rainforests and certain symbolic vertebrates. The geographic distribution of those concerned about diversity is not uniform and interest increases exponentially with the distance from where it is greatest and most threatened and peaks in places like Washington, DC and London. E.O. Wilson's excellent volume Biodiversity contains 57 chapters of which only two were written by experts from tropical countries. Those most knowledgeable and concerned about the importance of maintaining biodiversity also have their basic human needs well met, while those whose daily actions most drastically affect biodiversity do not. If we do not approach biodiversity maintenance through a process that also improves the economic and social well-being of rural people, the effort will most certainly fail. The small cadre of people who can effectively argue the case for biodiversity have to reach out and convince leaders in developing countries, their own national decision-makers, and the unconvinced experts in development agencies.

Why We are Losing Biodiversity

Failure to value and protect natural ecosystems (read 'biodiversity') is driven by deeply rooted cultural preference for disturbed landscapes, real or perceived necessity, and a lack of education of the public and decision-makers. Cultural values are reflected in poor and rich countries alike where subsidies tend to favour the conversion of natural ecosystems to provide for agriculture and pasture—regardless of the value of the goods and services offered by natural ecosystems that are lost. For example, in the United States subsidised grazing replaces poverty as an incentive to disturb arid areas and artificially low stumpage fees in publicly owned forests promotes their mismanagement.

Education on two fronts is needed to reverse the pressure on natural ecosystems. One emphasises economic evaluation of the goods and services offered by ecosystems, including both known marketable products and services (such as watershed protection) and future benefits yet undiscovered (new crops and medicines). The second is an ethical consideration which stresses that protecting nature is good and prudent, and that the loss of any species is bad. Scientists advocate preserving ecosystems because they contain unique and interesting information.

Both poverty and affluence take their toll on ecosystems and their associated species. However, without a measure of affluence, the loss of ecosystems and species will continue. Affluence provides the opportunity for choice on how ecosystems are to be used, rehabilitated and preserved. Running counter to the human drive to simplify is a love of life that also exists. But without education to enhance and focus innate human 'biophilia' into a moral and political force, increased income can increase destructiveness. In the Philippines, for example, increased affluence allows the dynamite fishermen to purchase larger boats and blow away more distant coral reefs.

An Evolving Strategy

For millennia simplification of ecosystems has been our most effective means of assuring that we can direct the sun's energy, water and nutrients to meet our own needs for food and fibre.

To persuade the great majority of the world's people to maintain biodiversity in apparent contradiction of their best interests is a formidable challenge. Efforts to date by dedicated conservationists have met surprising success, given the esoteric nature of the theme. The goal of the World Conservation Strategy has been modest—setting aside 10 per cent of each country in national parks and equivalent reserves. Success in this effort has absorbed much of the energy and funds of the conservation community. David Western points out that past efforts to conserve animal populations in Kenya's parks have been remarkably effective, but that 75 per cent of the wildlife is found outside the park boundaries. Probably an even higher percentage of the rainforest species in the Congo Basin are found outside of parks. The USAID-funded Biodiversity Support Program has established a commission made up predominantly of African professionals to develop a biodiversity-conservation agenda. Because of the African view of parks as a colonial legacy, they avoid linking biodiversity solely to parks, but advocate its conservation wherever it occurs—on farms, in parks, in hedgerows, etc. This approach closely parallels that advocated by Western. In Southern Africa, Zimbabwe is privatising wildlife management under the CAMPFIRE project. But we have to be careful in extrapolating from opportunities for the conservation of wildlife in East and Southern Africa: they have the highly visible 'big game' species that people will travel halfway around the world to see or shoot.

Biodiversity in a Life-zone Context

The IUCN interprets biodiversity to encompass all species of plants, animals and micro-organisms, and the ecosystems (including ecosystem processes) to which they belong. To be useful, this global perspective must be brought to a level, where public awareness and policies must eventually drive actions to maintain biodiversity. We have not identified all species, know little of the value of those we have catalogued, and even less about how to assure their survival. Given this reality, maintaining at least patches of as many ecosystems as possible is a prudent strategy. The Holdridge life-zone system provides a logical basis for defining local ecosystems in a globally comparable framework. All terrestrial ecosystems can be uniquely defined in terms of three parameters: precipitation and temperature (for which data are widely accessible) and, potential evapotranspiration, which is calculated using the first two parameters.

Life zones can be identified roughly using existing climatic and topographic data. Overlays of satellite imagery or maps showing land-use and vegetative cover can reveal whether actual cover matches the life-zone classification. Areas that have been substantially altered by human activity can be eliminated from consideration in the foreseeable future. Lands that are largely intact do not require immediate action. This leaves at-risk areas that are partially altered or undergoing change that may respond to management interventions. Further overlay of land-capability maps will reveal where conversion to agriculture or other uses may be more appropriate. In areas where the ecosystems are at least partially intact, policy initiatives can be implemented to maintain, rehabilitate and connect them. The identified ecosystems can then be stratified to assure that as many as possible are covered.

Forested life zones make up a majority of the blocks in dry forest, montane forest, humid forest, etc. All yield wood and other products of value and virtually all have been or are currently being heavily disturbed by fire and other forms of human intervention. A few sites in a variety of life zones have been managed sustainably. Intact forests represent a decreasing fraction of the potential cover. The loss or degradation of forest cover is highly uneven — among life zones, within life zones, and across countries. The drier, colder and wetter life zones in the three corners of the diagram are least attractive for crop agriculture and therefore tend to be less disturbed. Poor countries with high population densities or with a grazing tradition tend to push disturbance further into the arid, cold and damp corners of the diagram as well. Poverty causes the use of infertile soils and steep slopes that would be likely be left undisturbed in wealthier countries.

It is useful to note that most of the earth's surface is below 1000 m elevation. Conversely, as one proceeds upslope in essentially conically shaped mountains, the areal extent of each succeeding life zone is smaller and more isolated from similar ecosystems on neighbouring mountains. From the perspective of national planning for conservation of biodiversity this pattern is potentially important. Higher-elevation systems tend to be smaller in extent, more isolated, and potentially more threatened than some of the more extensive lowland systems and lowland rainforests may be the least threatened when compared to some upland ecosystems.

Land-capability Assessment

A strategy for assuring that land-use is compatible with the public interest is zoning based on land-capability assessment. To the extent that restraints limit intensity of use, biodiversity maintenance is well served. A common example is the delineation of river floodplain and coastal flood-prone areas with direct prohibitions on building or control of access to flood damage insurance and services. Costa Rica, for example, legally recognises a land-capability classification system that provides the biophysical basis for guiding credit, subsidy and colonisation programs. That legal recognition of the classification criteria is not accompanied by widespread application is a fact of political life, but does not detract from the utility of the system once the public chooses to demand enforcement of the regulations.

The system is rooted in the widely used Holdridge classification of life zones. Additional soil and slope parameters allow definition of land-use potential. Approval of credit and subsidies for agricultural and livestock production, including land clearing, can be based on objective criteria which include climatic regime, slope and soil depth, drainage, and fertility. Similarly, areas opened for colonisation under land-reform programs can be limited to those where appropriate technologies can be prescribed according to existing conditions and the capacity of the land to sustain the activity in question. The more intensive use categories—intensive annual crop production, permanent crops, grazing lands and tree crops—afford no protection to 'wild' biodiversity *per se*. Significant biodiversity conservation would occur in the following categories:

1. Production forest. Areas of high forest biomass production potential where best management practices will result in sustainable production of timber and other products. Conditions are not appropriate for other, more intensive agricultural uses.
2. Extensively managed forest: Areas with limitations not so severe as to be used solely for protection. Non-timber forest products as well as limited timber volumes can be extracted under tightly controlled conditions.
3. Protection: Areas lacking the minimal conditions for agriculture or like uses, generally steep slopes, swamps or areas of high precipitation. These areas have high value for watershed protection, aquifer recharge and wildlife habitat.

Financial institutions are concerned with the payment of principal and interest on credit granted. Thin, infertile soils on steep slopes with insufficient precipitation are unlikely to produce the crops and livestock needed to pay off a loan. The land-capability classification system limits government-subsidised incursions into areas that are agriculturally unproductive. This has a multiple appeal: the people and their fiscally responsible representatives have a tool to limit wasteful use of public funds on unprofitable land development schemes, the sustainable use of renewable natural resources is promoted, and the exponents of biodiversity maintenance find that these same lands remain under natural vegetative cover.

This land-capability classification system was not designed to conserve biodiversity, but rather to foster sustainable land-use. In most developing countries virtually all accessible areas classified for forestry or protection would be converted to cropland and pasture, often with the encouragement of subsidies and land-settlement policies. By encouraging adherence to the last three forestry-related categories, government would foster the conservation of biodiversity over large areas currently threatened with conversion.

The system does not address the need to conserve modest representatives of ecosystems with few constraints on intensive agricultural development. It is not likely that major reserves will be created on such favoured sites. Policies with associated incentives and penalties can be implemented to maintain hedgerows and stream corridors. Persuading people to set aside productive land for biodiversity conservation is fairly easy in a wealthy, motivated society; the options are extremely limited in poor, overcrowded countries.

Hunting and Fishing Reserves

Kenya's Masai Mara and Costa Rica's Monte Verde cloud forest both suffer the consequences of being overloaded with nature-oriented visitors, with resultant habitat degradation, negative effects on wildlife and loss of quality of the ecotourism experience. In contrast, the economic success of hunting reserves in such disparate locations as Zimbabwe, south Texas, and highland Ecuador attests to the appeal of this land-use, and the sustainability of the operation requires effective habitat maintenance, large areas and low numbers of tourists. The hunting of trophy animals on a well-managed reserve brings tourists who pay an order of magnitude more than the photo safarist sitting back in the tourist section of the plane. The Zimbabwe example proves that appeal to wealthy clients does not preclude significant local benefits, though in general the airlines, hotels and operators capture the bulk of the tourist's dollars. In Texas, hunting leases for deer, peccary and quail bring a higher return than cattle while maintaining a higher-quality, more diverse habitat for these and other species. In Ecuador, deer hunting in the paramo above 3600 meters offers a remunerative alternative or complement to the existing extensive grazing of sheep and cattle and the even more extensive uncontrolled annual burning, which creates a monotonous landscape dominated by unpalatable wire grass. Fire exclusion and management to create more browse for deer would simultaneously favour the re-establishment of a more diverse habitat for other animals as well.

Fly fishing has enjoyed (or suffered, in the eyes of the solitude-seeking angler) a boom in popularity, intensified by Maclean's A River Runs Through It, Redford's subsequent movie of the same name, and the more psychotherapeutic Fly Fishing Through the Midlife Crisis. What duck hunters through Ducks Unlimited have done for wetland habitat preservation, fly fishermen through Trout Unlimited and similar groups have accomplished for stream habitat maintenance and rehabilitation. Trout and salmon habitat maintenance is the primary tourism-oriented concern from Tierra del Fuego to Alaska, from New Zealand to Siberia, and in all cool flowing waters in between. Tiger fish in Lake Kariba, peacock bass in Amazonia, and bonefish on the mangrove-fringed flats support entrepreneurs with a vested interest in habitat maintenance. It is impressive to see a Belizean ex-commercial fisherman, now a bonefishing guide, gently but firmly insist that his client release his first bonefish.

Neither the safari outfitter nor the fishing-lodge operator has a primary interest in biodiversity conservation. The former has a two-dimensional interest in maintaining the length and breadth of the habitat for a narrow range of game animals; the primary concern of the latter is linear—the ribbon of stream or coastal habitat supporting his client's prey and its aesthetic surroundings. As for the average trout fisherman, the diverse array of benthic invertebrates anchoring the food chain supporting his beloved quarry is of but passing concern to him. In both cases indirect concerns are broader and activism extends to pressure on government to control land and water use affecting migrating game and fish.

If the operators have a vested interest in continued use of the resource, they will resist encroachment and be zealous pursuers of poachers, because a lost buffalo or a blasted reef represents lost profit. The long-term profitable management of any of these enterprises results in the maintenance of a good measure of the biodiversity that existed prior to human intervention. However, when profits are squeezed or intervening opportunities become more attractive, private operators will tend to intensify operations by bringing in more paying visitors, cutting corners on maintenance and protection, or simply selling out.

Private Conservation

Sherman Chickering practiced conservation of biodiversity by owning and protecting from intrusion 20,000 acres near Lake Tahoe in California. His action is exceptional only in the magnitude of the area set aside. Many private landowners evidence their appreciation of nature by setting aside and protecting natural areas. This has a legally defined analogue in the concept of the conservation easement, whereby the owner of land with natural vegetation cover can forgo in perpetuity the right to develop or improve the land (i.e. clear it for agriculture or urban-industrial use) in return for some form of tax relief.

Is the conservation-easement mechanism only applicable in economically developed countries? No, but it becomes easier to implement as a country's institutions become stronger and its population becomes more educated, more urban and economically better off. In the somewhat more conservation-minded Brazil of today, it is doubtful that the 1970s tax rebates in the south in return for investments in rainforest clearing for cattle ranches in Amazonia would be acceptable politically. An equally ambitious rebates-for-rainforest-conservation program could gain popular support, particularly if competitive income streams from sustainable management of timber, non-timber forest products, and ecotourism could be generated.

Coral-reef Parks, Tourists and Fishermen

Experience in Belize (personal observation) and the Philippines indicates that creation of reef parks or fishing exclusion zones can have a positive impact on biodiversity and fish populations in and around the area set aside. In Belize, the fishing cooperative agreed to respect the creation of the 256 ha Hol Chan Reserve, encompassing 1.6 km of barrier reef and shoreward seagrass beds and mangroves off the tourist destination town of San Pedro. Within one year after the establishment of the park, a dramatic increase in the number and diversity of fishes was observed and documented over time. Fishing success in the vicinity of the park was reported to have increased. Tourist dive boats were invariably concentrated along the reef within the park boundaries where intact reefs and the most fish could be seen. Income from increased catches and from guiding tourists to the reef benefited local people.

It appears that the intact reef in the exclusion area or park serves as a refuge and a breeding and growout habitat for reef fishes. Empirical evidence has convinced fishermen that the increase in fish catch in peripheral areas warrants respecting the park boundaries. For this system to function, the fishing practices in peripheral areas must respect the physical and biological health of the whole reef and other linked ecosystems. There can be no dynamiting, poisoning or overexploitation of the resource. Of critical importance are tenure arrangements that assure limited access to the inshore marine resource base. The economic viability of a reef park is greatly enhanced when ecotourism can draw divers and snorkelers under carefully controlled conditions.

In populated islands, coastal areas with fringing and barrier reefs, and associated ecosystems, it will be a challenge to the ecologist, fisheries biologist and resource economist to calculate the optimum proportion of protected to sustainably managed habitat. 'Optimum' depends on which stakeholder is consulted. In Belize there are at least three interested parties:

1. Conservation groups: The international and Belizean conservation groups interested primarily in establishing an extensive network of protected areas encompassing much of the coastline.
2. Developers: The tourist-hotel owners, dive shops and guides (some of whom are or were fishermen) to whom the Hol Chan Reserve represents a valuable attraction for their customers. Ironically, one developer who has devastated the mangroves near the park advertises the Reserve in the brochure for his coastal properties.
3. Fishermen: The fishermen themselves who for generations have exploited the area in and around the present reserve—trapping spiny lobster, collecting conchs and catching fish in accordance with a complex system of inherited fishing rights.

Neither the conservation community nor the government of Belize, even with the sympathetic support of the tourist industry, has the power to set aside and protect an extensive park system unless the fishermen are in agreement. It would appear that a practical solution is a system of protected areas that results in the maximum sustainable catch for the fisherman. The pattern of distribution and total area would be the focus of an applied research exercise involving participation by all stakeholders. There must be a pattern of alternating fishing areas with protected sites for breeding and refuge of the majority of the target species. This would not obviate the need to deal with special cases such as the concentrated breeding of grouper at a single site within a seemingly uniform habitat.

If fishermen are convinced their interests are being served, they can become a critical force in assuring that the ecological and spatial integrity of the protected areas is maintained and that the management of intervening areas is compatible with the overall production and protection goals. This assumes that the beneficiaries have exclusive access to the resource area being managed.

Alternatives to Timber Management for Biodiversity Conservation

Among the various uses of standing forests that have been proposed as being compatible with conservation are ecotourism (with or without an indigenous cultural component), extraction of non-timber forest products (fruits, medicinal materials, ornamentals, etc.), and forest management for timber.

Ecotourism

On flights carrying Costa Ricans from San Jose to Disney World and returning laden with North American ecotourists, the appeal of the exotic drives both streams—with the economic balance probably favouring Costa Rica. Ecotourism has great promise as a generator of foreign exchange while providing an economic incentive for conserving natural areas. The coming of ecotourists from abroad provides a high-profile demonstration to national decision-makers and the public that nature is valuable. The areas dedicated to ecotourism are attractive because the ecosystems are reasonably intact, the animals are relatively abundant, and the guides are effective at interpreting the landscape.

Generally a combination of primary and disturbed habitat affords the tourist the opportunity to observe the greatest diversity of landscapes, plants and animals. The greatest advantage and disadvantage of ecotourism are that it demands so little space. As long as guides keep groups a bend in the trail or river apart, large numbers can be accommodated and still assured a quality experience. Habituated animals afford thrill after thrill to passing tourists. A visit to a local community, a small zoo, videos and lectures make the tourism experience complete. Because ecotourism can be profitable within a few hundred or thousand hectares, there is little incentive to operators to buy or lease and protect large areas. The disadvantage is that ecotourism cannot be expected alone to justify the protection of large forest areas from conversion.

Perhaps Costa Rica provides the most striking example: this small country has the world's most developed forest-based ecotourism industry, even though only a small percentage of the original forest outside of parks is still standing and the rate of deforestation is relatively high. Still, even though ecotourists may not utilise large areas, awareness of the existence of extensive parks is a drawing card in countries like Ecuador.

Extraction of non-timber products

Historically, non-timber forest products have contributed to the meager cash income of people living in or on the fringes of forest areas, while loggers profited from timber exploitation. It is the position of some that non-timber forest products can offer an economically competitive alternative to logging and forest conversion; large extractive reserves are proposed as an effective mechanism for maintaining forest biodiversity, but this approach appears to be based more on ideology than on economics. What is provided by the extraction of non-timber forest products alone is sustained rural poverty; by any measure of well-being—income, housing, access to health care and education—the xate (ornamental palm) and allspice gatherers of the Guatemalan Petén are unrelievedly poor. Forgoing 50 to 90 per cent of the potential income stream that would derive from timber is a luxury few would willingly choose. Those engaged in extraction do not generate the surplus income needed either to pay for patrols to protect reserve borders or to pay concession fees that would allow government to protect the reserves from encroachment. These gleaners of non-timber products cover enormous areas of the forest every day, carrying their shotguns and accompanied by their dogs. If they are not to continue wreaking havoc on wildlife populations, the advocates of non-timber extraction will need to devise alternative sources of income and protein for these people.

There may be low-density populations of indigenous peoples with an extraction-based economy that have a genuine desire to live by their own traditional standards of well-being. Their reserves can effectively contribute to biodiversity conservation if they have political clout and sustained international support. Their case is totally different from that of migrants to the agricultural frontier who have socioeconomic aspirations similar to those of the rest of their dominant national culture.

Advocates of alternative management strategies should have the opportunity to compete for concessions on public forest lands. If governments establish concession fees based on the most remunerative sustainable use, presumably sustainable timber management, then all potential users have a common basis for bidding. The extractive bidders will presumably have to carry out international fund-raising in order to generate capital for concession fees and the costs of maintaining the integrity of the concession.

Ecotourism and non-timber forest product extraction are potentially remunerative and sustainable uses of forest systems that can contribute to biodiversity conservation. However, at the community level such uses need to be combined with timber management to assure that the overall forest-based enterprise is sustainable. An inescapable constraint is the need of every community to have long-term, well-funded NGO support, because most communities do not have the institutional structure, technical expertise, political clout or funds to engage in sustainable, market-oriented forest-resource management. And they simply are not enough NGOs or funds to meet the grass-roots needs of forest communities.

Commercial forest management

It is the basic hypothesis of this section that sustainable management of forest for timber and other products and services is a practical way to maintain forest-ecosystem biodiversity outside of parks. Most of the world's remaining forest ecosystems are found outside of parks and equivalent protected areas. Given that few of the existing parks receive adequate protection now because of lack of commitment and growing demographic and economic pressure, it is unrealistic to assume that the rest of the forest areas will be incorporated into any form of protected status.

Unfortunately, both mainstream and radical nature conservation groups find that anti-logging TV spots and literature, which reflect the perspective of many of their professional staff members, are effective for fund-raising. The stridency of these ads prejudices public opinion against all tree-cutting, whether sustainable or not. Actually, the conservationist opposition is to destructive logging, not sustainable timber management, which few have ever seen in practice. While preservation and management advocates wrangle over which strategy will save diversity, the poor farmer and cattleman are converting the forest to cropland and pasture.

It is also unfortunate that the sustainable management of forests for timber and other products and services is still being promoted primarily by other advocates of biodiversity conservation and not by the great majority of logging-company executives. Ironically, its most vociferous critics include both anti-timber advocates of biodiversity conservation and the timber and wood products industries themselves fearful that their supply of wood will be restricted. Both industry and conservationists must become convinced that sustainable timber management is profitable and one of the best available means of maintaining biodiversity outside of parks. Unless economically competitive uses for standing forests are found, they are likely to be converted to cornfields and pastures. The foremost enemy of biodiversity conservation is conversion to other uses.

A combination of compatible uses, with forest management for timber as the primary use both spatially and in terms of income generation, offers the highest potential for maintaining forest cover and a large measure of biodiversity in competition with conversion pressures. The potential for success will be far greater if: (i) the policies are neutral, or preferably favourable, to long-term use of suitable land for forest production, (ii) entrepreneurs and investors come to see sustainable forest management as good business within a favourable policy and regulatory setting (as they would demand if they were raising cattle or assembling computers), (iii) subsidies, both national and international, can be effectively directed to paying the opportunity cost, particularly to poor people who would otherwise be attracted to the conversion option for short-term survival, (iv) training and development programs are directed toward preparing people for productive involvement in forestry and complementary activities and (v) information programs convince the public and decision-makers that forest ecosystems are beautiful and economically valuable for the goods and services they provide.

The existing situation is not pretty. Both the United States and Canada have proved that having a competent forest service and an articulate and well-funded conservation community is not sufficient to assure sustainable management of their western coniferous forests. The situation in the mixed hardwood forest of eastern North America is more promising, with long-standing examples of sustainable management. In general, however, most developed country foresters study forest management and conservation in school and practice logging after graduation. Most loggers in developing countries never studied forestry in the first place. In this context logging is simply the removal of timber from the forest with no attention to the effects of the action on regeneration, erosion, service functions or biodiversity.

Sustained-yield management implies the removal of only the annual growth increment of the forest, extraction strategies that assure regeneration, and practices that promote maintenance of biodiversity both in the forest and downstream. The challenge is to persuade loggers to become dedicated experts in the sustainable management of forest ecosystems. This is likely to be accomplished when they become convinced that management is economically attractive and a legal condition of resource access. A critical first step is for the forest management operator to have confidence in long-term access to the resource, through either renewable concessions or secure ownership. Community industries must have confidence based on the same criteria. Becoming convinced that low-impact logging techniques are less costly than conventional practices is a relatively easy step toward voluntary sustainable management. Joint implementation agreements can actually result in the timber company's receiving a subsidy for low-impact logging.

Preferential access to 'green' markets is an added inducement, achievable only by conforming to all-encompassing certification criteria. As major wholesalers, and even political units, begin to require certification of sustainable management, the inducement becomes more coercive. Government verification of compliance with concession requirements can provide additional pressure for sustainable management of forest resources.

By comparison with the example of the western United States and Canada, the situation in the humid and wet tropical forests of the world is actually hopeful. These forests have fewer marketable species and infrequent occurrence of even-aged stands, two conditions that make devastating clearcuts attractive, the only major exception being the dipterocarp forests of Asia, which do have a high percentage of marketable species. Paradoxically, one of the long-greatest term threats to biodiversity in tropical forests, especially in the American tropics, is underutilisation. The extraction of only a tree or two per hectare leaves the forest virtually intact. This selective extraction has two distinct negative effects. First, the forest is devalued by high-grading, becoming less attractive to potential investors in sustainable timber management and, by default, more attractive to directed and spontaneous settlement. Second, most of the valuable species, like the mahoganies of America and Africa and the Asian dipterocarps, require larger gaps to reproduce than are produced in selective logging. Biodiversity is threatened if the economic competitiveness of the standing forest is reduced, making conversion a more attractive option.

Can biodiversity be preserved outside of officially designated parks and reserves? In developed countries the answer is a qualified yes. Well organised and funded nature conservation, fishing and hunting organisations support biodiversity conservation, at least indirectly. These countries can afford to pay the opportunity costs required to control urban sprawl, remove grazing subsidies, consume less and recycle more. They can afford incentives for conservation easements. These actions result in the conservation of more biodiversity. However, the general public and many politicians have only a modest interest in doing so.

In developing countries of the tropics, tiny Ecuador for example, the diversity of species is greater than in all of North America, yet the range of options for conserving this diversity is narrower. In most developing countries, the conservation movement is nascent and for the most part recently adopted and funded from abroad; biodiversity conservation is at best a slogan to a few politicians and an unknown concept among the general public. Building awareness and support is a critically important task in the long run. In the interim, pragmatic solutions must be sought. These usually do not involve overt championing of biodiversity conservation, but rather focus on making the case that the value of natural ecosystems to provide economically valuable goods and services-timber, non-timber products, clean water, etc. is greater than if the land were converted to alternative uses. Efforts to remove incentives and revoke policies that encourage ecosystem destruction can result in biodiversity conservation without competing with immediate development needs.

INTEGRATING PARK AND REGIONAL PLANNING THROUGH AN ECOSYSTEM APPROACH

Protected areas have been a feature of the landscape and of the resource and environmental manager's arsenal for over a century. Today systems of protected areas seek to preserve representative samples of ecological, geological and scenic wonders in most countries of the world. And there is growing urgency to 'complete' protected-areas systems before human pressures, land-use change, and political decisions eliminate all opportunity to preserve at least samples of all of the earth's species, habitats and ecosystems. Protected areas, especially biosphere reserves, national and provincial parks, and World Heritage Sites, are established in part to preserve natural, unaltered ecosystems and species as benchmarks and as areas for scientific study. In addition, parks are established for public use and experience of their intrinsic values, as well as means to demonstrate the potential for coexistence of nature and human activities.

These goals often conflict; park planners and managers face difficulties reconciling conflicting goals within parks and between the parks and their surrounding regions. Traditionally, park staff have turned to ecological sciences for guidance in making policy decisions. And, indeed, ecological understanding of protected areas tells us many important things about them. It can underscore that they are dynamic and complex systems of many interconnected and interacting components. The removal of one component, from a species to an entire ecosystem, can have unexpected, hard-to-predict consequences, including the numbers and distribution of species or ecosystems, or changes in physical processes and flows. Ecological understanding forces one to incorporate spatial and temporal dimensions into resource surveys, research, planning and management. The pattern of activities and ecosystems in space and time is of central importance to understanding and managing a protected area. Further, ecological understanding supports assessment of the impacts of different circumstances on a protected area: visitor activities, the refuse of old resource extraction activities, the effects of poaching, or the transport of pollutants in air and water from outside the boundaries. Ecological understanding highlights the fact that a protected area is subject to change and threats from both internal and external processes and activities, and that as a result management must be proactive.

Such issues are not new. But the more one applies 'lessons learned' to planning and management, the more one is pushed toward a focus on entire, functioning systems rather than arbitrarily limited protected areas. Technically, this is the domain of several rapidly developing areas of research that might collectively be referred to as ecosystem science.

Ecosystem Science

The holistic, interdisciplinary study of ecosystems has been around for twenty or thirty years. It gained early impetus from the International Biological and Man and the Biosphere programs and from the work of ecologists such as E.P. and H.T. Odum. Today there are many different but complementary approaches. Of particular relevance to parks are conservation biology, landscape ecology, ecosystem science, state-of-environment reporting and ecological integrity.

The lessons of conservation biology elaborate the implications of protected areas as islands in a sea of different land-uses and strongly altered ecosystems. Such islands may have difficulty maintaining species diversity, may not incorporate functional ecosystems, and, as a result, may require intensive management of populations because of small breeding populations. Conservation biology contributes to an understanding of the dynamics of small-scale population management within isolated ecosystems. It provides a view of the protected area as islands from the inside looking out.

Landscape ecology provides a view of the protected area as an island from the outside looking in. It deals with the protected area as the remnant of a once much larger landscape element, now isolated in an otherwise modified landscape. It identifies the dominant landscape elements, or matrix, and identifies other islands and corridor and network features that may link islands into functionally larger systems. Landscape ecology suggests quantitative measures of landscape structure and function, and provides a framework for outlining the processes of connection and change between protected areas and other landscape elements.

Ecosystem science is critical to an understanding of the actual processes within particular ecosystems at various scales. Such an understanding is what permits us to anticipate and mitigate alterations caused by internal or external threats. The idea of stress/response functions in ecosystems is a particularly useful one for park managers, whose lands are almost always stressed in some way and who can often improve their recovery responses through particular interventions.

A stress/response approach to park system management leads to a concern for the state of the environment in the protected area. What are the structural and functional features and characteristics of the protected area, and what is their current state? Such an assessment is critical for determining the effects of particular activities on the areas of the protected ecosystem that require more active intervention and protection. Such an approach emphasises the need for monitoring the protected area to track change as an aid to timely intervention. Many of these approaches can be used to collect and organise information for assessments of protected-area problems and to identify interventions needed for more effective management.

A related topic receiving much attention is 'ecological integrity'. It can be argued that the goal of ecosystem management should be to maintain their integrity. Indeed, since 1988 the Canadian National Parks Act makes the maintenance of ecological integrity of national parks the first priority of management. Yet ecological integrity is a difficult thing to define. The significant quantitative work done on freshwater ecosystems recognises that ecosystems are complex and interconnected and have their own inherent functional and organisational properties. They draw on a range of systems and other theories to emphasise the self-organising, self-maintaining abilities of intact ecosystems.

When we turn to the management of actual protected areas and their surrounding areas we are faced with other problems: 'large' ecosystems, a significant human presence and activities, and the need to integrate science with planning and management activities. This is where it may be useful to speak more generally of 'ecosystem approaches'.

Ecosystem Approaches

Protected-area management is never simply using science to understand the protected area. Science and the understanding it brings are of necessity, parts of planning and management. But planning and management also involve institutions, administrative hierarchies, organisations, and individuals with varied goals and perceptions, all of whose interests should be reflected in the planning and management processes. Yet these processes often reflect historical, political and disciplinary priorities and prejudices and are less inclusive and interdisciplinary than they should be. Such *a priori* narrowness creates problems for both scientific understanding and program implementation.

Over the last twenty years or so, in parallel with the growth of ecosystem science as described above, a number of disciplines have developed 'ecosystem approaches' based on ecological and systems principles that better integrate description, understanding and prescription in complex scientific and professional situations.

These ecosystem approaches use a holistic, interdisciplinary systems perspective and seek to place the system of primary interest in a larger context. The ecosystem is defined bioregionally or in terms of watersheds, and includes people and their activities. Ecosystem approaches focus on interactions and system behaviour, and take an ecological approach to changing patterns of structure and organisation. From traditions in human ecology and anthropology there is often an emphasis on linking biophysical and socioeconomic dimensions. When extended to planning and management, an ecosystem approach uses actor and institutional analyses to recommend or facilitate more consensual, participatory processes; cognitive or perceptual shifts; and institutional integration (Table 3.1).

Slocombe presents a review of theory and experience of ecosystem approaches in a range of disciplines. At their worst such approaches blend 1960s popular 'ecology' with a particular perspective on a problem. Ecosystem approaches are commonly criticised as being equilibrium-oriented, emphasising energy flow and functionalist approaches, and neglecting historical, evolutionary and individual factors. Yet with broad theoretical and empirical grounding, an ecosystem approach can provide a framework for organising and integrating research, planning, and management for protected and other areas. Although interest in ecosystem management for protected areas has been growing, the broader ecosystem approaches as described here are less common. But interest in transdisciplinary, integrative ecosystem approaches is growing. The Canadian Government recently announced a new fifty-million-dollar program for research on large ecosystems in which human activities are central. The next section briefly presents three case studies of national-park-centred regions where various initiatives suggest possible directions for ecosystem approaches.

Table 3.1. Core characteristics of ecosystem approaches.

Describing parts, systems, environments, and their interactions.

Holistic, comprehensive, transdisciplinary.

Including people and their activities in the ecosystem.

Describing system dynamics, e.g. through concepts of stability, feedback, etc.

Defining the ecosystem naturally, e.g. bioregionally, rather than arbitrarily.

Looking at different level/scales of system structure, process and function.

Recognising goals and taking an active management orientation.

Incorporating actor-system dynamics and institutional factors in the analysis.

Using an anticipatory, flexible research and planning process.

Entailing an implicit or explicit ethic of quality, well-being and integrity.

Recognising systemic limits to action—defining and seeking sustainability.

CONCLUSIONS AND RECOMMENDATIONS

In dealing with social concerns and biodiversity conservation in a regional context, we face the challenge of narrowing the definition of the problem, since all phases of planning involve the concerns of people. Thus, in discussions involving people from a number of disciplines the tendency is to expand the scope of the problem as linkages are noted—the result of which is a quagmire of advice which is far too broad to be effectively implemented. While the conclusions and recommendations presented here attempt to represent areas of concern routinely faced by planners they also represent concerns for biodiversity conservation. The chapter is organised into four main sections, all relating to the value of the new regional planning as a way to address the complex issues involved in conserving biodiversity: (i) social concerns, (ii) political will, (iii) information and (iv) development planning. Each of these is summarised in a series of 'principles' organised around a given concern of biodiversity conservation. Although they do not represent a comprehensive statement on the new regional planning, they do indicate that this methodology can contribute to the *in situ* conservation and *ex situ* maintenance of biodiversity. Their use will help ensure the long-term viability of these resources for all concerned parties-scientists, politicians, planners, managers, local residents and resource users.

Social Concerns and Biodiversity Conservation

The often-repeated principle that protected areas cannot be planned in isolation from their surroundings leads to the question of what constitutes a 'region'. Discussions center on the impact a protected area has on its surroundings and on the impact that economic activity often has on the integrity of a protected area. Two conclusions are relevant to this question:

1. Integrated regional development planning provides an effective framework for addressing social and economic concerns for biodiversity conservation regardless of scale or of the complexity of the region.

2. While the interactions between a protected area and its region are unique, the principles offered by integrated regional development planning can help to ensure that the benefits and costs of decisions concerning biodiversity conservation are fully evaluated in the planning process.

Questions are often voiced as to whether it would be preferable to establish a protected area first, and therefore influence the development process in its surrounding region or to formulate a regional plan that included projects to establish protected areas along with other development activities. In one sense, given the concern for biodiversity, the 'protected area first' choice is logical. On the other hand, since people are to be the ultimate beneficiaries of development planning they are the prime concern of the decision-making process, and therefore protected areas should not receive preferential treatment. However, the urgency in biodiversity conservation lies with those areas under pressure for unstructured development. Likewise, given the incipient state of our knowledge on biodiversity, planners may indeed have the opportunity to plan conservation areas more frequently if development planning places a priority on financing mechanisms to support basic field research as part of the planning process and if long-term public involvement in biodiversity conservation is based on an understanding of its benefits. Three principles are relevant here:

1. Resource planners and managers should identify areas that are important to biodiversity conservation and consider the potential development activities that would affect these areas of interest.
2. Community development activities identified in long-range plans should reflect an understanding of the direct relationship between a region's economic viability and maintenance of its biodiversity, whether included in a protected area or not.
3. Planners and decision-makers should be aware of the issues in the region of influence of a protected area as a basis for actively engaging and involving local communities in planning and implementing biodiversity conservation measures.

The long-term viability of local economies (as opposed to project financing) can depend on the benefits and costs of having a protected area nearby. The rights of indigenous peoples, for example, often suffer because of biodiversity conservation efforts that are neither locally initiated nor accepted. Two principles can be cited here:

1. Local resource users who live near protected areas often require support to establish effective institutions capable of influencing and participating in political, economic and conservation decisions that affect the viability of both the protected area and the local community.
2. Governments at all levels should consider institutional arrangements and policies to ensure information exchange, participation, and equal distribution of the economic costs and benefits to communities that depend directly on the resources available from a protected area—particularly when decisions are taken that would negatively affect a local community for the benefit of a larger population (region, nation, world). Such arrangements should include the creation of management structures having legal standing and authority that will allow full participation by populations that have historically used the protected area.

Problems for tourism and fisheries often center on the social and economic issues related to an influx of migrants to territories bordering a protected area. Sound commercial investments, though welcomed, can have direct negative impacts on the make-up of a community. Likewise, an unexpected influx of unskilled workers can affect both the local communities and the very viability of tourism based on natural amenities. A special case has to do with fishermen and the management of marine and coastal biodiversity and protected areas.

1. When development of a protected area results in an influx of migrant workers, their social welfare should be considered. State or national government must bear the responsibility of providing basic services for these people or the future of the protected area can easily be threatened.

2. The development of coastal and marine protected areas can affect access to established fishing areas and the use of traditional fishing methods. Compensation to local fishermen should be considered as part of the costs of implementation.

3. Often fisheries, tourism, and conservation of marine biodiversity cannot only coexist but be mutually beneficial as well. For example, local fishermen can be valuable sources of information and loyal employees of the conservation authority. Every effort should therefore be made to involve local fishermen in planning marine and coastal reserves.

Political Will and Biodiversity Conservation

Plans are often made that, no matter how well formulated in a technical sense, are not implemented. Many reasons can be given for the archives full of unimplemented plans, including those for what have come to be referred to as 'paper parks': a lack of funding, a change in development priorities, a change in political winds of a given nation or region. Sometimes, even after implementation begins, a project fails to fully produce what had been expected by the interest groups that supported it or is not fully accepted by the local community or even by the supposed beneficiaries. The reason often given for such failures is 'lack of political will'. Several observations, conclusions and recommendations can be mentioned concerning this phenomenon:

1. The desire and commitment (political will) of a decision-maker to support a proposed project is neither automatic nor predictable; it must be nurtured from the outset. Getting a decision-maker to accept a proposal is the responsibility of the planner, and therefore creating the atmosphere for its acceptance should be very high on the planning agenda.

2. Political decisions are complex and potentially conflictive with a number of things, including the conscience and aspirations of the decision-maker, the mandates from his or her constituencies, the desires of the groups who have given financial support, the needs of the opposition, and the beliefs of his or her peers, besides the extant legislative directives, policies and regulations. The decision will be taken according to what the decision-maker believes to be the route of minimal conflict with the more important of these.

3. Minimising conflicts while meeting stated goals and objectives is a function of the new regional planning. It is done through an iterative process that is integrated across sectors, is transparent and participatory, and seeks consensus from the affected parties.

4. Educating the decision-maker and the varying constituencies from the very outset greatly facilitates agreement on what a strategy or plan should contain. Building stakeholder pride and ownership helps to assure implementation and long-term viability.

5. Feedback mechanisms from local communities should be developed and nurtured. These provide valuable information to both the planning team and the decision-maker. Although this type of input is unwelcome in some cultures, its acceptance is increasing and should be cultivated.

6. There are inevitably 'winners' and 'losers' in any planning process. The decision-maker must be confident that the interests of all affected groups have been addressed in the strategy or plan.

7. Building a degree of flexibility into a strategy or plan can also help to ensure the support of the decision-maker. A 'perfect plan' may simply be unattainable in a political sense. Therefore,

both the planning team and the decision-maker must have alternative courses of action in the event that any one issue begins to threaten the viability of the project.

Information and Biodiversity Conservation

The availability of information is fundamental to any process that would lead to effective biodiversity conservation. Without comprehensive and credible information, constituencies cannot be built or objectives attained. However, although data gathering is a task that is never completed and planning must contend with incomplete information, every effort to keep it accurate, relevant and current must be made. Therefore,

1. Wide-ranging consultations should be held at an early stage in the planning process to identify interested parties and obtain their views on key issues that must be addressed by the planning team.
2. All involved and affected parties, including governments and corporations, have a responsibility to make relevant information available to the planning team as a means of ensuring the broadest possible debate on the decision as to how a given site should be managed.
3. All information collected by the planning team must be accessible to any group with legitimate interests in the outcome of the planning effort.
4. The information collected should be widely inclusive: technical, scientific, local and traditional.
5. Information used by the planning team should include the historical, natural and cultural aspects to learn from past experience in planning and managing for the future.

Development Planning and Biodiversity Conservation

Protected areas interact with their surrounding region in two ways: (i) they play an essential part in its economic development and (ii) the protection of their resources depends on their proper management in the widest possible regional context. Once a decision is made to establish a protected area, concern for its long-term health is fundamental to any community that hopes to realise long-term benefits from the arrangement. A number of conclusions and recommendations treat this subject.

1. Clear definition of the specific role that a protected area should play in a region or national development policy is of vital importance to ensure that it provides the wished-for range of goods and services over the long-term.
2. Realising the full potential of a protected area, regardless of its stated goals, requires the development of linkages with other sectors of society. Protected areas cannot exist without people.
3. Interactions between a protected area and other ongoing or proposed development activities, such as mining, forestry, agriculture, fisheries and urban development, need to be clearly identified.
4. Effective systems for gathering, storing and communicating information bring together the various partners and constituents of a planning exercise. They are essential for providing maximum benefits to the community.

Chapter 4

Species Extinction

INTRODUCTION

Individual organisms can usually be recognised, but the larger units we use to describe the diversity of life, such as populations, subspecies or species, are not so easily identifiable. Taxonomists further group species into genera, families, orders, and kingdoms, while ecologists group species into higher structures such as communities and ecosystems. The justification for these group terms is utility, rather than intrinsic naturalness, but as far as possible we attempt to delimit groups of organisms along natural fault lines, so that approximately the same groupings can be recovered by independent observers. However, there will be a virtually infinite number of different, albeit nested, ways of classifying the same organisms, given that life has evolved hierarchically.

Darwin felt that species were similar in kind to groupings at lower and higher taxonomic levels; in contrast, most recent authors suggest that species are more objectively identifiable, and thus more 'real' than, say, populations or genera. Today, much of ecology and biodiversity appears to depend on the idea that the species is the fundamental taxon, and many have argued that these fields could be undermined if, say, genera, or subspecies, had the same logical status as species.

Species concepts originate in taxonomy, where the species is 'the basic rank of classification' according to the International Commission of Zoological Nomenclature. The main use of species in taxonomy and derivative sciences is to order and retrieve information on individual specimens in collections or data banks. In evolution, we would like to delimit a particular kind of evolution, 'speciation', which produces a result qualitatively different from within-population evolution, although it may of course involve the same processes. In ecology, the species is a group of individuals within which variation can often be ignored for the purposes of studying local populations or communities, so that species can compete, for example, while subspecies or genera are not usually considered in this light. In biodiversity and conservation studies, and in environmental legislation, species are important as units, which we would like to be able to count both regionally and globally.

It would be enormously helpful if a single definition of species could satisfy all these uses, but a generally accepted definition has yet to be found, and indeed is believed by some to be an impossibility. A unitary definition should be possible, however, if species are more real, objectively definable and fundamental than, say, genera or subspecies. Conversely, even if species have no greater objectivity than other taxa, unitary nominalistic guidelines for defining species might be found, perhaps after much diplomacy, via international agreement among biologists; after all, if we can adopt meters and kilograms, perhaps we could agree on units of biodiversity in a similar way. In either case, knowledge of the full gamut of today's competing solutions to the species concept problem will probably be necessary for a universal species definition to be found.

DARWINIAN SPECIES CRITERIA

Darwin's Morphological Species Criterion

Before Darwin, it was often assumed that each species had an Aristotelian 'form' or 'essence', and that variation within a species was due to imperfections in the actualisation of this form. Each individual species was defined by its essence, which itself was unvarying and inherently different from all other species essences. This mode of thought of course precluded transformation of one species into another, and was associated with belief that each form was separately created by God. Darwin's extensive travels and knowledge of taxonomy led to a realisation that the distinction between intraspecific and interspecific variation was false. His abandonment of the essentialist philosophy and its species concept went hand in hand with his appreciation that variation itself was among the most important characteristics of living organisms, because it was this variation which allowed species to evolve.

Darwin guessed (correctly) that essentialist species would be hard to give up: '...we shall have to treat species in the same manner as those naturalists treat genera, who admit that genera are merely artificial combinations made for convenience. This may not be a cheering prospect; but we shall at least be freed from the vain search for the undiscovered and undiscoverable essence of the term species.' He argued that species were little more than varieties that acquired their claim to a greater reality only when intermediates died out leaving a morphological gap. This morphological gap criterion, which seems to have been accepted by most early evolutionists, has been called a 'morphological species concept' because Darwin used the gaps in morphology to delimit species; however, it is easy to extend his species criterion to ecology, behaviour or genetics.

Polytypic Species

A major revolution in zoological taxonomy occurred around 1900. As the great museum collections became more complete, it became obvious that apparently distinct species found in different areas frequently intergraded where they overlapped. These replacement species were usually combined as subspecies within a 'polytypic' species, an idea suggested for 'geographical varieties' by early systematists and Darwinists such as Wallace. The taxonomic clarification that followed, which allowed identifiable geographic varieties to be named below the species level as subspecies, was conceptually more or less complete by the 1920s and 1930s. At the same time, other infraspecific animal taxa such as local varieties or forms were deemed unnameable in the Linnaean taxonomy. These changes are now incorporated into the International Code of Zoological Nomenclature.

PHILOSOPHISATION OF SPECIES, THE 'INTERBREEDING' CONCEPT

In January 1904, E. B. Poulton read his famous presidential address—'What is a species?'—to the Entomological Society in London. Following up some ideas raised (but immediately dismissed) by Wallace, Poulton proposed 'syngamy' (i.e. interbreeding) as the true meaning of species. Poulton and Wallace were both particularly knowledgeable about swallowtail butterflies (Papilionidae). In swallowtails, there were strong sexual dimorphisms: the female colour pattern often mimics unrelated unpalatable butterflies while the male is nonmimetic. The females themselves are often polymorphic, each female form mimicking a different distasteful model. Under a morphological criterion each form could be designated as a different species, whereas mating observations in the wild showed that the forms were part of the same interbreeding group. Similar ideas were promoted by the botanist J.P. Lotsy, who termed the interbreeding species a 'syngameon'.

In the 1930s, T. Dobzhansky studied morphologically indistinguishable 'sibling species' of *Drosophila* fruit flies and concluded that Lotsy's approach had some value. A species will rarely, if ever, interbreed with its sibling; each chooses mates from within its own species. Dobzhansky proposed his own interbreeding species concept, later popularised by Mayr as the 'biological species concept', so named because interbreeding within species, coupled with reproductive isolation between species, was considered the single true biological meaning or reality of the term species.

A short definition of the biological species concept is: 'Species are groups of interbreeding natural populations that are reproductively isolated from other such groups.' This concept was not so much new as a clarification of two distinct threads: (i) a local component, the Poulton/Dobzhansky interbreeding concept and (ii) a global component which extended the interbreeding concept to cover geographical replacement series of actually or potentially interbreeding subspecies, as in the pre-existing idea of polytypic species.

This extended interbreeding concept was, until about 30 years ago, almost universally adopted by evolutionists. The species concept problem appeared to have been solved; species were interbreeding communities, each of which formed a 'gene pool' reproductively incompatible with other such communities. The new concept answered both perceived problems of Darwin's morphological approach: (i) that using a naive interpretation of morphological criteria, mutants and polymorphic variants within populations might be considered separate species and (ii) that sibling species might be misclassified morphologically as members of the same species.

The new approach was promoted in a long series of books and articles by Dobzhansky, Mayr and their followers. Mayr in particular was highly influential by justifying the taxonomic application of the polytypic species criteria in terms of the new concept of 'gene flow'. To adopt this change, it was necessary to see species in a new post-Darwinian light. Instead of species being defined simply, using man-made criteria based on demonstrable characters such as morphology, species became defined by characteristics important in their own maintenance, that is by means of their biological function. Significantly, the philosophical term 'concept' came into vogue along with these ideas about species, and the term species problem — which hitherto referred to the problem of how species arose, became instead the problem of defining what species were. The important features of species defined by the 'biological concept' were that they were protected from gene flow by what Dobzhansky termed 'isolating mechanisms', including prezygotic factors (ecological, mate choice and fertilisation incompatibilities) and postzygotic factors (hybrid inviability and sterility caused by genomic incompatibilities). Curiously, by going beyond simple character-based identification of species, the 'biological concept of species' became less universally applicable in biology; for example, Dobzhansky simply concluded that asexuals (between which no interbreeding is possible) could not have species.

Poulton, Mayr and Dobzhansky emphasised that their new concept was based on the reality that underlay species, rather than being merely a criterion useful in taxonomy. In this new philosophical approach, taxonomic criteria and conceptual issues of species became separate while taxonomic criteria took a more minor role. The concept was true from first principles, and was therefore untestable: difficulties such as hybridisation, intermediates or inapplicability to many plants and asexuals caused taxonomic problems, but did not disprove or even challenge the underlying truth of the concept itself. These imperfect actualisations of species' true reality were expected in nature. Mayr claimed that the biological concept would do away with 'typology' (his term for species definitions based on a fixed, unvarying type or Aristotelian essence), but in many ways it can be seen that the biological concept reverts to a new kind of essentialism, where evolutionary maintenance via interbreeding is the underlying reality or essence of species.

ALTERNATIVE SPECIES CONCEPTS

It is interesting that exactly this kind of search for the essence of species had been criticised by Darwin. In his chapter 'Hybridism', he specifically argued against using hybrid sterility and zygote inviability as a cut-and-dried characteristic of species. In this discussion, he made no mention of 'premating isolation', another component of the reproductive isolation that characterises species under the biological concept. However, we can infer that Darwin, the inventor of the term 'sexual selection', would almost certainly have argued that mate choice, like hybrid sterility and inviability, is likewise found within as well as between species. Oddly, Mayr claimed that Darwin treated species 'purely typologically (i.e. as an essentialist) as characterised by degree of difference', and also that Darwin 'had strong, even though perhaps unconscious, motivation to demonstrate that species lack the constancy and distinctiveness claimed for them by the creationists'.

Whether or not it is reasonable to criticise Darwin in such a contradictory way can be debated, but it is clear that Mayr's proposition that interbreeding is the true essence or reality of species immediately laid itself open to debate. Although the interbreeding concept had a long run (and still does), proposals for different kinds of biological reality of species were eventually forthcoming. By proposing a unified reality for species, Poulton, Dobzhansky, and Mayr opened the Pandora's box of alternative essences, deemed more important by other biologists.

Ecological Species Concept

Asexual organisms such as the bdelloid rotifers can clearly be clustered into groups recognisable as taxonomic species, very likely because competition made intermediates extinct. On the other hand, distinct forms such as oaks (*Quercus*), between which there are high rates of hybridisation, can remain recognisably distinct even where they co-occur. This suggested to van Valen and others that the true meaning of species was occupancy of an ecological niche rather than interbreeding. This ecological idea became known as the 'ecological species concept'. It became clear to Mayr during the 1970s also that gene flow could not unite every population in a polytypic, biological species' range, and that stabilisation of phenotype might be effected by ecologically mediated 'stabilising selection' rather than purely because of gene flow.

Recognition Concept of Species

An important attack on the biological species concept came from H.E.H. Paterson in the early 1980s. His claims were twofold: first, that the Dobzhansky/Mayr term isolating mechanisms implied that reproductive isolation was adaptive, which Paterson felt was unlikely; second, that the true reality underlying species was prezygotic compatibility, consisting of mating signals and fertilisation signals. According to Paterson, this compatibility is strongly conserved by stabilising selection, whereas isolating mechanisms such as hybrid sterility or inviability are nonadaptive and can be argued to be a result rather than a cause of species separateness. To Paterson, the true reality of species must be adaptive. He termed his idea of species the 'recognition concept' versus Mayr's 'isolation concept', and its important characteristics 'specific mate recognition systems' (SMRSs) instead of isolating mechanisms. Species were defined as 'that most inclusive population of individual biparental organisms which share a common fertilisation system'.

The idea is generally recognised as a useful critique and has gained strong currency in some circles. However, it has been pointed out that SMRSs are more or less the inverse of prezygotic isolating mechanisms, and that the recognition concept therefore differs from the biological species concept mainly by focusing on the subset of isolating mechanisms occurring before fertilisation. The interbreeding concept had always stressed a common gene pool and compatibility within a species, as well as isolation between species.

SPECIES CONCEPTS BASED ON HISTORY

Monophyly

The rise of 'cladistic' methods revolutionised systematics by proposing that all classification should be based on the idea of 'monophyly'. This new system formalised the principle that 'paraphyletic' and 'polyphyletic' taxa were unnatural groupings, which should not be used in taxonomy. It was natural to attempt to apply this idea throughout systematics, all the way down to the species level, leading to a monophyly criterion of species, a type of 'phylogenetic species concept'. Species were seen as forming when a single interbreeding population split into two branches or lineages that did not exchange genetic material. In a somewhat different formulation, the 'cladistic species concept', species are branch segments in the 'phylogeny', with every branching event leading to a new pair of species. Otherwise, if only one of the two branches were recognised as new, the other branch would become paraphyletic.

Perhaps the main criticism of this idea is that it could, if applied in taxonomy, cause great nomenclatural instability. Monophyly exhibits fractal self-similarity and can exist at very high or very low levels of the phylogeny, so the precise level at which species taxa exist becomes unclear. Suppose that a new 'monophyletic' form is discovered overlapping with, but remaining distinct from, a closely related local form in the terminal branches of an existing species. Recognition of this taxon as a species would leave the remaining branches within the original species paraphyletic. Many other branch segments would then need to be recognised at the species level, even if they interbreed and have reticulate, intermingling 'phylogenies'. Many 'phylogenetic' systematists therefore adopt a different phylogenetic concept, the diagnostic concept, which can allow 'paraphyly' at the species level.

Genealogy

Another problem with a monophyly concept is that a single, true phylogeny of taxa may rarely exist: an organismal phylogeny is in fact an abstraction of the actual genetic history, consisting of multiple gene genealogies, some of which may undergo genetic exchange with other taxa. There is now good evidence that occasional horizontal gene transfer and hybridisation may selectively cause genetic material to flow between unrelated forms. Furthermore, there are multiple gene lineages within any population, so that, if such a population were to become geographically or genetically split into two distinct forms, it would be some time before each branch became fixed for different, reciprocally monophyletic gene lineages at any single gene. The idea of monophyly for whole genomes then becomes hard to define, especially near the species boundary. However annoying, phylogenetic methods and evolutionary theory must face up to these facts. It has therefore been suggested that species should be defined when a consensus between multiple gene genealogies indicates reciprocal monophyly. This is called the 'genealogical species concept'.

Critics argue that this idea has many problems in common with other monophyly concepts of species. Geographic forms that have become isolated in small populations or on islands, say, could rapidly become fixed for gene lineages, and become viewed as separate species without any biologically important evolution taking place.

On the other hand, clearly distinct sister taxa such as humans and chimpanzees still share gene genealogy polymorphisms at some genes such as the human leukocyte antigen (HLA) complex involved in immunological defense, and might therefore be classified as the same species under genealogical considerations.

Diagnostic Species Concept

The motivation for the diagnostic concept, usually called the 'phylogenetic species concept' by its adherents, was again to incorporate phylogenetic thinking into species-level taxonomy. There are many cases of hybridisation between taxa on very different branches of species-level phylogenies, which suggests that interbreeding and 'phylogenetic realities'. conflict. Cracraft also noted that many bird taxa, normally thought of as subspecies, were far more recognisable and stable nomenclaturally than the polytypic species to which they supposedly belonged. Cracraft therefore argued that the polytypic/ interbreeding species concept should be rejected, and, in its place, we should use a diagnostic criterion in the form of fixed differences at one or more inherited characters. 'A phylogenetic species is an irreducible (basal) cluster of organisms, diagnosably distinct from other such clusters, and within which there is a parental pattern of ancestry and descent.' According to Cracraft, species defined in this way are the proper basal, real taxa suitable for phylogenetic analysis and evolutionary studies. Of course, if diagnostic criteria are applied strictly, rather small groups of individuals or even single specimens, might be defined as separate species, leading to unbridled 'taxonomic inflation'. Cracraft recognised this and argued that such diagnosable groups have no 'parental pattern of ancestry and descent', that is they are not proper populations.

However, this qualification appears similar to an interbreeding criterion of species, whereas the whole approach of using diagnostic characters was an attempt to get away from interbreeding. Most evolutionary biologists balk at the idea of speciation being merely the acquisition of a new geographically diagnostic character, a DNA base pair or colour pattern change perhaps.

Speciation is only a different, or special, kind of evolution if the new 'species' is a distinct population, which can coexist locally with its sibling or parent population without losing its integrity. Characters used to diagnose phylogenetic species may not be shared derived characters; they may be primitive (plesiomorphic) characters or they may have evolved several times. Therefore, phylogenetic species need not be monophyletic, and could presumably be paraphyletic and perhaps polyphyletic. Cracraft appears confused on this matter: on the one hand, he claims that phylogenetic species 'will never be nonmonophyletic, except through error', but on the other he recognises that 'their historical status may [sometimes] be unresolved because relative to their sister species they are primitive in all respects. Whether they ... [are] truly paraphyletic ... is probably unresolvable.'

It seems odd to allow a phylogenetic species even to be paraphyletic (let alone polyphyletic), because paraphyly and polyphyly contravene the basic tenets of phylogenetic systematics, and because one of the main justifications for a phylogenetic species concept is that species defined via other concepts might sometimes be paraphyletic: 'The biological species concept cannot be applied to the *Thomomys umbrinus* complex unless one is willing to accept paraphyletic species, and to do so would be a *de facto* admission that biological species are not units of evolution.' The phylogeneticists' resolution of that problem, using diagnostic characters, leads to the same difficulty all over again! This rather glaring logical inconsistency considerably undercuts the argument for a diagnostic species concept.

In spite of these logical problems, Cracraft highlighted some genuine and important practical problems with the polytypic application of the interbreeding concept, and as a result this phylogenetic species concept has been influential. Recently, many molecular systematists, including botanists, have taken up Cracraft's suggestion and used diagnostic differences between geographic populations, in some cases at single DNA base pairs, as evidence that two forms are separate species even if they intergrade freely at the boundaries of their distribution. Ornithologists and primatologists in particular have used diagnostic characters to reassign many taxa long thought of as subspecies to the level of full species, resulting in rather severe taxonomic inflation.

COMBINED SPECIES CONCEPTS

Evolutionary and Lineage Concepts

Faced with the problem of studying the evolution of species through time, the paleontologist Simpson proposed his 'evolutionary species concept', in which a species is 'a lineage (an ancestral–descendant sequence of populations) evolving separately from others and with its own unitary evolutionary role and tendencies'. In other words, Simpson combined the idea that species were historical lineages with the concept of their evolutionary and ecological role. The key essence here appears to be 'evolutionary independence'. This concept appeals to phylogenetic systematists and paleontologists alike, because of its historical dimension, and to neontologists because of its acknowledgement that biological mechanisms are what make the species real. de Queiroz is perhaps the most recent reviewer to propose that a single concept, which he calls 'the general lineage concept', under which 'species are segments of population-level lineages', underlies all other species concepts. According to de Queiroz, apparently competing species concepts merely emphasise different characters or criteria for species definition, but all acknowledge implicitly or explicitly that evolutionary separateness of lineage is the primary concept. This is a nice ideal, but evolutionary independence has little logical force in its application to actual forms that hybridise or undergo genetic exchange.

Cohesion Concept of Species

In similar vein, Templeton's 'cohesion concept' combines a number of competing ideas of species. He accepts criticisms of the interbreeding species concept, and fuses ideas from the ecological, recognition and genealogical concepts. Templeton argues that a combination of ecological and reproductive 'cohesion' is important for maintaining a species' evolutionary unity and integrity, thereby incorporating components of the evolutionary, ecological, recognition and interbreeding concepts. As well as applying to asexual taxa ('too little sex'), Templeton's idea also applied to species like oaks that undergo frequent hybridisation and gene flow ('too much sex' for the interbreeding concepts). He further argues that separateness of genealogy is another important characteristic of species.

We are perhaps nearing the apogee of the species debate with these combined concepts. By incorporating evolutionary and phylogenetic origins together with every possible biological means by which species are currently maintained, these combined concepts 'cover all the bases'. One can acknowledge that species evolve and are maintained as cohesive wholes by all of these multifarious processes; yet at the same time one can argue that species can, and perhaps should, be seen as separate from their histories of origin and from current reasons for their integrity. If groups with very different and conflicting biological and evolutionary characteristics are all considered species, there should exist a simpler criterion that unites them. It can also be argued that to conflate the origin and evolutionary role of a taxon with the definition of that taxon itself may lead to circularity, particularly in conservation or ecological studies or when investigating speciation.

DISSENT: MAYBE SPECIES ARE NOT REAL

Throughout the history of the species debate, starting with Darwin, there have been some who argue that species are not individual real objects, but should instead be considered merely as man-made constructs, merely useful in understanding biodiversity and its evolution. These people are not necessarily nihilists, who deny that species exist: they simply argue that actual morphological and genetic gaps between populations would be more useful for delimiting species than inferred processes underlying evolution or maintenance of these gaps. By their refusal to unite these ideas under a single named concept, this biologist 'silent majority' has rarely found a common voice.

Taxonomic Practice

Taxonomists are on the front line of the species battle, because it is they who ultimately decide whether to lump or split taxa, and at what level to name them as species. If the objectivity and individuality of species as the primary taxon exists, taxonomists' activities have not been made any easier; and many taxonomists have simply ignored or denied belief in the evolutionary reality of species. In general, it is probably true to say that at least 10 per cent of taxonomic species are subject to revision because of these practical difficulties in delimitation.

For this reason, since the rise of the polytypic/interbreeding species concept, there has been little impact of the postwar species concepts on practicing taxonomists, even while the debate raged around them, at least up until the late 1980s. Procedure, at least in zoology, was more or less as follows: geographic variants which blended (or were thought to be able to blend) together at their boundaries were united within a single, polytypic species, unless morphological or genetic differences were so great that it seemed necessary to recognise two species. On the other hand, whenever two divergent forms differing at several unrelated traits, overlapped spatially, they were recognised as separate species even if a few intermediates suggested some hybridisation or gene flow.

Some taxonomists regarded subspecies as artificial taxa to be avoided, and may either have ignored geographic variation, or elevated subspecies of polytypic species to the rank of full species. But good taxonomic practice on species remained broadly similar across most branches of systematics, and involved careful analysis of multiple, chiefly morphological character sets tested in large samples of specimens collected from as many geographic regions as possible.

This view on species and subspecies had led to a steady reduction in the numbers of recognised species in zoology, as more and more dubiously separated taxa, previously ranked as species, became inserted as subspecies into larger and larger polytypic species. Recently, however, the diagnostic version of the phylogenetic species concept has been making strong inroads into zoological nomenclature, with the result that counts of species on continents are again climbing as former subspecies are re-elevated to the species level, in spite of intergradation at their boundaries. However, the situation could get much worse; many *Heliconius* butterflies, for example, have over 30 geographic subspecies per species, all of which can be diagnosed easily. The numbers of bird and butterfly species could easily increase 2–10 times in some groups if the diagnostic criteria were generally adopted, and indeed in some well-known groups, for example primates, a doubling of species numbers has already been observed. Because most of the increase has come from reclassification of known subspecies or populations, rather than from discovery of new populations, the multiplication of species can be termed taxonomic inflation. The one reality that is clear in species-level taxonomy is that the species is not real enough to remain at the same taxonomic rank while fashions in species concept change.

This is surely good evidence that actual species taxa have been and still are purely man-made taxonomic units lacking in any objectively determined underlying biological or evolutionary essence, even if such an essence exists.

Populations are Evolutionary Units, Not Species

Botanists deal with geographically variable organisms with low powers of dispersal, and have therefore never been happy with the polytypic/interbreeding concept applied with such apparent success in zoology. Meanwhile, the strong surge in experimental population genetics and evolutionary studies that followed the books by Dobzhansky, Mayr and Stebbins has led to a greatly improved understanding of gene flow in natural populations. Gene flow, even in quite mobile animals such as birds or butterflies, may not unite local populations into a common gene pool. If local populations only rarely exchange genes, then gene flow across the range of a continental species is clearly insufficient to explain species integrity, because it would be outweighed easily by weak local patterns of adaptation or genetic drift.

This increasing input of population biology into systematics and evolution led to the proposal by Ehrlich and Raven, Levin and others that species are not real biological units at all; instead, local populations are the only real groupings united by gene flow within a common gene pool, and which adapt to local conditions, compete, and so on. Any homogeneity of ecological niche or genetics over the range of a species might be owing either to simple evolutionary inertia or to similar stabilising selection everywhere. To these authors, species exist and are real in local communities, but it is fallacious to treat distant populations in the same way.

This viewpoint is generally understood and respected by population biologists, but curiously has not been incorporated explicitly into current thinking on species in systematics and evolution. Perhaps there is a sneaking suspicion that even very weak levels of gene flow may explain the species integrity over wide areas.

Phenetic Species Concept

In the 1960s and 1970s, a major systematics movement proposed numerical methods in taxonomy now usually referred to as 'phenetics'. 'Pheneticists', as they were called, argued that taxonomy and systematics should be based on multivariate statistical analysis of characters rather than on underlying evolutionary or biological process information. If taxa were defined by nonevolutionary criteria, studies of evolution would be freed from the tautology of testing hypotheses about processes, when those same processes are used as assumptions in the definitions of taxa under study. Species, like other taxa, would be defined in numerical taxonomy on the basis of multivariate statistics, as clusters in phenotypic space.

Phenetics is reviled by those who believe that classifications should be phylogenetic. However, the approach is closely similar to the intuitive methods adopted by most actual taxonomists, who use multiple morphological or genetic characteristics to sort individual specimens into discrete groups between which there are few intermediates. Some large areas of practical taxonomy are based purely on this 'phenetic' approach. Bacterial systematists, for instance, use multiple biochemical tests to assign microbes to species taxa. The usefulness of this taxonomic method is attested by its success in hospitals for predicting pathogeneticity and antibiotic sensitivity.

Phenetic classifications based on morphology introduce the danger that, if convergent characters are used as data, one may group unrelated forms into paraphyletic or even polyphyletic taxa. In addition, single gene polymorphisms and sexual dimorphism can affect multiple morphological characters. This could lead to recognition of multiple species within polymorphic populations. Sibling species, on the other hand, could be lumped into the same species using a phenetic approach, unless a set of highly diagnostic characters could be found. Nonetheless, these problems are due mainly to the lack of characters found in morphological datasets. Phenetics has proved much more successful in distinguishing unrelated, although cryptic, taxa from polymorphic forms when coupled with molecular genetics techniques developed since the 1960s, including allozymes and DNA-based methods.

Genotypic Cluster or Genomic Cluster Criterion

For morphological or genetic gaps to exist between species, gene flow (if any) between species must be balanced by an opposing force of 'disruptive selection'. However, to define species by means of the gaps between them requires consideration of the nature of the gaps to avoid falling into the trap of defining polymorphic forms as separate species or of lumping sibling species. DNA has a digital, rather than analogue code, so there are genetic gaps between virtually any pair of individuals. Clearly, then, we cannot use just any discreteness at the genetic level to define species.

Separate sexes and polymorphic female forms of mimetic *Papilio* butterflies also have gaps between them in exactly this way. A genetic element, which may be a single base pair, an allele at a gene, the entire mitochondrial genome, a chromosomal rearrangement, or perhaps a sex chromosome, may determine the genetic or morphological differences between such polymorphic forms.

To be considered part of a single local population, and therefore part of the same local species, we expect that polymorphic genetic elements like mimicry genes and sex chromosomes will be approximately randomly combined with polymorphisms at genetic elements found on other chromosomes or extrachromosomal DNA. Each individual may be a distinct multilocus genotype, but we recognise a single grouping of genotypes because polymorphisms at one genetic element are independent of polymorphisms at others. Conversely, if alleles at one locus are strongly associated with alleles at other, unlinked elements (i.e. linkage disequilibrium or gametic disequilibrium), we have evidence for more than one separate population; if these two populations overlap spatially, the groupings are probably also separate species.

Several of us therefore proposed a 'genotypic cluster criterion' for species. The term 'genomic cluster' would perhaps be an apt synonym in today's postgenomic age. Species are recognised by morphological and genetic gaps between populations in a local area rather than by means of the phylogeny, cohesion, or reproductive isolation that are responsible for these gaps. 'In a local area, separate species are recognised if there are several clusters separated by multilocus phenotypic or genotypic gaps. A single species (the null hypothesis) is recognised if there is only a single cluster in the frequency distribution of multilocus phenotypes and genotypes.' The genotypic gaps may be entirely vacant or they may contain low frequencies of intermediate genotypes or hybrids (Fig. 4.1). The definition is useful because one avoids tautological thinking: hypotheses about speciation or phylogeny become independent of assumptions about the nature of reproductive isolation or phylogeny underlying the taxa studied.

Genotypic clusters are neither profound nor original. Many similar proposals have been made. The approach is essentially the same in most taxonomic decisions, like the phenetic concept, or a practical application of the biological species concept. Multilocus genotypic clusters are almost universally applied as a criterion of speciation in theoretical models of sympatric speciation: in these models, a bimodal genotypic distribution evolves via reproductive isolation, but it is the demonstration that a pair of genetically divergent groups of individuals emerge from a single population, rather than the mere existence of hybrid inviability or mate choice, that is required for an inference that speciation has occurred. This general use of direct morphological or genetic criteria in the definition of species, as opposed to reproductive or phylogenetic inferences made from such data, has apparently lacked widespread support due to the supposed need for a separation between 'concept' and taxonomic criterion. Most genotypic cluster species can be recognised morphologically; for example, minor pattern elements in *Papilio* can be used to unite the various polymorphic forms; however, with abundant molecular marker data, we could easily use the criterion to sort actual specimens.

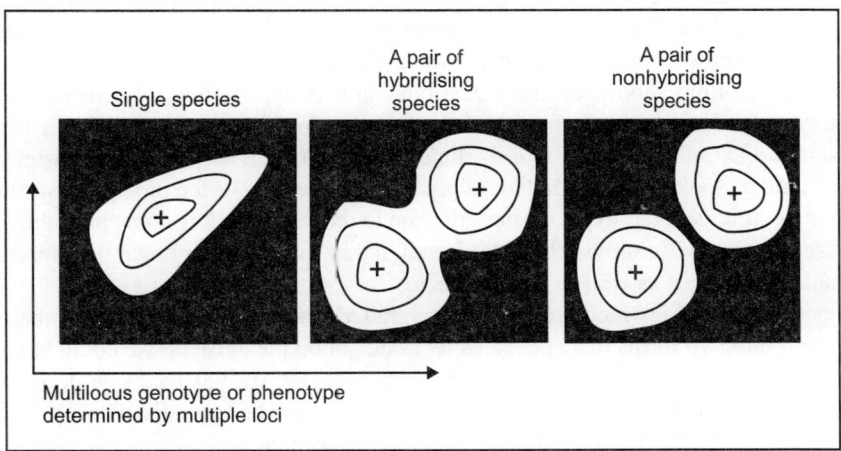

Fig. 4.1. Genotypic or genomic cluster criterion for species. A sample of individuals is made at a single place and time. Numbers of individuals are represented by the contours in multidimensional genotypic space. Peaks in the abundance are represented by '+'. Two species are detected if there are two peaks in the genotypic distribution (right, bimodal distribution). Otherwise the null hypothesis of a single species is not rejected (left, unimodal distribution). *Note*: The axes represent multidimensional morphological/genotypic space, not geographic space.

There is also every reason to conclude after seeing a male butterfly mating with an unlike female that they belong to the same species, but, because hybridisation does occur occasionally between forms normally thought of as different species, one is not so much using the mating behaviour itself to define species as inferring that the mating behaviour is a common enough event to cause homogenisation of genotypic frequencies between the male and female forms.

We infer that, if we were to analyse their genomes, the two forms would have similar genetic characteristics apart from those determining sexual dimorphism, that is they would belong to the same genotypic cluster. Instead of reproductive compatibility being the primary criterion of species, we can turn the argument on its head: we infer from limited data on reproductive compatibility that a single genotypic or genomic cluster is a likely result.

Asexual forms, unclassifiable under the interbreeding concept, and arbitrarily definable at any level under concepts depending on phylogeny, can be clustered and classified as a genotypic clusters in exactly the same way as sexual species. The precise taxonomic level of species clustering for asexuals is somewhat arbitrary, as in the phylogenetic concepts, but at least the method acknowledges this arbitrariness rather than purporting to use some higher evolutionary principle. However, many asexual forms such as bdelloid rotifers have easily distinguishable species taxa, probably due to ecological selection for distinct characteristics. In bacteria, competition is thought to structure promiscuous, but largely asexual, populations into recognisable genetic clusters. Thus, reproductive isolation is not required for genotypic clustering.

Critics have argued that the genotypic cluster criterion in sexual species is nothing other than a gene flow concept of species under a different guise. This is true for one specialised interpretation of gene flow in sexual populations. If we define gene flow as successful or effective, as opposed to actual input of genes, we can see that a 'gene flow criterion' becomes similar to the 'genotypic cluster' criterion. To find whether a hybridisation or gene flow event is successful, we must either follow every gene through all possible descendants for all time or we may examine the genotypic state of a population and determine if genes from one form are mixed randomly with genes from another form. Looking for random associations of genes within genotypes in the genotypic cluster approach will be methodologically the same as a genotypic analysis to determine whether a population is interbreeding, but the latter requires additional assumptions (and of course will exclude asexual taxa). The genotypic cluster criterion in sexual species could be looked upon simply as a practical application of the biological species concept. However, one may prefer the genotypic cluster criterion to the interbreeding concept, if only because its name emphasises that the definition is character-based, rather than actually based on interbreeding, and is thus applicable to asexuals as well as sexual species.

If a single geographic race, which previously intergraded at all its boundaries with other geographic races, were to split into two forms that coexist as separate genotypic clusters, we could have a situation that the original polytypic species became paraphyletic. The new species has been derived from only one of the component subspecies. Thus, paraphyly of species must be recognised as a possibility under this definition, as in both the interbreeding and diagnostic concepts.

Unreality of Species in Space and Time

Geographic races often form clusters differing at multiple loci from other races in the same species. The interbreeding concept or genotypic cluster criterion can be used to justify a classical polytypic species if the various geographic races are separated by zones, which contain abundant intermediates (hybrids). We sample multilocus genotypes or phenotypes in local areas of overlap and determine whether a single peak (one species, i.e. abundant hybrids) or two peaks (two different species, i.e. rare hybrids) are evident in the local genotypic distribution (Fig. 4.1). Hybridisation may occur, but if it is so rare that character and genotypic distributions remain distinctly bimodal in zones of overlap, we usually classify them as separate species, even under the interbreeding concept. Although this spatial extension of the local species is practical to apply to any pair of forms in contact, it is unlikely to lead to general agreement. The problem is that hybrid zones can be very narrow and may separate forms that are highly distinct at multiple characters or loci, in spite of complete unimodal blending in local areas of overlap. Even adherents of the interbreeding concept are reluctant to lump such geographic forms within the same species. Examples include North American swallowtail butterflies (*Papilio glaucus*/*P. canadensis*) and European toads (*Bombina bombina*/*B. variegata*).

An even worse problem is found in 'ring species', which form a continuous band of intergrading subspecies, but whose terminal taxa may be incompatible, and overlap without intergrading. A commonly cited example are the herring gulls and lesser black-backed gulls (*Larus argentatus* complex). Similarly, while most hybrid zones between European *Bombina* are unimodal, the same pair of taxa may have bimodal genotypic distributions in other zones of overlap. Thus, geographic forms may be apparently conspecific in some areas, but overlap as separate species in other areas. Finally, if distinct populations are geographically isolated and there is no area of overlap, one cannot disprove the null hypothesis of 'same species' under interbreeding or genotypic cluster criteria, but biologists are reluctant to unite such populations if they are very divergent. Laboratory hybridisation could be tried, but many overlapping species are known to hybridise freely in captivity, while remaining separate in nature. There are good examples even in our closest relatives, the great apes, for instance the bonobo (*Pan paniscus*) versus the chimpanzee (*P. troglodytes*), and among the gorillas, but similar decisions must be made in almost any animal or plant group.

The problem of extending local species criteria spatially is due to the way in which spatially separated lineages diverge: time since population divergence is correlated with distance. Paleontologists face a similar temporal problem when classifying fossils in different strata. Evolutionary rates may vary, but all lineages must ultimately be continuous, so there is no very logical place to put a species boundary in time any more than there is in space. Paleontologists, like neontologists, use operational species on the basis of morphological gaps between taxa from the same and different time periods.

These difficulties show why there is no easy way to tell whether related geographic or temporal forms belong to the same or different species. Species gaps can be verified only locally and at a point of time. One is forced to admit that Darwin's insight is correct: any local reality or integrity of species is greatly reduced over large geographic ranges and time periods.

IMPORTANCE OF SPECIES CONCEPTS FOR BIODIVERSITY AND CONSERVATION

Traditional: Species as Real Entities

Different species concepts seek to define species in mutually incompatible ways. Thus, a monophyletic species concept seems not very useful to evolutionary biologists because of difficulties with multiple gene genealogies and paraphyletic remnants. In contrast, the interbreeding concept and other concepts incorporating biological processes of species maintenance (e.g. recognition, ecological, evolutionary and cohesion concepts) suffer in the eyes of phylogenetic systematists because they lack phylogenetic coherence and produce paraphyletic taxa, or worse.

If we were to allow the basal unit of our taxonomy to incorporate paraphyly, it would be harder to justify a strict adherence to monophyly at other taxonomic levels. It is beyond the scope of this section to resolve these difficult issues, but these conceptual conflicts fuel the continued debate, and also highlight the fact that if species are indeed real, objective biological units, their unifying reality has been extremely difficult to verify.

Many ecological and biodiversity studies of actual organisms ignore these difficulties, and assume that species are objectively real basal units. Thus, in ecology, we have theories of global species diversity. In conservation, we have the Endangered Species Act in the United States, which prescribes the conservation of threatened taxa we call species. Populations not viewed as species, particularly putative hybrid taxa (like the red wolf, *Canis rufus*, of the southeast United States), maybe seen as less valuable, even if rare. How do we recognise that a taxon is hybridised? Obviously, to be a hybrid, it must be a mere intergrade between two, real, objectively identifiable entities. The Endangered Species Act viewed species as important real conservation units and hybrids as unimportant. It did this because it incorporated the species concept in vogue at the time of its enactment, that is the biological species concept, in which hybridisation is seen as a 'breakdown in isolating mechanisms'.

Alternatives: Genetic Differences More Valuable than Species Status

If the Endangered Species Act were to be rewritten now, what would it say? There is undoubtedly a greater realisation today that other levels in the taxonomic hierarchy are important elements of biodiversity, and indeed the infamous hybrid policy has now been removed. The diagnostic concept of species, while claiming to support the basal, objective nature of species, can at least have the beneficial effect of allowing its basal unit of biodiversity to be recognised at a lower level, in this case as subspecies within polytypic species. Some molecular geneticists have advocated conservation of 'evolutionary significant units', 'management units' or 'stocks' (a fisheries term) defined on criteria of continuous genetic differentiation at molecular markers as being more important than the species level. But the reality of spatiotemporally extended species eludes us, and biodiversity in terms of number of species, including endangered species, remains difficult to measure. If this is so, then it seems best to adopt some other measure of conservation value that relies purely on the degree of genetic differentiation, for instance, at molecular genetic markers.

Species Differences as Ecologically Important Markers

However, there are many who oppose using genetic divergence as a measure of biodiversity. Species within a local area such as a nature reserve are, for the most part, easily and objectively identifiable using morphology, behaviour, genetics or phylogeny. A pair of similar species must usually be ecologically distinct to coexist. Sexual species will need some prezygotic isolation, so their mating behaviour must also be different. Thus, counting species in a local area makes some ecological sense, and conserving species diversity in a local area would conserve actual ecological and behavioural diversity. Behavioural and morphological differences that cause speciesness seem more valuable evolutionarily, as well as interesting to conservationists, than the probably neutral genetic differences at molecular markers.

Biodiversity in Space and Time

As we have seen, this local view breaks down when we try to apply the term 'species' over large areas or geological timescales. In some cases, there is excellent homogeneity over large areas; for example, the painted lady butterfly (*Vanessa cardui*) and the barn owl (*Tyto alba*) have a virtually worldwide distribution and look similar everywhere. Other species are not so homogeneous: the familiar mallard (*Anas platyrhynchos*) group of ducks is as widespread, but has become highly differentiated into some 18 or so forms in farflung outposts of the world. Exactly how many mallard populations are good species, and how many are races or indeed, how many races there are in total, is a matter of taste. Current authorities recognise about 10 species, but there might easily be 5 or 15 in alternative treatments. One of the forms, the Mexican duck, *A. platyrhynchos diazi*, is threatened with hybridisation by the 'true' mallard, *A. platyrhynchos platyrhynchos*, which has been expanding from the north, and the American black duck (*A. rubripes*) also hybridises with the mallard, but appears to resist hybridisation somewhat better than the Mexican form—hence its species status.

Faced with these difficulties, should we worry about the species level when conserving endangered taxa over large areas? Whatever the answer to this question, it does not seem sensible to rely on the spatiotemporal reality of species as a guide. We might conform to taxonomic inflation, and upgrade the Mexican duck to a separate species instead of a subspecies, but this should surely have little effect on our view of its conservation value since there has been no actual change in the knowledge of biological characteristics that affect conservation value. Most conservationists now agree that the former fetish for species-level legislation was a mistake: conservation and legislation should now recognise that living, evolving populations form fractal continua with species, communities, and global ecosystems over time and space, rather than attempting a division into spurious 'fundamental' units.

Species are fundamental units of 'local' biodiversity, but they have this clarity only in a small zone of time and space, and so species counts become less and less meaningful as larger and larger areas are covered. Taxonomists might come to nominalistic agreements on a case-by-case basis, but even this shows little sign of happening yet.

Ecological theory, as well as conservation and biodiversity studies must however recognise that species counts over large expanses of space and time represent only a sketchy measure of biodiversity, a measure which owes more to taxonomic and metaphysical fashion than to science. Yet conservation still depends on lists of endangered species at both local and global levels. We clearly need either a better way than species lists to estimate conservation value or at the very least a more stable species criterion less prone to taxonomic inflation. However, it is the bleak truth that agreement on this matter has not yet been achieved.

SPECIES INFLUENCES UPON ECOSYSTEM FUNCTION

Species diversity may affect ecosystem function, and different species may have disproportional influences upon ecosystem processes. To start off with, the null hypothesis for the relationship between biodiversity and ecosystem processes would state: Species diversity does not affect ecosystem function. Given the varied plant assemblages and ecosystems on earth, however, diversity probably influences processes. Several alternative hypotheses have been proposed in order to elucidate the extent to which ecosystem function depends on diversity. Depending on the interpretation followed, the redundancy hypothesis implies that under existing conditions species richness is irrelevant or that a minimum level of diversity is necessary for proper system functioning, but most species overlap in their functional roles. A different view asserts that each and every species contributes to ecosystem functioning. Both the 'Type 1' linear relationship and the 'rivet' hypothesis are based on this idea of unique species contributions. The order of species deletions or additions does not matter in any of these hypotheses.

According to the Global Biodiversity Assessment (GBA), 'There is no evidence that each and every species plays a unique role such that its absence would immediately result in a dramatic change in the functioning of ecological system.' Moreover, in most ecosystems, species diversity is higher than the level required for efficient biogeochemical and trophic function. Also, some researchers assert that richness is not important *per se*, but a higher level of richness increases the likelihood of including a productive species. Scientists acknowledge that the relationship between diversity and function most likely sits between the two extreme hypotheses listed above. Instead, a few species probably dominate processes and species richness does not matter past a certain threshold. This differs from the redundancy hypothesis in that the order of deletions or additions matters. The idiosyncratic response hypothesis summarises the relationship: ecosystem function changes when diversity changes, but the magnitude and direction are unpredictable since the roles of individuals are complex and varied. The remainder of this section will consider the concepts invoked for the differential roles species may have in ecosystems. The focus will remain on autotrophs.

Species Traits

Species can individually affect ecosystem processes through unique traits. Examples of these unique traits include nitrogen (N) fixation, water redistribution, trace gas emission, high growth rates or unique secondary chemistry. A few species usually have values for traits quite different from the rest. These types of effects are often studied in the context of invasive species but can also occur in non-invasive situations. These functions may be predictable and may have profound effects if the species is added or deleted.

One framework for understanding species traits is that species modify available resources either through consumption or supply. The resource supply rate can also be affected by species. In particular, differences in tissue quality, which control litter decomposition, occur. Other environmental factors can be affected, including soil acidity and microenvironment (through evapotranspiration and insulation). These, of course, also will affect community processes. For example, in Asia Nepalese alder increases nitrogen (N) inputs while bamboo retains weathered potassium (K). Two processes affected by species traits—water redistribution and nutrient cycling—will be explored further below.

Hydraulic lift

Hydraulic lift denotes the phenomenon of water redistribution from deep, moist soil horizons to dry, shallow layers. Besides aiding the 'hydraulic lifter', water can also be provided to plants in the vicinity of the hydraulic lifter. The process involves passive movement of water from roots to soil water when the soil water potential is less than the xylem water potential (or, when one part of the soil has lower soil water potential than another part). One study found that groundwater lifted by sugar maple (*Acer saccharum*) was used by plants up to 2.5 metres (m) away from the tree, but no effect was seen further than 5 m from the tree. All neighbours used some fraction of the lifted water (3–60 per cent), and differential plant effects were noted. Since plant water deficit limits production, the existence of a hydraulic lifter in a community may be a crucial presence. This also may affect biogeochemical conditions for helping mineral availability, nutrient acquisition or microbial processes. To date, the phenomenon has been demonstrated in about 27 species of different life forms—grasses, shrubs, trees, and herbs—and scientists expect it to be more widespread.

Inverse hydraulic lift has also been demonstrated. Hydraulic lift may provide a significant amount of water needed for evapotranspiration and so contribute to this process at an ecosystem scale. On the other hand, demonstration of inverse lift may prevent neighbours/competitors from gaining access to water. These coarse-scale effects, in addition to the facilitation of neighbouring plants, make hydraulic lift a community—and ecosystem-level process. It also may be a general process, as additional studies of different taxa and ecosystems indicate.

Nutrient cycling

In general, individual plant species can play an important role in soil fertility. Usually they create positive feedbacks to plant persistence. Plants from low-nutrient systems use nutrients efficiently and grow slowly, so the cycling of nutrients is slow; the opposite is true for high-nutrient sites. Sometimes the vegetation is more important than abiotic effects in soil process control. Species traits elucidate whether a species loss or addition will affect nutrient cycling. Plants affect the nitrogen cycle through litter quality and its subsequent influence on transformation rates. Competitive displacement may occur after a change in nitrogen (N) supply due to deposition from anthropogenic sources. Community composition in moist meadow tundra was shown to have a potential effect on the nutrient cycling of the community. *Deschampsia caespitosa* had a positive influence on the rates of net soil N transformations, whereas *Acomastylis rossii* did not respond to an increased N supply and had a negative influence on the cycle. Therefore, the replacement of *Acomastylis rossii* by *Deschampsia caespitosa* had positive feedbacks, and the litter could reinforce the patchiness of species composition and available N.

Particular species do not always affect processes, though. In semi-arid grassland, researchers found that plant presence, rather than plant species identification, probably had the most effect on ecosystem processes. They do acknowledge that specific plant characteristics, such as life-span, biomass allocation, and tissue chemical allocation can affect nutrient dynamics and soil organic matter. The traits that best predict resource consumption are height (as related to an individual's ability to capture light and exploit a large soil volume) or biomass per individual and relative growth rate (as related to individual ability to capture carbon and nutrients). These traits vary continuously among organisms.

Dominance

Some species dominate ecosystem functioning by virtue of their abundance in a system. The most abundant species in a system usually have the most influence in terms of productivity, transpiration, decomposition and nutrient cycling. While many rare species may contribute to higher species richness, they might not contribute much to ecosystem function. Accordingly, the deletion of a dominant species would have a greater impact on processes than the deletion of a rare one. The mass ratio hypothesis derives from theoretical and experimental evidence that shows the extent to which a plant species affects ecosystem functions is likely to be predictable from its contribution to the total plant biomass.

Accordingly, the four standard models explained at the beginning of this section only work if the species are deleted in order of rarest to most abundant. If the most abundant species are deleted first, then the response of ecosystem function is wholly different. Although dominance is identified by a measure of biomass, the impact of dominance in a system is not due to biomass itself. Instead, the dominant species affects the ecosystem process most by being best adapted to the environmental conditions present in the system. Deleting a dominant species that comprises 40 per cent of the biomass of a system would have a greater impact upon the system than taking out 40 per cent of the biomass distributed proportionately among all the species present in the system. Since order of removal matters, the relationship between diversity and ecosystem process could be modelled many ways, depending on the order in which species are subtracted or deleted. This also supports criticism of experimental work in this field: species richness affects ecosystem properties only insofar as it increases the likelihood that a dominant species will be included in the assemblage.

Given that dominants dominate processes, it is important to figure out whether less common species play any role in ecosystem function. Although 'there is no *a priori* reason to suspect that such minor contributors must influence ecosystem functioning', Grime proposes considering how subordinates and transients affect ecosystem processes. Species are classified as belonging to one of three groups: dominant, subordinate, or transient. Dominants occupy a high proportion of biomass as well as usually have an expansive morphology and stature, and number only a few individuals. Dominants also occur in particular vegetation types. Subordinates usually co-occur with dominants, comprise less biomass, have smaller stature, and are more numerous. Transients, usually ignored by ecologists, have a heterogeneous distribution and no particular association.

They vary in number and functional traits, make a small contribution to biomass and have few individuals. They usually exist only as seedlings and juveniles. Ecosystem processes may be controlled by one dominant. Alternatively, there may be functional differences between 'co-dominants'. Ecosystem processes may benefit from this redundancy and complementarity. The association between dominants and subordinates might represent a fuller exploitation of resources; this may be a relationship of complementarity. Overall, changes in subordinates and transients will have less effect on ecosystem properties.

It is important to note that the response of a system after perturbation to species richness will depend on the rate of the process and the time at which response is studied. The longer a response to a deletion is measured, the lesser the magnitude will be due to the compensatory response by other species.

Succession

Whereas the mass ratio hypothesis tackles immediate effects on processes, some effects become apparent on longer time scales. This may be especially true of less abundant species. Grime terms longer-term effects 'filter and founder effects'. Often following disturbance, subordinates of the previous community establish first. Therefore, subordinates can influence the regeneration of dominants after disturbance events, thereby controlling the identity, functional diversity and relative abundance of dominants. Transients may indicate the effectiveness with which potential dominants are dispersed across the landscape into suitable ecosystems. A high presence of transients may also indicate low competition and a rich assortment of colonisers. This could allow the ingress of different plant functional types, some of which are capable of exploiting the new conditions created by a disturbance. So, whereas the mass ratio hypothesis argues against the importance of species richness for ecosystem function, the filter and founder effects argue for a need for diversity in a disturbance-laden world. In addition, species with a small effect on ecosystem processes can have large effect if they affect the abundance of other species with large ecosystem effect (e.g. a seed disperser or pollinator might be essential for persistence of canopy species with greater ecosystem impacts).

Keystones

The gain or loss of one or a few certain species sometimes seem to have amplifying effects on both community and ecosystem processes. Those species with this effect, but unpredictable traits, are termed 'keystone species'. First coined by Paine in 1966 to explain the trophic dynamics in a rocky intertidal pool, the concept has been extended to many relationships. The removal of the keystone species causes major changes in the rest of the community dynamics and ecosystem processes; usually, the species affected are functionally related. Keystones are distinguished from dominants in that they have a greater effect than would be expected given their abundance in the ecosystem. Typically this is still applied to trophic dynamics (e.g. the classic example of the cascade of effects after the removal of the sea otter from its environs), but it has also been applied to autotrophic situations. Since keystones do not have a set of specific traits but are defined upon their impacts, they are considered to be unpredictable.

Keystones have been organised into five categories: predators, herbivores, competitors, pathogens and mutualists. An example of a keystone plant species is a dominant tree outcompeting others during succession. Dominants, though, can be excluded by 'keystone weeds' that suppress the establishment of seedlings. Plants are also described as keystone mutualists, which provide critical support to pollinators and dispersers. Keystones can be a relative concept: a keystone from one area may not play such a crucial role in another due to the complexity of associations that occur.

Ecosystem Engineering

Ecosystem engineers are defined as species which directly or indirectly modulate the availability of resources (apart from themselves) to other species. Ecosystem engineers cause physical state changes in biotic or abiotic materials and thereby modify, maintain and/or create habitats and resources for other organisms. To some degree, the other organisms which make use of the changes are dependent upon the modulated structure. Autogenic engineers modify the ecosystem via their own physical structures (e.g. living and dead tissue of a tree). Allogenic engineers transform living or nonliving materials from one state to another. These activities differ from the direct provision of resources an organism may provide through living or dead tissue.

The supporters of the ecosystem engineer concept assert that every system is affected by engineering. An obvious example of an allogenic engineer is the beaver and its impact on riparian systems through the construction of dams. Earthworms have also been cited for moving soil, making casts and burying seeds. Trees and coral reefs represent autogenic engineers not because of the resources they provide through their tissues, but rather due to the effects their physical structures have on modulating the environment, such as altering hydrology, nutrient cycles, soil stability, and microclimate. According to the theory, early successional stages are 'plant-engineered environments'. Proponents of this theory incorporate other well-known processes. For example, the kelp which is disrupted by the loss of sea otters in the keystone example is considered itself to be an autogenic engineer. According to proponents of the ecosystem engineer story, the example should be retold: otters change the effect of an allogenic engineer—urchins—upon an autogenic engineer—kelp beds; the change in kelp bed structure causes the rest of the cascade of effects. The impact of engineers is scaled by: lifetime per capita activity of individual organisms; population density; spatial distribution of population (local and regional); the length of time the population is present on a site; the durability of the engineer's constructs; artifacts in absence of the engineer; the number and type of resource flows modulated by the engineer; and the number of other species dependent upon these flows. The gain or loss of a species will have its greatest impact on ecosystem processes when there are few species in the community, when the species gained or lost is a dominant species, and/or when the species differs strongly from other species in the community. The continuous distribution of traits means that species differ quantitatively rather than qualitatively in their effects on processes. The source of the change in biodiversity may also be important; grazing may affect a system differently than successional processes. According to the Global Biodiversity Assessment (GBA), 'An understanding of unique species' traits, overlap among species, and the possible functional significance of low or high numbers of species, apart from how they differ in traits, is clearly immediately relevant to understanding the conditions under which 'species matter.'

RARE SPECIES AND ECOSYSTEM FUNCTIONING

The role of diversity in the maintenance of ecosystems has been studied widely in the past decade. By correlating richness and diversity with basic ecosystem processes, these investigations lend support to the hypothesis that species diversity significantly influences ecosystem functioning and, in turn, provide support for the conservation of biodiversity.

Nonetheless, the majority of these investigations demonstrate that conservation of a relatively small number of generally dominant species is sufficient to maintain most processes. Indeed, there is remarkably little evidence to support the contention that less common species, those likely of highest conservation concern, are important in the maintenance of ecosystem functioning. Here we summarise studies, most employing alternative methodological strategies, wherein less common and rare species are demonstrated to make significant contributions to ecosystem functioning. Evidence exists among studies of keystone species, aggregate effects of less common species, and species turnover. The findings suggest that (i) less common species can make significant ecosystem contributions, (ii) further investigation into the effects of rare and less common species on ecosystem maintenance is sorely needed, (iii) further investigation should embrace a variety of approaches and (iv) until further research is conducted a prudent conservation approach is warranted wherein the contribution of less common species to ecosystem functioning is assumed.

Based on simple correlations between diversity and measures of ecosystem functioning, consensus is growing for the argument that biodiversity must be conserved to maintain ecosystems. Schwartz challenged this widespread conclusion in a review of empirical studies demonstrating that, in most cases, the relationship between diversity and ecosystem processes was not linear but asymptotic. In contrast to the prevailing interpretation, an asymptotic relationship suggests that ecosystem functioning may be maintained by conservation of relatively few, generally dominant species, not the whole of diversity.

In the majority of ecological communities most species exist in relatively low abundance and, as such, are expected to be at higher risk of extinction. Nonetheless, because of logistical, intellectual and financial constraints, studies on biodiversity and ecosystem functioning focus primarily on common species. In addition, it seems logical to assume that species with low abundances will have concomitantly low ecosystem impacts, as argued by Grime in his mass ratio theory. Consequently, the literature addressing the contribution of diversity to ecosystem functioning is chiefly limited to those species likely to make large ecosystem impacts because of their high relative abundance. The result is a lack of information to make the general conclusion that the maintenance of an entire suite of community species is necessary to maintain ecosystem functioning.

Schwartz and later Hector, argue that a lack of information regarding at-risk species requires that we rely on a precautionary approach, albeit scientifically unsatisfying. This is not a new concept. Leopold and Ehrlich and Mooney similarly advocate that, in lieu of evidence to the contrary, all species be assumed to contribute in important ways to the ecosystems in which they exist. This approach is justifiable, however, only if we find that less common or rare species can contribute significantly to ecosystem functioning. Given that we rarely study the specific effects of uncommon species on ecosystem properties, we know little about the role these species may play in ecosystems.

If, on occasion, we demonstrate that less common species do have strong impacts on ecosystem stability and functioning, a precautionary approach is warranted because of the possibility that many other not-yet-studied species may be important.

To substantiate the precautionary approach we located case studies providing evidence that less abundant species in a community, those often overlooked in conventional studies of the relationship between diversity and ecosystem functioning, make significant, measurable contributions to their ecosystems. We did not aim to quantify how often less common species are demonstrated to contribute to valued ecosystem attributes. We propose that a precautionary approach is warranted if we find that less common species, in those rare cases where they are the focus of a study, are demonstrated to make substantial contributions to ecosystem functioning. We grouped the studies into three categories: (i) less common species as keystone species, (ii) effects of less common species when considered in aggregate and (iii) spatial and temporal turnover of less common species.

Many measures have been used to examine the role of species in ecosystems. Among these metrics are measures of ecosystem processes and ecosystem services. Ecosystem processes are frequently used as a measure because they are assumed to be related in some way to ecosystem health or ecosystem functioning. The most commonly measured ecosystem processes include biomass production and nutrient cycling (typically measured by looking at metrics such as nitrogen retention or movement of carbon between different carbon pools). In general, ecosystem processes involve the movement of materials and energy through an ecosystem and include processes such as nutrient retention and cycling, water retention, productivity, community respiration and stability. When the ecosystem process in question has a direct effect on human health or economic well-being, it is also called an ecosystem service. Common ecosystem services include water filtration, oxygen production, timber production, and erosion control.

Rarity has several forms. For the purpose of this discussion, we focused on studies in which the authors suggest or explicitly define a species as less common, rare or subdominant. We were primarily concerned with species that represent a small component of ecosystem biomass (~1–5 per cent) within their trophic level and nonetheless demonstrate strong effects on ecosystem attributes.

Less Common Species as Keystone Species

Among studies of keystone species we expected to find many examples where relatively less common species had large ecosystem impacts because keystone species are defined as species with functional impacts disproportional to their abundance. In their review of keystone species, Power provide numerous examples of taxa found at relatively low abundance in their communities with major impacts [e.g. beaver (*Castor canadensis*), bass (*Micropterus* spp.), gophers (*Geomys bursarius, Thomomys bottae*), and snow geese (*Chen caerulescens*)]. Most species were too common for us to include here; we found appropriate examples among top predators, however, which tend to contribute relatively little to overall ecosystem biomass.

The loss of top predators can degrade ecosystems. One example of costs associated with removal of top predators is the frequent population explosions of white-tailed deer (*Odocoileus virginianus*) in the eastern United States during the past 50 years. Natural control through the actions of native predators is a function that has been lost in many ecosystems. An overabundance of deer causes loss of plant diversity and eventual habitat degradation through overbrowsing of preferred species. Large deer populations also increase the spread of Lyme disease in humans and human fatalities through night-time automobile-deer collisions. In this case, the loss of uncommon species (top predators) has resulted in a decline in ecosystem services.

Uncommon plant species can act as keystone species in soil resource dynamics. For example, a suite of *Equisetum* species in a cold-temperate Alaskan shrub wetland, constituting < 5 per cent of the above- and below-ground biomass, make substantial contributions to ecosystem processes. Despite their relatively low representation as a portion of the total community phytomass, *Equisetum* spp. contribute 29 per cent P and 39 per cent K to the total annual litter accumulation and 55 per cent P, 41 per cent K and 75 per cent Ca to 2-year-old soil nutrient pools. The roots of *Equisetum* spp. occupy the lower C soil horizons in this marsh system, and roots of the remaining species occupy the upper O horizons. Marsh suggest that *Equisetum* spp. act as nutrient pumps that move limiting resources such as P, and to a lesser extent K and C, from lower to higher soil horizons, making them available to plant community dominants such as *Myrica gale, Salix* spp. and *Carex* spp. The sequestration of copious amounts of nutrients in these relatively unpalatable species, which contain high tissue concentrations of silica and thiaminase, may also be important in the maintenance of nutrients in this system, which is highly affected by the grazing of moose and hare. Furthermore, although decomposition rates in this anaerobic and cold environment are slow, *Equisetum* spp. are relatively quickly mineralised and contribute to higher rates of nutrient cycling. The findings of Marsh may also explain the high concentrations of phosphorus in ponds dominated by *Equisetum* spp., suggesting that the ecosystem contributions of these uncommon species may be further reaching.

Aggregate effects of diversity

Ecologists have also documented the role of diversity in ecosystem functioning by aggregating the effects of less common species on the system as a whole. As is common among abundant taxa in a community, there may be key players among rare species for a given ecosystem process and an efficient method to detect these species is to treat them in aggregate.

Less common species can be important in the resistance of a community to new species invasions. Lyons and Schwartz reduced plant community diversity by removing the less dominant species. Once diversity was significantly lower than the controls, they introduced an exotic grass. The diversity-reduction treatment experienced significantly higher rates of colonisation than the control treatments. In addition, there was a significant positive correlation between the number of less common species removed and colonisation of the introduced species. The results of this study suggest that invasion resistance may be conveyed by the aggregate effect of less common species on available resources. More important, a linear relationship between richness and colonisation demonstrates that species removal results in incremental loss of ecosystem resistance to colonisation, emphasising the role of each species as a factor in the process.

Less common species in aggregate also can play an important role in nutrient cycling and retention. In an alpine meadow differences were found between early- and late-season nitrogen uptake rates among rare, subdominant and dominant species. It was hypothesised that dominant species were outcompeting less common species in this high-stress environment by usurping vital and limiting resources such as nitrogen. To determine seasonal uptake rates, a liquid solution of radioactively labelled nitrogen (^{15}N) was added at the beginning and middle of the growing season. Surprisingly, the highest per capita N concentrations were consistently found in rare and subdominant species, whereas dominant species had higher N-use efficiencies. Thus, elimination of less abundant taxa from this ecosystem could result in diminished retention of limited nutrients.

Temporal variability in species abundance

At any single time, the majority of species within an ecosystem are uncommon. Nonetheless, these uncommon species may play critical roles in the ecosystem by becoming dominant following particular environmental triggers. Thus, one way a species typically perceived as less common may play an important role in ecosystem functioning is through changes in its abundance over time. It is therefore critical to understand the potential impacts of fluctuations in abundance on ecosystem functioning. Variability in species abundance has been well documented in studies of succession; response to disturbance such as fire, flooding and hurricanes; and response to change in environmental conditions such as rainfall. Although these shifts in abundance have been studied with respect to their role in maintenance of diversity and species coexistence, they have yet to be examined with respect to the role that less common species play in maintaining ecosystem functioning.

Many less common plant species have been characterised as dependent on openings created by fire for persistence. Such plants often respond to fire by occupying recently burned sites before recolonisation by superior competitors. For instance, three narrowly endemic species have seed-banks that germinate following fire. This response is common in chaparral.

All three narrowly endemic species have higher survival, larger biomass and greater fecundity in recently burned areas. The species are generally found only in recently burned openings, and their abundance declines as the time since fire increases. Postfire, early germinating and otherwise uncommon plant species may stabilise soils and maintain vegetative cover until more common species recolonise.

Walker and others hypothesise that functionally similar minor and dominant species provide buffering against environmental perturbations such as increased herbivory. Specifically, minor species, functionally similar to dominant species but with different environmental requirements and tolerances, may increase resilience in ecosystem functioning under perturbations that favour them over the past dominants. Walker and others tested this theory in a lightly grazed versus heavily grazed site and found that species of low abundance in a lightly grazed site compensated for the functions provided by dominants in a more heavily grazed site. The findings of Walker and others are supported by theoretical work of Yachi and Loreau, who demonstrated that the replacement of dominants by previously rare species buffers and enhances ecosystem processes in the face of environmental stochasticity.

Although the tendency of insect populations to vary in size over time is well documented (e.g. the ecological literature on the potential role of outbreaks of insect herbivores on ecosystems is scarce). Carson and Root document six cases from the literature where outbreaks had large effects on vegetation. These cases include grasshoppers in prairies, spruce budworm (*Choristoneura fumiferana*) and ghost moths (*Hepialus humuli*) and suggest that the role of insect outbreaks in suppressing dominants may be much more common than was previously thought. During outbreak events herbivores cause long-term damage to hosts and increase light levels in the understory. These effects provide opportunities for growth and establishment of species in the understory and increasing tree seedling establishment. Thus, insects that are typically at low abundance within a system can have large ecosystem effects when their abundance increases during outbreaks.

Similarly, a study of pollination service in row crops by native bees demonstrates that the aggregate contribution of less common species is essential in sustaining a system with high yearly variation in species composition and abundance. In one year of the study, during which many species were low in abundance, the aggregate contribution of all species was required in order to reach a threshold necessary for sufficient crop pollination. Because of higher abundances in the second year, only a few species were required to reach the threshold. Temporal variability in species abundances was important in the maintenance of pollination service because in each year several of the most important contributors were either minor or entirely absent in the other year.

Thus, to assert that the whole of biological diversity should be conserved to maintain ecosystem properties, one must be able to demonstrate that less common species also make significant contributions. This is a challenging task within the current context of investigations into the role of biodiversity in ecosystem functioning because understandable logistical constraints to experimental research in this area limit our ability to explore the range and breadth of potential ways less common species contribute to ecosystem values. Indeed, with few exceptions, the studies reported on here fall outside the conventional approach in which experiments are designed to seek a strict correlation between diversity and ecosystem processes. Also noteworthy and of particular concern to the conventional approach is evidence that abundance can be an unreliable measure of impact.

EXTINCTION

In biology and ecology, extinction is the end of an organism or of a group of organisms (taxon), normally a species. The moment of extinction is generally considered to be the death of the last individual of the species, although the capacity to breed and recover may have been lost before this point. Because a species' potential range may be very large, determining this moment is difficult, and is usually done retrospectively. This difficulty leads to phenomena such as Lazarus taxa, where a species presumed extinct abruptly 're-appears' (typically in the fossil record) after a period of apparent absence.

Through evolution, new species arise through the process of speciation—where new varieties of organisms arise and thrive when they are able to find and exploit an ecological niche—and species become extinct when they are no longer able to survive in changing conditions or against superior competition. The relationship between animals and their ecological niches has been firmly established. A typical species becomes extinct within 10 million years of its first appearance, although some species, called living fossils, survive virtually unchanged for hundreds of millions of years. Most extinctions have occurred naturally, prior to Homo sapiens walking on earth: it is estimated that 99.9 per cent of all species that have ever existed are now extinct.

Mass extinctions are relatively rare events; however, isolated extinctions are quite common. Only recently have extinctions been recorded and scientists have become alarmed at the high rates of recent extinctions. Most species that become extinct are never scientifically documented. Some scientists estimate that up to half of presently existing species may become extinct by 2100. It is difficult to estimate the trajectory that biodiversity might have taken without human impact but scientists at the University of Bristol estimate that biodiversity might increase exponentially without human influence.

A species becomes extinct when the last existing member dies. Extinction therefore becomes a certainty when there are no surviving individuals that are able to reproduce and create a new generation. A species may become functionally extinct when only a handful of individuals survive, which are unable to reproduce due to poor health, age, sparse distribution over a large range, a lack of individuals of both sexes (in sexually reproducing species), or other reasons.

Pinpointing the extinction (or pseudoextinction) of a species requires a clear definition of that species. If it is to be declared extinct, the species in question must be uniquely identifiable from any ancestor or daughter species, or from other closely related species. Extinction of a species (or replacement by a daughter species) plays a key role in the punctuated equilibrium hypothesis of Stephen Jay Gould and Niles Eldredge.

In ecology, extinction is often used informally to refer to local extinction, in which a species ceases to exist in the chosen area of study, but still exists elsewhere. This phenomenon is also known as extirpation. Local extinctions may be followed by a replacement of the species taken from other locations; wolf reintroduction is an example of this. Species which are not extinct are termed extant. Those that are extant but threatened by extinction are referred to as threatened or endangered species. The extinction of one species' wild population can have knock-on effects, causing further extinctions. These are also called 'chains of extinction'. This is especially common with extinction of keystone species.

Pseudoextinction

Descendants may or may not exist for extinct species. Daughter species that evolve from a parent species carry on most of the parent species' genetic information, and even though the parent species may become extinct, the daughter species lives on. In other cases, species have produced no new variants, or none that are able to survive the parent species' extinction. Extinction of a parent species where daughter species or subspecies are still alive is also called pseudoextinction.

Pseudoextinction is difficult to demonstrate unless one has a strong chain of evidence linking a living species to members of a pre-existing species. For example, it is sometimes claimed that the extinct *Hyracotherium*, which was an early horse that shares a common ancestor with the modern horse, is pseudoextinct, rather than extinct, because there are several extant species of *Equus*, including zebra and donkeys. However, as fossil species typically leave no genetic material behind, it is not possible to say whether *Hyracotherium* actually evolved into more modern horse species or simply evolved from a common ancestor with modern horses. Pseudoextinction is much easier to demonstrate for larger taxonomic groups.

Causes

As long as species have been evolving, species have been going extinct. It is estimated that over 99.9 per cent of all species that ever lived are extinct. The average life-span of most species is 10 million years, although this varies widely between taxa. There are a variety of causes that can contribute directly or indirectly to the extinction of a species or group of species (Fig. 4.2).

Fig. 4.2. The Passenger Pigeon, one of hundreds of species of extinct birds, was hunted to extinction over the course of a few decades.

Most simply, any species that is unable to survive or reproduce in its environment, and unable to move to a new environment where it can do so, dies out and becomes extinct. Extinction of a species may come suddenly when an otherwise healthy species is wiped out completely, as when toxic pollution renders its entire habitat unliveable; or may occur gradually over thousands or millions of years, such as when a species gradually loses out in competition for food to better adapted competitors. Extinction may take place a long time after the events that set it in motion, a phenomenon known as extinction debt. Assessing the relative importance of genetic factors compared to environmental ones as the causes of extinction has been compared to the nature-nurture debate.

Currently, environmental groups and some governments are concerned with the extinction of species caused by humanity, and are attempting to combat further extinctions through a variety of conservation programs. Humans can cause extinction of a species through overharvesting, pollution, habitat destruction, introduction of new predators and food competitors, overhunting and other influences.

Explosive, unsustainable human population growth is an essential cause of the extinction crisis. According to the International Union for Conservation of Nature (IUCN), 784 extinctions have been recorded since the year 1500 (to the year 2004), the arbitrary date selected to define 'modern' extinctions, with many more likely to have gone unnoticed (several species have also been listed as extinct since the 2004 date).

Genetics and demographic phenomena

Population genetics and demographic phenomena affect the evolution, and therefore the risk of extinction, of species. Limited geographic range is the most important determinant of genus extinction at background rates but becomes increasingly irrelevant as mass extinction arises.

Natural selection acts to propagate beneficial genetic traits and eliminate weaknesses. It is nevertheless possible for a deleterious mutation to be spread throughout a population through the effect of genetic drift. Because traits are selected and not genes, the relationship between genetic diversity and extinction risk can be complex with factors such as balancing selection, cryptic genetic variation, phenotypic plasticity, and degeneracy all playing potential roles.

A diverse or deep gene pool gives a population a higher chance of surviving an adverse change in conditions. Effects that cause or reward a loss in genetic diversity can increase the chances of extinction of a species. Population bottlenecks can dramatically reduce genetic diversity by severely limiting the number of reproducing individuals and make inbreeding more frequent. The founder effect can cause rapid, individual-based speciation and is the most dramatic example of a population bottleneck.

Genetic pollution

Purebred wild species evolved to a specific ecology can be threatened with extinction through the process of genetic pollution, i.e. uncontrolled hybridisation, introgression genetic swamping which leads to homogenisation or out-competition from the introduced (or hybrid) species. Endemic populations can face such extinctions when new populations are imported or selectively bred by people, or when habitat modification brings previously isolated species into contact. Extinction is likeliest for rare species coming into contact with more abundant ones; interbreeding can swamp the rarer gene pool and create hybrids, depleting the purebred gene pool (for example, the endangered Wild water buffalo is most threatened with extinction by genetic pollution from the abundant domestic water buffalo). Such extinctions are not always apparent from morphological (non-genetic) observations. Some degree of gene flow is a normal evolutionarily process, nevertheless, hybridisation (with or without introgression) threatens rare species' existence.

The gene pool of a species or a population is the variety of genetic information in its living members. A large gene pool (extensive genetic diversity) is associated with robust populations that can survive bouts of intense selection. Meanwhile, low genetic diversity reduces the range of adaptions possible. Replacing native with alien genes narrows genetic diversity within the original population, thereby increasing the chance of extinction.

Habitat degradation

Habitat degradation is currently the main anthropogenic cause of species extinctions. The main cause of habitat degradation worldwide is agriculture, with urban sprawl, logging, mining and some fishing practices close behind. The degradation of a species' habitat may alter the fitness landscape to such an extent that the species is no longer able to survive and becomes extinct. This may occur by direct effects, such as the environment becoming toxic, or indirectly, by limiting a species' ability to compete effectively for diminished resources or against new competitor species.

Habitat degradation through toxicity can kill off a species very rapidly, by killing all living members through contamination or sterilising them. It can also occur over longer periods at lower toxicity levels by affecting life-span, reproductive capacity or competitiveness.

Habitat degradation can also take the form of a physical destruction of niche habitats. The widespread destruction of tropical rainforests and replacement with open pastureland is widely cited as an example of this; elimination of the dense forest eliminated the infrastructure needed by many species to survive. For example, a fern that depends on dense shade for protection from direct sunlight can no longer survive without forest to shelter it. Another example is the destruction of ocean floors by bottom trawling.

Diminished resources or introduction of new competitor species also often accompany habitat degradation. Global warming has allowed some species to expand their range, bringing unwelcome competition to other species that previously occupied that area. Sometimes these new competitors are predators and directly affect prey species, while at other times they may merely outcompete vulnerable species for limited resources. Vital resources including water and food can also be limited during habitat degradation, leading to extinction.

Predation, competition and disease

Before the evolution of hominids, life forms competed with each other and drove one another extinct. Recently in geologic time, humans have been transporting animals and plants from one part of the world to another for thousands of years, sometimes deliberately (e.g. livestock released by sailors onto islands as a source of food) and sometimes accidentally (e.g. rats escaping from boats). In most cases, such introductions are unsuccessful, but when they do become established as an invasive alien species, the consequences can be catastrophic. Invasive alien species can affect native species directly by eating them, competing with them, and introducing pathogens or parasites that sicken or kill them or, indirectly, by destroying or degrading their habitat. Human populations may themselves act as invasive predators.

Coextinction

Coextinction refers to the loss of a species due to the extinction of another; for example, the extinction of parasitic insects following the loss of their hosts. Coextinction can also occur when a species loses its pollinator, or to predators in a food chain who lose their prey. 'Species coextinction is a manifestation of the interconnectedness of organisms in complex ecosystems. While coextinction may not be the most important cause of species extinctions, it is certainly an insidious one.' Coextinction is especially common when a keystone species goes extinct. Models suggest that coextinction is the most common form of biodiversity loss. Coextinction allows for cascading effects across the trophic levels. Such effects are most severe in mutualistic and parasitic relationships. An example of coextinction would be the Haast's Eagle and the Moa. The Haast's Eagle is an example of a predator that became extinct because its food source became extinct. The moa were several species of flightless birds that served as a food source to the Haast's Eagle.

Mass Extinctions

There have been at least five mass extinctions in the history of life on earth, and four in the last 3.5 billion years in which many species have disappeared in a relatively short period of geological time. The massive eruptive event is considered to be one likely cause of the 'Great Dying' about 250 million years ago, which is estimated to have killed 90 per cent of species existing at the time. There is also evidence to suggest this event was preceded by another mass extinction known as Olson's Extinction. The Cretaceous-Tertiary extinction event occurred 65 million years ago at the end of the Cretaceous period and is best known for having wiped out non-avian dinosaurs, among many other species.

Modern extinctions

According to a 2007 survey of 400 biologists conducted by New York's American Museum of Natural History, nearly 70 per cent believed that they were currently in the early stages of a human-caused extinction, known as the Holocene extinction. In that survey, the same proportion of respondents agreed with the prediction that up to 20 per cent of all living populations could become extinct within 30 years (by 2028). Biologist E.O. Wilson estimated that if current rates of human destruction of the biosphere continue, one-half of all species of life on earth will be extinct in 100 years. More significantly the rate of species extinctions at present is estimated at 100 to 1000 times 'background' or average extinction rates in the evolutionary time scale of planet earth.

Chapter 5

Population Ecology

INTRODUCTION

Population ecology is the study of the processes that affect the distribution and abundance of animal and plant populations. A population is a subset of individuals of one species that occupies a particular geographic area and, in sexually reproducing species, interbreeds. The geographic boundaries of a population are easy to establish for some species but more difficult for others. For example, plants or animals occupying islands have a geographic range defined by the perimeter of the island. In contrast, some species are dispersed across vast expanses, and the boundaries of local populations are more difficult to determine. A continuum exists from closed populations that are geographically isolated from other populations of the same species to open populations that show varying degrees of connectedness.

POPULATION DENSITY AND GROWTH

Life histories and the Structure of Populations

A life history is the sequence and timing of events that occur between birth and death. Populations from different parts of the geographic range that a species inhabits may exhibit marked variations in their life histories. The patterns of variation seen within and among populations are referred to as the structure of populations. These variations include breeding frequency, age at which reproduction begins, the number of times an individual reproduces during its lifetime, the number of offspring produced at each reproductive episode (clutch size), the ratio of male to female offspring produced, and whether reproduction is sexual or asexual. These differences in life history characteristics can have profound effects on the dynamics, ecology, and evolution of populations. Of the many differences in life history that occur among populations, age at the time of first reproduction is one of the most important for understanding the dynamics and evolution of a population. All else being equal, natural selection will favour individuals that reproduce earlier than other individuals within a population, because by reproducing earlier an individual's genes enter the gene pool sooner than those of other individuals that were born at the same time but have not reproduced. The genes of the early reproducers then begin to spread throughout the population. Individuals whose genetic make-up allows them to reproduce earlier in life will come to dominate a population if there is no counterbalancing advantage to those individuals that delay reproduction until later in life. Not all populations, however, are made up of individuals that reproduce very early in life. In the course of a lifetime, an individual must devote energy and resources to physiological demands other than reproduction. To reproduce successfully, a plant first may have to grow to a certain height and outcompete its neighbours, and an animal may have to devote energy to growth so that it can reach a size at which it can fend off predators and successfully compete for mates. In many populations, individuals that delay reproduction have a better chance of surviving and leaving offspring than those that attempt to reproduce early. The opposing demands of growth, defense and reproduction are balanced within the constraints of different environments to produce populations that have a diverse range of life-history strategies. Populations often can be divided into one of two extreme types, based on their life-history strategy. Some populations, called r-selected, are considered opportunistic because their reproductive behaviour involves a high intrinsic rate of growth (r)—individuals give birth once at an early age to many offspring. Populations that exhibit this strategy often have been shaped by an extremely variable and uncertain environment. Because mortality occurs randomly in this setting, quantity of progeny rather than quality of care serves the species better. In another strategy, called K-selected, populations tend to remain near the carrying capacity (K), the maximum number of individuals that the environment can sustain. Individuals in a K-selected population give birth at a later age to fewer offspring. This equilibrial life history is exhibited in more stable environments where reproductive success depends more on the fitness of the offspring rather than on their numbers.

Life Tables and the Rate of Population Growth

Differences in life-history strategies greatly affect population dynamics. As stated above, populations in which individuals reproduce at an early age have the potential to grow much faster than populations in which individuals reproduce later. The effect of the age of first reproduction on population growth can be seen in the life tables for a particular species. Life tables were originally developed by insurance companies to provide a means of determining how long a person of a particular age could be expected to live. They are used not only by demographers of human populations but also by plant, animal and microbial ecologists to make projections about the life expectancies of nonhuman populations. The number of individuals in a closed population (a population in which neither immigration nor emigration occurs) is governed by the rates of birth (natality), growth, reproduction and death (mortality). Life tables are designed to evaluate how these rates influence the overall growth rate of a population.

Survivorship curves

Life tables follow the fate of a group of individuals all born in the same year. Of this group, or cohort, only a certain number of individuals will reach each age, and there is an age above which no individuals ever survive. Plotting the number of those members of the group that are still alive at each age results in a survivorship curve for the population. Survivorship curves are usually displayed on a semilogarithmic rather than an arithmetic scale. There are three general types of survivorship curves. Species such as humans and large mammals, which have fewer numbers of offspring but invest much time and energy in caring for their young (K-selected species), usually have a Type I curve. This relatively flat curve reflects low juvenile mortality, with most individuals living to old age. A constant probability of dying at any age, shown by the Type II curve, is evident as a straight line decreasing over time toward zero. Certain lizards, perching birds, and rodents exhibit this type of survivorship curve. In some species that produce many offspring but provide little care for them (r-selected species), mortality is greatest among the youngest individuals. The Type III survivorship curve indicative of this life history is initially very steep but flattens out as those individuals who reach maturity survive for a relatively longer time; it is exhibited by animals such as many insects or shellfish. Many populations have survivorship patterns that are more complex than, or fall in between, these three idealised curves. For example, passerine birds (perching birds such as finches) commonly suffer high mortality during the first year of life and a lower, more constant rate of death in subsequent years.

Calculating population growth

Life tables also are used to study population growth. The average number of offspring left by a female at each age and the proportion of individuals surviving to each age can be used to evaluate the rate at which the size of the population changes over time. These rates are used by demographers and population ecologists to estimate human population growth and to evaluate the effect that conservation efforts have on endangered species. The average number of offspring that a female leaves during her lifetime is called the net reproductive rate (R_0). If all females survived to the oldest possible age for that population, the net reproductive rate would simply be the sum of the average number of offspring left by females at each age. In real populations, however, some females die at every age. The net reproductive rate for a set cohort is obtained by multiplying the proportion of females surviving to each age (l_x) by the average number of offspring produced at each age (m_x) and then adding the products from all the age groups: $R_0 = \Sigma l_x m_x$. A net reproductive rate of 1.0 indicates that a population is neither increasing nor decreasing but replacing its numbers exactly. Any number below 1.0 indicates a decrease in population, any number above indicates an increase. In the example provided in the table, the net reproductive rate is 2.101, which means that the population of the Galapagos cactus finch (*Geospiza scandens*) can double its size each generation (Table 5.1).

Table 5.1. Life table for one Darwin finch, the Galapagos cactus finch (*Geospiza scandens*)*.

Age class** (x)	Probability of surviving to age x (l_x)	Average number of fledgling daughters (m_x)	Product of survival and reproduction ($\Sigma l_x m_x$)
0	1.0	0.0	0.0
1	0.512	0.364	0.186
2	0.279	0.187	0.052
3	0.279	1.438	0.401
4	0.209	0.833	0.174
5	0.209	0.500	0.104
6	0.209	0.833	0.174
7	0.209	0.250	0.052
8	0.209	3.333	0.696
9	0.139	0.125	0.017
10	0.070	0.0	0.0
11	0.070	0.0	0.0
12	0.070	3.500	0.245
13	0	–	–
			$R_0 = 2.101$

Net reproductive rate = $R_0 = \Sigma l_x m_x = 2.101$

Mean generation time = $T = (\Sigma x l_x m_x)/(R_0) = 6.08$ years

Intrinsic rate of natural increase of the population = r = approximately $\ln R_0/T = 2.101/6.08 = 0.346$

*The values are for the cohort of females born in 1975.

**Designated in years.

The other value needed to calculate the rate at which the population can grow is the mean generation time (T). Generation time is the average interval between the birth of an individual and the birth of its offspring.

To determine the mean generation time of a population, the age of the individuals (x) is multiplied by the proportion of females surviving to that age (l_x) and the average number of offspring left by females at that age (m_x). This calculation is performed for each age group, and the values are added together and divided by the net reproductive rate (R_0) to yield the result:

$$T = \frac{\sum x l_x m_x}{R_0}$$

The value that is used by population biologists to calculate the rate of increase of populations is the intrinsic rate of natural increase (r) or the Malthusian parameter. Very simply, this rate can be understood as number of births minus number of deaths per generation time—in other words, the reproduction rate less the death rate. To derive this value using a life table, the natural logarithm of the net reproductive rate is divided by the mean generation time:

$$r = \frac{\ln R_0}{T}$$

Values above zero indicate that the population is increasing; the higher the value, the faster the growth rate (as shown in the table). The Malthusian parameter can be used to compare growth rates of populations of a species that have different generation times. Some human populations have higher intrinsic rates of natural increase partially because individuals in those groups begin reproducing earlier than those in other groups.

Mice have higher intrinsic rates of natural increase than elephants because they reproduce at a much earlier age and have a much shorter mean generation time (Table 5.2).

Table 5.2. Intrinsic rate of increase (r)* calculated for populations of species that differ greatly in their potential for the rate of population growth.

Species	Intrinsic rate of increase (r)
Elephant seal	0.091
Ring-necked pheasant	1.02
Field vole	3.18
Flour beetle	23
Water flea	69

*Values above zero indicate that the population is increasing. The higher the value of r, the faster the intrinsic growth rate of the population.

If a population has an intrinsic rate of natural increase of 0, then it is said to have stable age distribution and is neither growing nor declining in numbers. A growing population has more individuals in the lower age classes than does a stable population, and a declining population has more individuals in the older age classes than does a stable population. Many human populations are currently undergoing population increase, far exceeding a stable age distribution. Although the human population has increased almost continuously throughout history, it has skyrocketed since the Industrial Revolution, primarily because of a drop in death rates. No other population has shown such steady growth.

NATALITY IN POPULATION ECOLOGY

Natality in Population Ecology is the scientific term for birth rate. Along with mortality rate, natality rate is used to calculate the dynamics of a population. They are the key factors in determining whether a population is increasing, decreasing or staying the same in size. Natality is the greatest influence on a population's increase. Natality is shown as a crude birth rate or specific birth rate. Crude birth rate is used when calculating population size (# of births per 1000 population/year), whereas specific birth rate is used relative to a specific criterion such as age. By calculating specific birth rate, the results are seen in an age-specific schedule of births.

Definitions

1. Natality rate—number of births per 1000 individuals per year.
2. Absolute natality—the number of births under ideal conditions (with no competition, abundance of resources such as food and water, etc.).
3. Realised natality—the number of births when environmental pressures come in to play.

Animal Natality

Specific birth rate is used when calculating animal natality. The criterion used is age. Animal natality is expressed as an age-specific schedule of births. This is represented by the quantity of young/unit of time by females in various age classes. The age-specific birth schedule will count females that have only given birth to females. Showing the number of females that have been born relative to the previous generation will show how much of that generation may have the ability to reproduce. To construct the Age-specific schedule, the average number of females born (m_x) must be calculated. A survivorship column must also be included to construct a fertility table. Taking the survivorship column and the m_x values from the life table, the number of offspring will be shown, giving us the natality rate.

Larger implications

Calculating natality for Animals has become an important part of the research for preservation of species. Studies have been conducted to determine whether a species may be going extinct because of climate or availability of resources. Measuring the natality of a species during extreme conditions such as drastic climate shift or a decrease in prey density will show scientists the measures that need to be taken to keep that species alive. Studies have been conducted about the Polar Bear population in Svalbard, Norway from 1988 to 2007.

As the average adult ages of both female and male increased, the natality rate decreased. Due to this observation, there was an interest in discovering why. The study was able to correlate the density of ringed seals, which are the polar bears' main prey, to the low natality rates.

Plant Natality

Plants' natality is more difficult to determine than animals. The factors that make it difficult to measure are:

1. Seed production of individual plants varying year to year.
2. Seed production will also vary age class to age class.
3. The seeds will become dormant for long periods of time before germinating.

Plant natality is an uncertain factor to measure. The time it takes for a plant to germinate its seeds may be extended over too long of a time period for accurate measurement.

Natality in Humans

Calculating and concluding the natality rate of humans is similar to animals. Age-specific schedules are constructed when determining growth rate. It can be used to find the effects that environmental chemicals/toxins have on women of a childbearing age. Birth rates are helpful in making government policies regarding population growth. The birth rate is an item of concern and policy for a number of national governments. Some, including those of Italy and Malaysia, seek to increase the national birth rate using measures such as financial incentives or provision of support services to new mothers. Conversely, other countries have policies to reduce the birth rate, for example, China's one child policy. Measures such as improved information about and availability of birth control have achieved similar results in countries such as Iran.

Calculations Where Natality is a Factor

1. Natality rate: Number of births/unit of time/Average population.
2. For wildlife management: $N_1 = N_0 + (B-D) + (I-E)$.
   where,

   N_1 = number of individuals at time 1.
   N_0 = number of individuals at time 0.
   B = number of individuals born.
   D = number of individuals that died.
   I = number of individuals that immigrated.
   E = number of individuals that emigrated between time 0 and time 1.
3. Intrinsic rate of increase: $(dN/dt)(1/N) = r$.
   r = Intrinsic rate of increase.
   (dN/dt) = Rate that population increases.
   N = Population size.

Fluctuation in Population Size

As surely as individuals are born and die, populations grow and decrease. At any time in its history, the size of a population is the product of births and deaths, and immigration and emigration. The rate at which individuals enter the population (births and immigration) and individuals leave (deaths and emigration) will determine whether the population is growing, shrinking or is stable, and if it is changing, how fast.

A population is define as a group of one species of organisms occupying the same general area, using the same resources, and acted upon by the same environmental factors. Populations cannot grow indefinitely, many populations will become stable over a period of time while others will show sharp increases followed by similar decreases. Population characteristics that are studied are its density and the spacing of its individuals. Population density is the number of individuals per unit area or volume. Population dispersion is the pattern of spacing among the parameters of the geographical boundaries of the population.

Density and dispersion

Every population has geographical barriers and a population size. Population density is the number of individuals per unit area or volume.

Measuring density

Since it is impractical to capture and/or count each individual in a given area, ecologists use a variety of methods to determine the density of various populations of organisms. Some of the techniques used are as follows: Counting the number of nests or burrows in a given area, analysing the number of tracks, examination of solid waste products left behind by a species, and an actual capture method used to tag and release the specimen studied.

Mark-recapture method—traps are placed within the boundaries of the population being studied and the captured animals are marked and released. After a few weeks, traps are set again. The proportions of marked to unmarked that are captured the second time give an estimate of the size of the entire population.

N = (Number of marked)(Total catch second time)/Number marked recaptured

Patterns of dispersion

Dispersion is the pattern of spacing among individuals within the geographical boundaries of the population. Local densities, within a population's range, may vary considerably due to differences in the limiting factors present. There are three types of patterns of dispersion in relationship to other individuals: clumped, uniform and random.

1. Clumped: Clumped is a pattern when individuals are aggregated in patches. This style is caused by a heterogeneous environment with resources concentrated in patches. Mating or social behaviour of the individuals may also contribute to this type of dispersion.
2. Uniform: Uniform is a pattern of equally spaced individuals. Competition between individuals may set up zones or territories for feeding, nesting or breeding.
3. Random: Random is a spacing pattern based on total unpredictability. This form of dispersal is highly uncommon in nature. If it does, it usually results from the absence of a strong competition among individuals.

Demography is the study of the vital statistics affecting a population size. This branch of science deals with the influence that immigration and emigration have on a given population, this does not usually include the new organisms leaving or entering the population birth and death rates are also studied.

Age structure and sex ratio states that each age group has a characterised birth and death rate (Juveniles and old people are more likely to die). Birth rate is the number of individuals produced during a certain amount of time (is greatest for individuals of intermediate age).

Age and sex ratios—several populations have gone beyond generations where individuals of more than one generation coexist. This situation produces an age structure in most populations. Every age group has a characteristic birth and death rate. A standard rule of thumb sets a high mortality for the lower and upper age groups and a low mortality rate for the intermediate age group.

Generations over time, is the average time span between the birth of individuals and the birth of their offspring and is strongly related to the body size over a broad range of organisms a shorter generation time will result in faster population growth. Body size has a major effect on generation time.

For example, Elephant takes 2 years to produce an offspring and several more years before the offspring can successfully reproduce. On the other hand, mice can produce a litter of individuals ever 21 days. Their offspring usually are ready to reproduce in 3 to 4 weeks after birth.

Females play a central role in a population since they are the ones to produce the offspring, but males may mate with several females each breeding season.

Density-dependent and density-independent factors

Density dependent factors: Population cycles is a couple of concepts—one idea is that crowding regulates cyclic populations and another is that population cycles are caused by a lag time in the response to the density-dependent factors, creating large fluctuations of population size above and below carrying capacity. Increasing population size reduces available resources and this eventually limits population growth. In restricting population growth, a density-dependent factor strengthens as the population size increases, affecting each individual more strongly.

On the other hand, the population growth declines because the death rate increases, birth rate decreases or both. If there is a reduction in the food supply or other major resources, could also restrict the reproduction cycle resulting in less offspring.

The animal species' battle to establish territories is a behavioural trait that may restrict population growth and available food supplies. Predators concentrate in areas where there is a high concentration of organisms. As long as the natural resources are available in sufficient quantity, the population will remain constant. As the population decreases so, do the predators.

Density-independent factors: They are unrelated to population size and will affect the same # of individuals regardless of weather and climate. Examples are weather, climate and natural disasters such as freezes, seasonal changes such as unusually hot or cold temperatures, hurricanes, and fires. All of these factors are unrelated to population size but affect everyone in the population regardless of population size or location.

Interspecific competition: As already discussed populations do not live alone, just as individuals do not. When individuals of the same species, or of two different species, depend on a common important resource then competition occurs. Competition can be defined as 'interactions between individuals brought about by a shared requirement for a resource in limited supply leading to a reduction in survivorship, growth, and reproduction of individuals'. Interspecific competition (inter = between) is the competition between two or more different species for a resource.

Modes of population growth

A population that begins at low levels in a favourable environment may increase rapidly for a while, but eventually the numbers must stop growing.

Exponential population growth

In an ideal environment, there are no restrictions on the abilities of individuals to harvest energy, grow and reproduce thus, the population will grow in size with every birth and with immigration, and the population will decrease with every death and emigration. In a population living under ideal conditions, the population grows fast, because all the members have access to abundant food and are free to reproduce.

Populations increase under these conditions and the size of the population increases rapidly, resulting in the population actually accumulating more new individuals per unit of time when it is large than when it was small, and it will pass the threshold point (the environment will not be able to sustain the population anymore).

Logistic population growth

In most populations, there is a limit to the number of individuals that can occupy a habitat. The carrying capacity is the maximum stable population that a particular environment can support of a relatively long period of time. Crowding and resource limitations can have a profound effect on the population growth. The logistic model is a model of species-specific competition: the competition between two or members of the same species. As the population sizes increases, the competition become more intense.

Territoriality is the defense of a well-bounded physical space. The model is the idea that even at low populations each individual added to the population has the same negative effect on population growth rate and that some populations show an effect in which individuals may have a time that is more difficult surviving and reproducing if the population size is too small.

Population Bioenergetics

Studies on population bioenergetics, i.e. energy flow in population give the true picture of a population and helpful especially when population comprising of individuals which differ greatly in size, e.g. ants, deer, lions are to be compared.

Biological Dispersal

Biological dispersal refers to species movement away from an existing population or away from the parent organism. Through simply moving from one habitat patch to another, the dispersal of an individual has consequences not only for individual fitness, but also for population dynamics, population genetics and species distribution. Understanding dispersal and the consequences both for evolutionary strategies at a species level, and for processes at an ecosystem level, requires understanding on the type of dispersal, the dispersal range of a given species and the dispersal mechanisms involved.

Biological dispersal may be contrasted with geodispersal, which is the mixing of previously isolated populations (or whole biotas) following the erosion of geographic barriers to dispersal or gene flow.

Types of dispersal

At some time during its life, an organism, whether animal or plant, moves, or is moved, so that it or its offspring do not die exactly where they were born. Such movement is called dispersal. Some organisms are motile throughout their lives, but others are adapted to move or be moved at precise, limited phases of their life-cycles. This is commonly called the dispersive phase of the life-cycle. The strategies of organisms' entire life-cycles often are predicated on the nature and circumstances of their dispersive phases (Fig. 5.1).

In general there are two basic types of dispersal:

1. Density independent dispersal: Organisms have evolved adaptations for dispersal that take advantage of various forms of kinetic energy occurring naturally in the environment. This is referred to as density independent or passive dispersal and operates on many groups of organisms (some invertebrates, fish, insects and sessile organisms such as plants) that depend on animal vectors, wind, gravity or current for dispersal.

2. Density dependent dispersal: Density dependent or active dispersal for many animals largely depends on factors such as local population size, resource competition, habitat quality, and habitat size. Due to population density, dispersal may relieve pressure for resources in an ecosystem, and competition for these resources may be a selection factor for dispersal mechanisms.

Dispersal of organisms is a critical process for understanding both geographic isolation in evolution through gene flow and the broad patterns of current geographic distributions (biogeography).

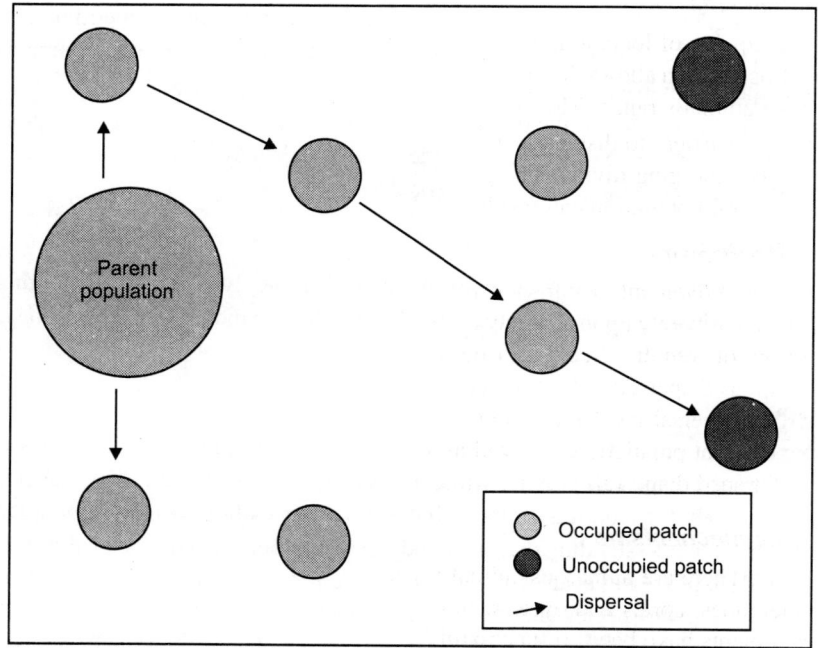

Fig. 5.1. Dispersal from parent population.

Dispersal range

'Dispersal range' refers to the distance a species can move from an existing population or the parent organism. An ecosystem depends critically on the ability of individuals and populations to disperse from one habitat patch to another. Therefore, biological dispersal is critical to the stability of ecosystems.

Environmental constraints

Few species are ever evenly or randomly distributed within or across landscapes. In general, species significantly vary across the landscape in association with environmental features that influence their reproductive success and population persistence. Spatial patterns in environmental features (e.g. resources) permit individuals to escape unfavourable conditions and seek out new locations. This allows the organism to 'test' new environments for their suitability, provided they are within animal's geographic range. In addition, the ability of a species to disperse over a gradually changing environment could enable a population to survive extreme conditions (i.e. climate change).

Dispersal barriers

A dispersal barrier may mean that the dispersal range of a species is much smaller than the species distribution. An artificial example is habitat fragmentation due to human land-use. Natural barriers to dispersal that limit species distribution include mountain ranges and rivers. An example is the separation of the ranges of the two species of chimpanzee by the Congo River.

On the other hand, human activities may also expand the dispersal range of a species by providing new dispersal methods (e.g. ships).

Dispersal mechanisms

Most animals are capable of locomotion and the basic mechanism of dispersal is movement from one place to another. Locomotion allows the organism to 'test' new environments for their suitability, provided they are within the animal's range. Movements are usually guided by inherited behaviours.

The formation of barriers to dispersal or gene flow between adjacent areas can isolate populations on either side of the emerging divide. The geographic separation and subsequent genetic isolation of portions of an ancestral population can result in speciation.

Plant dispersal mechanisms

Seed dispersal is the movement or transport of seeds away from the parent plant. Plants have limited mobility and consequently rely upon a variety of dispersal vectors to transport their propagules, including both abiotic and biotic vectors. Seeds can be dispersed away from the parent plant individually or collectively, as well as dispersed in both space and time. The patterns of seed dispersal are determined in large part by the dispersal mechanism and this has important implications for the demographic and genetic structure of plant populations, as well as migration patterns and species interactions. There are five main modes of seed dispersal: gravity, wind, ballistic, water and by animals.

Animal dispersal mechanisms

Nonmotile animals: There are numerous animal forms that are nonmotile, such as sponges, bryozoans, tunicates, sea anemones, corals and oysters. In common, they are all either marine or aquatic. It may seem curious that plants have been so successful at stationary life on land, while animals have not, but the answer lies in the food supply. Plants produce their own food from sunlight and carbon dioxide—both generally more abundant on land than in water. Animals fixed in place must rely on the surrounding medium to bring food at least close enough to grab, and this occurs in the three-dimensional water environment, but with much less abundance in the atmosphere.

All of the marine and aquatic invertebrates whose lives are spent fixed to the bottom (more or less; anemones are capable of getting up and moving to a new location if conditions warrant) produce dispersal units. These may be specialised 'buds' or motile sexual reproduction products, or even a sort of alteration of generations as in certain cnidaria.

Corals provide a good example of how sedentary species achieve dispersion. Corals reproduce by releasing sperm and eggs directly into the water. These release events are coordinated by lunar phase in certain warm months, such that all corals of one or many species on a given reef will release on the same single or several consecutive nights. The released eggs are fertilised, and the resulting zygote develops quickly into a multicellular planula. This motile stage then attempts to find a suitable substratum for settlement. Most are unsuccessful and die or are fed upon by zooplankton and bottom dwelling predators such as anemones and other corals. However, untold millions are produced, and a few do succeed in locating spots of bare limestone, where they settle and transform by growth into a *polyp*. All things being favourable, the single polyp grows into a coral head by budding off new polyps to form a colony.

Motile animals: The majority of all animals are motile. Although motile animals can, in theory, disperse themselves by their spontaneous and independent locomotive powers, a great many species utilise the existing kinetic energies in the environment, resulting in passive movement. Dispersal by water currents is especially associated with the physically small inhabitants of marine waters known as zooplankton.

Dispersal by dormant stages: Many animal species, especially freshwater invertebrates, are able to disperse by wind or by transfer with an aid of larger animals (birds, mammals or fishes) as dormant eggs, dormant embryos or, in some cases, dormant adult stages. Tardigrades, some rotifers and some copepods are able to withstand desiccation as adult dormant stages. Many other taxa (Cladocera, Bryozoa, Hydra, Copepoda and so on) can disperse as dormant eggs or embryos. Freshwater sponges usually have special dormant propagules called gemmulae for such a dispersal. Many kinds of dispersal dormant stages are able to withstand not only desiccation and low and high temperature, but also action of digestive enzymes during their transfer through digestive tracts of birds and other animals, high concentration of salts and many kinds of toxicants. Such dormant-resistant stages made possible the long-distance dispersal from one water body to another and broad distribution ranges of many freshwater animals.

ORIGIN OF SPECIES

Evolution of biosphere and organisms appear to be involved for origin of new species that results in species diversity. During beginning of life on earth, there was no free oxygen. In absence of oxygen, no ozone layer protected earth from deadly ultraviolet radiation of the sun. This radiation is believed to have triggered a chemical evaluation leading to formation of amino acids. However, life could develop only under protective cover of water as long as atmospheric oxygen and ozone remained scarce. Photosynthetic bacteria, especially Cyanobacteria, put oxygen into atmosphere 3.5 to 2 billion years ago. Origin of photosynthetic bacteria is still not clear.

However, gradual buildup of photosynthetically produced oxygen and its diffusion into stratosphere about 2 billion years ago brought about tremendous changes in geochemistry of earth and made possible rapid expansion of life and development of eukaryotic cell, which led to evolution of larger and more complex living organisms. Many minerals, such as iron were precipitated from water and formed characteristic geological formations. Fir nucleated cells appeared, when there was 0.6 per cent oxygen in atmosphere and scientists like Margulis has advocated that eukaryotic cell originated as a mutalistic coming together of once independent microbes. During Precambrian period, only single cell prokaryotes existed. Transition to eukaryotic cells appears to have occurred about 1.2 to 1.5 billion years ago. However, recent genetic studies suggest that eukaryotes diverged from prokaryotes closer to 2 billion years ago. Fossils do not yet agree with this date.

The old Kingdom Protista has been broken into many new kingdoms, reflecting new studies and techniques that help elucidate the true phylogenetic sequence of life on earth. Protists exhibit a great deal of variation in their life histories (life-cycles). They exhibit an alternation between diploid and haploid phases, that is similar to alternation of generations found in plants. Protist life-cycles vary from diploid dominant to haploid dominant. Diversity of form, habitat, mode of nutrition and life history exhibited by eukaryotes suggest that they evolved several times from various prokaryotes. This makes the protista a polyphyletic group. Eukaryotes are generally larger, have a variety of membrane-bound organelles, greater internal complexity than prokaryotes and have a specialised method of cell division (meiosis) that is a prelude to true sexual reproduction.

Multicellular organisms like sponges, corals, worms, shell, ancestor of seed plants and vertebrates appeared later. During Paleozoic era, green mantle of terrestrial vegetation provided more oxygen and food for subsequent evolution of large creatures such as dinosaurs, birds, mammals and finally human. Present day life forms including humans were evolved during the past 570 million years (Phenorozoic era) consisting of Paleozoic, Mesozoic and Cenozoic eras.

Direction of evolution has been mainly, with few exceptions, from simplicity to complexity and from imperfection to perfection. This ultimately led to highly evolved and specialised species of birds and mammals, which are perfectly matched with ecosystem that they live in. It seems most likely that life on this planet originated only once. Later on, it multiplied, diversified and colonised the already existing ecosystems. Kind of life that could survive and flourish in various 'virgin' ecosystems of world, at that point of time, would have largely been determined by abiotic components to begin with.

Isolating Mechanisms

Isolating Mechanisms are features of behaviour, morphology or genetics which serve to prevent breeding between species. Reproductive isolation of populations is established. It is particularly important to the biological species concept, as species are defined by reproductive isolation. Isolating mechanisms can be divided into two groups, Prezygotic isolating mechanisms and Postzygotic isolating mechanisms.

Prezygotic mechanisms

Factors which prevent individuals from mating:
1. Geographic isolation: Species occur in different areas, and are often separated by barriers.
2. Temporal isolation: Individuals do not mate because they are active at different times. This may be different times of the day or different seasons. The species mating periods may not match up. Individuals do not encounter one another during either their mating periods, or at all.
3. Ecological isolation: Individuals only mate in their preferred habitat. They do not encounter individuals of other species with different ecological preferences.
4. Behavioural isolation: Individuals of different species may meet, but one does not recognise any sexual cues that may be given. An individual chooses a member of its own species in most cases.
5. Mechanical isolation: Copulation may be attempted but transfer of sperm does not take place. The individuals may be incompatible due to size or morphology.
6. Gametic incompatibility: Sperm transfer takes place, but the egg is not fertilised.

Postzygotic isolating mechanisms

1. Genomic incompatibility, hybrid inviability or sterility.
2. Zygotic mortality: The egg is fertilised, but the zygote does not develop.
3. Hybrid inviability: Hybrid embryo forms, but is not viable.
4. Hybrid sterility: Hybrid is viable, but the resulting adult is sterile.
5. Hybrid breakdown: First generation (F1) hybrids are viable and fertile, but further hybrid generations (F2 and backcrosses) are inviable or sterile.

Reproductive Isolation

The mechanisms of reproductive isolation or hybridisation barriers are a collection of mechanisms, behaviours and physiological processes that prevent the members of two different species that cross or mate from producing offspring, or which ensure that any offspring that may be produced is not fertile. These barriers maintain the integrity of a species over time, reducing or directly impeding gene flow between individuals of different species, allowing the conservation of each species' characteristics.

The mechanisms of reproductive isolation have been classified in a number of ways. Zoologist Ernst Mayr classified the mechanisms of reproductive isolation in two broad categories: those that act before fertilisation (or before mating in the case of animals, which are called precopulatory) and those that act after. These have also been termed prezygotic and postzygotic mechanisms. The different mechanisms of reproductive isolation are genetically controlled and it has been demonstrated experimentally that they can evolve in species whose geographic distribution overlaps (sympatric speciation) or as the result of adaptive divergence that accompanies allopatric speciation.

Isolation mechanisms that occur before breeding or copulation (prezygotic isolation)

Prezygotic isolation mechanisms are the most economic in terms of the biological efficiency of a population, as resources are not wasted on the production of a descendent that is weak, nonviable or sterile.

Temporal or habitat isolation

Any of the factors that prevent potentially fertile individuals from meeting will reproductively isolate the members of distinct species. The types of barriers that can cause this isolation include: different habitats, physical barriers, and a difference in the time of sexual maturity or flowering. When factors change, especially physical barriers, often, species will branch off.

An example of the ecological or habitat differences that impede the meeting of potential pairs occurs in two fish species of the family *Gasterosteidae* (sticklebacks). One species lives all year round in freshwater, mainly in small streams. The other species lives in the sea during winter, but in spring and summer individuals migrate to river estuaries to reproduce. The members of the two populations are reproductively isolated due to their adaptations to distinct salt concentrations. An example of reproductive isolation due to differences in the mating season are found in the toad species *Bufo americanus* and *Bufo fowleri*. The members of these species can be successfully crossed in the laboratory producing healthy, fertile hybrids. However, mating does not occur in the wild even though the geographical distribution of the two species overlaps. The reason for the absence of interspecies mating is that *B. americanus* mates in early summer and *B. fowleri* in late summer. Certain plant species, such as *Tradescantia canaliculata* and *T. subaspera*, are sympatric throughout their geographic distribution yet they are reproductively isolated as they flower at different times of the year. In addition, one species grows in sunny areas and the other in deeply shaded areas.

Sexual isolation by behaviour or conduct

The different mating rituals of animal species creates extremely powerful reproductive barriers, termed sexual or behaviour isolation, that isolate apparently similar species in the majority of the groups of the animal kingdom. In dioecious species, males and females have to search for a partner, be in proximity to each other, carry out the complex mating rituals and finally copulate or release their gametes into the environment in order to breed.

Mating dances, the songs of males to attract females or the mutual grooming of pairs, are all examples of typical courtship behaviour that allows both recognition and reproductive isolation. This is because each of the stages of courtship depend on the behaviour of the partner. The male will only move onto the second stage of the exhibition if the female shows certain responses in her behaviour. He will only pass onto the third stage when she displays a second key behaviour. The behaviours of both interlink, are synchronised in time and lead finally to copulation or the liberation of gametes into the environment. No animal that is not physiologically suitable for fertilisation can complete this demanding chain of behaviour. In fact, the smallest difference in the courting patterns of two species is enough to prevent mating (for example, a specific song pattern acts as an isolation mechanism in distinct species of grasshopper of the genus *Chorthippus*). Even where there are minimal morphological differences between species, differences in behaviour can be enough to prevent mating. For example, *Drosophila melanogaster* and *D. simulans* which are considered twin species due to their morphological similarity, do not mate even if they are kept together in a laboratory. *Drosophila ananassae* and *D. pallidosa* are twin species from Melanesia. In the wild they rarely produce hybrids, although in the laboratory it is possible to produce fertile offspring. Studies of their sexual behaviour show that the males court the females of both species but the females show a marked preference for mating with males of their own species. A different regulator region has been found on Chromosome II of both species that affects the selection behaviour of the females.

Pheromones play an important role in the sexual isolation of insect species. These compounds serve to identify individuals of the same species and of the same or different sex. Evaporated molecules of volatile pheromones can serve as a wide-reaching chemical signal. In other cases, pheromones may be detected only at a short distance or by contact.

In species of the *melanogaster* group of *Drosophila*, the pheromones of the females are mixtures of different compounds, there is a clear dimorphism in the type and/or quantity of compounds present for each sex. In addition, there are differences in the quantity and quality of constituent compounds between related species, it is assumed that the pheromones serve to distinguish between individuals of each species. An example of the role of pheromones in sexual isolation is found in 'corn borers' in the genus *Ostrinia*. There are two twin species in Europe that occasionally cross. The females of both species produce pheromones that contain a volatile compound which has two isomers, E and Z; 99 per cent of the compound produced by the females of one species is in the E isomer form, while the females of the other produce 99 per cent isomer Z. The production of the compound is controlled by just one locus and the interspecific hybrid produces an equal mix of the two isomers. The males, for their part, almost exclusively detect the isomer emitted by the females of their species, such that the hybridisation although possible is scarce. The perception of the males is controlled by one gene, distinct from the one for the production of isomers, the heterozygous males show a moderate response to the odour of either type. In this case, just 2 'loci' produce the effect of ethological isolation between species that are genetically very similar.

Sexual isolation between two species can be asymmetrical. This can happen when the mating that produces descendants only allows one of the two species to function as the female progenitor and the other as the male, while the reciprocal cross does not occur. For instance, half of the wolves tested in the Great Lakes area of America show mitochondrial DNA sequences of coyotes. While mitochondrial DNA from wolves is never found in coyote populations. This probably reflects an asymmetry in inter-species mating due to the difference in size of the two species as male wolves take advantage of their greater size in order to mate with female coyotes, while female wolves and male coyotes do not mate.

Mechanical isolation

Mating pairs may not be able to couple successfully if their genitals are not compatible. The relationship between the reproductive isolation of species and the form of their genital organs was signalled for the first time in 1844 by the French entomologist Léon Dufour. Insects' rigid carapaces act in a manner analogous to a lock and key, as they will only allow mating between individuals with complementary structures, that is, males and females of the same species (termed co-specifics).

Evolution has led to the development of genital organs with increasingly complex and divergent characteristics, which will cause mechanical isolation between species. Certain characteristics of the genital organs will often have converted them into mechanisms of isolation. However, numerous studies show that organs that are anatomically very different can be functionally compatible, indicating that other factors also determine the form of these complicated structures.

Mechanical isolation also occurs in plants and this is related to the adaptation and coevolution of each species in the attraction of a certain type of pollinator (where pollination is zoophilic) through a collection of morphophysiological characteristics of the flowers (called floral syndrome), in such a way that the transport of pollen to other species does not occur.

Gametic isolation

The synchronous spawning of many species of coral in marine reefs means that inter-species hybridisation can take place as the gametes of hundreds of individuals of tens of species are liberated into the same water at the same time. Approximately a third of all the possible crosses between species are compatible, in the sense that the gametes will fuse and lead to individual hybrids. This hybridisation apparently plays a fundamental role in the evolution of coral species. However, the other two-thirds of possible crosses are incompatible. It has been observed that in sea urchins of the genus Strongylocentrotus the concentration of spermatocytes that allow 100 per cent fertilisation of the ovules of the same species is only able to fertilise 1.5 per cent of the ovules of other species. This inability to produce hybrid offspring, despite the fact that the gametes are found at the same time and in the same place, is due to a phenomenon known as gamete incompatibility, which is often found between marine invertebrates, and whose physiological causes are not fully understood.

In some *Drosophila* crosses, the swelling of the female's vagina has been noted following insemination. This has the effect of consequently, preventing the fertilisation of the ovule by sperm of a different species.

In plants the pollen grains of a species can germinate in the stigma and grow in the style of other species. However, the growth of the pollen tubes may be detained at some point between the stigma and the ovules, in such a way that fertilisation does not take place. This mechanism of reproductive isolation is common in the Angiosperms and is called cross-incompatibility or incongruence. A relationship exists between self-incompatibility and the phenomenon of cross-incompatibility. In general crosses between individuals of a self-compatible species (SC) with individuals of a self-incompatible (SI) species give hybrid offspring. On the other hand, a reciprocal cross (SI × SC) will not produce offspring, because the pollen tubes will not reach the ovules. This is known as unilateral incompatibility, which also occurs when two SC or two SI species are crossed.

Isolation mechanisms that occur after breeding or copulation (post-zygotic isolation)

A number of mechanisms which act after fertilisation preventing successful inter-population crossing are discussed below.

Zygote mortality and nonviability of hybrids

A type of incompatibility that is found as often in plants as in animals occurs when the ovule is fertilised but the zygote does not develop or it develops and the resulting individual has a reduced viability. This is the case for crosses between species of the frog genus, where widely differing results are observed depending of the species involved. In some crosses there is no segmentation of the zygote (or it may be that the hybrid is extremely nonviable and changes occur from the first mitosis). In others, normal segmentation occurs in the blastula but gastrulation fails. Finally, in other crosses, the initial stages are normal but errors occur in the final phases of embryo development. This indicates differentiation of the embryo development genes (or gene complexes) in these species and these differences determine the nonviability of the hybrids.

Similar results are observed in mosquitos of the *Culex* genus, but the differences are seen between reciprocal crosses, from which it is concluded that the same effect occurs in the interaction between the genes of the cell nucleus (inherited from both parents) as occurs in the genes of the cytoplasmic organelles which are inherited solely from the female progenitor through the cytoplasm of the ovule.

In Angiosperms, the successful development of the embryo depends on the normal functioning of its endosperm. The failure of endosperm development and its subsequent abortion has been observed in many interploidal crosses (that is, those between populations with a particular degree of intra, or interspecific ploidy, and in certain crosses in species with the same level of ploidy. The collapse of the endosperm, and the subsequent abortion of the hybrid embryo is one of the most common post-fertilisation reproductive isolation mechanism found in angiosperms.

Hybrid sterility

A hybrid has normal viability but is deficient in terms of reproduction or is sterile. This is demonstrated by the mule and in many other well known hybrids. In all of these cases sterility is due to the interaction between the genes of the two species involved; to chromosomal imbalances due to the different number of chromosomes in the parent species; or to nucleus-cytoplasmic interactions such as in the case of *Culex* described above.

Hinnies and mules are hybrids resulting from a cross between a horse and an ass or between a mare and a donkey, respectively. These animals are nearly always sterile due to the difference in the number of chromosomes between the two parent species. Both horses and donkeys belong to the genus *Equus*, but *Equus caballus* has 64 chromosomes, while *Equus asinus* only has 62. A cross will produce offspring (mule or hinny) with 63 chromosomes, that will not form pairs, which means that they do not divide in a balanced manner during meiosis. It is curious that they can cross with each other but the mule and the hinny are actually animals created by humans, as in the wild the species ignore each other and do not cross. In order to obtain mules or hinnies it is necessary to train the progenitors to accept copulation between the species or create them through artificial insemination.

The sterility of many of the interspecific hybrids among the angiosperms is a widely recognised and studied phenomenon. There are a variety of causes that can determine the interspecific sterility of hybrids in plants, these may be genetic, related to the genomes or the interaction between nuclear and cytoplasmic factors, as will be discussed in the corresponding section. Nevertheless, it should be pointed out that—on the contrary to the situation in animals—hybridisation in plants is a stimulus for the creation of new species. Indeed, although the hybrid may be sterile it can continue to multiply in the wild through the mechanisms of asexual reproduction, be they vegetative propagation or apomixis or the production of seeds. Indeed, interspecific hybridisation can be associated with polyploidia and, in this way, the origin of new species that are called allopolyploids. *Rosa canina*, for example, is the result of multiple hybridisations or there is a type of wheat that is an allohexaploid that contains the genomes of three different species.

Multiple mechanisms

In general, the barriers that separate species do not consist of just one mechanism. The twin species of *Drosophila*, *D. pseudoobscura* and *D. persimilis*, are isolated from each other by habitat (*persimilis* generally lives in colder regions at higher altitudes), by the timing of the mating season (*persimilis* is generally more active in the morning and *pseuoobscura* at night) and by behaviour during mating (the females of both species prefer the males of their respective species). In this way, although the distribution of these species overlaps in wide areas of the west of the United States of America, these isolation mechanisms are sufficient to keep the species separated. Such that, only a few fertile females have been found amongst the other species among the thousands that have been analysed. However, when hybrids are produced between both species, the gene flow between the two will continue to be impeded as the hybrid males are sterile. Also, and in contrast with the great vigour shown by the sterile males, the descendants of the backcrosses of the hybrid females with the parent species are weak and notoriously nonviable. This last mechanism restricts even more the genetic interchange between the two species of fly in the wild.

Hybrid gender: Haldane's rule

Haldane's rule states that when one of the two sexes is absent in interspecific hybrids between two specific species, then the gender that is not produced, is rare or is sterile is the heterozygous (or heterogametic) sex. In mammals, at least, there is growing evidence to suggest that this is due to high rates of mutation of the genes determining masculinity in the Y chromosome.

It has been suggested that Haldane's rule simply reflects the fact that the male gender is more sensitive than the female when the sex-determining genes are included in a hybrid genome. But there are also organisms in which the heterozygous sex is the female: birds and butterflies and the law is followed in these organisms. Therefore, it is not a problem related to sexual development, nor with the sex chromosomes. Haldane proposed that the stability of hybrid individual development requires the full gene complement of each parent species, so that the hybrid of the heterozygous sex is unbalanced (i.e. missing at least one chromosome from each of the parental species). For example, the hybrid male obtained by crossing *D. melanogaster* females with *D. simulans* males, which is nonviable, lacks the X chromosome of *D. simulans*.

Genetics of reproductive isolation barriers

Precopulatory isolation mechanisms in animals

The genetics of ethological isolation barriers will be discussed first. Precopulatory isolation occurs when the genes necessary for the sexual reproduction of one species differ from the equivalent genes of another species, such that if a male of species A and a female of species B are placed together they are unable to copulate. Study of the genetics involved in this reproductive barrier tries to identify the genes that govern distinct sexual behaviours in the two species. The males of *Drosophila melanogaster* and those of *D. simulans* conduct an elaborate courtship with their respective females, which are different for each species, but the differences between the species are more quantitative than qualitative. In fact the simulans males are able to hybridise with the melanogaster females. Although there are lines of the latter species that can easily cross there are others that are hardly able to. Using this difference, it is possible to assess the minimum number of genes involved in precopulatory isolation between the melanogaster and simulans species and their chromosomal location.

In experiments, flies of the *D. melanogaster* line, which hybridises readily with simulans, were crossed with another line that it does not hybridise with, or rarely. The females of the segregated populations obtained by this cross were placed next to simulans males and the percentage of hybridisation was recorded, which is a measure of the degree of reproductive isolation. It was concluded from this experiment that 3 of the 8 chromosomes of the haploid complement of *D. melanogaster* carry at least one gene that affects isolation, such that substituting one chromosome from a line of low isolation with another of high isolation reduces the hybridisation frequency. In addition, interactions between chromosomes are detected so that certain combinations of the chromosomes have a multiplying effect. Cross incompatibility or incongruence in plants is also determined by major genes that are not associated at the self-incompatibility *S locus*.

Postcopulation or fertilisation isolation mechanisms in animals

Reproductive isolation between species appears, in certain cases, a long time after fertilisation and the formation of the zygote, as happens, for example, in the twin species *Drosophila pavani* and *D. gaucha*. The hybrids between both species are not sterile, in the sense that they produce viable gametes, ovules and spermatozoa. However, they cannot produce offspring as the sperm of the hybrid male do not survive in the semen receptors of the females, be they hybrids or from the parent lines. In the same way, the sperm of the males of the two parent species do not survive in the reproductive tract of the hybrid female. This type of postcopulatory isolation appears as the most efficient system for maintaining reproductive isolation in many species.

In fact, the development of a zygote into an adult is a complex and delicate process of interactions between genes and the environment that must be carried out precisely, and if there is any alteration in the usual process, caused by the absence of a necessary gene or the presence of a different one, it can arrest the normal development causing the nonviability of the hybrid or its sterility. It should be borne in mind that half of the chromosomes and genes of a hybrid are from one species and the other half come from the other. If the two species are genetically different, there is little possibility that the genes from both will act harmoniously in the hybrid. From this perspective, only a few genes would be required in order to bring about postcopulatory isolation, as opposed to the situation described previously for precopulatory isolation.

In many species where precopulatory reproductive isolation does not exist, hybrids are produced but they are of only one sex. This is the case for the hybridisation between females of *Drosophila simulans* and *Drosophila melanogaster* males: the hybridised females die early in their development so that only males are seen among the offspring. However, populations of *D. simulans* have been recorded with genes that permit the development of adult hybrid females, that is, the viability of the females is 'rescued'. It is assumed that the normal activity of these speciation genes is to 'inhibit' the expression of the genes that allow the growth of the hybrid. There will also be regulator genes.

A number of these genes have been found in the *melanogaster* species group. The first to be discovered was 'Lhr' (Lethal hybrid rescue) located in Chromosome II of *D. simulans*. This dominant allele allows the development of hybrid females from the cross between *simulans* females and *melanogaster* males. A different gene, also located on Chromosome II of *D. simulans is* 'Shfr' that also allows the development of female hybrids, its activity being dependent on the temperature at which development occurs. Other similar genes have been located in distinct populations of species of this group. In short, only a few genes are needed for an effective postcopulatory isolation barrier mediated through the nonviability of the hybrids.

As important as identifying an isolation gene is knowing its function. The *Hmr* gene, linked to the X chromosome and implicated in the viability of male hybrids between *D. melanogaster* and *D. simulans*, is a gene from the proto-oncogene family *myb*, that codes for a transcriptional regulator. Two variants of this gene function perfectly well in each separate species, but in the hybrid they do not function correctly, possibly due to the different genetic background of each species. Examination of the allele sequence of the two species shows that change of direction substitutions are more abundant than synonymous substitutions, suggesting that this gene has been subject to intense natural selection.

The Dobzhansky–Muller model proposes that reproductive incompatibilities between species are caused by the interaction of the genes of the respective species. It has been demonstrated recently that *Lhr* has functionally diverged in *D. simulans* and will interact with *Hmr* which, in turn, has functionally diverged in *D. melanogaster* to cause the lethality of the male hybrids. *Lhr* is located in a heterochromatic region of the genome and its sequence has diverged between these two species in a manner consistent with the mechanisms of positive selection. An important unanswered question is whether the genes detected correspond to old genes that initiated the speciation favouring hybrid nonviability, or are modern genes that have appear post-speciation by mutation, that are not shared by the different populations and that suppress the effect of the primitive nonviability genes. The *OdsH* (abbreviation of *Odysseus*) gene causes partial sterility in the hybrid between *Drosophila simulans* and a related species, *D. mauritania*, which is only encountered on Mauritius, and is of recent origin. This gene shows monophyly in both species and also has been subject to natural selection. It is thought that it is a gene that intervenes in the initial stages of speciation, while other genes that differentiate the two species show polyphyly. *Odsh* originated by duplication in the genome of *Drosophila* and has evolved at very high rates in *D. mauritania*, while its paralogue, *unc-4*, is nearly identical between the species of the group *melanogaster*. Seemingly, all these cases illustrate the manner in which speciation mechanisms originated in nature, therefore they are collectively known as 'speciation genes' or possibly, gene sequences with a normal function within the populations of a species that diverge rapidly in response to positive selection thereby forming reproductive isolation barriers with other species. In general, all these genes have functions in the transcriptional regulation of other genes.

The *Nup96* gene is another example of the evolution of the genes implicated in postcopulatory isolation. It regulates the production of one of the approximately 30 proteins required to form a nuclear pore. In each of the *simulans* groups of *Drosophila* the protein from this gene interacts with the protein from another, as yet undiscovered, gene on the X chromosome in order to form a functioning pore. However, in a hybrid the pore that is formed is defective and causes sterility. The differences in the sequences of *Nup96* have been subject to adaptive selection, similar to the other examples of *speciation* genes described above.

Postcopulatory isolation can also arise between chromosomally differentiated populations due to chromosomal translocations and inversions. If, for example, a reciprocal translocation is fixed in a population, the hybrid produced between this population and one that does not carry the translocation will not have a complete meiosis. This will result in the production of unequal gametes containing unequal numbers of chromosomes with a reduced fertility. In certain cases, complete translocations exist that involve more than two chromosomes, so that the meiosis of the hybrids is irregular and their fertility is zero or nearly zero. Inversions can also give rise to abnormal gametes in heterozygous individuals but this effect has little importance compared to translocations. An example of chromosomal changes causing sterility in hybrids comes from the study of *Drosophila nasuta* and *D. albomicans* which are twin species from the Indo-Pacific region. There is no sexual isolation between them and the F1 hybrid is fertile. However, the F2 hybrids are relatively infertile and leave few descendants which have a skewed ratio of the sexes. The reason is that the X chromosome of *albomicans* is translocated and linked to an autosome which causes abnormal meiosis in hybrids. Robertsonian translocations are variations in the numbers of chromosomes that arise from either: the fusion of two acrocentric chromosomes into a single chromosome with two arms, causing a reduction in the haploid number or conversely; or the fission of one chromosome into two acrocentric chromosomes, in this case increasing the haploid number. The hybrids of two populations with differing numbers of chromosomes can experience a certain loss of fertility, and therefore a poor adaptation, because of irregular meiosis.

Postcopulation or fertilisation isolation mechanisms in plants

In plants, hybrids often suffer from an autoimmune syndrome known as hybrid necrosis. In the hybrids, specific gene products contributed by one of the parents may be inappropriately recognised as foreign and pathogenic, and thus trigger pervasive cell death throughout the plant. In at least one case, a pathogen receptor, encoded by the most variable gene family in plants, was identified as being responsible for hybrid necrosis.

Incompatibility caused by micro-organisms

In addition to the genetic causes of reproductive isolation between species there is another factor that can cause of postzygotic isolation: the presence of micro-organisms in the cytoplasm of certain species. The presence of these organisms in a species and their absence in another causes the nonviability of the corresponding hybrid. For example, in the semi-species of the group *D. paulistorum* the hybrid females are fertile but the males are sterile, this is due to the presence of a mycoplasma in the cytoplasm which alters spermatogenesis leading to sterility. It is interesting that incompatibility or isolation can also arise at an intraspecific level. Populations of *D. simulans* have been studied that show hybrid sterility according to the direction of the cross.

The factor determining sterility has been found to be the presence or absence of a micro-organism *Wolbachia* and the populations tolerance or susceptibility to these organisms. This interpopulation incompatibility can be eliminated in the laboratory through the administration of a specific antibiotic to kill the micro-organism. Similar situations are known in a number of insects, as around 15 per cent of species show infections caused by this symbiont. It has been suggested that, in some cases, the speciation process has taken place because of the incompatibility caused by this bacteria. Two wasp species *Nasonia giraulti* and *N. longicornis* carry two different strains of *Wolbachia*. Crosses between an infected population and one free from infection produces a nearly total reproductive isolation between the semi-species. However, if both species are free from the bacteria or both are treated with antibiotics there is no reproductive barrier. *Wolbachia* also induces incompatibility due to the weakness of the hybrids in populations of spider mites (*Tetranychus urticae*, between *Drosophila recens* and *D. subquinaria* and between species of *Diabrotica* (beetle) and *Gryllus* (cricket).

Selection for reproductive isolation

In 1950 K.F. Koopman reported results from experiments designed to examine the hypothesis that selection can increase reproductive isolation between populations. He used *D. pseudoobscura* and *D. persimilis* in these experiments. When the flies of these species are kept at 16°C approximately a third of the matings are interspecific. In the experiment equal numbers of males and females of both species were placed in containers suitable for their survival and reproduction. The progeny of each generation were examined in order to determine if there were any interspecific hybrids. These hybrids were then eliminated. An equal number of males and females of the resulting progeny were then chosen to act as progenitors of the next generation. As the hybrids were destroyed in each generation the flies that solely mated with members of their own species produced more surviving descendants than the flies that mated solely with individuals of the other species. In the Table 5.3 it can be seen that for each generation the number of hybrids continuously decreased up to the tenth generation when hardly any interspecific hybrids were produced. It is evident that selection against the hybrids was very effective in increasing reproductive isolation between these species. From the third generation, the proportions of the hybrids were less than 5 per cent. This confirmed that selection acts to reinforce the reproductive isolation of two genetically divergent populations if the hybrids formed by these species are less well adapted than their parents.

Table 5.3. Selection for reproductive isolation between two *Drosophila* species.

Generation	Percentage of hybrids
1	49
2	17.6
3	3.3
4	1.0
5	1.4
10	0.6

These discoveries allowed certain assumptions to be made regarding the origin of reproductive isolation mechanisms in nature. Namely, if selection reinforces the degree of reproductive isolation that exists between two species due to the poor adaptive value of the hybrids, it is expected that the populations of two species located in the same area will show a greater reproductive isolation than populations that are geographically separated.

This mechanism for 'reinforcing' hybridisation barriers in sympatric populations is called the 'Wallace effect', as it was first proposed by Alfred Russell Wallace at the end of the 19th century, and it has been experimentally demonstrated in both plants and animals.

The sexual isolation between *Drosophila miranda* and *D. pseudoobscura*, for example, is more or less pronounced according to the geographic origin of the flies being studied. Flies from regions where the distribution of the species is superimposed show a greater sexual isolation than exists between populations originating in distant regions (Fig. 5.2).

On the other hand, interspecific hybridisation barriers can also arise as a result of the adaptive divergence that accompanies allopatric speciation. This mechanism has been experimentally proved by an experiment carried out by Diane Dodd on *D. pseudoobscura*. A single population of flies was divided into two, with one of the populations fed with starch-based food and the other with maltose-based food. This meant that each sub-population was adapted to each food type over a number of generations. After the populations had diverged over many generations, the groups were again mixed; it was observed that the flies would mate only with others from their adapted population. This indicates that the mechanisms of reproductive isolation can arise even though the interspecific hybrids are not selected against.

Speciation

Speciation is the evolutionary process by which new biological species arise. The biologist Orator F. Cook seems to have been the first to coin the term 'speciation' for the splitting of lineages or 'cladogenesis', as opposed to 'anagenesis' or 'phyletic evolution' occurring within lineages. Whether genetic drift is a minor or major contributor to speciation is the subject matter of much ongoing discussion.

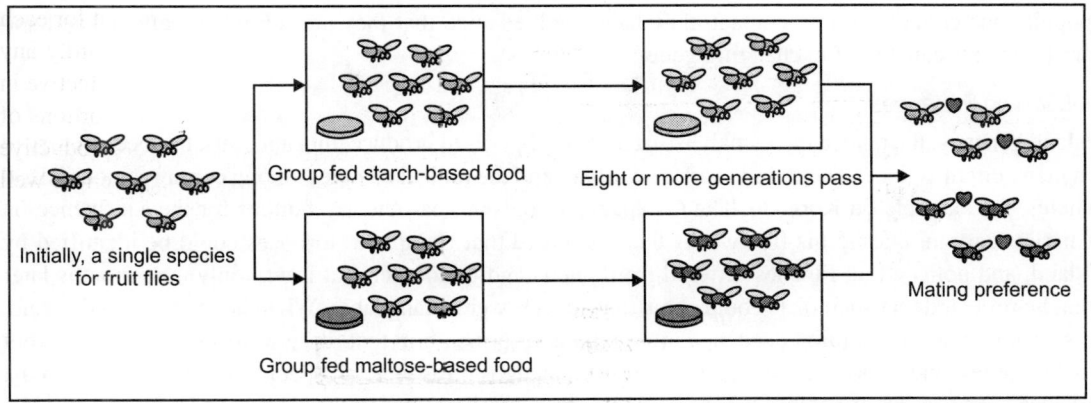

Fig. 5.2. Reproductive isolation mechanisms can be a consequence of allopatric speciation. A population of *Drosophila* was divided into sub-populations that were selected to adapt to different food types. After a number of generations the two sub-populations were mixed again. It was observed that the subsequent matings occurred between individuals belonging to the same adapted group.

There are four geographic modes of speciation in nature, based on the extent to which speciating populations are geographically isolated from one another: allopatric, peripatric, parapatric and sympatric. Speciation may also be induced artificially, through animal husbandry or laboratory experiments. Observed examples of each kind of speciation are provided throughout.

Natural speciation

All forms of natural speciation have taken place over the course of evolution; however it still remains a subject of debate as to the relative importance of each mechanism in driving biodiversity.

One example of natural speciation is the diversity of the three-spined stickleback, a marine fish that, after the last ice age, has undergone speciation into new freshwater colonies in isolated lakes and streams. Over an estimated 10,000 generations, the sticklebacks show structural differences that are greater than those seen between different genera of fish including variations in fins, changes in the number or size of their bony plates, variable jaw structure and colour differences.

Speciation rate

There is debate as to the rate at which speciation events occur over geologic time. While some evolutionary biologists claim that speciation events have remained relatively constant over time, some palaeontologists such as Niles Eldredge and Stephen Jay Gould have argued that species usually remain unchanged over long stretches of time, and that speciation occurs only over relatively brief intervals, a view known as punctuated equilibrium.

Allopatric

During allopatric (from the ancient Greek allos, 'other' + Greek patra-, 'fatherland') speciation, a population splits into two geographically isolated populations (for example, by habitat fragmentation due to geographical change such as mountain building). The isolated populations then undergo genotypic and/or phenotypic divergence as: (i) they become subjected to dissimilar selective pressures, (ii) they independently undergo genetic drift and (iii) different mutations arise in the two populations. When the populations come back into contact, they have evolved such that they are reproductively isolated and are no longer capable of exchanging genes.

Observed instances

Island genetics, the tendency of small, isolated genetic pools to produce unusual traits, has been observed in many circumstances, including insular dwarfism and the radical changes among certain famous island chains, for example on Komodo. The Galápagos islands are particularly famous for their influence on Charles Darwin. During his five weeks there he heard that Galápagos tortoises could be identified by island, and noticed that Finches differed from one island to another, but it was only nine months later that he reflected that such facts could show that species were changeable. When he returned to England, his speculation on evolution deepened after experts informed him that these were separate species, not just varieties, and famously that other differing Galápagos birds were all species of finches. Though the finches were less important for Darwin, more recent research has shown the birds now known as Darwin's finches to be a classic case of adaptive evolutionary radiation.

Peripatric

In peripatric speciation, a subform of allopatric speciation, new species are formed in isolated, smaller peripheral populations that are prevented from exchanging genes with the main population. It is related to the concept of a founder effect, since small populations often undergo bottlenecks. Genetic drift is often proposed to play a significant role in peripatric speciation.

Observed instances:

1. Mayr bird fauna.
2. The Australian bird *Petroica multicolour*.
3. Reproductive isolation occurs in populations of *Drosophila* subject to population bottlenecking.

Parapatric

In parapatric speciation, there is only partial separation of the zones of two diverging populations afforded by geography; individuals of each species may come in contact or cross habitats from time to time, but reduced fitness of the heterozygote leads to selection for behaviours or mechanisms that prevent their interbreeding. Parapatric speciation is modelled on continuous variation within a 'single', connected habitat acting as a source of natural selection rather than the effects of isolation of habitats produced in peripatric and allopatric speciation.

Ecologists refer to parapatric and peripatric speciation in terms of ecological niches. A niche must be available in order for a new species to be successful.

Sympatric

Sympatric speciation refers to the formation of two or more descendant species from a single ancestral species all occupying the same geographic location.

In sympatric speciation, species diverge while inhabiting the same place. Often-cited examples of sympatric speciation are found in insects that become dependent on different host plants in the same area. However, the existence of sympatric speciation as a mechanism of speciation is still hotly contested. People have argued that the evidences of sympatric speciation are in fact examples of micro-allopatric, or heteropatric speciation. The most widely accepted example of sympatric speciation is that of the cichlids of Lake Nabugabo in East Africa, which is thought to be due to sexual selection.

Artificial speciation

New species have been created by domesticated animal husbandry, but the initial dates and methods of the initiation of such species are not clear. For example, domestic sheep were created by hybridisation, and no longer produce viable offspring with *Ovis orientalis*, one species from which they are descended. Domestic cattle, on the other hand, can be considered the same species as several varieties of wild ox, gaur, yak, etc. as they readily produce fertile offspring with them.

The best-documented creations of new species in the laboratory were performed in the late 1980s. William Rice and GW Salt bred fruit flies, *Drosophila melanogaster*, using a maze with three different choices of habitat such as light/dark and wet/dry. Each generation was placed into the maze, and the groups of flies that came out of two of the eight exits were set apart to breed with each other in their respective groups. After thirty-five generations, the two groups and their offspring were isolated reproductively because of their strong habitat preferences: they mated only within the areas they preferred, and so did not mate with flies that preferred the other areas. The history of such attempts is described in Rice and Hostert.

Diane Dodd was also able to show how reproductive isolation can develop from mating preferences in *Drosophila pseudoobscura* fruit flies after only eight generations using different food types, starch and maltose. Dodd's experiment has been easy for many others to replicate, including with other kinds of fruit flies and foods.

Genetics

Few speciation genes have been found. They usually involve the reinforcement process of late stages of speciation. In 2008 a speciation gene causing reproductive isolation was reported. It causes hybrid sterility between related subspecies.

Hybrid speciation

Hybridisation between two different species sometimes leads to a distinct phenotype. This phenotype can also be fitter than the parental lineage and as such natural selection may then favour these individuals. Eventually, if reproductive isolation is achieved, it may lead to a separate species. However, reproductive isolation between hybrids and their parents is particularly difficult to achieve and thus hybrid speciation is considered an extremely rare event. The Mariana Mallard is known to have arisen from hybrid speciation.

Hybridisation is an important means of speciation in plants, since polyploidy (having more than two copies of each chromosome) is tolerated in plants more readily than in animals. Polyploidy is important in hybrids as it allows reproduction, with the two different sets of chromosomes each being able to pair with an identical partner during meiosis. Polyploids also have more genetic diversity, which allows them to avoid inbreeding depression in small populations.

Hybridisation without change in chromosome number is called homoploid hybrid speciation. It is considered very rare but has been shown in *Heliconius* butterflies and sunflowers. Polyploid speciation, which involves changes in chromosome number, is a more common phenomenon, especially in plant species.

Gene transposition as a cause

Theodosius Dobzhansky, who studied fruit flies in the early days of genetic research in 1930s, speculated that parts of chromosomes that switch from one location to another might cause a species to split into two different species. He mapped out how it might be possible for sections of chromosomes to relocate themselves in a genome. Those mobile sections can cause sterility in interspecies hybrids, which can act as a speciation pressure. In theory, his idea was sound, but scientists long debated whether it actually happened in nature. Eventually a competing theory involving the gradual accumulation of mutations was shown to occur in nature so often that geneticists largely dismissed the moving gene hypothesis.

However, 2006 research shows that jumping of a gene from one chromosome to another can contribute to the birth of new species. This validates the reproductive isolation mechanism, a key component of speciation.

Interspersed repeats

Interspersed repetitive DNA sequences function as isolating mechanisms. These repeats protect newly evolving gene sequences from being overwritten by gene conversion, due to the creation of non-homologies between otherwise homologous DNA sequences. The non-homologies create barriers to gene conversion. This barrier allows nascent novel genes to evolve without being overwritten by the progenitors of these genes. This uncoupling allows the evolution of new genes, both within gene families and also allelic forms of a gene. The importance is that this allows the splitting of a gene pool without requiring physical isolation of the organisms harbouring those gene sequences.

In 2011, it has been proposed that rapid amplification of repetitive DNA by genetic drift in small sub-populations can help them to 'drift apart' which may lead to the origin of new species.

Human speciation

Humans have genetic similarities with chimpanzees and bonobos, their closest relatives, suggesting common ancestors. Analysis of genetic drift and recombination using a Markov model suggests humans and chimpanzees speciated apart 4.1 million years ago.

Isolation of Population

Two populations becomes sufficiently distinct, when interbreeding is difficult or impossible between them. However, then there must be relatively little gene flow (migration) between them. Genetic changes in one population will soon become widespread in other as well, when there is a great deal of gene flow.

Genetic Divergence

Genetic divergence is the process in which two or more populations of an ancestral species accumulate independent genetic changes (mutations) through time, often after the populations have become reproductively isolated for some period of time. In some cases, sub-populations living in ecologically distinct peripheral environments can exhibit genetic divergence from the remainder of a population, especially where the range of a population is very large (see parapatric speciation). The genetic differences among divergent populations can involve silent mutations (that have no effect on the phenotype) or give rise to significant morphological and/or physiological changes.

Genetic divergence will always accompany reproductive isolation, either due to novel adaptations via selection or due to genetic drift, and is the principal mechanism underlying speciation.

Allopatric speciation

Allopatric speciation or geographic speciation is speciation that occurs when biological populations of the same species become isolated due to geographical changes such as mountain building or social changes such as emigration. The isolated populations then undergo genotypic and/or phenotypic divergence as: (i) they become subjected to different selective pressures, (ii) they independently undergo genetic drift and (iii) different mutations arise in the populations' gene pools.

The separate populations over time may evolve distinctly different characteristics. If the geographical barriers are later removed, members of the two populations may be unable to successfully mate with each other, at which point, the genetically isolated groups have emerged as different species. Allopatric isolation is a key factor in speciation and a common process by which new species arise. Adaptive radiation, as observed by Charles Darwin in Galapagos finches, is a consequence of allopatric speciation among island populations.

Isolating mechanisms

Allopatric speciation may occur when a species is subdivided into two genetically isolated populations. Allopatric and allopatry are terms from biogeography, referring to organisms whose ranges are entirely separate such that they do not occur in any one place together. If these organisms are closely related (e.g. sister species), such a distribution is usually the result of allopatric speciation. Separation may be attributed to either geological processes or population dispersal.

Geographical isolation: Geological processes can fragment a population through such events as emergence of mountain ranges, canyon formation, glacial processes, the formation or destruction of land bridges, or the subsidence of large bodies of water. On a global scale, plate tectonics are major geological factors leading to separation of populations and the resulting distribution of species.

Approximately 50,000 years ago, the Death Valley region of the western United States had a rainy climate which produced an interconnecting system of freshwater rivers and lakes. Climatic changes resulted in a drying trend that has continued for the last 10,000 years. As the lakes and rivers shrank, fish populations became geologically isolated. The few remaining (separated) springs are currently home to a variety of fish, many sharing a close common ancestor; yet each has uniquely adapted to its own particular pool. The extent to which a geological barrier can effectively isolate a population correlates to the mobility of the organism or its offspring. For example physical barriers such as canyons may effectively block migration and dispersal of small mammals; however, have little impact on flying birds or wind-borne seeds.

Population dispersal: Population dispersal is used to describe migratory events, either in the form of range expansion (natural movement away from parents) or jump dispersal (crossing of barriers), which may lead to genetic isolation. If the smaller population fragment becomes genetically isolated from the parental group, it may be subjected to its own unique mutations, selection forces, and genetic drift effects; thus, it will follow its own evolutionary pathway. Migrations or accidental relocations (such as birds being blown off course) may lead to population fragments; whereby groups merely become separated by distance. Once gene flow between the two groups is disrupted, speciation becomes a possibility.

Figure 5.3 shows the comparison of allopatric, peripatric, parapatric and sympatric speciation.

Allopatric speciation in peripheral populations

When populations become genetically isolated, heritable variations may accumulate so that they become different from the parental population. Given sufficient time, these variations may lead to reproductive isolation.

Portions of a populations that exist along the edges of the parent population's geographic territory have higher likelihood of developing reproductive isolation. Such peripheral populations are likely to possess genes that are different from the parental population. After isolation, the founding population is less likely to represent the gene pool of the parent population. In addition, peripheral isolates are likely to represent a small number of individuals, meaning their gene pool is more susceptible to the effects of genetic drift (random chance). Furthermore, it is likely that the peripheral population will inhabit an environment different from its ancestral gene pool, likely causing it to be subjected to different selective pressures as it colonises new areas. The outer periphery of a population's habitat tends to be extreme; hence, the reason range expansion is kept in check. For most peripheral isolates, it is more likely that they die off rather than survive and speciate.

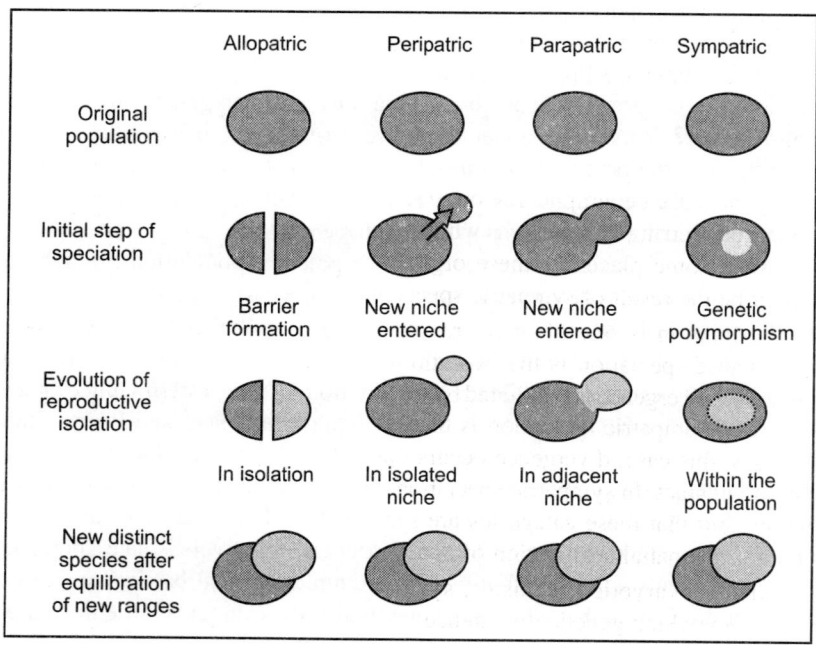

Fig. 5.3. Comparison of allopatric, peripatric, parapatric and sympatric speciation.

Genesis of reproductive barriers

Adaptive divergence may occur when a population becomes geographically divided: followed by an accumulation of genetic differences as they adapt to their own unique environments. Reproductive barriers do not evolve as a consequence of external forces that drive populations toward speciation. Rather, the evolution of reproductive isolation, leading to speciation, is generally thought to be an incidental by-product of genetic divergence, particularly adaptive changes that evolve through natural selection in response to different environmental conditions in separate geographic areas.

The frequency of other types of speciation, such as sympatric speciation, parapatric speciation and heteropatric speciation, is debated. Proponents of peripatric speciation contend that small population size in the peripheral isolate (sometimes referred to as a 'splinter population') increases genetic drift, which can be a more powerful force than natural selection in small populations. It deconstructs complex genotypes, allowing the creation of novel gene combinations. Both forms need not be mutually exclusive. In practice, passive isolation or fragmentation as well as active dispersal seem to play a role in many cases of speciation.

Alternative modes of speciation

Sympatric speciation represents an alternative method of speciation that does not require physical separation; instead speciation occurs within a population sharing the same geographic boundaries. For example, the development of polyploidy in plant species can lead to a new species arising within the geographic range of its parent population.

In parapatric speciation there is no physical barrier to gene exchange within the population. Instead, the population is continuous; however, mating is not random. Individuals mate with their closest neighbours rather than with individuals in a more distant location. Divergence may occur as a consequence of both reduced gene flow and natural selection, imposed by the large distance between individuals within a population's habitat. Allopatric speciation is thought to be the dominant mode of speciation.

Sympatric speciation

Sympatric speciation is the process through which new species evolve from a single ancestral species while inhabiting the same geographic region. In evolutionary biology and biogeography, sympatric and sympatry are terms referring to organisms whose ranges overlap or are even identical, so that they occur together at least in some places. If these organisms are closely related (e.g. sister species), such a distribution may be the result of sympatric speciation.

Sympatric speciation is one of three traditional geographic categories for the phenomenon of speciation. Allopatric speciation is the evolution of geographically isolated populations into distinct species. In this case, divergence is facilitated by the absence of gene flow, which tends to keep populations genetically similar. Parapatric speciation is the evolution of geographically adjacent populations into distinct species. In this case, divergence occurs despite limited interbreeding where the two diverging groups come into contact. In sympatric speciation, there is no geographic constraint to interbreeding. It has been pointed out that these categories are special cases of a continuum from zero (sympatric) to complete (allopatric) spatial segregation of diverging groups.

In multicellular eukaryotic organisms, sympatric speciation is thought to be an uncommon but plausible process by which genetic divergence (through reproductive isolation) of various populations from a single parent species and inhabiting the same geographic region leads to the creation of new species. In bacteria, however, the analogous process (defined as 'the origin of new bacterial species that occupy definable ecological niches') might be more common because bacteria are less constrained by the homogenising effects of sexual reproduction and prone to comparatively dramatic and rapid genetic change through horizontal gene transfer.

Evidence

A number of models have been proposed to account for this mode of speciation. The most popular, which invokes the disruptive selection model, was first put forward by John Maynard Smith. Maynard Smith suggested that homozygous individuals may, under particular environmental conditions, have a greater fitness than those with alleles heterozygous for a certain trait. Under the mechanism of natural selection, therefore, homozygosity would be favoured over heterozygosity, eventually leading to speciation. Sympatric divergence could also result from the sexual conflict.

Disruption may also occur in multiple-gene traits. The Medium Ground Finch (*Geospiza fortis*) is showing gene pool divergence in a population on Santa Cruz Island. Beak morphology conforms to two different size ideals, while intermediate individuals are selected against. Some characteristics (termed magic traits) such as beak morphology may drive speciation because they also affect mating signals. In this case, different beak phenotypes may result in different bird calls, providing a barrier to exchange between the gene pools.

A well studied circumstance of sympatric speciation is when insects feed on more than one species of host plant. In this case insects become specialised as they struggle to overcome the various plants' defense mechanisms. Sympatric speciation events are vastly more common in plants, as they are prone to developing multiple homologous sets of chromosomes, resulting in a condition called polyploidy. The polyploidal offspring occupy the same environment as the parent plants (hence sympatry), but are reproductively isolated.

A rare example of sympatric speciation in animals is the divergence of 'resident' and 'transient' Orca forms in the northeast Pacific. Resident and transient orcas inhabit the same waters, but avoid each other and do not interbreed. The two forms hunt different prey species and have different diets, vocal behaviour and social structures. Some divergences between species could also result from contrasts in microhabitats.

Parapatric speciation

Parapatry is a term from biogeography, referring to organisms whose ranges do not significantly overlap but are immediately adjacent to each other; they only occur together in the narrow contact zone, if at all. This geographical distribution is opposed to sympatry (same area) and allopatry or peripatry (2 cases of distinct areas). This distribution may along time cause speciation into sister species, a process called parapatric speciation.

Parapatric speciation is a form of speciation that occurs due to apparition of dimorphism between populations of a species, and simultaneous variation in the mating habits, within a continuous geographical area. In this model, the parent species lives in a continuous habitat, in contrast with allopatric speciation and peripatric speciation where sub-populations become geographically isolated, and sympatric speciation inside the same area (and still contested).

Niches in this habitat can differ along an environmental gradient, hampering gene flow, and thus creating a cline. In parapatric speciation there is no specific extrinsic barrier to gene flow. The population is continuous, but nonetheless, it does not mate randomly. Individuals are more likely to mate with their geographic neighbours than with individuals in a different part of the population's range. In this mode, divergence may happen because of reduced gene flow within the population as a whole and varying selection pressures across the population's range.

Examples

An example of this is the grass *Anthoxanthum*, which has been known to undergo parapatric speciation in such cases as mine contamination of an area. This creates a selection pressure for tolerance to those metals. Flowering time generally changes (tending toward character displacement—strong selection against interbreeding—as the hybrids are generally ill-suited to the environment) and often plants will become self-pollinating. Similarly, a recent study provided evidence for parapatric speciation in Tennessee cave salamanders, involving divergence with gene flow between cave and surface populations. Another example are ring species.

Operation of Reproductive Isolation

Genetic divergence during period of isolation is a necessary condition for origin of a new species, but it is not sufficient unless part of that genetic divergence cause the development of something that ensures reproductive isolation. Structural and/or behavioural modifications that prevent interbreeding are isolating mechanisms. Once such mechanisms have arisen, they have a clear adaptive value. Separate species tend to be genetically distinct in ways that adapt them to different environments. Any individual that mates with a member of another species probably produce unfit or sterile offspring, thereby, 'wasting' its genes and contributing nothing to future generations. Thus, there is strong environmental pressure to avoid mating between different species.

Ring Species

In biology, a ring species is a connected series of neighbouring populations, each of which can interbreed with closely sited related populations, but for which there exist at least two 'end' populations in the series, which are too distantly related to interbreed, though there is a potential gene flow between each 'linked' species. Such nonbreeding, though genetically connected, 'end' populations may coexist in the same region thus closing a 'ring'. Ring species provide important evidence of evolution in that they illustrate what happens over time as populations genetically diverge, and are special because they represent in living populations what normally happens over time between long deceased ancestor populations and living populations, in which the intermediates have become extinct. Richard Dawkins observes that ring species 'are only showing us in the spatial dimension something that must always happen in the time dimension'. Ring species also present an interesting case of the species problem, for those who seek to divide the living world into discrete species. After all, all that distinguishes a ring species from two separate species is the existence of the connecting populations—if enough of the connecting populations within the ring perish to sever the breeding connection, the ring species' distal populations will be recognised as two distinct species.

Formally, the issue is that interfertile 'able to interbreed' is not a transitive relation—if A can breed with B, and B can breed with C, it does not follow that A can breed with C—and thus does not define an equivalence relation. A ring species is a species that exhibits a counterexample to transitivity (Fig. 5.4).

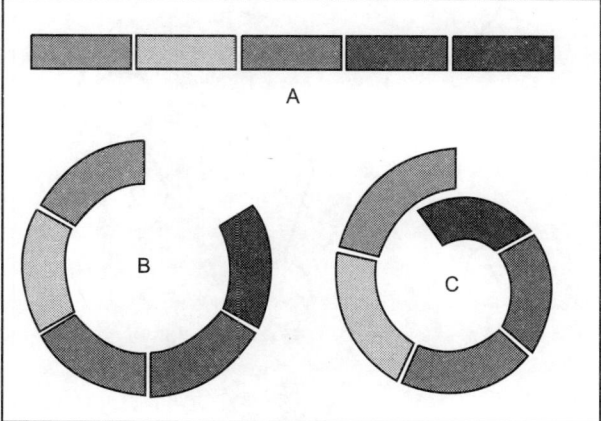

Fig. 5.4. In this diagram, interbreeding populations are represented by different shaded blocks. Variation along a cline may bend right around, forming a ring.

The problem, then, is whether to quantify the whole ring as a single species (despite the fact that not all individuals can interbreed) or to classify each population as a distinct species (despite the fact that it can interbreed with its near neighbours). Ring species illustrate that the species concept is not as clear-cut as it is often thought to be.

Larus gulls

A classic example of ring species is the *Larus* gulls' circumpolar species 'ring'. The range of these gulls forms a ring around the North Pole, which is not normally transited by individual gulls. The Herring Gull *L. argentatus*, which lives primarily in Great Britain and Ireland, can hybridise with the American Herring Gull *L. smithsonianus*, (living in North America), which can also hybridise with the Vega or East Siberian Herring Gull *L. vegae*, the western subspecies of which, Birula's Gull *L. vegae birulai*, can hybridise with Heuglin's gull *L. heuglini*, which in turn can hybridise with the Siberian Lesser Black-backed Gull *L. fuscus*. All four of these live across the north of Siberia. The last is the eastern representative of the Lesser Black-backed Gulls back in northwestern Europe, including Great Britain.

The Lesser Black-backed Gulls and Herring Gulls are sufficiently different that they do not normally hybridise; thus the group of gulls forms a continuum except where the two lineages meet in Europe. However, a recent genetic study entitled The herring gull complex is not a ring species has shown that this example is far more complicated than presented here: this example only speaks to the complex of species from the classical Herring Gull through Lesser Black-backed Gull. There are several other taxonomically unclear examples which belong in the same superspecies complex, such as Yellow-legged Gull *L. michahellis*, Glaucous Gull *L. hyperboreus* and Caspian Gull *Larus cachinnans* (Fig. 5.5).

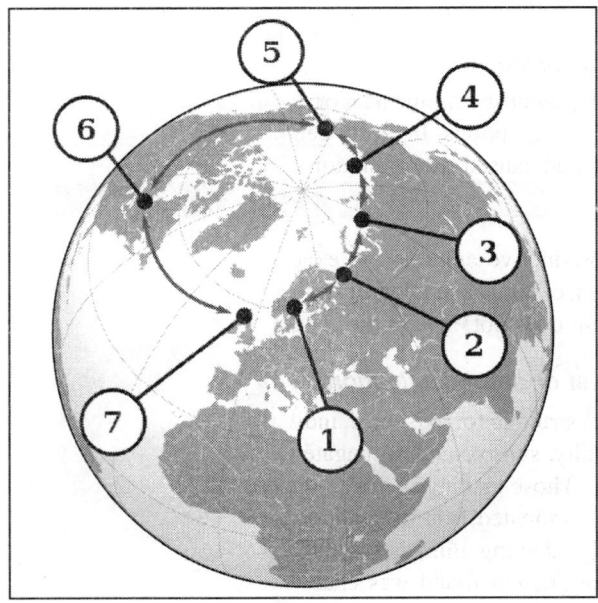

Fig. 5.5. The *Larus* gulls interbreed in a ring around the arctic: (1) *Larus argentatus argentatus*, (2) *Larus fuscus sensu stricto*, (3) *Larus fuscus heuglini*, (4) *Larus argentatus birulai*, (5) *Larus argentatus vegae*, (6) *Larus argentatus smithsonianus*, (7) *Larus argentatus argenteus*.

Darwinism

Darwinism is a set of movements and concepts related to ideas of transmutation of species or of evolution, including some ideas with no connection to the work of Charles Darwin. The meaning of 'Darwinism' has changed over time, and varies depending on who is using the term. In the United States, the term 'Darwinism' is often used by creationists as a pejorative term in reference to beliefs such as atheistic naturalism, but in the United Kingdom the term has no negative connotations, being freely used as a shorthand for the body of theory dealing with evolution, and in particular, evolution by natural selection.

Prodigality of production

Organisms have pontentiality of rapid increase. *Paramoecium* divide by fission about 600 times per year. If all survive and continue to divide, after some months, they would exceed the carrying capacity of earth. But practically that does not happen.

Struggle for existence

Individuals multiply in geometric ratio. As space and food remain almost constant, three-fold struggle occur in nature as follow.

Intraspecific struggle

Competition among individuals of same species or in closely related forms. This is very severe, as need and approach of competiting organism is precisely same.

Interspecific struggle

Struggle between organisms of different species living together for food, shelter and breeding places.

Struggle against environment

Struggle of living forms against extreme heat or cold, excess of moisture or drought, lightening, storms, earthquakes and volcanic eruptions. In North central State of America, a severe winter with late snow cuts off food supplies and caused extermination of quail.

Variation

No two sexually reproductive animals are exactly alike. Individuals of each species vary in size, proportion, colouration, external and internal structure and physiology. Darwin recognised this variation. His theory assumes but does not explain the origin of variation.

Survival of the fittest or natural selection

Darwin thought that in struggle for existence, individuals with favourable variations meet the conditions of life more successfully, survive and propagate their kind. Herbert Spencer termed such process 'the survival of the fittest'. Those lacking such variations perish or fail to breed, and characters, which they possessed, would be eliminated from population. In consequent generations, process would continue and result in gradually adapting animals suitable to their environment. Natural selection was based on inherent variation that Darwin found was characteristic of species. In a natural population, selective forces, like environmental factors (such as food availability, temperature or moisture conditions) or biotic factors (such as predation or competition between individuals), would favour those fit individuals than less fit individuals in prevailing conditions. Darwin clearly observed the variation conferring a selective advantage would be passed to future generations. Unfavourable variations would be diminished in future generations and ultimately eliminated. Central of Darwin's theory lies the idea that evolution proceeded by accumulation of small, heritable changes, not large, sudden changes, and selective forces acted on individual. It was Darwin's contention that evolution acted without design heritable traits accumulated randomly and natural selection depended on prevailing conditions.

Sexual selection

The sexual form of selection:

> "... depends, not on a struggle for existence, but on a struggle between the males for possession of the females; the result is not death to the unsuccessful competitor, but few or no offspring."

> "... when the males and females of any animal have the same general habits ... but differ in structure, colour or ornament, such differences have been mainly caused by sexual selection."

The sexual selection concept arises from the observation that many animals develop features whose function is not to help individuals survive, but help them to maximise their reproductive success. This can be realised in two different ways:

1. By making themselves attractive to the opposite sex (intersexual selection, between the sexes).
2. By intimidating, deterring or defeating same-sex rivals (intrasexual selection, within a given sex).

Thus, sexual selection takes two major forms: intersexual selection (also known as 'mate choice' or 'female choice') in which males compete with each other to be chosen by females; and intrasexual selection (also known as 'male–male competition') in which members of the less limited sex (typically males) compete aggressively among themselves for access to the limiting sex. The limiting sex is the sex which has the higher parental investment, which therefore faces the most pressure to make a good mate decision.

For intersexual selection to work, one sex must evolve a feature alluring to the opposite sex, sometimes resulting a 'fashion fad' of intense selection in an arbitrary direction. Or, in the second case, while natural selection can help animals develop ways of killing or escaping from other species, intrasexual selection drives the selection of attributes that allow alpha males to dominate their own breeding partners and rivals.

Sexual selection sometimes creates monstrously absurd features that, in harder times, could help cause a species' extinction, as has been suggested for the giant antlers of the Irish Elk (*Megaloceros giganteus*) that became extinct in Pleistocene Europe. However, sexual selection can also do the opposite, driving species divergence—sometimes through elaborate changes in genitalia—such that new species emerge.

Although the driving force for both sexes is reproductive success, the two genders have different concerns: males may seek to monopolise access to a group of fertile females, while the females want to maximise return on the energy they invest in reproduction, seeing their offspring grow into healthy adults and especially into alpha males with well-developed, sexually attractive features that sire them many descendants. Because of their limited number of breeding opportunities (due to seasonal breeding cycles, limited litter sizes, and the amount of food available to bring up the offspring) females have much more reason to be 'picky'. They need a way to choose the males that are most capable. Male and female investments in rearing offspring are not equal, females' energy expenditures on gestation and parental care being much higher. In contrast, males often use every opportunity they have to mate, as they are less invested in each individual offspring.

Modern interpretation

Today, biologists would say that certain evolutionary traits can be explained by intraspecific competition—competition between members of the same species—distinguishing between competition before or after sexual intercourse.

Before copulation, intrasexual selection—usually between males—may take the form of male-to-male combat. Also, intersexual selection, or mate choice, occurs when females choose between male mates. Traits selected by male combat are called secondary sexual characteristics (including horns, antlers, etc.), which Darwin described as 'weapons', while traits selected by mate (usually female) choice are called 'ornaments'.

After copulation, male–male competition distinct from conventional aggression may take the form of sperm competition, as described by Parker in 1970. More recently, interest has arisen in *cryptic female choice*, a phenomenon of internally fertilised animals such as mammals and birds, where a female will get rid of a male's sperm without his knowledge.

Mate choice

Mate choice or intersexual selection, is an evolutionary process in which selection of a mate depends on attractiveness of its traits. It is one of two components of sexual selection (the other is male-male competition or intrasexual selection). Darwin first introduced his ideas on sexual selection in 1871 but advances in genetic and molecular techniques have led to major progress in this field recently.

Five mechanisms that explain the evolution of mate choice are currently recognised. They are direct phenotypic benefits, sensory bias, Fisherian runaway, indicator traits and genetic compatibility. These mechanisms can co-occur and there are many examples of each.

In systems where mate choice exists, one sex is competitive with same-sex members and the other sex is choosy (selective when it comes to picking individuals to mate with). In most species, females are the choosy sex that discriminate amongst competitive males but there are several examples of reversed roles.

Direct and indirect benefits

Direct benefits are those that increase the fitness of the choosy sex through direct material advantages. These benefits include but are not limited to increased territory quality, increased parental care and protection from predators. There is much support for maintenance of mate choice by direct benefits and it is the least controversial model to explain discriminate mating.

Indirect benefits increase genetic fitness for the offspring. When it appears that the choosy sex does not receive direct benefits from his or her mate, indirect benefits may be the payoff for being selective. Examples of indirect benefits include better genetic quality and more attractive offspring. R.A. Fisher described this less obvious model in a book called *The Genetical Theory of Natural Selection*. Fisher explained that, through indirect selection, fitter individuals inherit both the genes and the mating preference for some indicator trait. This linkage of an indicator trait and the preference for such trait results in exaggerated phenotypes and is known as Fisherian runaway selection.

Indicator traits

Indicator traits are those that signal good overall quality of the individual. Traits that are perceived as attractive must reliably indicate broad genetic quality in order for selection to favour them and for preference to evolve. This is an example of indirect genetic benefits received by the choosy sex because mating with such individuals will result in high quality offspring. The indicator traits hypothesis is split into three highly related subtopics: the handicap theory of sexual selection, the good genes hypothesis, and the Hamilton-Zuk hypothesis.

Indicator traits are condition-dependent and have associated costs. Therefore, individuals that can handle these costs well should be desired by the choosy sex for their superior genetic quality. This is known as the handicap theory of sexual selection.

The good genes hypothesis states that the choosy sex will mate with individuals who possess traits that signify overall genetic quality. In doing so, they gain an evolutionary advantage for their offspring through indirect benefit.

Sperm competition

Sperm competition is a term used to refer to the competitive process between spermatozoa of two different males to fertilise an egg of a lone female. Competition occurs whenever females engage in promiscuous mating to increase their chances in producing more viable offspring. Sperm competition is an evolutionary pressure on males, and has led to the development of adaptations to increase males' chance of reproductive success. Sperm competition results from sexual conflict between males and females. Males have evolved several defensive tactics including: mate-guarding, mating plugs, and releasing toxic seminal substances to reduce female re-mating tendencies to cope with sperm competition. Offensive tactics of sperm competition involve direct interference by one male on the reproductive success of another male, for instance by physically removing another male's sperm prior to mating with a female.

Sperm competition is often compared to having tickets in a raffle; a male has a better chance of winning (i.e. fathering offspring) the more tickets he has (i.e. the more sperm he inseminates a female with). However, sperm are not free to produce, and as such males are predicted to produce sperm of a size and number that will maximise their success in sperm competition. By making many spermatozoa, males can buy more 'raffle tickets', and it is thought that selection for numerous sperm has contributed greatly to the evolution of anisogamy (because of the energy trade-off between sperm size and number). Alternatively males may evolve faster sperm to enable their sperm to reach and fertilise the ovum first. Dozens of adaptations have been documented in males that help them succeed in sperm competition.

Defensive adaptations

Mate-guarding is a defensive behavioural trait that occurs in response to sperm competition; males try to prevent other males from approaching the female (and/or *vice versa*) thus preventing their mate from engaging in further copulations. Precopulatory and postcopulatory mate-guarding occurs in birds, lizards, insects and primates. In these instances the males guard their female by keeping them in close enough proximity so that if an opponent male shows up in his territory he will be able to fight off the rival male which will prevent the female from engaging in extra-pair copulation with the rival male.

Strategic mate-guarding occurs when the male only guards the female during her fertile periods. This stragegy can be more effective because it may allow the male to engage in both extra-pair paternity and within-pair paternity.

Copulatory plugs are frequently observed in insects, reptiles, some mammals and spiders. Copulatory plugs are inserted immediately after a male copulates with a female, which reduce the possibility of fertilisation by subsequent copulations from another male, by physically blocking the transfer of sperm. The Indian Meal Moth (*P. interpunctella*) are restricted in engaging in further mating activities because the spermatacore serves as a copulatory plug immediately after copulation. Bumblebee mating plugs, in addition to providing a physical barrier to further copulations, contain linoleic acid, which reduces re-mating tendencies of females.

Similarly, *Drosophila melanogaster* males release toxic seminal fluids, known as ACPs (accessory gland proteins), from their accessory glands to impede the female from participating in future copulations. These substances act as an anti-aphrodisiac causing a dejection of subsequent copulations, and also stimulate ovulation and oogenesis. Seminal proteins can have a strong influence on reproduction, sufficient to manipulate female behaviour and physiology.

Another strategy, known as sperm partitioning, occurs when males conserve their limited supply of sperm by reducing the quantity of sperm ejected. In drosophila, ejaculation amount during sequential copulations is reduced; this results in half filled female sperm reserves following a single copulatory event, but allows the male to mate with a larger number of females without exhausting his supply of sperm. To facilitate sperm partitioning, some males have developed complex ways to store and deliver their sperm. In the blue headed wrasse thalassoma bifsciatum the sperm duct is sectioned into several small chambers that are surrounded by a muscle that allows the male to regulate how much sperm is released in one copulatory event. A strategy common among insects is for males to participate in prolonged copulations. By engaging in prolonged copulations a male has an increased opportunity to place more sperm within the female's reproductive tract and prevent the female from copulating with other males.

Offensive adaptations

Offensive adaptation behaviour differs from defensive behaviour because it involves an attempt to ruin the chances of another male's opportunity in succeeding in copulation by engaging in an act that tries to terminate the fertilisation success of the previous male. A male on the offensive side of mate-guarding may terminate the guarding male's chances at a successful insemination by brawling with the guarding male to gain access to the female. In *Drosophila*, males release seminal fluids that contain additional toxins like pheromones and modified enzymes that are secreted by their accessory glands intended to destroy the sperm that have already made their way into the female's reproductive tract from a recent copulation. Based on the 'last male precedence' idea, some males can remove sperm from previous males by ejaculating new sperm into the female; hindering successful insemination opportunities of the previous male.

Sexual selection in primates

Sexual selection in multi-male, multi-female primate groups is intense because social context of mating is complex and dynamic. Sexes compete, both sexes are choosy, both sexes have dominant relations and both sexes form alliances. Sexual relationships develop over weeks and years rather than minutes. Under these relentlessly social conditions, reproductive success came to depend on mental capacity for 'Chimpanzee politics', 'Machiavellian intelligence', 'special friendships' and creative courtship rather than simple physical ornaments and short-term courtship behaviours as in most other animals. Primates and especially hominids are extremely 'K-selected taxa' having slower development, larger bodies, fewer offspring, higher survival rates and longer life-spans than more 'r-selected taxa' such as insects, fish or rodents. The more K-selected the species, the more important sexual selection usually becomes compared to natural selection. K-Selection usually reduces relative energetic demand of reproduction on female and almost eliminates need for male help, because slow gestation spread maternal investment over a longer period and small litters of large, well-developed offspring are easier to care for.

Altruism in animals

Altruism is a well-documented animal behaviour, which appears most obviously in kin relationships but may also be evident amongst wider social groups, in which an animal sacrifices its own well-being for the benefit of another animal.

In the science of ethology (the study of behaviour), and more generally in the study of social evolution, on occasion, some animals do behave in ways that reduce their individual fitness but increase the fitness of other individuals in the population; this is a functional definition of altruism. Research in evolutionary theory has been applied to social behaviour, including altruism. Cases of animals helping individuals to whom they are closely related can be explained by kin selection, and are not considered true altruism. Beyond the physical exertions that mothers, and in some species fathers, undertake to protect their young, extreme examples of sacrifice may occur. One example is matriphagy (the consumption of the mother by her offspring) in the spider *Stegodyphus*. Hamilton's rule describes the benefit of such altruism in terms of Wright's coefficient of relationship to the beneficiary and the benefit granted to the beneficiary minus the cost to the sacrificer. Should this sum be greater than zero a fitness gain will result from the sacrifice.

When apparent altruism is not between kin, it may be based on reciprocity. A monkey will present its back to another monkey, who will pick out parasites; after a time the roles will be reversed. Such reciprocity will pay off, in evolutionary terms, as long as the costs of helping are less than the benefits of being helped and as long as animals will not gain in the long run by 'cheating'—that is to say, by receiving favours without returning them. This is elaborated on in evolutionary game theory and specifically the prisoner's dilemma as social theory.

Kin selection

Kin selection refers to apparent strategies in evolution that favour the reproductive success of an organism's relatives, even at a cost to the organism's own survival and reproduction. Charles Darwin was the first to discuss the concept of group/kin selection.

Kin selection refers to changes in gene frequency across generations that are driven at least in part by interactions between related individuals, and this forms much of the conceptual basis of the theory of social evolution. Indeed, some cases of evolution by natural selection can only be understood by considering how biological relatives influence one another's fitness. Under natural selection, a gene encoding a trait that enhances the fitness of each individual carrying it should increase in frequency within the population; and conversely, a gene that lowers the individual fitness of its carriers should be eliminated. However, a hypothetical gene that prompts behaviour which enhances the fitness of relatives but lowers that of the individual displaying the behaviour, may nonetheless increase in frequency, because relatives often carry the same gene; this is the fundamental principle behind the theory of kin selection. According to the theory, the enhanced fitness of relatives can at times more than compensate for the fitness loss incurred by the individuals displaying the behaviour. As such, this is a special case of a more general model, called inclusive fitness (in that inclusive fitness refers simply to gene copies in other individuals, without requiring that they be kin). However the validity of this analysis has recently been challenged.

MODERN EVOLUTIONARY SYNTHESIS

The modern evolutionary synthesis is a union of ideas from several biological specialities which provides a widely accepted account of evolution. It is also referred to as the new synthesis, the modern synthesis, the evolutionary synthesis, millennium synthesis and the neo-darwinian synthesis. The synthesis, produced between 1936 and 1947, reflects the current consensus. The previous development of population genetics, between 1918 and 1932, was a stimulus, as it showed that Mendelian genetics was consistent with natural selection and gradual evolution. The synthesis is still, to a large extent, the current paradigm in evolutionary biology.

The modern synthesis solved difficulties and confusions caused by the specialisation and poor communication between biologists in the early years of the 20th century. At its heart was the question of whether Mendelian genetics could be reconciled with gradual evolution by means of natural selection. A second issue was whether the broad-scale changes (macroevolution) seen by palaeontologists could be explained by changes seen in local populations (microevolution).

The synthesis included evidence from biologists, trained in genetics, who studied populations in the field and in the laboratory. These studies were crucial to evolutionary theory. The synthesis drew together ideas from several branches of biology which had become separated, particularly genetics, cytology, systematics, botany, morphology, ecology and paleontology.

Summary of the Modern Synthesis

The modern synthesis bridged the gap between experimental geneticists and naturalists, and between palaeontologists. It states that:

1. All evolutionary phenomena can be explained in a way consistent with known genetic mechanisms and the observational evidence of naturalists.
2. Evolution is gradual: small genetic changes regulated by natural selection accumulate over long periods. Discontinuities amongst species (or other taxa) are explained as originating gradually through geographical separation and extinction (not saltation).
3. Natural selection is by far the main mechanism of change; even slight advantages are important when continued. The object of selection is the phenotype in its surrounding environment.
4. The role of genetic drift is equivocal. Though strongly supported initially by Dobzhansky, it was downgraded later as results from ecological genetics were obtained.
5. Thinking in terms of populations, rather than individuals, is primary: the genetic diversity existing in natural populations is a key factor in evolution. The strength of natural selection in the wild is greater than previously expected; the effect of ecological factors such as niche occupation and the significance of barriers to gene flow are all important.
6. In palaeontology, the ability to explain historical observations by extrapolation from microevolution to macroevolution is proposed. Historical contingency means explanations at different levels may exist. Gradualism does not mean constant rate of change.

The idea that speciation occurs after populations are reproductively isolated has been much debated. In plants, polyploidy must be included in any view of speciation. Formulations such as 'evolution consists primarily of changes in the frequencies of alleles between one generation and another' were proposed rather later. The traditional view is that developmental biology (evo-devo) played little part in the synthesis, but an account of Gavin de Beer's work by Stephen J. Gould suggests he may be an exception.

CONTINENTAL DRIFT

Continental drift is the movement of the earth's continents relative to each other. Evidence for continental drift is now extensive. Similar plant and animal fossils are found around different continent shores, suggesting that they were once joined. The fossils of Mesosaurus, a freshwater reptile rather like a small crocodile, found both in Brazil and South Africa, are one example; another is the discovery of fossils of the land reptile *Lystrosaurus* from rocks of the same age from locations in South America, Africa and Antarctica. There is also living evidence—the same animals being found on two continents. Some earthworm families (e.g. Ocnerodrilidae, Acanthodrilidae, Octochaetidae) are found in South America and Africa, for instance.

The complementary arrangement of the facing sides of South America and Africa is obvious, but is a temporary coincidence. In millions of years, slab pull and ridge-push, and other forces of tectonophysics will further separate and rotate those two continents. It was this temporary feature which inspired Wegener to study what he defined as continental drift, although he did not live to see his hypothesis become generally accepted.

Widespread distribution of Permo-Carboniferous glacial sediments in South America, Africa, Madagascar, Arabia, India, Antarctica and Australia was one of the major pieces of evidence for the theory of continental drift. The continuity of glaciers, inferred from oriented glacial striations and deposits called tillites, suggested the existence of the supercontinent of Gondwana, which became a central element of the concept of continental drift. Striations indicated glacial flow away from the equator and toward the poles, in modern coordinates, and supported the idea that the southern continents had previously been in dramatically different locations, as well as contiguous with each other.

It is now known that there are two kinds of crust, continental crust and oceanic crust. Continental crust is inherently lighter and of a different composition to oceanic crust, but both kinds reside above a much deeper fluid mantle. Oceanic crust is created at spreading centers, and this, along with subduction, drives the system of plates in a chaotic manner, resulting in continuous orogeny and areas of isostatic imbalance. The theory of plate tectonics explains all this, including the movement of the continents, better than Wegener's theory.

GAIA HYPOTHESIS

The Gaia hypothesis, also known as Gaia theory or Gaia principle, proposes that all organisms and their inorganic surroundings on earth are closely integrated to form a single and self-regulating complex system, maintaining the conditions for life on the planet.

The scientific investigation of the Gaia hypothesis focuses on observing how the biosphere and the evolution of life forms contribute to the stability of global temperature, ocean salinity, oxygen in the atmosphere and other factors of habitability in a preferred homoeostasis. The Gaia hypothesis was formulated by the chemist James Lovelock and co-developed by the microbiologist Lynn Margulis in the 1970s. Initially received with hostility by the scientific community, it is now studied in the disciplines of geophysiology and earth system science, and some of its principles have been adopted in fields like biogeochemistry and systems ecology. This ecological hypothesis has also inspired analogies and various interpretations in social sciences, politics and religion under a vague philosophy and movement.

The Gaia theory posits that the earth is a self-regulating complex system involving the biosphere, the atmosphere, the hydrospheres and the pedosphere, tightly coupled as an evolving system. The theory sustains that this system as a whole, called Gaia, seeks a physical and chemical environment optimal for contemporary life.

Gaia evolves through a cybernetic feedback system operated unconsciously by the biota, leading to broad stabilisation of the conditions of habitability in a full homeostasis. Many processes in the earth's surface essential for the conditions of life depend on the interaction of living forms, especially micro-organisms, with inorganic elements. These processes establish a global control system that regulates earth's surface temperature, atmosphere composition and ocean salinity, powered by the global thermodynamic desequilibrium state of the earth system.

The existence of a planetary homeostasis influenced by living forms had been observed previously in the field of biogeochemistry, and it is being investigated also in other fields like earth system science. The originality of the Gaia theory relies on the assessment that such homoeostatic balance is actively pursued with the goal of keeping the optimal conditions for life, even when terrestrial or external events menace them.

Regulation of the salinity in the oceans: Ocean salinity has been constant at about 3.4 per cent for a very long time. Salinity stability in oceanic environments is important as most cells require a rather constant salinity and do not generally tolerate values above 5 per cent. Ocean salinity constancy was a long-standing mystery, because river salts should have raised the ocean salinity much higher than observed. Recently it was suggested that salinity may also be strongly influenced by seawater circulation through hot basaltic rocks, and emerging as hot water vents on mid-ocean ridges. However, the composition of seawater is far from equilibrium, and it is difficult to explain this fact without the influence of organic processes. One suggested explanation lies in the formation of salt plains throughout earth's history. It is hypothesised that these are created by bacteria colonies that fix ions and heavy metals during life processes.

Regulation of oxygen in the atmosphere: The atmospheric composition remains fairly constant providing the ideal conditions for contemporary life. All the atmospheric gases other than noble gases present in the atmosphere are either made by organisms or processed by them. The Gaia theory states that the earth's atmospheric composition is kept at a dynamically steady state by the presence of life.

The stability of the atmosphere in earth is not a consequence of chemical equilibrium like in planets without life. Oxygen is the second most reactive element after fluorine, and should combine with gases and minerals of the earth's atmosphere and crust. Traces of methane (at an amount of 1,00,000 tons produced per annum) should not exist, as methane is combustible in an oxygen atmosphere.

Dry air in the atmosphere of earth contains roughly (by volume) 78.09 per cent nitrogen, 20.95 per cent oxygen, 0.93 per cent argon, 0.039 per cent carbon dioxide, and small amounts of other gases including methane. While air content and atmospheric pressure varies at different layers, air suitable for the survival of terrestrial plants and terrestrial animals is currently known only to be found in earth's troposphere and artificial atmospheres. Oxygen is a crucial element for the life of organisms, who require it at stable concentrations.

Regulation of the global surface temperature: Since life started on earth, the energy provided by the Sun has increased by 25 to 30 per cent; however, the surface temperature of the planet has remained within the levels of habitability, reaching quite regular low and high margins. Lovelock has also hypothesised that methanogens produced elevated levels of methane in the early atmosphere, giving a view similar to that found in petrochemical smog, similar in some respects to the atmosphere on Titan. This, he suggests tended to screen out ultraviolet until the formation of the ozone screen, maintaining a degree of homoeostasis.

Biodiversity and stability of ecosystems: The importance of the large number of species in an ecosystem, led to two sets of views about the role played by biodiversity in the stability of ecosystems in Gaia theory. In one school of thought labelled the 'species redundancy' hypothesis, proposed by Australian ecologist Brian Walker, most species are seen as having little contribution overall in the stability, comparable to the passengers in an aeroplane who play little role in its successful flight. The hypothesis leads to the conclusion that only a few key species are necessary for a healthy ecosystem. The 'rivet-popper' hypothesis put forth by Paul R. Ehrlich and his wife Anne H. Ehrlich, compares each species forming part of an ecosystem as a rivet on the aeroplane (represented by the ecosystem). The progressive loss of species mirrors the progressive loss of rivets from the plane, weakening it till it is no longer sustainable and crashes.

Later extensions of the Daisyworld simulation which included rabbits, foxes and other species, led to a surprising finding that the larger the number of species, the greater the improving effects on the entire planet (i.e. the temperature regulation was improved). It also showed that the system was robust and stable even when perturbed. Daisyworld simulations where environmental changes were stable gradually became less diverse over time; in contrast gentle perturbations led to bursts of species richness. These findings lent support to the idea that biodiversity is valuable.

This finding was later proved in a eleven-year old study of the factors species composition, dynamics and diversity in successional and native grasslands in Minnesota by David Tilman and John A. Downing wherein they discovered that 'primary productivity in more diverse plant communities is more resistant to, and recovers more fully from, a major drought'. They go on to add 'Our results support the diversity stability hypothesis but not the alternative hypothesis that most species are functionally redundant.'

Processing of CO_2

Gaia scientists see the participation of living organisms in the carbon cycle as one of the complex processes that maintain conditions suitable for life. The only significant natural source of atmospheric carbon dioxide (CO_2) is volcanic activity, while the only significant removal is through the precipitation of carbonate rocks. Carbon precipitation, solution and fixation are influenced by the bacteria and plant roots in soils, where they improve gaseous circulation or in coral reefs, where calcium carbonate is deposited as a solid on the sea floor. Calcium carbonate is used by living organisms to manufacture carbonaceous tests and shells. Once dead, the living organisms' shells fall to the bottom of the oceans where they generate deposits of chalk and limestone.

One of these organisms is *Emiliania huxleyi*, an abundant coccolithophore algae which also has a role in the formation of clouds. CO_2 excess is compensated by an increase of coccolithophoride life, increasing the amount of CO_2 locked in the ocean floor. Coccolithophorides increase the cloud cover, hence control the surface temperature, help cool the whole planet and favour precipitations necessary for terrestrial plants. Lately the atmospheric CO_2 concentration has increased and there is some evidence that concentrations of ocean algal blooms are also increasing.

Lichen and other organisms accelerate the weathering of rocks in the surface, while the decomposition of rocks also happens faster in the soil, thanks to the activity of roots, fungi, bacteria and subterranean animals. The flow of carbon dioxide from the atmosphere to the soil is therefore regulated with the help of living beings. When CO_2 levels rise in the atmosphere the temperature increases and plants grow. This growth brings higher consumption of CO_2 by the plants, who process it into the soil, removing it from the atmosphere.

GEOLOGIC TIME SCALE

The geologic time scale provides a system of chronologic measurement relating stratigraphy to time that is used by geologists, paleontologists and other earth scientists to describe the timing and relationships between events that have occurred during the history of the earth. The table of geologic time spans presented here agrees with the dates and nomenclature proposed by the International Commission on

Stratigraphy, and uses the standard colour codes of the United States Geological Survey. Evidence from radiometric dating indicates that the earth is about 4.570 billion years old. The geological or deep time of earth's past has been organised into various units according to events which took place in each period. Different spans of time on the time scale are usually delimited by major geological or paleontological events, such as mass extinctions.

For example, the boundary between the Cretaceous period and the Paleogene period is defined by the Cretaceous–Tertiary extinction event, which marked the demise of the dinosaurs and of many marine species. Older periods which predate the reliable fossil record are defined by absolute age. Each era on the scale is separated from the next by a major event or change.

The largest defined unit of time is the supereon, composed of eons. Eons are divided into eras, which are in turn divided into periods, epochs and ages. The terms eonothem, erathem, system, series and stage are used to refer to the layers of rock that correspond to these periods of geologic time.

Geologists qualify these units as Early, Mid and Late when referring to time, and Lower, Middle, and Upper when referring to the corresponding rocks. For example, the Lower Jurassic Series in chronostratigraphy corresponds to the Early Jurassic Epoch in geochronology. The adjectives are capitalised when the subdivision is formally recognised, and lower case when not; thus 'early Miocene' but 'Early Jurassic'.

Geologic units from the same time but different parts of the world often look different and contain different fossils, so the same period was historically given different names in different locales. For example, in North America the Lower Cambrian is called the Waucoban series that is then subdivided into zones based on succession of trilobites. In East Asia and Siberia, the same unit is split into Alexian, Atdabanian and Botomian stages. A key aspect of the work of the International Commission on Stratigraphy is to reconcile this conflicting terminology and define universal horizons that can be used around the world.

MOLECULAR CLOCK

The molecular clock (based on the molecular clock hypothesis (MCH)) is a technique in molecular evolution that uses fossil constraints and rates of molecular change to deduce the time in geologic history when two species or other taxa diverged. It is used to estimate the time of occurrence of events called speciation or radiation. The molecular data used for such calculations is usually nucleotide sequences for DNA or amino acid sequences for proteins. It is sometimes called a gene clock or evolutionary clock.

The molecular clock alone can only say that one time period is twice as long as another: it cannot assign concrete dates. To achieve this, the molecular clock must first be calibrated against independent evidence about dates, such as the fossil record. Alternatively, for viral phylogenetics and ancient DNA studies, two areas of evolutionary biology where it is possible to sample sequences over an evolutionary timescale, the dates of the samples themselves can be used to calibrate the molecular clock.

The molecular clock technique is an important tool in molecular systematics, the use of molecular genetics information to determine the correct scientific classification of organisms or to study variation in selective forces. Knowledge of approximately-constant rate of molecular evolution in particular sets of lineages also facilitates establishing the dates of phylogenetic events, including those not documented by fossils, such as the divergence of living taxa and the formation of the phylogenetic tree. But in these cases—especially over long stretches of time—the limitations of MCH (above) must be considered; such estimates may be off by 50 per cent or more.

Chapter 6

Biodiversity in India

INTRODUCTION

Biodiversity is defined as 'the variability among living organisms from all sources, including terrestrial, marine and other aquatic ecosystems and the ecological complexes of which they are a part; this includes diversity within species, between species and of ecosystems'. Conservation and sustainable use of biodiversity is fundamental to ecologically sustainable development. Biodiversity is part of our daily lives and livelihood, and constitutes resources upon which families, communities, nations and future generations depend. Every country has the responsibility to conserve, restore and sustainably use the biological diversity within its jurisdiction. Biological diversity is fundamental to the fulfilment of human needs. An environment rich in biological diversity offers the broadest array of options for sustainable economic activity, for sustaining human welfare and for adapting to change. Loss of biodiversity has serious economic and social costs for any country. The experience of the past few decades has shown that as industrialisation and economic development in the classical sense takes place, patterns of consumption, production and needs, change, straining, altering and even destroying ecosystems. India, a megabiodiversity country, while following the path of development, has been sensitive to needs of conservation and hence is still rich in biological resources. Ethos of conservation and harmonious living with nature is very much ingrained in the lifestyles of India's people.

India is one of 12 megadiversity countries of the world. The innumerable life forms harboured by the forests, deserts, mountains, other land, air and oceans provide food, fodder, fuel, medicine, textiles, etc. There are innumerable species, the potential of which is not as yet known. It would therefore be prudent to not only conserve the species we already have information about, but also species we have not yet identified and described from economic point of view. *Taxus baccata*, a tree found in the Sub-Himalayan regions, once believe to be of no value is now considered to be effective in the treatment of certain types of cancer. The diversity of genes, species and ecosystem is a valuable resource that can be tapped as human needs and demands change, the still more basic reasons for conservation are the moral, cultural and religious values. The importance of biodiversity can be understood, it is not easy to define the value of biodiversity, and very often difficult to estimate it. The value of biodiversity is classified into direct and indirect values. Biodiversity has direct consumptive value in agriculture, medicine and industry. Approximately 80,000 edible plants have been used at one time or another in human history, of which only about 150 have even been cultivated on a large scale. Today a mere 10 to 20 species provide 80–90 per cent food requirements of the world. The indirect values imply the functions performed by biodiversity which are not of any direct use such as ecological processes, etc. In India, many rural communities particularly the tribals obtain considerable part of their daily food from the wild plants. Some examples are: *Ceropegia bubosa* in Central India and Western Ghats; *Codonopisis ovata* in Himalayan region; *Ardisia* and *Meliosma pinnata* in the North-east; *Eremurus himalaicus, Origanum vulgare* and *Urtica hyperborea* in Lahul-Spiti and Ladakh; *Allium carolinianum* and *Cicer microphyllum* in Kashmir and *Sesuvium portulacastrum* in Coastal areas. Similarly, a variety of faunal species, e.g. insects, molluscs, spiders, wild herbivores are consumed by many tribal and non-tribal communities in India.

At one time, nearly all medicines were derived from biological resources. Even today they remain vital and as much as 67–70 per cent of modern medicine are derived from natural products. In developing countries, a large majority of the people rely on traditional medicines for their primary health care, most of which involve the use of plant extracts.

Around 20,000 plant species are believed to be used medicinally in the third world. In India, almost 95 per cent of the prescriptions are plant-based in the traditional systems of Unani, Ayurveda and Siddha. Many indigenous medicines also utilise animals and their parts or extracts as remedies for various diseases. Diverse habitats and species also have non-consumptive use-value. Tourism, recreation and scientific research are the major examples. The indirect use-value of biodiversity includes ecosystem process of biological diversity, which provides valuable ecological services to the biosphere; some examples are the ecosystem's ability to absorb pollution, maintain soil fertility and microclimates, recharge groundwater, and provide other invaluable services. Many plants, animals and their parts are used in rituals all over the country. To name a few: flowers of *Hibiscus, Datura* and *Euphorbia*; leaves of *Aegle marmelos* (bel), *Eragrostis cynasuroides* (kusa grass), rice til, chenopods, odourous roots of *Dolomiaea macrocephala* (dhup). Further, sacred values are attached to entire ecosystems, for example patches of forests were believed to be the abode of gods, and are used only for prayers and rituals. Many sacred groves still exist in different parts of India.

PRESSURE

Habitat destruction, overexploitation, pollution and species introduction are the major causes of biodiversity loss in India. Other factors included fires, which adversely affect regeneration in some cases, and such natural calamities as droughts, diseases, cyclones and floods. Habitat destruction, decimation of species, and the fragmentation of large contiguous populations into isolated, small and scattered ones has rendered them increasingly vulnerable to inbreeding depression, high infant mortality and susceptibility to environmental stochasticity and, in the long run, possibly to extinction. Besides these, the failure to stem this tide of destruction results from an amalgamation of lacunae in economic, policy, institutional and governance systems. Among others, these include:

1. Management with limited local community participation and involvement and inadequate implementation of ecodevelopment programs; poor implementation of the Wildlife (Protection) Act of 1972 as amended in 1991.
2. Poor conviction rates of wildlife cases due to inadequate legal competence in the forest department, and the lackadaisical approach of courts with cases pending for years (Fig. 6.1).

Protected Area Network comprises National Parks and Sanctuaries which covers a mere 4.2 per cent of the land area and is inadequate in protecting such ecologically important and fragile ecosystems such as wetlands, mangroves and grasslands that lie outside such protected areas. The protected areas themselves are susceptible to denotification and further reduction in extent due to other pressures emanating from the industrial-commercial-political combine.

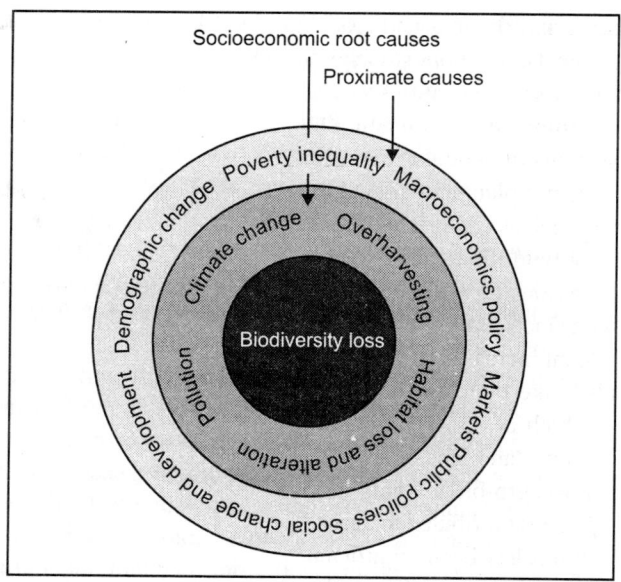

Fig. 6.1. Biodiversity loss: proximate and socioeconomic root causes.

Biodiversity conservation in India is also impeded by a lack of knowledge of the magnitude, patterns, causes and rates of deforestation and biodiversity loss at the ecosystem and landscape level. Poaching and trade in wildlife species are among the most important concerns in the management of protected areas today but information on poaching, trade and trade routes is sketchy and current wildlife protection and law enforcement measures are inadequate and inefficient.

Major Problems with Biodiversity Conservation

1. Low priority for conservation of living natural resources.
2. Exploitation of living natural resources for monetary gain.
3. Values and knowledge about the species and ecosystem inadequately known.
4. Unplanned urbanisation and uncontrolled industrialisation.

Major biodiversity threats

1. Habitat destruction.
2. Extension of agriculture.
3. Filling up of wetlands.
4. Conversion of rich biodiversity site for human settlement and industrial development.
5. Destruction of coastal areas.
6. Uncontrolled commercial exploitation.

This erosion of biodiversity is largely due to habitat loss caused by the expansion of various development projects such as mines, dams and road and canal construction. It is estimated that, after Independence, the country has lost 4696 million hectares of forestland to non-forestry purposes. While 0.07 million ha of forest land has been illegally encroached upon, 4.37 million ha has been subjected to cultivation, 0.52 million ha given to river valley projects, 0.14 million ha to industries and townships,

0.06 million ha for transmission lines and roads; and the rest for miscellaneous purposes. Habitat loss leads to the fragmentation of continuous stretches of land and consequently fragments wildlife populations inhabiting them. These small populations are increasingly vulnerable to inbreeding depression, high infant mortality, susceptibility to environmental stochasticity, and, in the long run, possibly to extinction. Apart from the primary loss of habitats, there are numerous other problems contributing to the loss and endangered status of several plant and animal species.

Habitat degradation such as changes in forest composition and quality can in turn lead to declines in primary food species for wildlife. Poaching is another insidious threat that has emerged in recent years as one of the primary reasons for extinction of species such as the tiger. Poaching pressures, however, are unevenly distributed since certain selected species are more heavily targeted than others. Population pressures and concomitant increases in the collection of fuelwood and fodder, and grazing in forests by local communities also take their toll on the forests and consequently its biodiversity. Other minor factors include fires, which adversely affect regeneration in some cases, and natural calamities like droughts, diseases, cyclones and landslides.

India's contribution to agro-biodiversity has been impressive. India stands seventh in the world as far as the number of species contributed to agriculture and animal husbandry is concerned. In qualitative terms too, the contribution has been significant, as it has contributed such useful animal species as water buffalo and camel and plant species such as rice and sugarcane. India has also been a secondary centre of domestication for animal species such as horse and goat, and such plant species as potato and maize. Animal species, which are reported to be threatened in India, have been listed in Table 6.1.

Table 6.1. Comparative statement of recorded number of animal species in India and the World (endemic and threatened animals species for India are also shown).

Taxa	India			World	Percentage of India to the world
	Species	Endemic species	Threatened species		
Protista	2577			31259	8.24
Mollusca	5060	967		66535	7.62
Arthropoda	68389	16214 (Insects)		987949	6.90
Other Invertebrates	8329		22	87121	9.56
Protochordata	119			2106	5.65
Pisces	2546		4	21723	11.72
Amphibia	209	110	3	5150	4.06
Reptilia	456	214	16	5817	7.84
Aves	1232	69	73	9026	13.66
Mamalia	390	38	75	4629	8.42

India has 47,000 species of flowering and non-flowering plants representing about 12 per cent of the recorded world's flora. Out of 47,000 species of plants, 5150 are endemic and 2532 species are found in the Himalayas and adjoining regions and 1782 in the peninsular India. India is also rich in the number of endemic faunal species it possesses, while its record in agro-biodiversity is very impressive as well. There are 166 crop species and 320 wild relatives along with numerous wild relatives of domesticated animals. Overall India ranks seventh in terms of contribution to world agriculture.

STATE IMPACT

Status of Biodiversity in India

India occupies only 2.4 per cent of the world's land area but its contribution to the world's biodiversity is approximately 8 per cent of the total number of species, which is estimated to be 1.75 million (As per Global Biodiversity Assessment of UNEP of 1995, described number of species so far is 1.75 million). Of these, 1,26,188 have been described in India. The species recorded includes flowering plants (angiosperms), mammals, fish, birds, reptiles and amphibians, constitute 17.3 per cent of the total whereas nearly 60 per cent of India's bio-wealth is contributed by fungi and insects. Such a distribution is similar to that found in the tropics and the subtropics. Biogeographically, India is situated at the trijunction of three realms namely afrotropical, Indo-Malayan and Paleo-Arctic realms, and therefore, has characteristic elements from each of them. This assemblage of three distinct realms probably is a fact which is believed to partly account for its rich and unique in biological diversity. Based on the available data, India ranks tenth in the world and fourth in Asia in plant diversity, and ranks tenth in the number of endemic species of higher vertebrates in the world. There are 10 biogeographical zones in India. They can be classified as under:

Biogeographic Classification of India

1. Trans-Himalayas: An extension of the Tibetan plateau, harbouring high-altitude cold desert in Laddakh (J&K) and Lahaul Spiti (HP) comprising 5.7 per cent of the country's landmass.
2. Himalayas: The entire mountain chain running from north-western to northeastern India, comprising a diverse range of biotic provinces and biomes, 7.2 per cent of the country's landmass.
3. Desert: The extremely arid area west of the Aravalli hill range, comprising both the salty desert of Gujarat and the sand desert of Rajasthan. 6.9 per cent of the country's landmass.
4. Semi-arid: The zone between the desert and the Deccan plateau, including the Aravalli hill range. 15.6 per cent of the country's landmass.
5. Western ghats: The hill ranges and plains running along the western coastline, south of the Tapti river, covering an extremely diverse range of biotic provinces and biomes. 5.8 per cent of the country's landmass.
6. Deccan peninsula: The largest of the zones, covering much of the southern and southcentral plateau with a predominantly deciduous vegetation. 4.3 per cent of the country's landmass.
7. Gangetic plain: Defined by the Ganges river system, these plains are relatively homogenous. 11 per cent of the country's landmass.
8. North-east India: The plains and non-Himalayan hill ranges of northeastern India, with a wide variation of vegetation. 5.2 per cent of the country's landmass.
9. Islands: The Andaman and Nicobar Islands in the Bay of Bengal, with a highly diverse set of biomes. 0.03 per cent of the country's landmass.
10. Coasts: A large coastline distributed both to the west and east, with distinct differences between the two; Lakshadeep islands are included in this with the per cent area being negligible.

Apart from the biogeographic classifications described above ecosystems can also be demarcated on the basis of purely geographical or geological features like mountains, islands, valleys, plateaux, oceans; on the basis of vegetative cover like forests, grasslands, mangroves and deserts; on the basis of climatic conditions like arid and semi-arid areas, permanently snow-bound areas, high rainfall areas; on the basis of soil characteristic and other such criteria.

In some descriptions the biomes/ecosystems are clubbed together into very general habitat classifications. The main natural habitat types are:

1. Forests.
2. Grasslands.
3. Wetlands.
4. Mangroves.
5. Coral reefs.
6. Deserts.

Forests

The forest cover of the country is placed at 6,33,397 sq km according to the forest survey of India assessment. This presents 19.27 per cent of India's total geographical areas. India is endowed with diverse forest types ranging from the Tropical wet evergreen forests in North-Eastern to the Tropical thorn forests in the Central and Western India. The forests of the country can be divided into 16 major groups comprising 221 types. The distribution of these groups, and the percentage of total forest area covered by each are given in Table 6.2.

Table 6.2. Forest types—distribution and percentage.

Forest type	Distribution	% of forest area
Tropical forest		
Tropical wet evergreen	North East and South, Andaman and Nicobar island	5.8
Tropical semi evergreen	South and East	2.5
Tropical moist deciduous	Central and East	30.3
Tropical littoral and swamp	Along the coast	0.9
Tropical dry deciduous	West and Central	38.2
Tropical thorn	West and Central	6.7
Tropical dry evergreen	Central and South	0.1
Subtropical forests		
Subtropical broad leaved hill forests	South	0.4
Subtropical pine	Sub-Himalayan tract	5.0
Subtropical dry evergreen	North-East and South	0.2
Temperate forests		
Montane wet temperate	Himalaya and Nilgiris (in Western Ghats)	2.0
Himalayan moist temperate	Temperate areas of Himalayas	3.4
Himalayan dry temperate	Dry temperate areas of Himalayas	0.2
Sub-alpine and alpine forests		
Sub-alpine	Himalaya	4.3
Moist alpine shrub	Himalaya	4.3
Dry alpine shrub	Himalaya	4.3

Grasslands

In India the spread of grassland and shrubland is put at 12 per cent of the total landmass while the planning commission and Grasslands and Fodder research Institute, Jhansi gives an estimate of about

3.7 to 3.9 per cent. The diversity of grasslands in India is high ranging from semi-arid pastures of the western part of the Deccan peninsula, the humid, semi-waterlogged tall grassland of the Terai belt, the rolling shola grasslands of the western ghat hilltops, and the high-altitude alpine pastures of the Himalayas.

The grass flora in India is also quite diverse, consisting of about 1256 species in 245 genera and an estimated 370 endemic species reported. Unfortunately due to greater neglect than Forests the status of grasslands is not so well known or documented.

Wetlands

Wetlands cover 3 per cent of the Indian landmass or nearly 1,00,000 sq km. Wetlands in India harbour a vast variety of life forms that are a part of the complex food of these transitional ecosystems. About 320 species of birds are associated with the Indian Wetlands. Apart from birds, the wetlands support a diverse population of plants and animals including 150 species of amphibians. Wetlands are the habitat of some of the world's endangered and threatened flora and fauna. The Western and Central flock of Siberian crane, one of the most endangered cranes in the world, uses Keoladeo as its winter site. The brown antlered deer (*Cervus eldi eldi*) or 'sangai' is found only in *phumadis* (floating landmasses) of Lok Tak Lake. Gahirmatha beach is a major breeding site of olive ridley turtles. Chilka is the habitat of many threatened species such as green sea turtle, Hawksbill turtle, dugong and blackbuck.

Mangroves

Government of India estimated mangrove cover of 6,74,000 ha, which is about 7 per cent of the world's mangrove.

Mangroves are salt-tolerant ecosystems in tropical and subtropical regions. These ecosystems are largely characterised by assemblage of unrelated tree genera that share the common ability to grow in saline tidal zone. India harbours some of the best mangroves swamps in the world, located in the alluvial deltas of Ganga, Mahanadi, Godavari, Krishna and Cauveri rivers and on the Andaman and Nicobar group of Islands. The total area covered by mangroves in India is estimated at about 6700 sq km. amounting to about 7 per cent of the worlds mangroves.

The largest stretch of mangroves in the country lies in the Sunderbans in West Bengal covering an area of about 4200 sq km. The predominant mangroves species are *Avicennia* officinalis Excoecaria agallocha, Heritiera fomes, Bruguiera parviflora, Ceriops decandra, *Rhizophora mucronata* and *Xylocarpus granatum*. Mangroves also harbour a number of molluscs, polychaetes and honeybees. The Indian mangroves are host to 105 species of fish, 20 kinds of shellfish and 229 crustacean species. The Royal Bengal tiger is found in the Sunderban mangroves. Different species of monkeys, otters, deer, fishing cats, snakes and wild pigs are common. A total of 117 species of migratory and residential birds have been reported. The most common birds are flamingos, storks, sea eagles, kites, kingfishers, sandpipers, bulbuls and whistlers.

Coral reefs

Accurate estimates of coral reef extent in the world are not available. A rough estimate puts it at 6,00,000 sq km out of which 60 per cent occurs in the Indian Ocean region and most of it in south-east Asia.

The coral reef cover in Indian waters is roughly estimated upto 19,000 sq km. Indian reefs belong to the following categories:

PalkBay and Gulf of Mannar	:	Fringing
Gulf of Kachchh	:	Fringing, Patchy

Andaman and Nicobar Islands	:	Fringing
Lakshadeep Islands	:	Atolls
Central West coast	:	Patchy

The diversity of the Indian coral reefs is very impressive with about 200 coral species belonging to 71 genera. The richest being Andaman and Nicobar Islands which alone harbours 179 species.

Deserts

In India, deserts extend over about 2 per cent of the landmass. Three kinds of deserts are noticeable in India:

1. The sand desert of western Rajasthan and neighbouring areas.
2. The vast salt desert of Gujarat.
3. The high-altitude cold desert of Jammu and Kashmir and Himachal Pradesh.

Desert fauna in India is also quite diverse, with about 1200 sp. of animals reported from Thar region of which 440 are vertebrates and 755 are invertebrates. Desert fox, Desert cat, Houbara Bustard and some Sandgrouse species are restricted to the Thar area. In the remote part of Great Rann, Gujarat lies the nesting ground of Flamingoes and the only known population of Asiatic wild ass.

The cold deserts in India cover a vast area of 1,09,990 sq km, about 87,780 sq km in Laddakh (Jammu and Kashmir) and 22,210 sq km in Lahaul-Spiti (Himachal Pradesh). The diversity of the high altitude cold deserts has been studied only recently with many insect species being endemic. Interestingly the cold desert harbours *Kiang* a close relative of the Indian wild ass found the Rann of Kachchh. Other distinctive animals include Snow leopard, Yak, Tibetan antelope, Ibex, Blue sheep, Tibetan gazelle, Woolly hare, etc.

Biodiversity Hotspots

Biodiversity hotspots are areas that are unusually rich in species, most of which are endemic, and are under a constant threat of being overexploited. Among the 18 hotspots in the world, two are found in India. These are two distinct areas: the Eastern Himalayas and the Western Ghats and are also depicted in the National forest vegetation map of India. Together these 18 sites contain approximately 49,955 endemic plant species or 20 per cent of the world's recorded plants species, in only 7,46,400 sq km or 0.5 per cent of the earth's land surface.

Eastern Himalayas

Phytogeographically, the Eastern Himalayas forms a distinct floral region and comprises Nepal, Bhutan, neighbouring states of east and north-east India, and a contiguous sector Yunnan province in south western China. In the whole of Eastern Himalayas, there are an estimated 9000 plant species, with 3500 (i.e. 39 per cent) of them being endemic. In India's sector of the area, there occur some 5800 plant species, roughly 2000 (i.e. 36 per cent) of them being endemic.

At least 55 flowering plants endemic to this area are recognised as rare, for example, the pitcher plant (*Nepenthes khasiana*).

The area has long been recognised as a rich centre of primitive flowering plants and the area is recognised as '*Cradle of Speciation*'. Species of several families of monocotyledons, Orchidaceae, Zingiberaceae and Arecaceae abound in the area. Gymnosperms and pteridophytes (ferns) are also well represented in the area.

The area is also rich in wild relatives of plants of economic significance, e.g. rice banana, citrus, ginger, chilli, jute and sugarcane. The region is regarded as the centre of origin and diversification of five palms of commercial importance namely, coconut, arecanut, palmyra palm, sugar palm and wild date palm.

Tea (*Thea sinensis*) is reported to be in cultivation in this region for the last 40,000 years. Many wild and allied species of tea, the leaves of which are used as substitute of tea, are found growing in the North East in the natural habitats.

The 'taxol' plant *Taxus wallichiana* is sparsely distributed in the region and has come under red data category due to its over exploitation for extraction of a drug effectively used against cancer.

As regards faunal diversity, 63 per cent of the genera of land mammals in India are known from this area. During the last four decades, two new mammals have been discovered from the region: Golden Langur from Assam–Bhutan region, and Namdapha flying squirrel from Arunachal Pradesh indicating the species richness of the region.

The area is also a rich centre of avian diversity—more than 60 per cent of the Indian birds are recorded in the North East. The region also has two endemic genera of lizards, and 35 endemic reptilian species, including two turtle. Of the 204 Indian amphibians, at least 68 species are known from North East, 20 of which are endemic.

From Namdapha National Park itself, a new genus of mammal, a new subspecies of bird, 6 new species of amphibia, four new species of fish, at least 15 new species of beetles and 6 new species of flies have been discovered.

Western ghats

The Western Ghats region is considered as one of the most important biogeographic zones of India, as it is one of the richest centres of endemism. Due to varied topography and micro-climatic regimes, some areas within the region are considered to be active zones of speciation.

The region has 490 arborescent taxa, of which as many as 308 are endemics this endemism of tree species shows a distinct trend, being the highest (43 per cent) in 8N–10°30′ N location and declining to 11 per cent in 16N–16°30′N location.

About 1500 endemic species of dicotyledonous plants are reported from the Western Ghats. 245 species of orchids belonging to 75 genera are found here, of which 112 species in 10 genera are endemic to the region.

As regards the fauna, as many as 315 species of vertebrates belonging to 22 genera are endemic, these include 12 species of mammals, 13 species of birds, 89 species of reptiles, 87 species of amphibians and 104 species of fish.

The extent of endemism is high in amphibian and reptiles. There occur 117 species of amphibians in the region, of which 89 species (i.e. 76 per cent) are endemic. Of the 165 species of reptiles found in Western Ghats, 88 species are endemic. Many of the endemics and other species are listed as threatened. Nearly 235 species of endemic flowering plants are considered endangered. Rare fauna of the region includes: Lion Tailed Macaque, Nilgiri Langur, Nilgiri Tahr, Flying Squirrel and Malabar Gray Hornbill.

Biodiversity Contribution to Indian Economy

Biodiversity products have obtained a commercial value and have been increasingly exchanged in the markets having a monetary value, from which their share in the national economy can be judged. In the Indian context it is difficult to put a value on diversity as such because the marketable products are of various kinds both legal and illegal, e.g. wood and non-wood products from forests where wood comprises the major commercial produce is both legally exported as well as illegally smuggled out of the country. Many non-wood forest produce and the illegal produce is not accounted for in the official documents.

The contribution of natural and agricultural biodiversity in terms of crops, livestock, fisheries, etc. is very substantial in terms of commercial value.

Such biodiversity has a major contribution to make to the Indian GDP (gross domestic product). The large economic implications of biodiversity in its wild and domesticated forms is the rice improvement program. Rice accounts for 22 per cent of the total cropped area and 39 per cent of the total area under cereals, which reflects its importance in the country's struggle to attain self-sufficiency in food. When the rice crop was doomed due to the grassy stunt virus in the 1970s, one single gene from the wild strain of rice, namely *Oryza nivara* from Uttar Pradesh, showed resistance to this virus and proved vital in the fight against the virus.

With respect to the commercial value of the plant species of medicinal value, the world trade is of several billion dollars and this is growing. The export market for medicinal plants has also increased. India's foreign exchange reserves from horticultural products are from high yielding varieties. Increased production of oilseeds also helped in saving large amounts of foreign exchange spent on edible oil import.

The aforesaid pressures will lead to loss of biodiversity in India and will also result in considerable drop in Indian GDP and foreign exchange earnings from horticultural products, oil seeds, oil meal and oil cake will drop down to a great extent.

RESPONSE

The Ministry of Environment and Forests (MoEF) is the nodal agency in the Government of India for planning, promotion, coordination and overseeing the implementation of the environmental and forestry programs. The MoEF is also the focal point for implementation of the Convention on Biological Diversity. The mandates of the Ministry *inter alia* include survey of flora, fauna, forests and wildlife, and conservation of natural resources (Fig. 6.2). These objectives are supported by legislative and regulatory measures. A number of institutions affiliated with the Ministry are involved in the work related to various aspects of biological diversity. Survey and inventorisation of the floral and faunal resources are carried out by the Botanical Survey of India (BSI) established in 1890, and the Zoological Survey of India (ZSI) established in 1916. The Forest Survey of India established in 1981 assesses the forest cover, with a view to develop an accurate database for planning and monitoring purposes. The Wildlife Institute of India undertakes studies of endangered species of animals and critical ecosystems. Over 47,000 species of plants and 89,000 animals species have been recorded by the BSI and ZSI respectively.

The Survey organisations have published over the years, documents on flora and fauna at country, state and in some cases district levels and for selected ecosystems. Besides, extensive reports on inventories of resources indicating level of biodiversity in selected areas have also been brought out. The Surveys have also published Red Data Books on endangered species (Fig. 6.3). The voucher specimens are preserved in Central National Herbarium (CNH) of BSI and National Zoological Collection (NZC) of ZSI.

The Forest Survey of India publishes every three years, a State of Forest in India report based on remote sensing and ground truth data.

Fig. 6.2. Black buck: needs conservation.

Fig. 6.3. Pitcher plant: an endangered species.

Existing Policy Response

In situ conservation (within natural habitat)

Some important measures taken are as follows:

1. Approximately 4.2 per cent of the total geographical area of the country has been earmarked for extensive *in situ* conservation of habitats and ecosystems. A protected area network of 85 National Parks and 448 Wildlife Sanctuaries have been created. The results of this network have been

significant in restoring viable population of large mammals such as tiger, lion, rhinoceros, crocodiles, elephants, etc.

2. The Indian Council of Forestry Research and Education (ICFRE) has identified 309 forest preservation plots of representative forest types for conservation of viable and representative areas of biodiversity. 187 of these plots are in natural forests and 112 in plantations, covering a total area of 8500 hectares.

3. A program entitled 'Eco-development' for *in situ* conservation of biological diversity involving local communities has been initiated in recent years. The concept of eco-development integrates the ecological and economic parameters for sustained conservation of ecosystems by involving the local communities with the maintenance of earmarked regions surrounding protected areas. The economic needs of the local communities are taken care of under this programs through provision of alternative sources of income and a steady availability of forest and related produce.

4. To conserve the respective ecosystems, a Biosphere Reserve Program is being implemented. Twelve biodiversity rich areas of the country have been designated as Biosphere Reserves (Table 6.3) applying the diversity and genetic integrity of plants, animals and micro-organisms in their totality as part of the natural ecosystems, so as to ensure their self-perpetuation and unhindered evolution of the living resources.

Table 6.3. Biosphere reserves set up.

Name of the site	Date of notification	Location (State)
Nilgiri	01-08-86	Part of Wynad, Nagarhole, Bandipur and Madumalai, Nilambur, Silent Valley and Siruvani hills (Tamil Nadu)
Nanda Devi	18-01-88	Part of Chamoli, Pithoragarh, Almora districts (Uttarakhand)
Nokrerk	01-09-88	Part of Gora Hills (Meghalaya)
Manas	14-03-89	Part of Kokrajhar, Bongaigaon, Barpeta, Nalbari, Kamprup and Darang district (Assam)
Sunderbans	29-03-89	Part of delta of Ganga and Brahmaputra river system (West Bengal)
Gulf of Mannar	18-02-89	Indian part of Gulf of Mannar between India and Sri Lanka (Tamil Nadu)
Great Nicobar	06-01-89	Southern most islands of Andaman and Nicobar (A&N islands)
Similpal	21-06-94	Part of Mayurbhanj district (Orissa)
Dibru-Saikhowa	28-07-97	Part of Dibrugarh and Tinsukia district (Assam)
Dehang Debang	02-09-98	Part of Siang and Debang valley in Arunachal Pradesh
Pachmarhi	03-03-99	Parts of Betul, Hoshangabad and Chhindwara districts of Madhya Pradesh
Kanchanjanga	07-02-2000	Part of Kanchanjanga Hills and Sikkim

5. Programs have also been launched for scientific management and wise use of fragile ecosystem. Specific programs for management and conservation of wetlands, mangroves and coral reef systems are also being implemented. 21 wetlands, 15 mangrove areas and 4 coral reef areas have been identified for management. National and sub-national level committees oversee and guide these programs to ensure strong policy and strategic support.

6. Six internationally significant wetlands of India have been declared as 'Ramsar Sites' under the Ramsar Convention. To focus attention on urban wetlands threatened by pollution and other anthropogenic activities, State Governments were requested to identify lakes that could be include the National Lake Conservation Plan. The activities of the NLCP include formulation of

perspective plans for conservation based on resource survey using remote sensing technology and GIS studies on biodiversity and related ecological matters, prevention of pollution from point and non-point sources, treatment of catchment, desilting and weed control.

7. Wild Life Protection Act is in the final stage of revision and provisions have been made for conservation reserves and community reserves to allow restrictive use to make it more people oriented. Presently Biodiversity Act which is in the final stage, has got the component of National Biodiversity Authority to control access to genetic resources from international community. There will also be State Biodiversity Boards to control access to domestic consumers (Table 6.4).

Table 6.4. World heritage sites.

Site	Location
Kaziranga Nation Park	Assam
Keoladeo Ghana National Park	Rajasthan
Manas Wildlife Sanctuary	Assam
Nanda Devi National Park	Uttarakhand
Sundarbans National Park	West Bengal

8. Under the World Heritage Convention, five natural sites have been declared as 'World Heritage Sites'.

9. The Tura Range in Gora Hills of Meghalaya is a gene sanctuary for preserving the rich native diversity of wild Citrus and Musa species.

10. Sanctuaries for rhododendrons and orchids have been established in Sikkim.

11. Large mammal species targeted protection based on the perception of threat to them have been under implementation.

12. Project Tiger: A potential example of an highly endangered species is the Indian Tiger (*Panthera tigris*). The fall and rise in the number of Tigers in India is an index of the extent and nature of conservation efforts. It is estimated that India had about 40,000 tigers in 1900, and the number declined to a mere about 1800 in 1972. Hence, Project Tiger was launched in 1973 with the following objectives:

(a) To ensure maintenance of available population of Tigers in India for scientific, economic, aesthetic, cultural and ecological value.

(b) To preserve, for all times, the areas of such biological importance as a national heritage for the benefit, education and enjoyment of the people.

(c) At present there are 25 Tiger Reserves spreading over in 14 states and covering an area of about 33,875 sq km and the Tiger population has more than doubled now due to a total ban on hunting and trading tiger products at national and international levels and the implementation of habitat improvement and anti-poaching measures.

13. Project Elephant was launched in 1991–92 to assist States having free ranging population of wild elephants to ensure long-term survival of identified viable populations of elephants in their natural habitats (Fig. 6.4). Major activities of Project Elephant are:

(a) Ecological restoration of existing natural habitats and migratory routes of elephants.

(b) Development of scientific and planned management for conservation of elephants habitats and value population of wild Asiatic elephants in India.

(c) Promotion of measures for mitigation of man-elephant conflict in crucial habitats and moderating pressures of human and domestic stock activities in crucial elephant habitats.

(d) Strengthening of measures for protection of wild elephants from poachers and unnatural cause of death.

(e) Research on Project Elephant management related issues.

(f) Public education and awareness programs.

(g) Eco-development.

(h) Veterinary care.

14. Rhinos have been given special attention in selected sanctuaries and national parks in the North East and North-west India. All these programs, though focused on a single species, have a wider impact as they conserve habitats and a variety of other species in those habitats.

15. The Ministry of Environment and Forests constituted the National Afforestation and Eco-development Board (NAEB) in August 1992. National Afforestation and Eco-development Board has evolved specific schemes for promoting afforestation and management strategies, which help the states in developing specific afforestation and management strategies and eco-development packages for augmenting biomass production through a participatory planning process of Joint Forest Management and microplanning.

Fig. 6.4. Herd of elephants of North-east India.

Ex situ conservation (outside natural habitats)

To complement *in situ* conservation, attention has been paid to *ex situ* conservation measures. According to currently available survey, Central Government and State Government together run and manage 33 Botanical Gardens. Universities have their own botanic gardens. There are 275 zoos, deer parks, safari parks, aquaria, etc. A Central Zoo Authority was set up to secure better management of zoos. A scheme entitled Assistance to Botanic Gardens provides one-time assistance to botanic gardens to strengthen and institute measure for *ex situ* conservation of threatened and endangered species in their respective regions.

Recent conservation initiatives

Several recent initiatives of the Indian Government have focused on wetland, mangroves and coral reef management. In 1998–99, an amount of Rs. 140 lakhs were released to the State Governments for the preparation of management action plans for Pongdam in Himachal Pradesh, Wullar in Kashmir, Loktak in Manipur, Rudrasagar in Tripura and Kolleru in Andhra Pradesh. Additionally, one more wetland has been identified for conservation, i.e. Rudrasagar from Tripura, thus increasing the list to 20 wetlands for intensive conservation in the country. Additionally, a wetland strategy has been drafted.

The National Committee on Conservation and Management of Mangroves and Coral Reefs in September 1998 recommended the establishment of an Indian Coral Reef Monitoring Network to develop Action Plans for important coral reefs of the country. Preparation of these plans is already underway. Moreover, financial assistance from UNDP/GEF has led to a PDF-B project on strengthening the Gulf of Mannar Biosphere Reserve. The ZSI has initiated another UNDP/GEF project relating to management of Andaman's coral reefs.

Policy Gaps

1. Lack of policies for protection of wetlands, grasslands, sacred groves and other areas significant from the point of view of biodiversity.
2. Lacunae in economic policy, institutional and governance system.
3. Inadequate enforcement of existing laws.
4. Poor implementation of wildlife protection act 1972 as amended in 1991.
5. Inadequate implementation of eco-development programs.
6. Need for enhanced role of NGOs and other institutions.
7. Need for political commitment and goodwill.
8. Need for providing Institutional Structure.
9. Need for more sectoral financial outlay.
10. Human resource development—limited local community participation.

Knowledge/Information/Data

1. Documentation of biodiversity is an urgent requirement as latest statistics and data on floral and faunal biodiversity of India has not been compiled and documented.
2. The information and data should be made available to the scientific and socioeconomic agencies to support the evaluation/revision of the policies.
3. Lack of knowledge of the magnitude, patterns, causes and rates of deforestation and biodiversity laws at the ecosystem and landscape level.
4. Information on poaching trade and trade routes is sketchy and current wildlife protection and law enforcement measures are inadequate and inefficient procedure.
5. Biodiversity Act/Bill should not override the provisions of Wildlife Protection Act.

Policy Recommendations

1. Most of the legal provisions pertain mainly to use/exploitation of biological resources, rather than their conservation. Even Wild Life Protection Act 1972, focuses on protection rather than conservation. Protection under Wild Life Protection Act is largely directed towards large animal species (charismatic terrestrial species) rather than the large spectrum of fauna and flora also found in the marine realm.
2. Hence the existing laws relating to biodiversity shall be examined in order to bring them in tune with the provisions of convention to reflect current understanding of biodiversity conservation.
3. Need for comprehensive legislation on biodiversity conservation and use especially fisheries policies, which is generally ignored.

4. Formulation of policies for protection of wetlands, grasslands, sacred groves, marine flora and fauna and other areas significant from the point of view of biodiversity.
5. Improving policy environment.
6. Passage of biodiversity bill.
7. A presence of a biodiversity cell in all development departments impinging on land and water.
8. Documentation of biodiversity.
9. Increase allocation of financial resources for conservation of biodiversity.
10. Integrating conservation with development.
11. Incentives and disincentives for improper use of biodiversity.
12. Biodiversity Act/Bill should not override the provisions of Wildlife Protection act.
13. There should be continuous monitoring of biodiversity use for review of results of implementation of policies and programs.

Chapter 7

Species Diversity

INTRODUCTION

Species diversity is the effective number of different species that are represented in a collection of individuals (a dataset). The effective number of species refers to the number of equally-abundant species needed to obtain the same mean proportional species abundance as that observed in the dataset of interest (where all species may not be equally abundant). Species diversity consists of two components, species richness and species evenness. Species richness is a simple count of species, whereas species evenness quantifies how equal the abundances of the species are.

CALCULATION OF DIVERSITY

Species diversity in a dataset can be calculated by first taking the weighted average of species proportional abundances in the dataset, and then taking the inverse of this. The equation is:

$$^{q}D = \frac{1}{\sqrt[q-1]{\sum_{i=1}^{S} p_i p_i^{q-1}}}$$

The denominator equals mean proportional species abundance in the dataset as calculated with the weighted generalised mean with exponent $q - 1$. In the equation, S is the total number of species (species richness) in the dataset, and the proportional abundance of the ith species is p_i. The proportional abundances themselves are used as weights. The equation is often written in the equivalent form:

$$^{q}D = \left(\sum_{i=1}^{S} p_i^{q} \right)^{1/(1-q)}$$

The value of q defines which kind of mean is used. $q = 0$ corresponds to the harmonic mean, and gives values equal to species richness because the p_i values cancel out. $q = 1$ corresponds to the geometric mean and $q = 2$ to the arithmetic mean. As q approaches infinity, the generalised mean approaches the maximum p_i value. In practice, q modifies species weighting, such that increasing q increases the weight given to the most abundant species, and fewer equally-abundant species are hence needed to reach mean proportional abundance. Consequently, large values of q lead to smaller species diversity than small values of q for the same dataset. If all species are equally abundant in the dataset, changing the value of q has no effect, but species diversity at any value of q equals species richness.

Negative values of q are not used, because then the effective number of species (diversity) would exceed the actual number of species (richness). As q approaches negative infinity, the generalised mean approaches the minimum p_i value. In many real datasets, the least abundant species is represented by a single individual, and then the effective number of species would equal the number of individuals in the dataset. The same equation can be used to calculate the diversity in relation to any classification, not only species. If the individuals are classified into genera or functional types, p_i represents the proportional abundance of the ith genus or functional type, and ^qD equals genus diversity or functional type diversity, respectively.

DIVERSITY INDICES

Often researchers have used the values given by one or more diversity indices to quantify species diversity. Such indices include species richness, the Shannon index, the Simpson index and the complement of the Simpson index (also known as the Gini-Simpson index).

When interpreted in ecological terms, each one of these indices corresponds to a different thing, and their values are therefore not directly comparable. Species richness quantifies the actual rather than effective number of species. The Shannon index equals $\log(^qD)$, and in practice quantifies the uncertainty in the species identity of an individual that is taken at random from the dataset. The Simpson index equals $1/^qD$ and quantifies the probability that two individuals taken at random from the dataset (with replacement of the first individual before taking the second) represent the same species. The Gini-Simpson index equals $1 - 1/^qD$ and quantifies the probability that the two randomly taken individuals represent different species.

SAMPLING CONSIDERATIONS

Depending on the purposes of quantifying species diversity, the dataset used for the calculations can be obtained in different ways. Although species diversity can be calculated for any dataset where individuals have been identified to species, meaningful ecological interpretations require that the dataset is appropriate for the questions at hand. In practice, the interest is usually in the species diversity of areas so large that not all individuals in them can be observed and identified to species, but a sample of the relevant individuals has to be obtained. Extrapolation from the sample to the underlying population of interest is not straightforward, because the species diversity of the available sample generally gives an underestimation of the species diversity in the entire population. Applying different sampling methods will lead to different sets of individuals being observed for the same area of interest, and the species diversity of each set may be different. When a new individual is added to a dataset, it may introduce a species that was not yet represented. How much this increases species diversity depends on the value of q: when $q = 0$, each new actual species causes species diversity to increase by one effective species, but when q is large, adding a rare species to a dataset has little effect on its species diversity.

In general, sets with many individuals can be expected to have higher species diversity than sets with fewer individuals. When species diversity values are compared among sets, sampling efforts need to be standardised in an appropriate way for the comparisons to yield ecologically meaningful results. Resampling methods can be used to bring samples of different sizes to a common footing. Species accumulation curves and the number of species only represented by one or a few individuals can be used to help in estimating how representative the available sample is of the population from which it was drawn.

TRENDS IN SPECIES DIVERSITY

The observed species diversity is affected not only by the number of individuals but also by the heterogeneity of the sample. If individuals are drawn from different environmental conditions (or different habitats), the species diversity of the resulting set can be expected to be higher than if all individuals are drawn from a similar environment. Increasing the area sampled increases observed species diversity both because more individuals get included in the sample and because large areas are environmentally more heterogeneous than small areas.

SPECIES RICHNESS

Species richness is the number of different species represented in a set or collection of individuals. Species richness is simply a count of species, and it does not take into account the abundances of the species or their relative abundance distributions. In contrast, species diversity takes into account both species richness and species evenness.

Sampling Considerations

Depending on the purposes of quantifying species richness, the individuals can be selected in different ways. They can be, for example, trees found in an inventory plot, birds observed from a monitoring point, or beetles collected in a pitfall trap. Once the set of individuals has been defined, its species richness can be exactly quantified, provided the species-level taxonomy of the organisms of interest is well enough known. Applying different species delimitations will lead to different species richness values for the same set of individuals.

In practice, people are usually interested in the species richness of areas so large that not all individuals in them can be observed and identified to species. Then applying different sampling methods will lead to different sets of individuals being observed for the same area of interest, and the species richness of each set may be different. When a new individual is added to a set, it may introduce a species that was not yet represented in the set, and thereby increase the species richness of the set. For this reason, sets with many individuals can be expected to contain more species than sets with fewer individuals.

If species richness of the obtained sample is taken to represent species richness of the underlying habitat or other larger unit, values are only comparable if sampling efforts are standardised in an appropriate way. Resampling methods can be used to bring samples of different sizes to a common footing. Properties of the sample, especially the number of species only represented by one or a few individuals, can be used to help estimating the species richness in the population from which the sample was drawn.

Trends in Species Richness

The observed species richness is affected not only by the number of individuals but also by the heterogeneity of the sample. If individuals are drawn from different environmental conditions (or different habitats), the species richness of the resulting set can be expected to be higher than if all individuals are drawn from similar environments. The accumulation of new species with increasing sampling effort can be visualised with a species accumulation curve. Such curves can be constructed in different ways. Increasing the area sampled increases observed species richness both because more individuals get included in the sample and because large areas are environmentally more heterogeneous than small areas. Many organism groups have most species in the tropics, which leads to latitudinal gradients in species richness. There has been much discussion about the relationship between productivity and species richness. Results have varied among studies, such that no global consensus on either the pattern or its possible causes has emerged.

Applications

Species richness is often used as a criterion when assessing the relative conservation values of habitats or landscapes. However, species richness is blind to the identity of the species. An area with many endemic or rare species is generally considered to have higher conservation value than another area where species richness is similar, but all the species are common and widespread.

SPECIES EVENNESS

Species evenness refers to how close in numbers each species in an environment are. Mathematically it is defined as a diversity index, a measure of biodiversity which quantifies how equal the community is numerically. So if there are 40 foxes, and 1000 dogs, the community is not very even. But if there are 40 foxes and 42 dogs, the community is quite even. The evenness of a community can be represented by Pielou's evenness index:

$$J' = \frac{H'}{H'_{max}}$$

where, H' is the number derived from the Shannon diversity index and H'_{max} is the maximum value of H', equal to:

$$H_{max} = -\sum_{i-1}^{S} \frac{1}{S} \ln \frac{1}{S} = \ln S.$$

E is constrained between 0 and 1. The less variation in communities between the species, the higher E is. Other indices have been proposed by authors where $H'_{min} > 0$, e.g. Hurlburt's evenness index. S is the total number of species.

TYPES OF DIVERSITY

Alpha Diversity

The term alpha diversity (α-diversity) was introduced by R.H. Whittaker together with the terms beta diversity (β-diversity) and gamma diversity (γ-diversity). Whittaker's idea was that the total species diversity in a landscape (gamma diversity) is determined by two different things, the mean species diversity in sites or habitats at a more local scale (alpha diversity) and the differentiation among those habitats (beta diversity).

Scale considerations

Both the area or landscape of interest and the sites or habitats within it may be of very different sizes in different situations, and no consensus has been reached on what spatial scales are appropriate to quantify alpha diversity. It has therefore been proposed that the definition of alpha diversity does not need to be tied to a specific spatial scale: alpha diversity can be measured for an existing dataset that consists of subunits at any scale. The subunits can be, for example, sampling units that were already used in the field when carrying out the inventory or grid cells that are delimited just for the purpose of analysis. If results are extrapolated beyond the actual observations, it needs to be taken into account that the species diversity in the subunits generally gives an underestimation of the species diversity in larger areas.

Different alpha diversity concepts

Ecologists have used several slightly different definitions of alpha diversity. Whittaker himself used the term both for the species diversity in a single subunit and for the mean species diversity in a collection of subunits. It has been argued that defining alpha diversity as a mean across all relevant subunits is preferable, because it agrees better with Whittaker's idea that total species diversity consists of alpha and beta components.

Definitions of alpha diversity can also differ in what they assume diversity to be. Often researchers use the values given by one or more diversity indices, such as species richness, the Shannon index or the Simpson index. However, it has been argued that it would be better to use the effective number of species as the universal measure of species diversity. This measure allows weighting rare and abundant species in different ways, just as the diversity indices collectively do, but its meaning is intuitively easier to understand. The effective number of species is the number of equally-abundant species needed to obtain the same mean proportional species abundance as that observed in the dataset of interest (where all species may not be equally abundant).

Calculating alpha diversity

Suppose species diversity is equated with the effective number of species, and alpha diversity with the mean species diversity per subunit. Then alpha diversity can be calculated in two different ways that give the same result. The first approach is to calculate a weighted mean of the within-subunit species proportional abundances, and then take the inverse of this mean. The second approach is to calculate the species diversity for each subunit separately, and then take a weighted mean of these.

If the first approach is used, the equation is:

$$^qD_\alpha = \frac{1}{\sqrt[q-1]{\sum_{j=1}^{N}\sum_{i=1}^{S}p_i p_{i/j}^{q-1}}}$$

The denominator equals mean proportional species abundance within the subunits as calculated with the weighted generalised mean with exponent $q - 1$. In the equation, N is the total number of subunits and S is the total number of species (species richness) in the dataset. The proportional abundance of the ith species in the jth subunit is $p_{i/j}$. These proportional abundances are weighted by the proportion of data that each contributes to the dataset, which equals p_{ij}.

If the second approach is used, the equation is:

$$^qD_\alpha = \sqrt[1-q]{\sum_{j=1}^{N}w_j (^qD_{\alpha j})^{1-q}}$$

This also equals a weighted generalised mean but with exponent $1 - q$. Here a mean is taken of the $^qD_{\alpha j}$ values, which represent the effective species density (species diversity per subunit) in the jth subunit. Here the weight of subunit j is w_j, which equals the proportion of data that the subunit contributes to the dataset. Large values of q lead to smaller alpha diversity than small values of q, because increasing q increases the weight given to those species with the highest proportional abundance and to those subunits with the lowest species diversity.

Beta Diversity

The term beta diversity (β-diversity) was introduced by R.H. Whittaker together with the terms alpha diversity (α-diversity) and gamma diversity (γ-diversity). The idea was that the total species diversity in a landscape (γ) is determined by two different things, the mean species diversity at the habitat level (α) and the differentiation among habitats (β). Whittaker proposed several ways of quantifying differentiation, and subsequent generations of ecologists have invented more. As a result, the definition of beta diversity has become quite contentious.

Some use beta diversity as a broad umbrella term that can refer to any of several indices related to compositional heterogeneity. Others argue that such broad usage should be avoided because it leads to confusion, and that when beta diversity indices correspond to different phenomena, they should also be called by different names.

Beta diversity in the strict sense

Gamma diversity and alpha diversity can be calculated directly from species inventory data. The simplest of Whittaker's original definitions of beta diversity is:

$$\beta = \gamma/\alpha$$

Here gamma diversity is the total species diversity of a landscape, and alpha diversity is the mean species diversity per habitat. Because the limits among habitats and landscapes are diffuse and to some degree subjective, it has been proposed that gamma diversity can be quantified for any inventory dataset, and that alpha and beta diversity can be quantified whenever the dataset is divided into subunits. Then gamma diversity is the total species diversity in the dataset and alpha diversity the mean species diversity per subunit. Beta diversity quantifies how many subunits there would be if the total species diversity of the dataset and the mean species diversity per subunit remained the same, but the subunits shared no species.

Other meanings of beta diversity

Absolute species turnover

Some researchers have preferred to partition gamma diversity into additive rather than multiplicative components. Then beta diversity becomes

$$\beta_A = \gamma - \alpha$$

This quantifies how much more species diversity the entire dataset contains than an average subunit within the dataset. This can also be interpreted as the total amount of species turnover among the subunits in the dataset.

When there are two subunits, and presence-absence data are used, this can be calculated with the following equation:

$$\beta_A = (S_1 - c) + (S_2 - c)$$

where, S_1 = the total number of species recorded in the first community, S_2 = the total number of species recorded in the second community, and c = the number of species common to both communities.

Whittaker's species turnover

If absolute species turnover is divided by alpha diversity, a measure is obtained that quantifies how many times the species composition changes completely among the subunits of the dataset. This measure was proposed by Whittaker, so it has been called Whittaker's species turnover.

It is calculated as:
$$\beta_W = (\gamma - \alpha)/\alpha = \gamma/\alpha - 1$$

When there are two subunits, and presence-absence data are used, this equals the one-complement of the Sørensen similarity index.

Proportional species turnover

If absolute species turnover is divided by gamma diversity, a measure is obtained that quantifies what proportion of the species diversity in the dataset is not contained in an average subunit. It is calculated as:
$$\beta_P = (\gamma - \alpha)/\gamma = 1 - \alpha/\gamma$$

When there are two subunits, and presence-absence data are used, this measure as ranged to the interval [0, 1] equals the one-complement of the Jaccard similarity index.

Gamma Diversity

The term gamma diversity (γ-diversity) was introduced by R.H. Whittaker together with the terms alpha diversity (α-diversity) and beta diversity (β-diversity). Whittaker's idea was that the total species diversity in a landscape (γ) is determined by two different things, the mean species diversity in sites or habitats at a more local scale (α) and the differentiation among those habitats (β). According to this reasoning, alpha diversity and beta diversity constitute independent components of gamma diversity:
$$\gamma = \alpha \times \beta$$

Scale considerations

The area or landscape of interest may be of very different sizes in different situations, and no consensus has been reached on what spatial scales are appropriate to quantify gamma diversity. It has therefore been proposed that the definition of gamma diversity does not need to be tied to a specific spatial scale, but gamma diversity can be measured for an existing dataset at any scale of interest. If results are extrapolated beyond the actual observations, it needs to be taken into account that the species diversity in the dataset generally gives an underestimation of the species diversity in a larger area. The smaller the available sample in relation to the area of interest, the more species that actually exist in the area are not found in the sample. The degree of underestimation can be estimated from a species-area curve.

Different gamma diversity concepts

Researchers have used different ways to define diversity, which in practice has led to different definitions of gamma diversity as well. Often researchers use the values given by one or more diversity indices, such as species richness, the Shannon index or the Simpson index. However, it has been argued that it would be better to use the effective number of species as the universal measure of species diversity. This measure allows weighting rare and abundant species in different ways, just as the diversity indices collectively do, but its meaning is intuitively easier to understand. The effective number of species is the number of equally-abundant species needed to obtain the same mean proportional species abundance as that observed in the dataset of interest (where all species may not be equally abundant).

Calculating gamma diversity

Suppose species diversity is equated with the effective number of species in a dataset. Then gamma diversity can be calculated by first taking the weighted mean of species proportional abundances in the dataset, and then taking the inverse of this mean.

The equation is:

$$^{q}D_{\gamma} = \frac{1}{\sqrt[q-1]{\sum_{i=1}^{S} p_i p_i^{q-1}}}$$

The denominator equals mean proportional species abundance in the dataset as calculated with the weighted generalised mean with exponent $q - 1$. In the equation, S is the total number of species (species richness) in the dataset, and the proportional abundance of the ith species is p_i.

Large values of q lead to smaller gamma diversity than small values of q, because increasing q increases the weight given to those species with the highest proportional abundance, and fewer equally-abundant species are hence needed to obtain this proportional abundance.

RELATIVE SPECIES ABUNDANCE

Relative species abundance is a component of biodiversity and refers to how common or rare a species is relative to other species in a defined location or community. Relative species abundances tend to conform to specific patterns that are among the best-known and most-studied patterns in macroecology.

Relative species abundance and species richness describe key elements of biodiversity. Relative species abundance refers to how common or rare a species is relative to other species in a given location or community.

Usually relative species abundances are described for a single trophic level. Because such species occupy the same trophic level they will potentially or actually compete for similar resources. For example, relative species abundances might describe all terrestrial birds in a forest community or all planktonic copepods in a particular marine environment.

Relative species abundances follow very similar patterns over a wide range of ecological communities. When plotted as a histogram of the number of species represented by 1, 2, 3, …,n individuals usually fit a hollow curve, such that most species are rare (represented by a single individual in a community sample), and relatively few species are abundant (represented by a large number of individuals in a community sample). This pattern has been long-recognised and can be broadly summarised with the statement that 'most species are rare'. For example, Charles Darwin noted in 1859 in the Origin of Species that '…rarity is the attribute of vast numbers of species in all classes…'.

Species abundance patterns can be best visualised in the form of relative abundance distribution plots. The consistency of relative species abundance patterns suggests that some common macroecological 'rule' or process determines the distribution of individuals among species within a trophic level.

Distribution Plots

Relative species abundance distributions are usually graphed as frequency histograms ('Preston Plots'; Fig. 7.1) or rank-abundance diagrams ('Whittaker plots'; Fig. 7.2).

Frequency histogram (Preston plot):
1. x-axis: Logarithm of abundance bins [usually $\log_2$ (because this was historically a simple way to approximate the natural log)].
2. y-axis: Number of species at given abundance.

Rank-abundance diagram (Whittaker plot):
1. x-axis: Species list, ranked in order of descending abundance (i.e. from common to rare).
2. y-axis: Logarithm of % relative abundance.

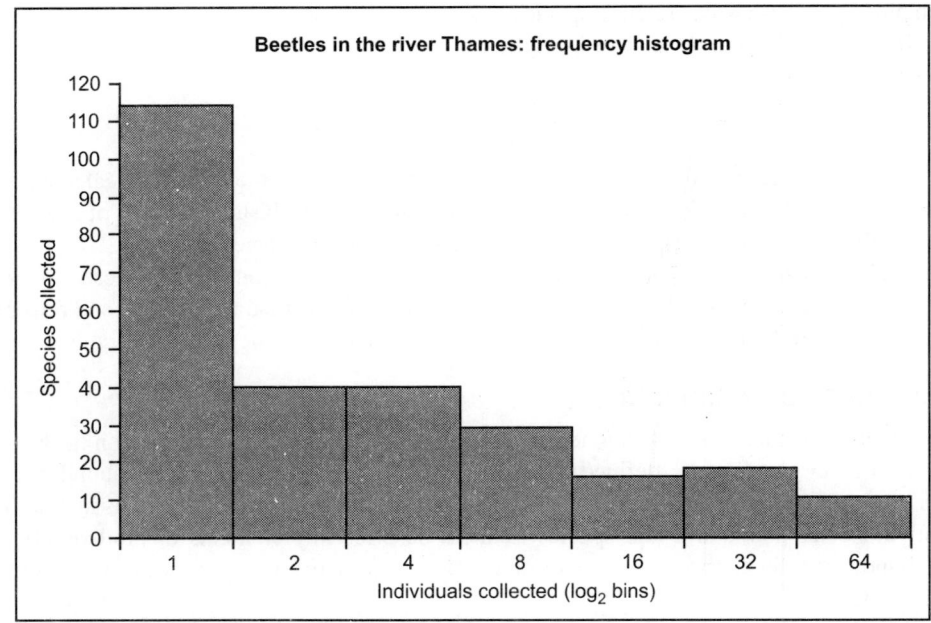

Fig. 7.1. Preston plot of beetles sampled from the river Thames showing a strong right-skew.

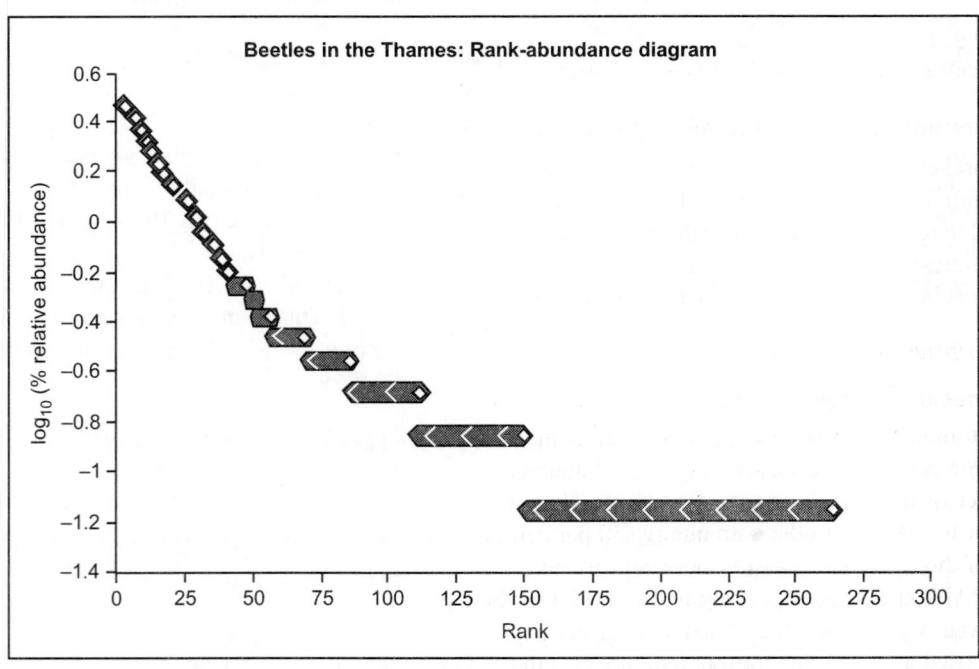

Fig. 7.2. Whittaker plot of beetles sampled from the river Thames showing a slight 's'-shape.

When plotted in these ways, relative species abundances from wildly different data sets show similar patterns: frequency histograms tend to be right-skewed (e.g. Fig. 7.1) and rank-abundance diagrams tend to conform to the curves illustrated in Fig. 7.3.

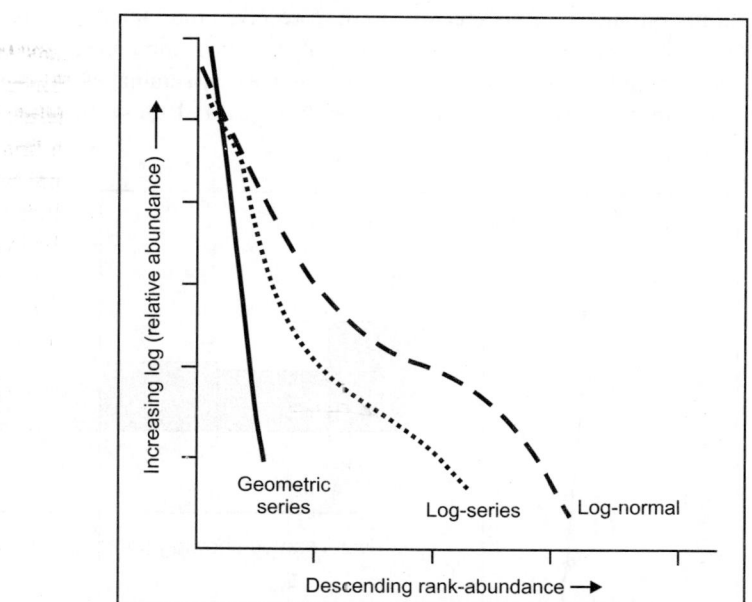

Fig. 7.3. Generic Rank-abundance diagram of three common mathematical models used to fit species abundance distributions: Motomura's geometric series, Fisher's logseries and Preston's log-normal series.

Understanding Relative Species Abundance Patterns

Researchers attempting to understand relative species abundance patterns usually approach them in a descriptive or mechanistic way. Using a descriptive approach biologists attempt to fit a mathematical model to real data sets and infer the underlying biological principles at work from the model parameters. By contrast, mechanistic approaches create a mathematical model based on biological principles and then test how well these models fit real data sets.

Descriptive approaches

Geometric series (Motomura 1932)

I. Motomura developed the geometric series model based on benthic community data in a lake. Within the geometric series each species' level of abundance is a sequential, constant proportion (k) of the total number of individuals in the community. Thus if k is 0.5, the most common species would represent half of individuals in the community (50 per cent), the second most common species would represent half of the remaining half (25 per cent), the third, half of the remaining quarter (12.5 per cent) and so forth. Although Motomura originally developed the model as a statistical (descriptive) means to plot observed abundances, the 'discovery' of his paper by Western researchers in 1965 led to the model being used as a niche apportionment model—the 'niche-preemption model'. In a mechanistic model k represents the proportion of the resource base acquired by a given species.

The geometric series rank-abundance diagram is linear with a slope of $-k$, and reflects a rapid decrease in species abundances by rank (Fig. 7.3). The geometric series does not explicitly assume that species colonise an area sequentially, however, the model fits the concept of niche preemption, where species sequentially colonise a region and the first species to arrive receives the majority of resources. The geometric series model fits observed species abundances in highly uneven communities with low diversity. This is expected to occur in terrestrial plant communities (as these assemblages often show strong dominance) as well as communities at early successional stages and those in harsh or isolated environments (Fig. 7.4).

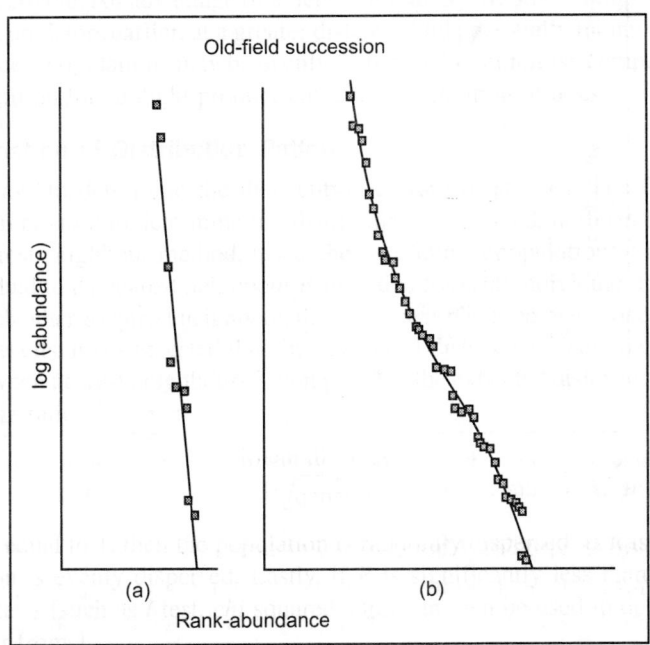

Fig. 7.4. Plant succession in abandoned fields within Brookhaven National Laboratory, NY. Species abundances conform to the geometric series during early succession but approach lognormal as the community ages.

Logseries (Fisher 1943)

$$S = \alpha \ln\left(1 + \frac{N}{\alpha}\right)$$

where,

  S = the number of species in the sampled community.

  N = the number of individuals sampled.

  α = a constant derived from the sample data set.

The logseries was developed by Ronald Fisher to fit two different abundance data sets: British moth species (collected by Carrington Williams) and Malaya butterflies (collected by Alexander Steven Corbet). The logic behind the derivation of the logseries is varied however Fisher proposed that sampled species abundances would follow a negative binomial from which the zero abundance class (species too rare to be sampled) was eliminated. He also assumed that the total number of species in a community was infinite. Together, this produced the logseries distribution (Fig. 7.3). The logseries predicts the number of species at different levels of abundance (n individuals) with the formula:

$$S_n = \frac{\alpha x^n}{n}$$

where,

S = the number of species with an abundance of n.

x = a positive constant ($0 < x < 1$) which is derived from the sample data set and generally approaches 1 in value.

The number of species with 1, 2, 3,..., n individuals are therefore:

$$\alpha, \frac{\alpha x^2}{2}, \frac{\alpha x^3}{3}, ..., \frac{\alpha x^n}{n}$$

Fisher's constants

The constants α and x can be estimated through iteration from a given species data set using the values S and N. Fisher's dimensionless α is often used as a measure of biodiversity, and indeed has recently been found to represent the fundamental biodiversity parameter, θ, from neutral theory.

Lognormal

Using several data sets (including breeding bird surveys from New York and Pennsylvania and moth collections from Maine, Alberta and Saskatchewan) Frank W. Preston argued that species abundances follow a Normal (Gaussian) distribution, partly as a result of the Central Limit Theorem (Fig. 7.3). According to his argument, the right-skew observed in species abundance frequency histograms (including those described by Fisher) was, in fact, a sampling artifact. Given that species toward the left side of the x-axis are increasingly rare, they may be missed in a random species sample. As the sample size increases however, the likelihood of collecting rare species in a way that accurately represents their abundance also increases, and more of the normal distribution becomes visible. The point at which rare species cease to be sampled has been termed Preston's veil line. As the sample size increases Preston's veil is pushed farther to the left and more of the normal curve becomes visible. Interestingly, Williams' moth data, originally used by Fisher to develop the logseries distribution, became increasingly lognormal as more years of sampling were completed.

Calculating theoretical species richness

Preston's theory has an interesting application: if a community is truly lognormal yet under-sampled, the lognormal distribution can be used to estimate the true species richness of a community. Assuming the shape of the total distribution can be confidently predicted from the collected data, the normal curve can be fit via statistical software or by completing the Gaussian formula:

$$n = n_0 e^{-(\alpha R)^2}$$

where,

n_0 is the number of species in the modal bin (the peak of the curve).

n is the number of species in bins R distant from the modal bin.

a is a constant derived from the data.

It is then possible to predict how many species are in the community by calculating the total area under the curve (*N*):

$$N = \frac{n_0 \sqrt{\pi}}{a}$$

The number of species missing from the data set (the missing area to the left of the veil line) is simply *N* minus the number of species sampled. Preston did this for two lepidopteran data sets, predicting that, even after 22 years of collection, only 72 and 88 per cent of the species present had been sampled.

Yule model (Nee 2003)

The Yule model is based on a much earlier, Galton–Watson model which was used to describe the distribution of species among genera. The Yule model assumes random branching of species trees, with each species (branch tip) having the equivalent probability of giving rise to new species or becoming extinct. As the number of species within a genus, within a clade, has a similar distribution to the number of individuals within a species, within a community (i.e. the 'hollow curve'), Sean Nee used the model to describe relative species abundances. In many ways this model is similar to niche apportionment models, however, Nee intentionally did not propose a biological mechanism for the model behaviour, arguing that any distribution can be produced by a variety of mechanisms.

Mechanistic approaches: niche apportionment

Note: This section provides a general summary of niche apportionment theory, more information can be found under niche apportionment models.

Most mechanistic approaches to species abundance distributions use niche-space, i.e. available resources, as the mechanism driving abundances. If species in the same trophic level consume the same resources (such as nutrients or sunlight in plant communities, prey in carnivore communities, nesting locations or food in bird communities) and these resources are limited, how the resource 'pie' is divided among species determines how many individuals of each species can exist in the community. Species with access to lots of resources will have higher carrying capacities than those with little access. Mutsunori Tokeshi later elaborated niche apportionment theory to include niche filling in unexploited resource space. Thus, a species may survive in the community by carving out a portion of another species' niche (slicing up the pie into smaller pieces) or by moving into a vacant niche (essentially making the pie larger, for example, by being the first to arrive in a newly available location or through the development of a novel trait that allows access previously unavailable resources). Numerous niche apportionment models have been developed. Each make different assumptions about how species carve up niche-space.

Unified neutral theory (Hubbell 1979/2001)

The Unified Neutral Theory of Biodiversity and Biogeography (UNTB) is a special form of mechanistic model that takes an entirely different approach to community composition than the niche apportionment models. Instead of species populations reaching equilibrium within a community, the UNTB model is dynamic, allowing for continuing changes in relative species abundances through drift.

A community in the UNTB model can be best visualised as a grid with a certain number of spaces, each occupied with individuals of different species. The model is zero-sum as there are a limited number of spaces that can be occupied: an increase in the number of individuals of one species in the grid must result in corresponding decrease in the number of individuals of other species in the grid. The model then uses birth, death, immigration, extinction and speciation to modify community composition over time.

Hubbell's theta

The UNTB model produces a dimensionless 'fundamental biodiversity' number, θ, which is derived using the formula:

$$\theta = 2J_m v$$

where,

J_m is the size of the metacommunity (the outside source of immigrants to the local community).

v is the speciation rate in the model.

Relative species abundances in the UNTB model follow a zero-sum multinomial distribution. The shape of this distribution is a function of the immigration rate, the size of the sampled community (grid) and θ. When the value of θ is small, the relative species abundance distribution is similar to the geometric series (high dominance). As θ gets larger, the distribution becomes increasingly s-shaped (log-normal) and, as it approaches infinity, the curve becomes flat (the community has infinite diversity and species abundances of one). Finally, when $\theta = 0$ the community described consists of only one species (extreme dominance).

Fisher's alpha and Hubbell's theta—an interesting convergence

An unexpected result of the UNTB is that at very large sample sizes, predicted relative species abundance curves describe the metacommunity and become identical to Fisher's logseries. At this point θ also becomes identical to Fisher's α for the equivalent distribution and Fisher's constant x is equal to the ratio of birthrate:deathrate. Thus, the UNTB unintentionally offers a mechanistic explanation of the logseries 50 years after Fisher first developed his descriptive model.

SPECIES-AREA CURVE

In ecology, a species-area curve is a relationship between the area of a habitat, or of part of a habitat, and the number of species found within that area. Larger areas tend to contain larger numbers of species, and empirically, the relative numbers seem to follow systematic mathematical relationships. The species-area relationship is usually constructed for a single type of organism, such as all vascular plants or all species of a specific trophic level within a particular site. It is rarely, if ever, constructed for all types of organisms if simply because of the prodigious data requirements. It is related to, but not identical with, the species discovery curve (Fig. 7.5).

Ecologists have proposed a wide range of factors determining the slope and elevation of the species-area relationship. These factors include the relative balance between immigration and extinction, rate and magnitude of disturbance on small vs. large areas, predator-prey dynamics and clustering of individuals of the same species as a result of dispersal limitation or habitat heterogeneity. The species-area relationship has been reputed to follow from the 2nd law of thermodynamics. In contrast to these 'mechanistic' explanations, others assert the need to test whether the pattern is simply the result of a random sampling process. Authors have classified the species-area relationship according to the type of habitats being sampled and the census design used. Frank Preston, an early investigator of the theory of the species-area relationship, divided it into two types: samples (a census of a contiguous habitat that grows in census area, also called 'mainland' species-area relationships), and isolates (a census of discontiguous habitats, such as islands, also called 'island' species-area relationships). Michael

Rosenzweig also notes that species-area relationships for very large areas—those collecting different biogeographic provinces or continents—behave differently from species-area relationships from islands or smaller contiguous areas. It has been presumed that 'island'-like species-area relationships have higher slopes (in log-log space) than 'mainland' relationships, but a recent meta-analysis of almost 700 species-area relationships found the former had lower slopes than the latter.

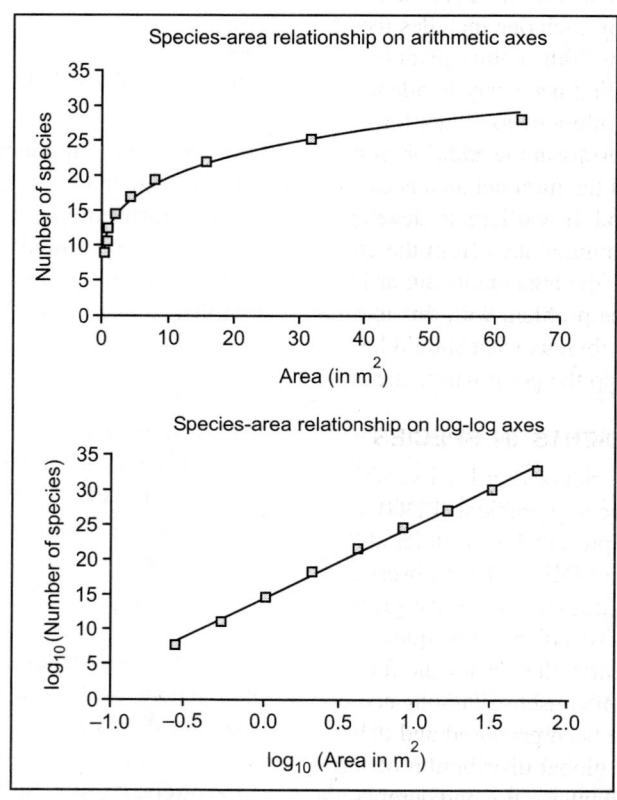

Fig. 7.5. The species-area relationship for a contiguous habitat.

Regardless of census design and habitat type, species-area relationships are often fit with a simple function. Frank Preston advocated the power function based on his investigation of the lognormal species-abundance distribution. If S is the number of species, A is area, c is a constant (number of species in the smallest sampling area) and z is the slope of the species area relationship in log-log space, then the power function species-area relationship goes as:

$$S = cA^z$$

which looks like a straight line on log-log axes. In contrast, Henry Gleason championed the semilog model:

$$S = c + z \log(A)$$

which looks like a straight line on semilog axes, where area is logged and the number of species is arithmetic. In either case, the species-area relationship is almost always decelerating (has a negative second derivative) when plotted arithmetically. Species-area relationships are often graphed for islands (or habitats that are otherwise isolated from one another, such as woodlots in an agricultural landscape) of different sizes. Although larger islands tend to have more species, it is possible that a smaller island will have more than a larger one. In contrast, species-area relationships for contiguous habitats will always rise as areas increases, provided that the sample plots are nested within one another.

The species-area relationship for mainland areas (contiguous habitats) will differ according to the census design used to construct it. A common method is to use quadrats of successively larger size, so that the area enclosed by each one includes the area enclosed by the smaller one (i.e. areas are nested).

In the first part of the 20th century plant ecologists often used the species-area curve to estimate the minimum size of a quadrat necessary to adequately characterise a community. This is done by plotting the curve (usually on arithmetic axes, not log-log or semilog axes), and estimating the area after which using larger quadrats results in the addition of only a few more species. This is called the minimal area. A quadrat that encloses the minimal area is called a relevé, and using species-area curves in this way is called the relevé method. It was largely developed by the Swiss ecologist Josias Braun-Blanquet.

Estimation of the minimal area from the curve is necessarily subjective, so some authors prefer to define minimal area as the area enclosing at least 95 per cent (or some other large proportion) of the total species found. The problem with this is that the species area curve does not usually approach an asymptote, so it is not obvious what should be taken as the total. In fact, the number of species always increases with area up to the point where the area of the entire world has been accumulated.

LATITUDINAL GRADIENTS IN SPECIES DIVERSITY

The increase in species richness or biodiversity that occurs from the poles to the tropics, often referred to as the latitudinal diversity gradient (LDG), is one of the most widely recognised patterns in ecology. Put another way, in the present day localities at lower latitudes generally have more species than localities at higher latitudes. The LDG has been observed to varying degrees in earth's past.

Explaining the latitudinal diversity gradient is one of the great contemporary challenges of biogeography and macroecology. The question 'What determines patterns of species diversity?' was among the 25 key research themes for the future identified in 125th Anniversary issue of Science (July 2005). There is a lack of consensus among ecologists about the mechanisms underlying the pattern, and many hypotheses have been proposed and debated.

Understanding the global distribution of biodiversity is one of the most significant objectives for ecologists and biogeographers. Beyond purely scientific goals and satisfying curiosity, this understanding is essential for applied issues of major concern to humankind, such as the spread of invasive species, the control of diseases and their vectors, and the likely effects of global climate change on the maintenance of biodiversity. Tropical areas play a prominent role in the understanding of the distribution of biodiversity, as their rates of habitat degradation and biodiversity loss are exceptionally high.

Patterns in the Past

The LDG is a noticeable pattern among modern organisms that has been described qualitatively and quantitatively. It has been studied at various taxonomic levels, through different time periods and across many geographic regions. The LDG has been observed to varying degrees in earth's past, possibly due to differences in climate during various phases of earth's history. Some studies indicate that the LDG was strong, particularity among marine taxa, while other studies of terrestrial taxa indicate the LDG had little effect on the distribution of animals.

Hypotheses for Pattern

Although many of the hypotheses exploring the latitudinal diversity gradient are closely related and interdependent, most of the major hypotheses can be split into four general hypotheses.

Spatial/Area hypotheses

There are two major hypotheses that depend solely on the spatial and areal characteristics of the tropics.

Mid-domain effect

Using computer simulations, Colwell and Hurtt and Willig and Lyons first pointed out that if species' latitudinal ranges were randomly shuffled within the geometric constraints of a bounded biogeographical domain (e.g. the continents of the New World, for terrestrial species), species' ranges would tend to overlap more toward the center of the domain than towards its limits, forcing a mid-domain peak in species richness.

Colwell and Lees called this stochastic phenomenon the mid-domain effect (MDE), presented several alternative analytical formulations for one-dimensional MDE, and suggested the hypothesis that MDE might contribute to the latitudinal gradient in species richness, together with other explanatory factors considered here, including climatic and historical ones. Because 'pure' mid-domain models attempt to exclude any direct environmental or evolutionary influences on species richness, they have been claimed to be null models.

On this view, if latitudinal gradients of species richness were determined solely by MDE, observed richness patterns at the biogeographic level would not be distinguishable from patterns produced by random placement of observed ranges. Others object that MDE models so far fail to exclude the role of environment at the population level and in setting domain boundaries, and therefore cannot be considered null models. Mid-domain effects have proven controversial.

While some studies have found evidence of a potential role for MDE in latitudinal gradients of species richness, particularly for wide-ranging species others report little correspondence between predicted and observed latitudinal diversity patterns.

Geographical area hypothesis

Another spatial hypothesis is the geographical area hypothesis. It asserts that the tropics are the largest biome and that large tropical areas can support more species. More area in the tropics allows species to have larger ranges, and consequently larger population sizes. Thus, species with larger ranges are likely to have lower extinction rates. Additionally, species with larger ranges may be more likely to undergo allopatric speciation, which would increase rates of speciation. The combination of lower extinction rates and high rates of speciation leads to the high levels of species richness in the tropics.

A critique of the geographical area hypothesis is that even if the tropics is the most extensive of the biomes, successive biomes north of the tropics all have about the same area. Thus, if the geographical area hypothesis is correct these regions should all have approximately the same species richness, which is not true, as is referenced by the fact that polar regions contain fewer species than temperate regions. To explain this, suggested that if species with partly tropical distributions were excluded, the richness gradient north of the tropics should disappear. Blackburn and Gaston 1997 tested the effect of removing tropical species on latitudinal patterns in avian species richness in the New World and found there is indeed a relationship between the land area and the species richness of a biome once predominantly tropical species are excluded. Perhaps a more serious flaw in this hypothesis is some biogeographers suggest that the terrestrial tropics are not, in fact, the largest biome, and thus this hypothesis is not a valid explanation for the latitudinal species diversity gradient. In any event, it would be difficult to defend the tropics as a 'biome' rather than the geographically diverse and disjunct regions that they truly include.

Species-energy hypothesis

The species energy hypothesis suggests the amount of available energy sets limits to the richness of the system. Thus, increased solar energy (with an abundance of water) at low latitudes causes increased net primary productivity (or photosynthesis). This hypothesis proposes the higher the net primary productivity the more individuals can be supported and the more individuals the more species there will be in an area. Put another way, this hypothesis suggests that extinction rates are reduced towards the equator as a result of the higher populations sustainable by the greater amount of available energy in the tropics. Lower extinction rates lead to more species in the tropics.

One critique of this hypothesis has been that increased species richness over broad spatial scales is not necessarily linked to increased number of individuals, which in turn is not necessarily related to increased productivity. Additionally, the observed changes in the number of individuals in an area with latitude or productivity are either too small (or in the wrong direction) to account for the observed changes in species richness. The potential mechanisms underlying the species-energy hypothesis, their unique predictions and empirical support have been assessed in a major review by Currie.

Climate harshness hypothesis

Another climate-related hypothesis is the climate harshness hypothesis, which states the latitudinal diversity gradient may exist simply because fewer species can physiologically tolerate conditions at higher latitudes than at low latitudes because higher latitudes are often colder and drier than tropical latitudes. Again, Cardillo find fault with this hypothesis, stating that although it is clear that climatic tolerance can limit species distributions, it appears that species are often absent from areas whose climate they can tolerate.

Climate stability hypothesis

Similarly to the climate harshness hypothesis, climate stability is suggested to be the reason for the latitudinal diversity gradient. The mechanism for this hypothesis is that while a fluctuating environment may increase the extinction rate or preclude specialisation, a constant environment can allow species to specialise on predictable resources, allowing them to have narrower niches and facilitating speciation. The fact that temperate regions are more variable both seasonally and over geological timescales suggests that temperate regions are thus expected to have less species diversity than the tropics.

Critiques for this hypothesis include the fact that there are many exceptions to the assumption that climate stability means higher species diversity. For example, low species diversity is known to occur often in stable environments such as tropical mountaintops. Additionally, many habitats with high species diversity do experience seasonal climates, including many tropical regions that have highly seasonal rainfall.

Historical/Evolutionary hypotheses

There are three main hypotheses that are related to historical and evolutionary explanations for the increase of species diversity towards the equator.

The historical perturbation hypothesis

The historical perturbation hypothesis proposes the low species richness of higher latitudes is a consequence of an insufficient time period available for species to colonise or recolonise areas because of historical perturbations such as glaciation. This hypothesis suggests that diversity in the temperate regions have not yet reached equilibrium, and that the number of species in temperate areas will continue to increase until saturated.

The evolutionary rate hypothesis

The evolutionary rate hypothesis argues higher evolutionary rates in the tropics have caused higher speciation rates and thus increased diversity at low latitudes. Higher evolutionary rates in the tropics have been attributed to higher ambient temperatures, higher mutation rates, shorter generation time and/or faster physiological processes. Faster rates of microevolution in warm climates (i.e. low latitudes and altitudes) have been shown for plants, mammals and amphibians. Based on the expectation that faster rates of microevolution result in faster rates of speciation, these results suggest that faster evolutionary rates in warm climates almost certainly have a strong influence on the latitudinal diversity gradient. More research needs to be done to determine whether or not speciation rates actually are higher in the tropics. Understanding whether extinction rate varies with latitude will also be important to whether or not this hypothesis is supported.

The hypothesis of effective evolutionary time

This hypothesis assumes that diversity is determined by the evolutionary time under which ecosystems have existed under relatively unchanged conditions, and by evolutionary speed directly determined by effects of environmental energy (temperature) on mutation rates, generation times and speed of selection. It differs from most other hypotheses in not postulating an upper limit to species richness set by various abiotic and biotic factors, i.e., it is a nonequilibrium hypothesis assuming a largely non-saturated niche space. It does accept that many other factors may play a role in causing latitudinal gradients in species richness as well. The hypothesis is supported by much recent evidence, in particular the studies of Allen and Wright.

Biotic hypotheses

Biotic hypotheses claim ecological species interactions such as competition, predation, mutualism and parasitism are stronger in the tropics and these interactions promote species coexistence and specialisation of species, leading to greater speciation in the tropics. These hypotheses are problematic because they cannot be proximate cause of the latitudinal diversity gradient as they fail to explain why species interactions might be stronger in the tropics. An example of one such hypothesis is the greater intensity of predation and more specialised predators in the tropics has contributed to the increase of diversity in the tropics. This intense predation could reduce the importance of competition and permit greater niche overlap and promote higher richness of prey.

However, as discussed above, even if predation is more intense in the tropics (which is not certain), as it cannot be the ultimate cause of species diversity in the tropics because it fails to explain what gives rise to the richness of the predators in the tropics.

Several recent studies have failed to observe consistent changes in ecological interactions with latitude. These studies suggest the intensity of species interactions are not correlated with the change in species richness with latitude. Ecology is the study of life around you.

Synthesis and Conclusions

There are many other hypotheses related to the latitudinal diversity gradient, but the above hypotheses are a good overview of the major ones still cited today. It is important to note that many of these hypotheses are similar to and dependent on one another. For example, the evolutionary hypotheses are closely dependent on the historical climate characteristics of the tropics.

Generality of the latitudinal diversity gradient

Recently, Hillebrand performed an extensive meta-analysis of nearly 600 latitudinal gradients from published literature to determine the generality of the latitudinal diversity gradient across different organismal, habitat and regional characteristics. Hillebrand found the latitudinal gradient occurs in marine, terrestrial and freshwater ecosystems, in both hemispheres. The gradient is steeper and more pronounced in richer taxa (i.e. taxa with more species), larger organisms, in marine and terrestrial versus freshwater ecosystems, and at regional versus local scales.

The gradient steepness (the amount of change in species richness with latitude) is not influenced by dispersal, animal physiology (homeothermic or ectothermic) trophic level, hemisphere or the latitudinal range of study. The study could not directly falsify or support any of the above hypotheses, however results do suggest a combination of energy/climate and area processes likely contribute to the latitudinal species gradient. Notable exceptions to the trend include the ichneumonidae, shorebirds, penguins and freshwater zooplankton.

Thus, the fundamental macroecological question that the latitudinal diversity gradient depends on is 'What causes patterns in species richness'? Species richness ultimately depends on whatever proximate factors are found to affect processes of speciation, extinction, immigration and emigration. While some ecologists continue to search for the ultimate primary mechanism that causes the latitudinal richness gradient, many ecologists suggest instead this ecological pattern is likely to be generated by several contributory mechanisms. For now the debate over the cause of the latitudinal diversity gradient will continue until a groundbreaking study provides conclusive evidence or there is general consensus that multiple factors contribute to the pattern.

DIVERSITY INDEX

A diversity index is a statistic that increases when the number of types into which a set of entities has been classified increases, and obtains its maximum value for a given number of types when all types are represented by the same number of entities. When diversity indices are used in ecology, the entities of interest are usually individual plants or animals, and the types of interest are species or other taxa. In demography, the entities of interest can be people, and the types of interest various demographic groups, and in information science, the entities can be characters and the types the different letters of the alphabet. The most commonly used diversity indices are simple transformations of the effective number of types (also known as 'true diversity'), but each diversity index can also be interpreted in its own right as a measure corresponding to some real phenomenon (but a different one for each diversity index).

Terms

Species richness

The species richness S is simply the number of species present in an ecosystem. This index makes no use of relative abundances. In practice, measuring the total species richness in an ecosystem is impossible, except in very depauperate systems. The observed number of species in the system is a biased estimator of the true species richness in the system, and the observed species number increases non-linearly with sampling effort. Thus S, if indicating the observed species richness in an ecosystem, is usually referred to as species density.

Species evenness

The species evenness is the relative abundance or proportion of individuals among the species.

Concentration ratio

Concentration ratio is a crude indicator of the extent to which a few groups such as species, demographic groups or companies dominate an environment, the total share taken by the top n species or firms. However by itself the concentration ratio does not indicate how much that share is divided between those top n firms or species.

Indices that Measure Diversity

Simpson's diversity index

If p_i is the fraction of all organisms which belong to the ith species, then Simpson's diversity index is most commonly defined as the statistic:

$$D = \sum_{i=1}^{S} p_i^2$$

This quantity was introduced by Edward Hugh Simpson in 1949. The Herfindahl index in competition economics is essentially the same.

If n_i is the number of individuals of species i which are counted, and N is the total number of all individuals counted, then:

$$\frac{\sum_{i=1}^{S} n_i(n_i - 1)}{N(N-1)}$$

is an estimator for Simpson's index for sampling without replacement.

Note that $0 \leq D \leq 1$, with values near zero corresponding to highly diverse or heterogeneous ecosystems and values near one corresponding to more homogeneous ecosystems. Biologists who find this confusing sometimes use 1/D instead; confusingly, this reciprocal quantity is also called Simpson's index. Another response is to redefine Simpson's index as:

$$\tilde{D} = 1 - D = 1 - \sum_{i=1}^{S} p_i^2$$

This quantity is called by statisticians the index of diversity.

In sociology, psychology and management studies the index is often known as Blau's Index, as it was introduced into the literature by the sociologist Peter Blau.

In economics essentially the same quantity is called the Hirschman-Herfindahl index (HHI), defined as the sum of the squares of the shares in the population across groups (with E as the group size, that is, the number of employees or the number of specimina):

$$D = \sum_{i=1}^{} \left(\frac{E_i}{E} \right)^2.$$

Note that a HHI is also used within sectors, to measure competition.

The index of diversity (also referred to as the Index of variability) is a commonly used measure, in demographic research, to determine the variation in categorical data. Gibbs and Martin defined the Simpson's diversity index for use in sociology as:

$$D = 1 - \sum_{i=1}^{N} p_i^2$$

where,

p = proportion of individuals or objects in a category.

N = number of categories.

A perfectly homogeneous population would have a diversity index score of 0. A perfectly heterogeneous population would have a diversity index score of 1 (assuming infinite categories with equal representation in each category). As the number of categories increases, the maximum value of the diversity index score also increases (e.g. 4 categories at 25 per cent = 0.75, 5 categories with 20 per cent = 0.8, etc.).

An example of the use of the index of diversity would be a measure of racial diversity in a city. Thus, if Sunflower City was 85 per cent white and 15 per cent black, the index of diversity would be: 0.255.

The interpretation of the diversity index score would be that the population of Sunflower City is not very heterogeneous but is also not homogeneous.

Shannon's diversity index

Shannon's diversity index is simply the ecologist's name for the communication entropy introduced by Claude Shannon:

$$H' = -\sum_{i=1}^{S} p_i \ln p_i$$

where, p_i is the fraction of individuals belonging to the ith species. This is by far the most widely used diversity index. The intuitive significance of this index can be described as follows. Suppose we devise binary codewords for each species in our ecosystem, with short codewords used for the most abundant species, and longer codewords for rare species. As we walk around and observe individual organisms, we call out the corresponding codeword. This gives a binary sequence. If we have used an efficient code, we will be able to save some breath by calling out a shorter sequence than would otherwise be the case. If so, the average codeword length we call out as we wander around will be close to the Shannon diversity index.

It is possible to write down estimators which attempt to correct for bias in finite sample sizes, but this would be misleading since communication entropy does not really fit expectations based upon parametric statistics. Differences arising from using two different estimators are likely to be overwhelmed by errors arising from other sources. Current best practice tends to use bootstrapping procedures to estimate communication entropy.

Shannon himself showed that his communication entropy enjoys some powerful formal properties, and furthermore, it is the unique quantity which does so. These observations are the foundation of its interpretation as a measure of statistical diversity (or 'surprise', in the arena of communications). The applications of this quantity go far beyond the one discussed here.

Berger-Parker index

The Berger-Parker diversity index is simply:

$$\max_{1 \leq i \leq S} p_i$$

This is an example of an index which uses only partial information about the relative abundances of the various species in its definition.

Rényi entropy

The species richness, the Shannon index, Simpson's index and the Berger-Parker index can all be identified as particular examples of quantities bearing a simple relation to the Rényi entropy,

$$H_\alpha = \frac{1}{1-\alpha} \log \sum_{i=1}^{S} p_i^\alpha$$

for α approaching 0, 1, 2, ∞ respectively.

Unfortunately, the powerful formal properties of communication entropy do not generalise to Rényi entropy, which largely explains the much greater power and popularity of Shannon's index with respect to its competitors.

Income inequality

Related to diversity indices are many income inequality indices, such as the Gini index and the Theil index. Generally these measure a lack of diversity, but the only difference with the measures mentioned above is a minus sign.

The Theil index in particular is the maximum possible diversity $\log(N)$ minus Shannon's diversity index. It is the maximum possible entropy of the data minus the observed entropy. The Theil index is called redundancy in information theory.

SPECIES DISTRIBUTION

Species distribution is the manner in which a biological taxon is spatially arranged. Species distribution is not to be confused with dispersal, which is the movement of individuals away from their area of origin or from centers of high population density. A similar concept is the species range. A species range is often represented with a species range map. Biogeographers try to understand the factors determining a species' distribution. The pattern of distribution is not permanent for each species. Distribution patterns can change seasonally, in response to the availability of resources, and also depending on the scale at which they are viewed.

Dispersion usually takes place at the time of reproduction. Populations within a species are translocated through many methods, including dispersal by people, wind, water and animals. People are one of the largest distributors due to the current trends in globalisation and the expanse of the transportation industry. For example, large tankers often fill their ballasts with water at one port and empty them in another, causing a wider distribution of aquatic species (Fig. 7.6).

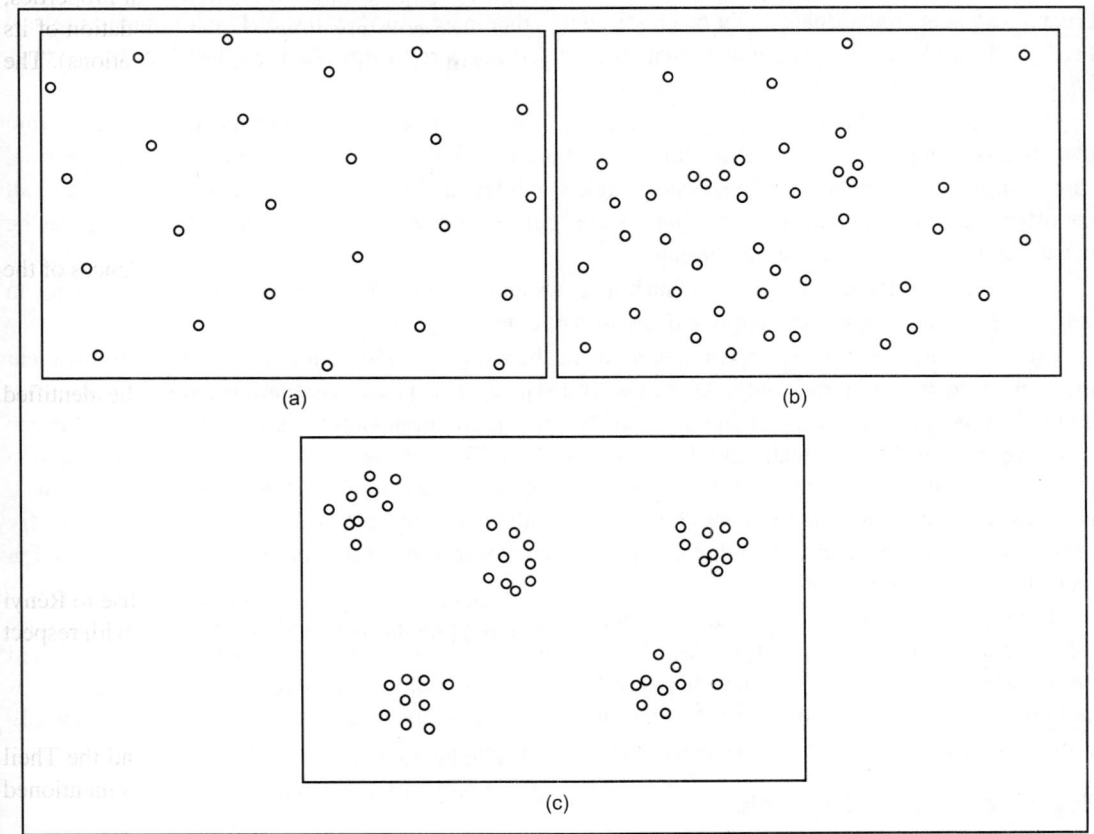

Fig. 7.6. There are three basic types of population distribution within an area: (a) uniform, (b) random and (c) clumped.

Biogeography is the study of the distribution of biodiversity over space and time. It is very useful in understanding species distribution through factors such as speciation, extinction, continental drift, glaciation, variation of sea levels, river capture and available resources. This branch of study not only gives a description of the species distribution, but also a geographical explanation for the distribution of particular species. The traditional biogeographic regions were first modelled by Alfred Wallace in The Geographical Distribution of Animals. These were based on the work of Sclater's terrestrial biogeographic regions. Wallace's system was based on both birds and vertebrates, including non-flying mammals, which better reflect the natural divisions of the earth due to their limited dispersal abilities.

Clumped Distribution

Clumped distribution is the most common type of dispersion found in nature. In clumped distribution, the distance between neighbouring individuals is minimised. This type of distribution is found in environments that are characterised by patchy resources. Clumped distribution is the most common type of dispersion found in nature because animals need certain resources to survive, and when these resources become rare during certain parts of the year animals tend to 'clump' together around these crucial resources. Individuals might be clustered together in an area due to social factors such as selfish herds and family groups. Organisms that usually serve as prey form clumped distributions in areas where they can hide and detect predators easily.

Other causes of clumped distributions are the inability of offspring to independently move from their habitat. This is seen in juvenile animals that are immobile and strongly dependent upon parental care. For example, the bald eagle's nest of eaglets exhibits a clumped species distribution because all the offspring are in a small subset of a survey area before they learn to fly. Clumped distribution can be beneficial to the individuals in that group.

However, in some herbivore cases, such as cows and wildebeests, the vegetation around them can suffer, especially if animals target one plane in particular.

Clumped distribution in species acts as a mechanism against predation as well as an efficient mechanism to trap or corner prey. African wild dogs, *Lycaon pictus*, use the technique of communal hunting to increase their success rate at catching prey. It has been shown that larger packs of African wild dogs tend to have a greater number of successful kills. A prime example of clumped distribution due to patchy resources is the wildlife in Africa during the dry season; lions, hyenas, giraffes, elephants, gazelles, and many more animals are clumped by small water sources that are present in the severe dry season. It has also been observed that extinct and threatened species are more likely to be clumped in their distribution on a phylogeny.

The reasoning behind this is that they share traits that increase vulnerability to extinction because related taxa are often located within the same broad geographical or habitat types where human-induced threats are concentrated. Using recently developed complete phylogenies for mammalian carnivores and primates it has been shown that the majority of instances threatened species are far from randomly distributed among taxa and phylogenetic clades and display clumped distribution.

Regular or Uniform Distribution

Less common than clumped distribution, uniform distribution, also known as even distribution, is evenly spaced. Uniform distributions are found in populations in which the distance between neighbouring individuals is maximised. The need to maximise the space between individuals generally arises from competition for a resource such as moisture or nutrients, or as a result of direct social interactions between individuals within the population, such as territoriality. For example, penguins often exhibit uniform spacing by aggressively defending their territory among their neighbours. Plants also exhibit uniform distributions, like the creosote bushes in the southwestern region of the United States. *Salvia leucophylla* is a species in California that naturally grows in uniform spacing. This flower releases chemicals called terpenes which inhibit the growth of other plants around it and results in uniform distribution. This is an example of allelopathy, which is the release of chemicals from plant parts by leaching, root exudation, volatilisation, residue decomposition and other processes.

Allelopathy can have beneficial, harmful or neutral effects on surrounding organisms. Some allelochemicals even have selective affects on surrounding organisms; for example, the tree species *Leucaena leucocephala* exudes a chemical that inhibits the growth of other plants but not those of its own species, and thus can affect the distribution of specific rival species. Allelopathy usually results in uniform distributions, and its potential to suppress weeds is being researched. Farming and agricultural practices often create uniform distribution in areas where it would not previously exist, for example, orange trees growing in rows on a plantation.

Random Distribution

Random distribution, also known as unpredictable spacing, is the least common form of distribution in nature and occurs when the members of a given species are found in homogeneous environments in which the position of each individual is independent of the other individuals: they neither attract nor repel one another. Random distribution is rare in nature as biotic factors, such as the interactions with neighbouring individuals, and abiotic factors, such as climate or soil conditions, generally cause organisms to be either clustered or spread apart. Random distribution usually occurs in habitats where environmental conditions and resources are consistent. This pattern of dispersion is characterised by the lack of any strong social interactions between species. For example, when dandelion seeds are dispersed by wind, random distribution will often occur as the seedlings land in random places determined by uncontrollable factors. Tropical fig trees exhibit random distribution as well because of wind pollination. In addition to tropical fig trees and dandelion seeds, oyster larvae can travel hundreds of kilometres powered by sea currents, which causes random distribution when the larvae land in random places. Although random is thought to be unpredictable, it is the only dispersion that has a mathematical equation to represent it. This is due to the individualistic characteristics of random dispersion based on the idea that every species has equal opportunity and access to resources.

Species Distribution Model

Species distribution can now be potentially predicted based on pattern of biodiversity at spatial scales. A general hierarchical model can integrate disturbance, dispersal and population dynamics. Based on factors of dispersal, disturbance, resources limiting climate, and other species distribution, predictions of species distribution can create a bioclimate range or bioclimate envelope. The envelope can range from a local to a global scale or a density independence to density dependence. The hierarchical model takes into consideration of requirements and impacts or resources as well as local extinctions in disturbance factors. Models can integrate the dispersal/migration model, the disturbance model, and abundance model. SDM's can be used to assess climate change impacts and conservation management issues. Species distribution models include, presence/absence models, the dispersal/migration models, disturbance models and abundance models.

A prevalent way of creating predicted distribution maps for different species is to reclassify a land cover layer depending on whether or not the species in question would be predicted to habit each cover type. This simple SDM is often modified through the use of range data or ancillary information—such as elevation or water distance.

Recent studies have indicated that the grid size used can have an effect on the output of these species distribution models. The standard 50×50 km grid size can select up to 2.89 times more area than when modelled with a 1×1 km grid for the same specie. This has several effects on the species conservation planning under climate change predictions (global climate models which are frequently used in the creation of species distribution models—usually consists of 50–100 km size grids) which could lead to over-prediction of future ranges in species distribution modelling. This can result in the misidentification of protected areas intended for a specie's future habitat.

Abiotic and Biotic Factors

The distribution of species into clumped, uniform or random depends on different abiotic and biotic factors. Any nonliving chemical or physical factor in the environment is considered an abiotic factor. There are three main types of abiotic factors: climatic factors consist of sunlight, atmosphere, humidity, temperature and salinity; edaphic factors are abiotic factors regarding soil, such as the coarseness of soil, local geology, soil pH and aeration; and social factors include-land use and water availability. An example of the effects of abiotic factors on species distribution can be seen in drier areas, where most individuals of a species will gather around water sources, forming a clumped distribution.

Biotic factors, such as predation, disease, and competition for resources such as food, water and mates, can also affect how a species is distributed. A biotic factor is any behaviour of an organism that affects another organism, such as a predator consuming its prey. For example, biotic factors in a quail's environment would include their prey (insects and seeds), competition from other quail, and their predators, such as the coyote. An advantage of a herd, community, or other clumped distribution allows a population to detect predators earlier, at a greater distance, and potentially mount an effective defense. Due to limited resources, populations may be evenly distributed to minimise competition, as is found in forests, where competition for sunlight produces an even distribution of trees.

Statistical Determination of Distribution Patterns

There are various ways to determine the distribution pattern of species. The Clark–Evans nearest neighbour method can be used to determine if a distribution is clumped, uniform or random. To utilise the Clark–Evans nearest neighbour method, researchers examine a population of a single species. The distance of an individual to its nearest neighbour is recorded for each individual in the sample. For two individual that are each other's nearest neighbour, the distance is recorded twice, once for each individual. To receive accurate results, it is suggested that the number of distance measurements is at least 50. The average distance between nearest neighbours is compared to the expected distance in the case of random distribution to give the ratio:

$$\frac{\text{Mean distance}}{\frac{1}{2}\sqrt{\text{density}}}$$

If this ratio (R) is equal to 1, then the population is randomly dispersed. If R is significantly greater than 1, the population is evenly dispersed. Lastly, if R is significantly less than 1, the population is clumped. Statistical tests (such as t-test, chi squared, etc.) can then be used to determine whether R is significantly different from 1.

The Variance/Mean ratio method focuses mainly on determining whether a species fits a randomly spaced distribution, but can also be used as evidence for either an even or clumped distribution. To utilise the Variance/Mean ratio method, data is collected from several random samples of a given population. In this analysis, it is imperative that data from at least 50 sample plots is considered. The number of individuals present in each sample is compared to the expected counts in the case of random distribution. The expected distribution can be found using Poisson distribution. If the variance/mean ratio is equal to 1, the population is found to be randomly distributed. If it is significantly greater than 1, the population is found to be clumped distribution. Finally, if the ratio is significantly less than 1, the population is found to be evenly distributed. Typical statistical tests used to find the significance of the variance/mean ratio include Student's t-test and chi squared.

However, many researchers believe that species distribution models based on statistical analysis, without including ecological models and theories, are too incomplete for prediction. Instead of conclusions based on presence-absence data, probabilities that convey the likelihood a species will occupy a given area are more preferred because these models include an estimate of confidence in the likelihood of the species being present/absent. Additionally, they are also more valuable than data collected based on simple presence or absence because models based on probability allow the formation of spatial maps that indicates how likely a species is to be found in a particular area. Similar areas can then be compared to see how likely it is that a species will occur there also; this leads to a relationship between habitat suitability and species occurrence.

KEYSTONE SPECIES

A keystone species is a species that has a disproportionately large effect on its environment relative to its abundance. Such species play a critical role in maintaining the structure of an ecological community, affecting many other organisms in an ecosystem and helping to determine the types and numbers of various other species in the community.

The role that a keystone species plays in its ecosystem is analogous to the role of a keystone in an arch. While the keystone is under the least pressure of any of the stones in an arch, the arch still collapses without it. Similarly, an ecosystem may experience a dramatic shift if a keystone species is removed, even though that species was a small part of the ecosystem by measures of biomass or productivity. It has become a very popular concept in conservation biology.

A classic keystone species is a small predator that prevents a particular herbivorous species from eliminating dominant plant species. Since the prey numbers are low, the keystone predator numbers can be even lower and still be effective. Yet without the predators, the herbivorous prey would explode in numbers, wipe out the dominant plants, and dramatically alter the character of the ecosystem. The exact scenario changes in each example, but the central idea remains that through a chain of interactions, a non-abundant species has an out-sized impact on ecosystem functions. One example is the weevil and its suggested keystone effects on aquatic plant species diversity by prey activities on nuisance Eurasian Watermilfoil.

Predators

As was described by Dr. Robert Paine in his classic 1966 paper, some sea stars may prey on sea urchins, mussels, and other shellfish that have no other natural predators. If the sea star is removed from the ecosystem, the mussel population explodes uncontrollably, driving out most other species, while the urchin population annihilates coral reefs. Similarly, sea otters protect kelp forests from damage by sea urchins. Kelp 'roots', called holdfasts, are merely anchors, and not the vast nutrient gathering networks of land plants. Thus the sea urchins only need to eat the roots of the kelp, a tiny fraction of the plant's biomass, to remove it from the ecosystem. These creatures need not be apex predators. Sea stars are prey for sharks, rays and sea anemones. Sea otters are prey for orca.

The jaguar, whose numbers in Central and South America have been classified as Near Threatened, acts as a keystone predator by its widely varied diet, helping to balance the mammalian jungle ecosystem with its consumption of 87 different species of prey.

Mutualists

Keystone mutualists are organisms that participate in mutually beneficial interactions, the loss of which would have a profound impact upon the ecosystem as a whole. For example, in the Avon Wheatbelt region of Western Australia, there is a period of each year when *Banksia prionotes* (Acorn Banksia) is the sole source of nectar for honeyeaters, which play an important role in pollination of numerous plant species. Therefore the loss of this one species of tree would probably cause the honeyeater population to collapse, with profound implications for the entire ecosystem. Another example is frugivores such as the cassowary, which spreads the seeds of many different trees, and some will not grow unless they have been through a cassowary.

Engineers

In North America, the grizzly bear is a keystone species—not as a predator but as ecosystem engineers. They transfer nutrients from the oceanic ecosystem to the forest ecosystem. The first stage of the transfer is performed by salmon, rich in nitrogen, sulphur, carbon, and phosphorus, who swim up rivers, sometimes for hundreds of miles. The bears then capture the salmon and carry them onto dry land, dispersing nutrient-rich feces and partially eaten carcasses. It has been estimated that the bears leave up to half of the salmon they harvest on the forest floor.

The prairie dog is also an ecosystem engineer. Prairie dog burrows provide the nesting areas for Mountain Plovers and Burrowing Owls. Prairie dog tunnel systems also help channel rainwater into the water table to prevent runoff and erosion, and can also serve to change the composition of the soil in a region by increasing aeration and reversing soil compaction that can be a result of cattle grazing. Prairie dogs also trim the vegetation around their colonies, perhaps to remove any cover for predators. Even grazing species such as Plains bison, pronghorn, and Mule deer have shown a proclivity for grazing on the same land used by prairie dogs. It is believed that they prefer the vegetative conditions after prairie dogs have foraged through the area.

Another ecosystem engineering keystone species is the beaver, which transforms its territory from a stream to a pond or swamp. In the African savannah, the larger herbivores, especially the elephants, shape their environment. The elephants destroy trees, making room for the grass species. Without these animals, much of the savannah would turn into woodland.

BIODIVERSITY HOTSPOT

A biodiversity hotspot is a biogeographic region with a significant reservoir of biodiversity that is under threat from humans. The concept of biodiversity hotspots was originated by Norman Myers in two articles in 'The Environmentalist', revised after thorough analysis by Myers and others in 'Hotspots: Earth's Biologically Richest and Most Endangered Terrestrial Ecoregions'.

To qualify as a biodiversity hotspot on Myers 2000 edition of the hotspot-map, a region must meet two strict criteria: it must contain at least 0.5 per cent or 1500 species of vascular plants as endemics, and it has to have lost at least 70 per cent of its primary vegetation.

Around the world, at least 25 areas qualify under this definition, with nine others possible candidates. These sites support nearly 60 per cent of the world's plant, bird, mammal, reptile and amphibian species, with a very high share of endemic species.

Hotspot Conservation Initiatives

Only a small percentage of the total land area within biodiversity hotspots is now protected. Several international organisations are working in many ways to conserve biodiversity hotspots.

1. Critical Ecosystem Partnership Fund (CEPF) is a global program that provides funding and technical assistance to nongovernmental organisations and participation to protect the earth's richest regions of plant and animal diversity including: biodiversity hotspots, high-biodiversity wilderness areas and important marine regions. CI works in more than 40 countries on four continents, with headquarters near Washington, DC.

2. The World Wildlife Fund has derived a system called the 'Global 200 Ecoregions', the aim of which is to select priority Ecoregions for conservation within each of 14 terrestrial, 3 freshwater and 4 marine habitat types. They are chosen for their species richness, endemism, taxonomic uniqueness, unusual ecological or evolutionary phenomena and global rarity. All biodiversity hotspots contain at least one Global 200 Ecoregion.
3. Birdlife International has identified 218 'Endemic Bird Areas' (EBAs) each of which hold two or more bird species found nowhere else. Birdlife International has identified more than 11,000 Important Bird Areas all over the world.
4. Plantlife International coordinates several projects around the world aiming to identify Important Plant Areas.
5. Alliance for Zero Extinction is an initiative of a large number of scientific organisations and conservation groups who co-operate to focus on the most threatened endemic species of the world. They have identified 595 sites, including a large number of Birdlife's Important Bird Areas.
6. The National Geographic Society has prepared A World map of the hotspots and ArcView shapefile and metadata for the Biodiversity Hotspots including details of the individual endangered fauna in each hotspot, which is available from Conservation International.

These initiatives are all based on scientific criteria and quantitative thresholds.

Biodiversity Hotspots by Region

North and Central America

1. California Floristic Province.
2. Caribbean Islands.
3. Madrean pine-oak woodlands.
4. Mesoamerica.

South America

1. Atlantic Forest.
2. Cerrado.
3. Chilean Winter Rainfall—aldivian Forests.
4. Tumbes-Chocó-Magdalena.
5. Tropical Andes.

Europe and Central Asia

1. Caucasus.
2. Irano-Anatolian.
3. Mediterranean Basin.
4. Mountains of Central Asia.

Africa

1. Cape Floristic Region.
2. Coastal Forests of Eastern Africa.
3. Eastern Afromontane.

4. Guinean Forests of West Africa.
5. Horn of Africa.
6. Madagascar and the Indian Ocean Islands.
7. Maputaland-Pondoland-Albany.
8. Succulent Karoo.

South Asia

1. Eastern Himalaya, India.
2. Indo-Burma, India and Myanmar.
3. Western Ghats, India.
4. Sri Lanka.

East Asia and Asia-Pacific

1. East Melanesian Islands.
2. Japan.
3. Mountains of Southwest China.
4. New Caledonia.
5. New Zealand.
6. Philippines.
7. Polynesia-Micronesia.
8. Southwest Australia.
9. Sundaland.
10. Wallacea.

Critiques of Hotspots

The high profile of the biodiversity hotspots approach has resulted in considerable criticism. Papers such as Kareiva and Marvier have argued that the biodiversity hotspots:

1. Do not adequately represent other forms of species richness (e.g. total species richness or threatened species richness).
2. Do not adequately represent taxa other than vascular plants (e.g. vertebrates or fungi).
3. Do not protect smaller scale richness hotspots.
4. Do not make allowances for changing land use patterns. Hotspots represent regions that have experienced considerable habitat loss, but this does not mean they are experiencing ongoing habitat loss. On the other hand, regions that are relatively intact (e.g. the Amazon Basin) have experienced relatively little land loss, but are currently losing habitat at tremendous rates.
5. Do not protect ecosystem services.
6. Do not consider phylogenetic diversity.

A recent series of papers has pointed out that biodiversity hotspots (and many other priority region sets) do not address the concept of cost. The purpose of biodiversity hotspots is not simply to identify regions that are of high biodiversity value, but to prioritise conservation spending. The regions identified include regions in the developed world (e.g. the California Floristic Province), alongside regions in the developing world (e.g. Madagascar). The cost of land is likely to vary between these regions by an order of magnitude or more, but the biodiversity hotspots do not consider the conservation importance of this difference.

MEGADIVERSE COUNTRIES

The megadiverse countries are a group of countries that harbour the majority of the earth's species and are therefore considered extremely biodiverse. Conservation International identified 17 megadiverse countries in 1998, most are located in the tropics.

In 2006, Mexico formed a separate organisation focusing on Like-Minded Megadiverse Countries, consisting of countries rich in biological diversity and associated traditional knowledge. This organisation does not include all the megadiverse countries as identified by Conservation International.

In alphabetical order, the 17 Megadiverse countries are:

Australia	India	Peru
Brazil	Indonesia	Philippines
China	Madagascar	South Africa
Colombia	Malaysia	United States
Democratic Republic of the Congo	Mexico	Venezuela
Ecuador	Papua New Guinea	

Cancún Initiative and Declaration of Like-minded Megadiverse Countries

On 18 February 2006, the Ministers in charge of the Environment and the Delegates of Brazil, China, Colombia, Costa Rica, India, Indonesia, Kenya, Philippines, Mexico, Peru, South Africa and Venezuela assembled in the Mexican city of Cancún. These countries declared to set up a Group of Like-Minded Megadiverse Countries as a mechanism for consultation and cooperation so that their interests and priorities related to the preservation and sustainable use of biological diversity could be promoted. They also declared that they would call on those countries that had not become Parties to the Convention on Biological Diversity, the Cartagena Protocol on Biosafety, and the Kyoto Protocol on climate change to become parties to these agreements.

At the same time, they agreed to meet periodically, at the ministerial and expert levels, and decided that upon the conclusion of each annual Ministerial Meeting, the next rotating host country would take on the role of Secretary of the group, to ensure its continuity, the further development of cooperation among these countries and to reach the agreements and objectives set forth herein.

The current member countries of the Like-Minded Megadiverse Countries organisation are as follows, in alphabetical order:

Bolivia	Ecuador	Mexico
Brazil	India	Peru
China	Indonesia	Philippines
Colombia	Kenya	South Africa
Costa Rica	Madagascar	Venezuela
Democratic Republic of the Congo	Malaysia	

Chapter 8

Impact of Biotechnology on Biodiversity

INTRODUCTION

The convention on biological diversity requires that all member states take measures to preserve both native and agricultural biodiversity. The intrinsic value of species and ecosystems, in addition to their value as starting material for finding new products, is the basis for these measures.

The biggest threat to biodiversity is habitat destruction. The ever-increasing spread of cities and the accompanying expansion of agriculture must be held largely responsible. Humid tropical forests are particularly valuable reservoirs of biodiversity and are currently being seriously threatened. As the human population expands, the need for food is expected to double in the next 30 years, with the ensuing threat of massive habitat destruction particularly in the less developed countries. Increasing crop productivity on the land already under cultivation would prevent or at least reduce habitat destruction. One of several measures aimed at increasing yields is the use of better seeds, including those enhanced by modern biotechnology. Many other measures from the technical, socioeconomic and political fields need to be taken at the same time in order to balance intensification and sustainability of modern agriculture.

Modern biotechnology offers new means of improving rather than threatening biodiversity. If properly tested for both risks and benefits to humans and the environment, transgenic crops are more likely to increase agricultural biodiversity and help maintain native biodiversity rather than to endanger it. Such applications need to be judged by the criteria of improved sustainability and compared to current as well as alternative farming practices.

ESSENCE OF BIODIVERSITY

Biodiversity is the multitude of different living beings in a particular ecosystem or on the whole earth. Biodiversity can be seen and studied at different organisational levels: genetic, organismal and ecological. It touches upon both native environments on land and sea as well as agricultural and other man-made surroundings.

Native Biodiversity

The biodiversity we observe today is the result of 3.5 billion years of evolution. Through the processes of mutation and selection all living organisms we know today, as well as those that ever lived before, developed from one single-cell micro-organism. How this first living being arose 3.5 billion years ago is still a matter for speculation. This unitary origin explains why all organisms share the same basic chemistry: DNA is always the storage molecule of genetic information and the complex process of protein biosynthesis is virtually the same in all organisms. Metabolic pathways are also similar in all organisms (see also cell metabolism), e.g. the reactions by which energy is generated or the way in which fatty acids, sugars and amino acids are made. Separate species arose when mutations between relatives no longer allowed for interbreeding, for instance after geographic or reproductive separation.

Dinosaurs are by no means the only creatures of the past that became extinct: far more organisms lived at some time on our earth than are living here today. The vast majority, probably more than 99 per cent, of species that arose on this globe, disappeared again. This shows that evolution is an ongoing process, with species coming and going. This dynamic situation is important when discussing the conservation of species living here today. In the long-term view, there has never been any stability in life on earth—only change. However, these changes were very slow compared to the length of a human life, or even compared to the time humans have existed. Clearly, today, with the massive amount of human interference on the globe, changes are much faster than at any other time in the last 65 million years, the point at which the trilobites and later the dinosaurs and many other creatures vanished from the surface of the globe in a relatively short time period.

As suggested by Raven in 1992, the number of species of plants, animals and eukaryotic micro-organisms is probably around 10 million today, but only 1.4 million have been characterised and given a name by scientists. There is much variation in what is known about the different groups. Virtually all of the 40,000 vertebrate animals are known and most of the 3,00,000 vascular plant species as well. On the other hand, there are likely to be over a million species each of fungi and nematodes, of which only 70,000 and 13,000 have been named. There are thought to be far more than a million different insect species as well. With prokaryotes, the situation is even more extreme: about 5000 bacteria and viruses have been named individually, yet the total number in both of these two groups may well according to Bull be in excess of one million. Since many micro-organisms and viruses are associated with specific plants and animals, Staley considers their own biodiversity will depend on the biodiversity of their hosts, as well as the micro-organisms own host range.

All the different species of plants and animals do not live an independent existence, but are associated in specific communities and ecosystems to form more or less stable associations. One such association is, for instance, the humid (and dry) tropical forest, which is generally thought to have the highest degree of biodiversity, with more tree species per km^2 than there are tree species in North America or in Europe, as discussed by Burslem. Another example comes from specific types of alpine meadows or specific sorts of rivers or ponds. Biodiversity needs to be considered not only in qualitative terms of the species present, but also in quantitative terms, considering how many individuals of certain species of plants and animals are present. With large mammals this is quite easy to determine, but it is impossible with micro-organisms. Often the number of species found in a given ecosystem is taken as a measure of the biodiversity of that system: other criteria are more difficult to apply.

In this section, we will concentrate on terrestrial biodiversity, although it is clear that streams, lakes and oceans provide habitats for a vast biodiversity of animals, plants and micro-organisms. In addition, as calculated by Naylor, they are an important source of protein as food for humans and feed for farm animals, with about 120 millions tons used per year.

Agricultural Biodiversity

In addition to biodiversity in the wild, there is the biodiversity of organisms used for farming and other human activities. In agriculture, 7000 species of plants are used by farmers somewhere in the world, but only 30 species provide 90 per cent of our calorific intake. Within these dominant crop species, there are many hundred thousands of varieties (landraces, cultivars) adapted to local climates, farming practices, and cultural predilections like taste, colour, structure, ability to store the products, etc. Much of this large crop diversity is important for providing the initial material for breeding. However, it must be recalled that the genetic diversity found in crops is much less broad than the genetic diversity observed in plants or animals living in the wild, which points to the importance of wild species for agricultural breeding programs.

The top three crops are wheat, rice and maize (corn) with around 500 million tons annual production each. Traditional breeding led us in to the trap of narrowing down genomes, and perhaps wisely used biotechnology could bring back at least that part of genetic diversity which enhances pest resistance and also yield.

There are many indications that mixtures of varieties of a crop or of different crops may give higher yields and be more resistant to pests and diseases than monocultures, as reported recently by Zhu for rice in China. However, even in mixed cultures, high quality, well defined varieties and pure seeds are required and the sustainability of mixed cropping related to pest management has still to be proven. In addition, it is still not clear, whether or not, in natural, non-agricultural habitats yield is dependent on biodiversity.

Based on experimental results, some researchers claim that the loss of species leads to a reduction in biomass, while others disagree. There simply may not be valid correlations between biodiversity and biomass yield, in either agricultural nor non-agricultural settings.

Human Population Expansion

The one species that is still globally expanding in numbers are humans. The world population has gone up from 2.5 billion in 1950 to 6 billion today; it is expected, by the UN, to reach 8 billion in 2015 and 9–10 billion in 2050. Over 95 per cent of the expected population increase will be in the developing countries. In those countries, most of the population growth will occur in the cities. The additional population will require more space to live in, more water, more energy, more food and more services. For the years 1995–2020 the largest relative population increase (80 per cent) is expected in sub-Saharan Africa: in absolute numbers it is expected to go from 500 to 900 million. It is not clear whether the HIV/AIDS epidemic, with an estimated 25 million infected in this region in 2000, will substantially affect population dynamics within the next 25 years, as most of the infected are of reproductive age, as pointed out by Cohen in 2006. It is important to realise that the FAO estimates that there are over 800 million people globally who do not have enough to eat today, and it is imperative that more food be produced for them where it is needed, namely in the developing countries. As pointed out by Lipton in 2006, poverty reduction in rural, agricultural areas of developing countries is also likely to have an indirect effect in favour of the urban poor. Improved seeds will have a positive effect, but many other factors such as education, access to microcredit, reduction in farm subsidies in the North, etc. will be at least as important.

INTERNATIONAL AGREEMENTS

In view of the importance of biodiversity for the future of mankind, several international agreements have been reached. Since this has occurred only in the last few years, the long-term impact of these agreements cannot yet be estimated.

Convention on Biological Diversity (CBD)

Recognising that the biodiversity of organisms in the wild should be maintained both for their own intrinsic value, but also on practical grounds, the United Nations prepared the Convention on Biological Diversity (CBD) and succeeded in having it adopted in 1992. It entered into force in 1993. This is the first time that a large majority of States, have agreed to a legally binding instrument for biodiversity conservation and the sustainable use of biological resources. A radical change brought about by the CBD is the recognition that States have a sovereign right over biodiversity within their own territory: previously organisms were considered to be the common heritage of mankind. Living organisms or their products may, under the terms of the CBD, only be removed from a country under mutually agreed conditions. The CBD is a comprehensive approach to biodiversity conservation of both wild and domesticated species. It aims at conservation at the genetic, species and ecosystem levels. As reviewed by Buhenne-Guilmin, action is delegated to the national level obliging States to assess biodiversity, enact legislation for its conservation *in situ* and *ex situ*, and to enforce legislation within national boundaries.

The field of biotechnology is particularly affected by articles 16 and 19 of the CBD, since they require a fair and equitable sharing of benefits derived from the use of genetic resources. This includes providing facilities and financial means for technology transfer and open access to scientific and technical information. The sovereignty over biological resources means that no one can remove specimens of plants, animals or micro-organisms from a country without the prior consent of that country. One example of a joint effort in 'bioprospecting' is a search for specific active ingredients of plants by the Merck Company in the tropical forests of Costa Rica. This brought the country 2 million dollars over a five-year period, in addition to the potential for royalties, should profitable products emerge. A small number of similar agreements have been undertaken elsewhere in the world.

The regulations of the CBD have only been in operation for a few years. It is too early to assess their long-term effects. As far as biotechnology and 'bioprospecting' is concerned, it will take more time to establish smooth administrative procedures to allow simple routine implementation of close collaborations. This system will only be perpetuated if both the national authorities from countries rich in biodiversity, and pharmaceutical or other companies see the mutual advantages to be gained from such collaboration. One of the real obstacles to this are the exorbitantly high costs of drug testing that needs to be done to meet strict and justified regulation before marketing. The expectations of some LDCs to make rapid earnings may have been too optimistic. The search for natural, highly active pharmaceuticals in wild plants may often be more cumbersome than laboratory searches using genomics, rational drug design and combinatorial chemistry.

Cartagena Protocol on Biosafety

The CBD provided a basis for developing and formulating a further international agreement, namely one primarily regulating the trans-boundary movement of living GMOs. After much debate in 1999, the new protocol was agreed upon in early 2000 under the name of the 'cartagena protocol on biosafety'. The 'Intergovernmental Committee for the Cartagena Protocol on Biosafety had its first meeting in Montpellier in December 2000. It paved the way for launching a pilot phase of the 'biosafety clearing house', a centre for information exchange, as explained on the internet.

The Protocol is a worldwide regulation for the transfer, handling and use of GMOs that may have an adverse effect on biodiversity, also taking into account risks to human health and focusing on transboundary movements. It makes explicit reference to the precautionary approach. It establishes an advance informed agreement (AIA) procedure for imports of GMOs intended for introduction into the environment and an alternative procedure for mass movements of GMOs intended for food, feed and for processing (commodities). The permit for transboundary movement will or will not be issued on the basis of a risk assessment procedure performed by the national competent authority. The Protocol does not pertain to pharmaceuticals or other non-living products made by genetic modification.

In practice, the Protocol will be most important for the import of transgenic seeds. It requires a risk assessment by the national authorities and allows countries to reject GMOs. The protocol specifies that 'lack of scientific certainty due to insufficient scientific information and knowledge regarding the extent of the potential adverse effects shall not prevent the party from taking a decision.' This may or may not be in agreement with the WTO rules, but will need to be tested in the courts. If there are disputes between the interpretation of the Protocol and the WTO regulations, which allow for trade barriers virtually only if there are scientific reasons to do so, the outcome of such disputes will depend to a considerable degree on the quality of scientific data used for assessing the benefits and risks of GMOs.

The Protocol become operational in the year 2002, after ratification by 50 nations. Early in 2001 it had been signed by 86 nations and ratified by two. The Protocol will, as pointed out by Mahoney, be a challenge to the scientific community to provide solid scientific data to convince national authorities of the benefits of transgenic crops and their relatively low and manageable risks. It is far too early to make any assessment of the Protocol's effect. Hopefully it will allow (and not prevent) the planting of GM crops that may have real advantages for developing countries, while at the same time minimising their risks to humans and their environment. The Protocol may contribute to a better public understanding of what transgenic crops are, but it may on the other hand, by treating GMO crops as a special group, increase fears of a negative impact, even in the absence of any science-based data showing such an impact. It will be difficult to move forward with the Precautionary Approach, if it is taken as a stringent principle. It will be more advisable to see the Precautionary Approach as guiding a case by case decision-making tree, in the sense of a systems approach as described by Verma Niraj of the Churchman School from Berkeley, California.

LOSS OF BIODIVERSITY AND CONSERVATION

Losses of biodiversity are undoubtedly occurring in many parts of the globe, often at a rapid pace. These losses require countermeasures such as an increased effort towards conservation by many different means.

Reduction of Biodiversity

The loss of biodiversity can be measured by a loss of individual species, groups of species or decreases in numbers of individual organisms. In a given location the loss will often reflect a degradation or a destruction of a whole ecosystem. Recently the Subsidiary Body on Scientific, Technical and Technological Advice (SBSTTA) of the CBD ranked the priority of threats to global biodiversity in the following manner: first comes habitat loss (most of it through the expansion of cultivated land), second comes the introduction of exotic species. Habitat loss comes not only from taking more land under the plough, but also from expanding cities and road building. In addition, habitats can be damaged by flooding, lack of water, climate changes, salination, etc. all of which phenomena that may be both natural or man-made.

Since tropical humid forests are particularly rich in biodiversity, their destruction is disproportionately damaging to biodiversity. It is estimated by Pimm and Raven that of the original 16 million km^2 of these forests known a century ago, only half are left, with about one million km^2 being destroyed every 5 to 10 years. Burning and selective logging may damage an even greater area. Biodiversity is not homogeneously distributed over the humid tropical forests, rather there are hotspots with a particularly high level of biodiversity. These hotspots are of particular interest for the implementation of conservation measures.

The second most important reason for loss of biodiversity is invasion by exotic plants and animals. Knowingly, or unknowingly, imported plant species threaten the native ones by being highly competitive and often by lacking local predators, such as insects or birds. One of the most extreme examples is seen in the pampas of Argentina, a flat grassland with a moderate climate, from which nearly all the native grasses have disappeared and have been replaced by European plants. This invasion was brought about by European farmers who introduced animals and crops, in addition to accidentally spreading many different weeds. This phenomenon was already noted in 1833 by Charles Darwin. Today, still, droves of gardeners transport seeds all over the globe and never think of the possible threat to biodiversity, as suggested by Ammann in 1997. It has been estimated that one in ten imported plants may spread in a modest way and that one in a hundred may turn into a nuisance weed. Even in today's Europe, invasion by exotics may threaten ecosystems. In the Ticino region of Southern Switzerland *Robinia pseudoacacia*, a native of North America, is displacing chestnut and oak trees, whilst in the Northern regions of the country *Solidago canadensis* is replacing native irises in swampy areas. Islands are particularly threatened by invaders, as is well documented for Hawaii, New Zealand and the Galapagos Islands. For North America, it has been estimated that the damage caused by exotics amounts to 137 billion dollars a year. Although such calculations are fraught with uncertainties, there is no doubt that the costs of exotics are tremendous.

Exotic biological control agents are often introduced to agricultural ecosystems on purpose, in order to control pests or weeds without resorting to chemical control agents. One example is the introduction of the seven-spot ladybird, which was intended to fight the Russian wheat aphid. The consequence, however, was the disappearance of the native ladybirds, for which the seven-spot import was a competitor and an actual predator.

Another example is the decimation of the large American moths, which are killed by European *Compsilura* flies, introduced nearly a century ago to control the gypsy moth. Field experiments recently done by Jensen showed that caterpillars of the American moth *Cecropia* were killed by massive infestations of *Compsilura* maggots.

It cannot be said today, on the basis of experimental evidence, whether transgenic plants are specifically prone to spreading in the long-term. However, one would not expect this to be the case unless the transgenic plant had an increased fitness. There is no good argument why crops that have for centuries depended for survival on human care should become weeds, just because of the addition of one or a few well characterised genes, in addition to the many thousands of genes they already carry. However, this issue needs to be carefully studied on a case by case basis, keeping in mind that the absence of a negative effect can never be proven with absolute certainty. The results of a fairly long-term study of the performance of transgenic crops in natural habitats were recently presented by Crawley in 2001. Four different crops (oilseed rape, potato, maize and sugar beet) were grown in 12 different habitats and monitored over a period of 10 years. In no case were transgenic plants found to be more invasive or more persistent than their conventional counterparts, in agreement with the general hypothesis put forward above.

Conservation Strategies

Conservation may be *in situ* or *ex situ*, either in the natural or semi-natural habitat, or in some purpose-built environment. The choice of one or the other technique, or a combination of both, will depend on the particular case. *In situ* conservation will involve the maintenance and protection of natural habitats, while botanical gardens and seed banks are used for the *ex situ* conservation. Both of the latter require precise knowledge of taxonomy.

Conserving a substantial, but selected fraction, of the humid tropical forests would still allow half or so of their indigenous species to be preserved. This would require a selection of the most appropriate areas, called the hotspots. Protecting huge tracts of land will pose major socioeconomic and political problems. It has been asked how forests destined to be protected from human encroachment can be kept free of hungry people in search of potential farmland. Jennings thinks that a viable strategy may be to find a sustainable livelihood for rural populations in connection with conserving tropical humid forests. Policing alone will not be successful over vast territories, as seen today in the war on drugs in South America and Asia. Today, conservation also embraces various components of agro-biodiversity like crop varieties, land races, semi-domesticates and crop relatives. The role of indigenous communities in maintaining agro-biodiversity is stressed by the Global Biodiversity Assessment and the Leipzig Plan of Action, two recently concluded international agreements.

APPLICATIONS OF BIOTECHNOLOGY AND ITS EFFECT ON BIODIVERSITY

The methods of biotechnology can be applied to the study of virtually any biological phenomenon and will in some cases have practical applications for maintaining biodiversity. Conversely, threats to biodiversity by biotechnology also need to be considered.

Biotechnology for the Acquisition of Knowledge

In the context of this section there are two quite different applications of biotechnology, or of molecular biology, that are relevant. The first is the use of biotechnology as a tool for acquiring knowledge, whilst the second is the use of biotechnology to directly intervene in plant and animal breeding, in particular to transfer genetic information from one sort of organism to a particular crop or to a farm animal to make it transgenic (see also *Transgenic plants* and *Transgenic animals*).

Today, biological research can hardly be conducted without using biotechnology in one way or another. Taxonomy uses molecular markers to identify individual strains of organisms or to identify species, much in the same way as in forensic medicine to identify criminals. This is useful for *ex situ* conservation of plants and micro-organisms. In seed banks, genetic fingerprints are used to establish the origin of a seed or the relatedness of one plant variety to another. A rational classification of most micro-organisms has only become possible with these biotechnological methods. This is important to the many collections of micro-organisms that exist around the world.

Biotechnology has also proven useful for following genetic markers in plant and animal breeding. Here, animal or plant varieties are crossed by conventional, sexual means. By analysing a few cells of the newly born calf or of the newly sprouted crop, one can predict some of the expected properties of the progeny, by looking at the presence or absence of certain forms of genes. This enables one to predict a phenotypic property, which will only show up later in life, for instance certain characteristics of a cow's milk or the crop's expected resistance to an infectious plant disease. Using *in vitro* fertilisation of animals, the laboratory test can be done even before the embryo is implanted. This is called pre-implantation diagnostics.

These applications of biotechnology are in general not controversial. Here, the public perception is similar to the application of modern biotechnology in medicine, for instance to industrially produced pharmaceuticals, vaccines and diagnostics. It may be noted that the worldwide sales of biotechnology products in 2008 was worth around 18 billion dollars in the medical field and only 3.2 billion in agriculture.

The availability of genome sequences will be a boost to research. The first two complete plant genome sequences determined were those of *Arabidopsis* and rice. The 120 million base pairs (MBP) of the small brassica *Arabidopsis* were sequenced by an international academic consortium and the data made public. The 430 MBP sequence of rice was completed only a few weeks later by an industrial group lead by Syngenta, and will be available by contract to other researchers. Syngenta intends to make the data available free of charge for research directly benefiting subsistence farmers. It will hopefully become common practice for companies to make their basic discoveries publicly available, to everyone's benefit. The Monsanto company has also opened up some of its rice sequencing data.

Direct Gene Transfer to Crops and Farm Animals

Since all genes consist of DNA, and the information in this DNA molecule is read in the same way in all organisms in order to make proteins, it is in principle possible to take any (single) gene from any organism and transfer it into any other organism so that the recipient produces a protein normally only made in the donor. The resulting organism is called transgenic. From the time this simple strategy was devised, it took molecular biologists about twenty years until the first transgenic plants were made in 1985. *Ten* years later, the first transgenic crop appeared in supermarkets in the USA, the 'FlavrSavr' tomato. In 2000 there were, worldwide, about 45 million hectares planted with commercial transgenic crops. Most transgenic crops planted commercially in 2000 were in the US, Canada, Argentina, with smaller amounts in China, Australia, South Africa, México and Spain. Soyabean and corn ranked first and second, making up 57 per cent and 22 per cent of the total area planted with GMOs. Cotton and canola accounted for about 5.3 and 2.8 million ha each, whilst only small areas of transgenic potato, squash and papaya were grown commercially. With regards to the genetically-modified traits, herbicide tolerance was dominant with 74 per cent, while insect resistance was 19 per cent. According to the ISAAA, the amount of virus resistant crops was quite small. Compared to 1999 there was an increase of 10 per cent in the area planted with GMOs. For corn there was a decrease, presumably because of an anti-GMO-wave that started in Europe in early 1999; soya and cotton showed an increase.

The reason that US farmers have adopted the transgenic crops surprisingly quickly is because of the economic benefits they offer. In most surveys done by different researchers in different parts of the US, the yields were the same or somewhat higher with the new seeds (www.internutrition.ch). The most noticeable difference to the farmers was the saving on herbicides. The US National Centre for Food and Agricultural Policy cited an annual saving of US$ 220 million to soyabean farmers. It was also found that the new crops needed less frequent sprayings and allowed 'no till' management. These benefits mostly offset the initial higher cost of the transgenic seeds, although farm profits, with or without modern biotechnology, vary a great deal from year to year and region to region. Clearly these economic considerations only hold for countries with economic structures similar to those of the US, not to developing countries.

It is important to remember that a large number of transgenic crops are still in the development stage and will only come onto the market in a few years from now. They are likely to show benefits for the consumers and some may be of particular interest to farmers in tropical countries. Two rice varieties, with anticipated consumer benefits, are those containing Vitamin A or an increased level of iron in the product, which were developed by Potrykus and Beyer last year. Despite traditional preventative measures (distribution of free vitamin A, encouragement to eat more fruit and vegetables), worldwide there are 130 million young people who are vitamin A-deficient, one to two million die annually as a consequence of vitamin A-deficiency and 5,00,000 turn irreversibly blind every year. A bowl of 300 g of this cooked rice is thought to be enough to overcome the vitamin A-deficiency to a significant degree. Similarly iron-deficiency, particularly prevalent in pregnant women, can potentially be alleviated by rice containing an increased amount of iron in its endosperm. Such rice varieties have been successfully developed in the laboratory, but are far from commercialisation, for both scientific and political reasons.

For farmers in developing countries, the following GMOs may be of interest:

1. Virus-resistant cassava.
2. Virus-resistant sweet potatoes.
3. Virus-resistant papaya (already on the market in Hawaii).
4. Rice with an increased rate of photosynthesis, and therefore with a potential yield increase of up to 25 per cent.
5. Rice with increased salt tolerance.
6. Diverse varieties that are partially aluminium-resistant and have the potential to grow in degraded tropical soils.
7. Diverse crops that are more drought-resistant than the usual varieties.

All of these and many more crops have been proven to work, in principle, in laboratory and glasshouse trials. The practical benefits and risks of the crops need to be assayed in the field and their products scrutinised, like any other novel food. Several lines of transgenic farm animals have been produced, but none have been made commercial. Some lines are made for the pharmaceutical industry to produce drugs in their milk. Others may show improved resistance towards certain infections. Transgenic salmon that grow faster than normal have been developed and have roused considerable concern amongst ecologists. As pointed out by Reichhardt, many environmental issues still need to be clarified in this context. However, it is clear that the transgenic crops that have been commercialised so far have not been seen to have done any harm to either the environment or consumers.

NATIVE BIODIVERSITY AND BIOTECHNOLOGY

Biodiversity in the wild has been massively reduced in the industrialised countries over a long period of time, in Europe, for example, over several millennia. Hardly any ecosystem is the same here as it was before humans started to clear forests and develop farming. When we look out of the window we see houses, roads and meadows, whilst three thousand years ago the area was covered in beech and oak forests. Even Europe's forests are more like manicured gardens than virgin forests, despite the mystical 'naturalness' attributed to our forests, at least in Germanic countries. North America still has far more native, untouched ecosystems, either in the form of protected areas or in less hospitable regions like the North of Canada. Biodiversity has already diminished on a massive scale in the industrialised countries.

Despite many conservation efforts, about half of the tropical humid forests have already been destroyed. How then can the rest be preserved, given that the regional population is increasing rapidly, as for instance in sub-Saharan Africa, where a near doubling of the population is expected in the next 25 years? Already today, the per capita food supply in sub-Saharan countries is only 2100 kilocalories compared to 3500 kilocalories in the industrialised countries, so there will be a huge pressure to provide more food. Even if some food is imported, the vast majority has to be produced regionally or nationally, both for reasons of economy and ecology (e.g. poverty reduction, transport and distribution costs).

Yields of cereals have gone up very considerably in the last forty years. In the developing countries, this is primarily a result of the green revolution. However, the annual growth increases in cereal yields have slowed down from about 3 per cent to 1 per cent per year as shown by Pinstrup-Andersen and his collaborators in 2007. For the developing countries they were 2.8 per cent in 1967–1982, 1.9 per cent in 1982–1994 and only 1.2 per cent in 1993–2020.

These lower yield increases of recent years mean that productivity will probably not keep up with demand in the developing countries. The consequence on biodiversity is devastating and means that more land will be required for farming. This land will primarily come from areas with high native biodiversity, in particular the aforementioned tropical humid and dry forests or from marginal land. Whilst the green revolution has also had negative consequences, such as salination, excessive water use and soil degradation, the increased productivity it achieved allowed the maintenance of large tracts of native untouched land not used for farming.

Conway concludes that the single most promising way to avoid habitat destruction is to increase farm yields in a process that has been called the second green revolution. Several components will be required to increase productivity: better training and education of farmers (in particular women), more favourable economic and political climate, availability of microcredits, etc. In addition, technical contributions will also be necessary. One such contribution is improved seed, produced either by traditional crop breeding or by modern biotechnology.

There will have to be more reliance on the latter, since traditional breeding seems to have reached a yield plateau. So agricultural biotechnology, which is viewed controversially in the public debate, may contribute markedly to conserve biodiversity by preventing the appropriation of native biodiversity-rich land for farming purposes. It should be noted, however, that this technology—like all others—is no panacea. Pinstup-Andersen and Cohen believe that each application needs to be studied carefully on a case-by-case basis, like any other new technology.

A valid concern is the possible effect of *Bt*-crops and similar plants on non-target insects. The *Bt*-crops contain a gene coding for an insecticidal protein originally produced by the soil bacterium *Bacillus thuringensis*. They were developed to make the plants resistant to a particular, highly damaging pest and have been quite successful in reducing pesticide input when infestation rates are high. In laboratory studies Losey showed in 2007 that the pollen from *Bt*-corn could kill larvae of the Monarch butterfly when a large amount of pollen was sprayed on the larvae's favourite food plant, milkweed. Subsequent field studies by Sears showed that the *Bt*-corn caused little or no damage to the Monarch in real agricultural settings. This shows that the impact of transgenic crops on non-target organisms cannot be studied solely in the laboratory, but also requires farmland experimentation.

A more limited concern, that largely touches Northern Europe, is the conservation of native plants and animals, in particular birds, in farmed areas. The birds' habitats are fields, hedges, roadsides and fallow land where they depend for food on insects and seeds produced by weeds in or near the crops. These seeds are particularly important in the winter months. Computer models by Watkinson suggest that more intense weed control measures may lead to smaller amounts of seeds being available to birds. This effect seems plausible under certain conditions, but depends on weed management regimes rather than on the presence or absence of transgenic plants, and therefore is not an issue of biotechnology. Herbicide tolerant beets may allow farmers to tolerate weeds for a longer time and fight them only after sowing. This is made possible by a post-emergence herbicide treatment. More efficient weed management may also make it possible to set aside more land. If the lack of food results in a reduction in the bird population, this may lead to an increased number of harmful insects in the fields. It must be remembered that farming the land serves quite different purposes, particularly in Northern Europe. The primary goal is obviously the production of food, but secondary goals, such as conservation of biodiversity and giving city dwellers opportunities for outdoor activities, are also important. For the latter purpose, setting aside more farmland would be helpful. In order to do this, political will and financial incentives are a prerequisite.

AGRICULTURAL BIODIVERSITY AND BIOTECHNOLOGY

In addition to wild plants, old landraces might be threatened by transgenic crops. It should be remembered that vertical gene transfer by pollen has always occurred between different old landraces and between different new varieties of crops. Despite this, varieties of apples or cereals have been stable over many years and specific traits have not disappeared. Pollen has always flown. What has become far more precise is the method of analysis. Thanks to gene probes and to GMOs, it has become much easier to follow gene flow, since one is no longer dependent on visible traits, but can follow a specific gene. Nevertheless, it is important to preserve landraces and native relatives of crops for their intrinsic value as well as for having starting material for future crop breeding.

Can newly introduced transgenic crops transfer genes vertically to native wild plants and thereby change important characteristics of the wild plants? Vertical gene transfer between cultivars and wild plants has always occurred within the limits of species, if the two types of plants were in close proximity and flowered at the same time. No new problems can be expected from transgenic plants, except if the gene transferred from the GMO to the wild plant significantly increased the fitness of the recipient. This seems rather unlikely *a priori*, but needs to be studied experimentally, as suggested by Ammann and collaborators in 2007, both in the laboratory and the field. Herbicide resistance transfer from a herbicide tolerant, transgenic crop to a close relative can occur in the field, as has been shown for canola by Mikkelsen in Denmark. However, this would be of major significance only if the recipient weed was controlled by this herbicide in this farm setting.

Small farmers in many developing countries use a remarkable number of races of many different crops. These are often well adapted to the local climate and topography, and are used to produce foods for different cultural purposes. In the Andes region of South America, dozens of different varieties of potatoes are grown, often side-by-side on small plots. In Europe, different varieties of potatoes are used for the industrial production of chips, and for preparing rösti in the home. Will the traditional races disappear when, and if, transgenic crops are introduced by farmers in developing countries? To judge from the past in Europe, the answer will be largely yes, not because of any biological hazard emanating from the GMOs, but because farmers need to produce their products and sell them economically. Many old varieties of apples have disappeared in Europe because of the preferences of the food retailers and consumers, who buy just a few varieties.

The rapid consolidation in the global seed market is also a concern for the maintenance of agricultural biodiversity: a decreasing number of companies control the global seed and agrochemical markets. Anti-trust legislation needs to prevent the formation of overly powerful monopolies. This concern is not a consequence of the introduction of agricultural biotechnology, but a consequence of globalisation. New technologies have, not only in biotechnology, but also in many other areas, made large companies more powerful and more visible. It may be possible that modern breeding methods based on molecular genetics will *per se* reverse the trend of variety losses, since the insight in genomics may bring back respect for races and the need to conserve them.

SOCIAL CONSEQUENCES

The introduction and spread of new technologies generally have social consequences with winners and losers. For biotechnology, this has led to intense public debate across many different facets, among them ethical, economic, legal and emotional (for details readers can also see *Biotechnology in the environment: potential effects on biodiversity*, and *Why genetic engineering causes concern—social, cultural and political impacts*).

Economic Considerations

One may wonder under what economic conditions biotechnology will benefit agricultural productivity, prevent the expansion of farmland and thereby help the preservation of biodiversity. It is worth remembering that the green revolution, which improved wheat and rice production in Asia, was initiated through work done in the public sector, namely in the Philippines at the International.

Rice research institute (IRRI) and in México at the Centro Internacional de Mejoramiento de Maiz y Trigo (CIMMYT), both institutions which belong to the Consultative Group of International Agricultural Research (CGIAR). Today, innovative research in agricultural biotechnology is largely done by a few large companies, with smaller contributions from the public sector. That companies strive for intellectual property rights (IPR) through patenting or other protective mechanisms is understandable and largely unavoidable, if innovation is to be maintained. However, provisions need to be made so that the agricultural research institutions of developing countries can obtain easy access to the patented materials and procedures as well as to the knowledge they need in order to improve the position of poorer farmers, as proposed for instance by IRRI.

A very limited number of transfers of specific patents have already occurred, such as from Monsanto to the Kenyan Agricultural Research Institute for producing virus-resistant sweet potatoes or another to IRRI for the promoter genes used in golden rice. This approach, including the possibility of forced licensing, is more practicable than wanting to overthrow the world's patenting system with the slogan 'no patents on life'. The patenting system does, however, need to be adapted in order to do justice to the complexity of living organisms. It may be noted in passing that the first patent granted for a living organism was given to Louis Pasteur in 1873 for a specific type of yeast.

A new issue arising in the field of IPR is the attempt to protect traditional, indigenous knowledge and preserve biodiversity through new international legislation. This legal instrument is still in an embryonic state of development.

At the level of the individual human being, the lack of food appears to be, and is, largely a problem of distribution: there is not sufficient food available where it is needed. In addition, the poor simply do not have the money to buy sufficient food. Since in developing countries most of the poor still live on the land, it is there that more income has to be generated, so that there is not only enough for survival, but a surplus which can be sold on the market. Increasing agricultural productivity is the most effective way to do this and should, according to economists, embrace modern biotechnology as an important component.

Ethical Considerations

The issues involved in the interaction between biodiversity and biotechnology have far-reaching consequences and need to be subject to an open and knowledge-based dialogue in society. The heated public debate, seen primarily in Europe in the last couple of years, and which in some circles led to near hysteria, is according to Leisinger not sufficient to solve the underlying problems. The dialogue needs to include many different stakeholders, including farmers of developing countries, diverse scientists, policy-makers and communicators. Cultural values involved in farming and food production need to be taken into consideration, just as much as the emotional side of eating and drinking.

A significant aspect of ethical behaviour is openness. Transparent information is required both from scientists in non-commercial settings as well as from industry. Although competition from fellow researchers and from industries does not allow the immediate divulgence of all research results, both groups should make concessions to the interested public and to policy-makers. This should be done by scientists personally and would help improve mutual confidence.

Several large projects in biotechnology like HUGO, the Human Genome Organisation, have an 'ELSI' component, dealing with ethical, legal and societal implications of biotechnology. This shows that some scientists realise that science cannot be seen as a human activity taking place in a void, without any connection to social and political realities. Ethicists stress that it is the responsibility of scientists to be actively concerned with these issues and that they should take special responsibility in communicating with non-specialists to explain what they know as well as what they don't know. Matrices of criteria have been set up for instance by Mepham to make it possible to ask relevant questions, such as who and what may be affected by agricultural biotechnology, and what properties like well-being, autonomy or justice may or may not be infringed upon. The questions put forward by ethicists need to answered on the basis of concrete scientific knowledge, using comparisons with other, established agricultural practices. Ethicists will have critical and thought provoking questions. However, their answers and attitudes may vary considerably. If they are positive as Comstock, ethicists will suggest pursuing research and application with appropriate caution. There is no easy answer to what the word 'appropriate' means, beyond answering questions on a case-by-case basis to balance the benefits and risks. Actively shaping the future and attempting to solve problems always involves risk.

It will be important to see ethics not as a stable set of principles, but rather as a discursive process, which will be able to cope with the growing speed of new developments. The Nuffield council of bioethics recently came to the conclusion that 'the moral imperative for making GM crops readily and economically available is compelling', if well-informed governments of developing countries want to introduce them. This imperative should be taken seriously and treated with urgency: every day 20,000 children die of malnutrition and 30,000 ha of humid tropical forest are destroyed.

Chapter 9

Biotechnology and Agrobiodiversity

INTRODUCTION

Agrobiodiversity has become an increasingly useful source for crop and animal improvement since biotechnology has widened and facilitated the potential use of genetic resources. Useful traits can be much more easily linked to specific genes and gene complexes, and such genes can be transferred across traditional genetic barriers. These options have raised awareness about the value of genetic resources, but not on the need to maintain our genetic diversity *in situ* as part of wider agrobiodiversity.

Awareness raising and public discussions will continue to be needed to correct for the current imbalance between biotechnology and agrobiodiversity.

For centuries traditional biotechnological processes have been used to contribute to agricultural production and food processing. However, as a result of the recent advances in molecular biology and genetics, existing biotechnological applications have been refined and many new biotechnological tools have been developed. Modern biotechnology has left the laboratory and reached the market place. The impact of biotechnology on agricultural production and food processing has accordingly increased. Traditional and modern biotechnology can be regarded as a continuum, with modern biotechnology characterised as laboratory-based and resource-intensive. As a consequence of its biological basis, the use of traditional and modern biotechnology shows strong interactions with agricultural biodiversity: biotechnological applications are based on biodiversity, and in turn influence biodiversity.

Various definitions of biological diversity or biodiversity, agricultural biodiversity or agrobiodiversity, and biotechnology have been given. Now widely accepted are the definitions in the Convention on Biological Diversity, agreed in 1992, for biological diversity and biotechnology. Biotechnology is defined as 'any technological application that uses biological systems, living organisms or derivatives thereof, to make or modify products or processes for specific use'. It is clear that this definition encompasses both traditional and modern biotechnological applications. Biological diversity is defined as 'the variability among living organisms from all sources including, *inter alia*, terrestrial, marine and other aquatic ecosystems and the ecological complexes of which they are part; this includes diversity within species, between species and of ecosystems'.

Questions of diversity within species, i.e. genetic diversity, can be addressed at the species, population, and within-population levels. Thus, this definition distinguishes three levels of integration with increasing complexity. Agricultural biodiversity, also known as agrobiodiversity, forms a subset of total biodiversity. It refers to biodiversity related to agriculture and can be described as 'the variety and variability amongst living organisms (of animals, plants and micro-organisms) that are important to food and agriculture in the broad sense and associated with cultivating crops and rearing animals and the ecological complexes of which they form a part'.

It includes the diversity found in farming systems as well as their surroundings to the extent that the latter influences agriculture. Food processing technologies — strictly speaking — are not covered by this definition, and will not be discussed, although the relationship between biotechnology and agrobiodiversity also largely holds for these technologies.

The interactions between biotechnology and agrobiodiversity are manifold. Traditional biotechnological applications in particular make use of microbial organisms. They include:

1. Composting, the accelerated microbial degradation of organic matter.
2. Nitrogen fixation, based on the ability of bacterial symbionts of leguminous plants to fix atmospheric nitrogen.
3. Fermentation, generating alcoholic beverages and dairy products and preserving a large variety of meat and plant products.
4. Ethnoveterinary practices aimed at protection of domestic animals against infectious diseases (vaccine production).

For these applications, specific microbial strains have coevolved and often been selected. In other words, biodiversity has been exploited to allow for these traditional biotechnological applications. Modern biotechnology has developed much more refined tools to select and generate optimal microbial organisms for a larger set of applications, and each of these involves the use of specific traits generated and expressed in the microbial domain.

These applications, as well as applications making use of plant and animal biodiversity, have contributed to the shaping of agricultural production. Cereals and pulses have been intercropped in all Centers of Origin of agricultural crops, an essential combination to guarantee high yields. Other crops, such as soyabean and enset (Ethiopia), have gained importance because protocols were developed to ferment the harvested products and allow for prolonged food storage. Ruminants (cattle as well as sheep and goats) have been selected for milk production concomitantly with the development of milk fermentation technologies.

Modern biotechnology has already influenced the genetic diversity of crops and animals cultivated and raised in the fields by allowing for the rapid and widespread introduction of desired starting material through *in vitro* technology. It is now on the brink of more profoundly changing our agriculture, in particular by the generation of novel crop varieties containing traits which could not be incorporated before and altering our farming practices. In this way, biotechnology has also changed and will further change our agricultural biodiversity. It can improve our production systems and the diversity of products we desire. The question to answer is whether agricultural biotechnology will be able to change the current, often negative image of modern agriculture into a new image which stands for a more sustainable way of dealing with resources, people and problems or is going to intensify current problems in agriculture concerning sustainability and biodiversity. Modern biotechnology also has a potential or actual influence on natural ecosystems, either by allowing changes in agriculture which affect natural biodiversity or by directly influencing natural biodiversity itself. These developments are now often discussed in the framework of biosafety policies.

This introduction so far attempts to stress the continuum in the interactions between traditional and modern biotechnology and biodiversity. Also from another perspective, a continuum characterises the influence of biotechnology on (agro) biodiversity in particular and agricultural practices in general.

From ancient times onwards, farming practices, including animal and plant breeding, have profoundly influenced our agroecosystems and the biodiversity these contain, and thus this influence is not the prerogative of biotechnology but of technologies applied in agriculture in general. Impacts on biodiversity may not be different in nature, only in degree, from those of our traditional or conventional practices.

The term genetic resources refer to a resource-centred view on biodiversity. Accordingly, it is often stated that loss of biodiversity means loss of capital and potentially useful resources. The immediate relevance of agrobiodiversity concerns food production and the prime incentive to preserve agrobiodiversity is based on economics. Although agriculture is both the predominant land-use system on earth and the largest global user of biodiversity, agrobiodiversity is a surprisingly minor topic in the global biodiversity debate.

The following text further elaborates the interactions between biodiversity and the applications of biotechnology. In particular, it focuses on the effects of the rapid developments in modern biotechnology on these interactions, both beneficial and detrimental. It deals with the use of biodiversity in biotechnological applications, and with the subsequent effects of these applications on the biodiversity in farming systems and natural environments. Although a substantial part of research is currently devoted to the diversity of beneficial and noxious insects, micro-organisms and other soil organisms, we have limited ourselves to the interface between agrobiotechnology and plant and animal biodiversity relevant to agroecosystems.

TECHNICAL ASPECTS OF AGRICULTURAL BIOTECHNOLOGY AT THE INTERFACE WITH AGROBIODIVERSITY

Modern biotechnology can be defined as all biotechnology that involves a laboratory phase in its development or application. The term stands for several different laboratory-based technologies that are used either independently or in combination with each other. These technologies are *in vitro* technology, marker technology and gene technology. More recent technological advances, commonly referred to as 'genomics', allow the application of former technologies in an (ultra) high-throughput context. As a consequence, genomics will further speed up the pace of (agro) biotechnological innovations. We will give a brief overview of the tools of modern agricultural biotechnology as applied to plants and animals. Although microbes are an integral part of agricultural systems, and microbial biotechnology is well developed, its application is predominantly in processing and food production.

In Vitro Technologies

Modern plant biotechnology is based on hydroponics, the capacity to grow plants without a soil substrate in an aqueous solution of macro and micronutrients and light. Hydroponics has developed into a major technology for the mass propagation of high-value vegetable crops. It also formed the basis for the development of plant cell and tissue culture techniques. Micropropagation allows the regeneration of complete new plants from leaves or shoot apexes; free from diseases, notably viruses, and thus a rapid clonal propagation of elite plant material. This is usually but not necessarily carried out in specialised laboratories. In particular the micropropagation of ornamentals and trees has become an important economic enterprise. The capacity of a single cell to regenerate into whole plants, is known as cell totipotency. It signifies the ultimate form of micropropagation. As an alternative application, *in vitro* somatic embryogenesis can be used for rapid propagation of plant material, and in various cases encapsulated embryos are replacing seeds. Also, the culturing of specialised plant cells in the form of mass cell suspensions has developed into a technology for the production of specialty chemicals. Finally, *in vitro* technology has been instrumental in the development of technologies as protoplast fusion, embryo rescue, mutagenesis and others that are part of modern plant breeding, as well as in the development of cryopreservation techniques, the storage of living cells at ultra-low temperatures.

Storage of cells, tissues and organs of animal origin has also been achieved by cryopreservation. Micropropagation of plant material allows the production of millions of clonal progeny that are essential identical. Nowadays, more than a thousand plant species are being propagated in tissue culture. The obvious advantage is that elite material can be quickly multiplied without changing the desirable genetic constitution of the material. Obtaining disease-free plants is another major advantage of the tissue culture approach. However, micropropagation in tissue culture also suffers from somaclonal variation, due to rearrangements and mutation of the DNA during the maintenance of cells in tissue culture. Tissue culture-induced somaclonal variation can also be exploited as a new source of genetic variation for crop improvement. This way, tissue culture, as well as mutagenic treatments, can contribute to a limited extent to an increase of diversity.

In contrast to plant cells, animal cells are generally not totipotent. Although individual cells can be propagated in culture, regeneration of such cells into a new, fully developed animal is not known. What has been realised is the induction of directed differentiation (based on multipotentiality) of somatic mammalian and other animal cells.

In its effects, artificial insemination can be regarded as the animal equivalent of plant micropropagation. More recent and advanced developments have resulted *in vitro* fertilisation combined with embryo transfer, widening the options for rapid propagation of desired genotypes.

Genetic Modification and the Sourcing of Genes

In vitro technology formed a prerequisite for the development of genetic modification of plants. Genetic modification of plants combines techniques for plant tissue culture, techniques for cloning, *in vitro* amplification (PCR), and transfer of DNA, either by the use of the soil bacterium *Agrobacterium tumefaciens* as a vector, through electroporation or by the use of a particle gun. It is based on the ability to change the genetic constitution of a single cell and regenerate a new plant from that single cell. The aims of plant genetic modification are manifold. The technology can be used to transfer a gene from a wild relative simply to speed up breeding, but it can also be used for the transfer of genes into the plant gene pool that could not be introduced into plant genomes by other means.

In 1999 more than 70 transgenic crop varieties were registered for commercial cultivation. The crops involved include corn, soyabean, rapeseed, tomato, tobacco, potato, chicory, papaya, pumpkin and clover. In 2005, biotech soyabean continued to be the principal biotech crop, occupying 54.4 million hectares (60 per cent of global biotech area), followed by maize (21.2 million hectares at 24 per cent), cotton (9.8 million hectares at 11 per cent) and canola (4.6 million hectares at 5 per cent of global biotech crop area). During the first decade, 1996 to 2005, herbicide tolerance has consistently been the dominant trait followed by insect resistance and stacked genes for the two traits. In the future diversification may be expected not only from the private sector, but also from the (international) public sector. The CGIAR Challenge program Generation focuses on drought tolerance in cereals, legumes and clonal crops. Genetic modification efforts of the M.S. Swaminathan Foundation regard combating drought and salinity in locally adapted rice and wheat varieties, whereas the Indian Initiative on Crop Biofortification focuses on bio-availability and bio-efficacy with respect to iron, zinc and phosphor in rice, wheat and maize.

Beneficial applications of genetic modification in crop breeding and food production may include the following short-term options:

1. Harnessing natural fungal resistance expressed in the seed phase for resistance in other phases of the plant growth cycle.
2. Developing non-toxin based insect resistance traits (e.g. hairiness).
3. Changing seasonal conditions for plant growth, allowing the shift from winter crops to spring crops in moderate climates and allowing fallow land periods.
4. Developing crops with higher levels of insect tolerance.

In order to improve or change a plant variety by genetic modification, a biotechnologist requires genes to code for the new character. As a consequence, biotechnology has a great need for well-characterised genes, preferably available as pieces of DNA or cDNA (= cloned RNA). Characteristic for genetic modification is that the origin of the gene of interest is not important. DNA is virtually independent of the organism from which it originates. The attitude of plant biotechnologists to plant improvement is often as straightforward as ambitious: for any trait desired, somewhere there will be an organism providing the genes that are able to do the job. The challenge is to find those organisms, to identify the gene or genes involved, and to transfer those genes into a suitable plant. Thus bioprospecting, the identification and evaluation of properties in organisms for potential exploitation, is currently becoming a major input in (plant) biotechnology. The suitable organism may be a plant, but may also be a bacterium or a whale: the concept of 'wild relatives' is of no relevance to biotechnology. The interest of biotechnology in genetic diversity is therefore unlimited and transcends the scope of gene bank collections or populations being conserved on-farm. Botanical gardens, national parks or pristine nature may be as useful, hence valuable, as the diversity in local farming communities. Any organism in any ecosystem may contain genes that are or may become of use for any application. All organisms should therefore be conserved. Biotechnologists are in a prime position to recognise the importance of genetic resources. Here the key issue is documentation. Which genes or properties or compounds are present in which organism; what is the function of these genes and compounds? And where is the organism and the relevant DNA or cDNA likely to be found?

Breeders and curators of plant genetic resources collections traditionally distinguish the primary, secondary and tertiary gene pools, based on the ease by which germplasm of different plants can be recombined. In line with this terminology, biotechnology relies on what could be considered the 'quaternary gene pool', encompassing all genetic material that occurs in nature but cannot be introduced into a plant by any sexual means. To fulfil a role in current plant biotechnology and plant genetic modification, genebanks might consider storing DNAs and cDNAs from a much wider range of species, forming part of this quaternary genepool. Furthermore, biotechnological tools might enable the much wider exploitation of indigenous knowledge on the properties of plants and microbial organisms which form part of the quaternary genepool, in food production and beyond. In the near future, biotechnology may also be able to create new diversity that could be considered a 'fifth-level gene pool'. Examples of novel diversity created in the laboratory include the generation of new genes that encode proteins with improved functions by DNA shuffling, the random recombination of existing genes. Random shuffling of genes can be considered as the molecular equivalent of processes resulting in heterosis. However, this technique creates genetic diversity that does not necessarily have a counterpart in nature.

Genetic modification of animals is based on a combination of technologies, as in the case of plants. It involves the genetic information and reproductive tissues that generate new organisms: nuclei, fertilised eggs, embryos, gonads or embryogenic stem cells, as well as the totipotent cells originating from the inner cell mass of the early blastula stage. Embryonic stem cells can be cultured *in vitro*, engineered and introduced in the inner cell mass of a suitable mammalian embryo. Gene transfer in animal genetic engineering is achieved either by microinjection, lipofection or electroporation.

In principle, new techniques involving inter-species somatic cell cloning and inter-species nuclear transfer technology allow for more efficient rescue of endangered animal species. Discussions about the application of animal biotechnology run parallel to those on the application of plant biotechnology, but bio-ethical considerations are much more prominent in applications involving animals.

LOSS OF AGROBIODIVERSITY: PLANTS AND ANIMALS FOR FOOD AND AGRICULTURE

During the last 10,000 years, human civilizations have benefited greatly from the domestication, conservation and use of a group of animals and plants species used for agriculture and food production. The genetic diversity within these species, breeds and varieties, which have been largely exchanged through history and readapted to local conditions, have evolved adaptations that allow production in a wide range of situations, including some of the most stressful environments inhabited by man and that now provide a coherent basket of sustainable solutions to disease resistance, survival and efficient production. We all greatly depend on this diversity but especially the rural poorest, which depend on the intensive management of biodiversity to support their livelihoods.

While the international community now acknowledges the very essential role of agricultural biodiversity, the loss of this biological treasure is still increasing around the world. The following facts and figures may bring some light:

1. 120 cultivated plant species provide 90 per cent of human food supply from plants.
2. 14 mammalian and bird species provide 90 per cent of human food supply from animals.
3. 12 plant species and 5 terrestrial animal species provide 70 per cent of all human food supply.
4. Potatoes, rice, maize and wheat together with cattle, swine and chicken provide over 50 per cent of all human food supply.
5. More than 1300 aquatic species are farmed or collected from the wild; almost 80 per cent of this production tales place in developing countries or countries in transition.
6. Over the past 15 years, 300 of 6000 animal breeds identified by FAO have become extinct.
7. 1350 breeds currently face extinction, two breeds are lost each week.

Genetic erosion continues at an increasing rate. The most prominent threats to populations are:

1. Wars, pest and disease outbreaks (animal and human), and other natural disasters (drought, floods, earthquakes, etc.).
2. Social and economic changes, urbanisation, market changes and intensification leading to 'farmer extinction', 'habitat extinction'.
3. Global marketing of breeding material.
4. Breed/variety substitution or absorption, crossbreeding with exotic breeds/varieties.
5. Short-term goals, lack of recognition of current or future value of GRFA.
6. Poor monitoring and management, lack of sustainable breeding programs.
7. Poor policies: development, re-stocking.
8. Land-use policies that annex common grazing grounds displace pastoral societies and lead to loss of animal breeds.

Animals and plants genetically adapted to their environment need to be further developed and conserved because they will:

1. Form an insurance against unforeseen events and provide genetic resources for future generations.

2. Be most effective in achieving local food security objectives because they are more resilient to climatic stress and to local parasites and diseases.
3. Be more productive at lower costs and in low-input systems, and sustainable in the long-term.
4. Support food, agriculture and cultural diversity, including supply of special products and cultural values.
5. Constitute an unique source of genes for improving health and performance of industrial breeds/varieties.
6. May offer new business opportunities.

The International Framework

International awareness of the essential role played by agricultural biodiversity in food and agriculture is gradually increasing. The intergovernmental discussions were started in the 1980's by the FAO, when an intergovernmental body was created to address the policy questions regarding the management and exchange of plant genetic resources for food and agriculture. The Commission on Genetic Resources for Food and Agriculture (CGRFA), hosted by the Food and Agriculture Organisation of the United Nations (FAO), is still the most important permanent forum for governments to discuss and negotiate matters relevant to genetic resources for food and agriculture. In 1995, its mandate was broadened to cover all components of agrobiodiversity of relevance to food and agriculture, although the broadening has not yet been implemented for forestry and fisheries genetic resources. At present 164 countries and the European Community are members of the CGRFA.

Other forums have contributed to advanced awareness and international cooperation in this area. In 1992, the United Nations Conference on Environment and Development, the Convention on Biological Diversity (CBD) and the Agenda 21 stared to provide a formal framework for dealing with global biodiversity. In 1996 and 2002 the FAO World Food Summits recognised the contribution of crop and animal genetic resources to food security, poverty alleviation and rural development. The Commission on Sustainable Development strongly emphasised the importance of promoting sustainable agriculture and rural development (SARD), and underlined the essential need to ensure the conservation and sustainable use of genetic resources in achieving sustainable agriculture. For this to happen, holistic value has to be re-evaluated and the cultural heritage appreciated. Ecosystem approaches to management, particularly of agroecosystems, must be focused, however, not only on the biological organisation but also on the human interactions that shape and influence them.

The International Community has now recognised, through the CBD and the FAO governing bodies, the special nature of agricultural biodiversity as it has some distinctive features that include the following:

1. Agricultural biodiversity is essential to satisfy basic human needs for food and livelihood security.
2. Agricultural biodiversity is managed by farmers; many components of agricultural biodiversity depend on this human influence; indigenous knowledge and culture are integral parts of the management of agricultural biodiversity.
3. There is a great interdependence between countries for the genetic resources for food and agriculture.
4. For crops and domestic animals, diversity within species is at least as important as diversity between species and has been greatly expanded through agriculture.
5. Because of the degree of human management of agricultural biodiversity, its conservation and development in production systems is inherently linked to sustainable use.
6. Nonetheless, much biological diversity is now conserved *ex situ* in gene banks or breeders' materials.

7. The interaction between the environment, genetic resources and management practices that occurs *in situ* within agroecosystems often contributes to maintaining a dynamic portfolio of agricultural biodiversity.

In recognising the distinctive features and problems of agricultural biodiversity, the International Community also recognised that these need to be solved through distinctive solutions. The member countries of the CGRFA have therefore developed and continue to monitor the Global Strategy for the Management of Farm Animal Genetic Resources; and the Global System for Plant Genetic Resources. A brief introduction of these two instruments follows in the next sections. Furthermore, the CGRFA reviews and advises FAO on policy, sectorial and cross sectorial matters, programs and activities related to the conservation, sustainable use and equitable sharing of benefits derived from the utilisation of genetic resources of relevance to food and agriculture, for present and future generations. The CGRFA facilitates and oversees cooperation between FAO and other relevant intergovernmental and non-governmental bodies, including, the Conference of Parties to the Convention of Biological Diversity (CBD) and its subsidiary bodies, and the institutes of the CGIAR.

International Treaty for Plant Genetic Resources for Food and Agriculture

A major achievement in this area has been the adoption by the FAO Conference in November 2001, of the International Treaty on Plant Genetic Resources for Food and Agriculture. This new and unique international agreement underwrites the conservation and sustainable use of all plant genetic resources for food and agriculture, and the fair and equitable sharing of the benefits arising out of their use. This is essential for agricultural development and for ensuring world food security, for this and future generations. The International Treaty will enter into force this year after 40 governments have ratified it.

Through the International Treaty, countries agree to establish a multilateral system to facilitate access to plant genetic resources for food and agriculture, and to share the benefits in a fair and equitable way. The Multilateral System applies to a set of crops and forages selected according to criteria of food security and interdependence. Establishing an effective and transparent Multilateral System will be vital in ensuring the continued availability of plant genetic resources that countries need to feed their people.

An other unique element of the International Treaty is the article on Farmers' Rights, a recognition of the contribution that farmers and their communities have made and continue to make to the conservation and development of plant genetic resources.

Finally the International Treaty contains a series of supporting components, different instruments that along with the establishment of a funding strategy will hopefully facilitate the ultimate objective of the Treaty, to make diversity work for sustainable development and food security.

Global System on Plant Genetic Resources

The objectives of the Global System are to ensure the safe conservation, and promote the availability and sustainable use of plant genetic resources by providing a flexible framework for sharing the benefits and burdens. The CGRFA, with its Intergovernmental Technical Working Group on Plant Genetic Resources for Food and Agriculture, monitors and coordinates the development of the Global System. Two key elements of the Global System are:

1. The Report on the State of the World's Plant Genetic Resources.
2. The Global Plan of Action for the Conservation and Sustainable Utilisation of Plant Genetic Resources for Food and Agriculture.

The first Report on the State of the World's Plant Genetic Resources was prepared through a participatory, country-driven process. It assessed the state of plant genetic diversity, and capacities at the local and global levels for *in situ* and *ex situ* management, conservation and utilisation of plant genetic resources. The Report was presented to the Fourth International Technical Conference held in Leipzig, Germany, in June 1996. A Global Plan of Action (GPA) was formally adopted by 150 countries through the *Leipzig Declaration*. The GPA comprises a set of activities covering capacity building, and the *in situ* and *ex situ* conservation of plant genetic resources. It is a rolling plan that is monitored, reviewed and updated by the CGRFA. The second Report on the State of the World's Plant Genetic Resources is under preparation.

In addition to these two elements, the Global System comprises international agreements, a variety of codes of conduct, scientific standards, technical mechanisms and global instruments for plant genetic resources for food and agriculture.

Global Strategy for the Management of Farm Animal Genetic Resources

The Global Strategy for the Management of Farm Animal Genetic Resources provides a technical and operational framework for assisting countries, comprising:

1. An intergovernmental mechanism for direct government involvement and policy development.
2. A country-based global infrastructure to help countries cost-effectively plan, implement and maintain national strategies for the management of animal genetic resources.
3. A technical program aimed at supporting effective action at the country level in the sustainable intensification, conservation, characterisation and access to Animal Genetic Resources.
4. A reporting and evaluation system to guide the Strategy's implementation, facilitate collaboration, coordination and policy development and maximise cost-effectiveness of activity.

The First Report on the State of the World's Animal Genetic Resources based on country reports and regional syntheses, and including a Report on Strategic Priorities for Action, is expected in 2006. The draft Report on Strategic Priorities for Action will be discussed by the CGRFA in its 10th session.

Documents of the Commission on Genetic Resources for Food and Agriculture http://www.fao.org/ag/cgrfa/docs.htm and http://dad.fao.org/en/Home.htm.

SUSTAINABLE LIVELIHOODS APPROACH

The sustainable livelihoods framework can help to explore the linkages between agrobiodiversity, gender and local knowledge. Moreover, it will help us broaden our perspective and apply a more holistic view to these issues. Recent research, on traditional crops and livestock species, suggests there is a significant gap between development and research priorities and farmers' needs. Two main perspectives can be identified, which are compared in the Table 9.1.

Table 9.1. Comparison of different perspectives on agrobiodiversity.

Livelihoods perspective	Natural resource management perspective
Focus is on local people and their livelihood strategies	Focus is on genetic resources and their production potential and use
Holistic in terms of understanding the purposes and functions played by agrobiodiversity in livelihood strategies	Narrow in terms of understanding and strengthening different purposes and functions of agrobiodiversity

(Contd ...)

Livelihoods perspective	Natural resource management perspective
Dynamic in terms of changing priorities and needs of different people at different times	Static resulting from the pre-selection of priority species for improvement and conservation
Builds on people's strength, e.g. local knowledge for species selection and *in situ* conservation practices	Draws heavily on external knowledge and technologies for species improvement, including *ex situ* conservation practices
Macro-micro linkages, e.g. policy lobbying for Farmers' Rights to secure local access to genetic diversity	Tends to focus more on either natural resource level or policy level
Sustainability related to improved local capacities and empowerment of local people	Sustainability questionable because little attention is given to building local capacities

What is the departure point of the livelihoods perspective? The people themselves must be the main entry point for analysing the management of agrobiodiversity. If people are not the starting point, it will be difficult to come up with research and development priorities that are in line with the views of the local people. The merits of using a livelihoods perspective to understand the management of agrobiodiversity are described in more detail below:

People-Centred

The entry point to agrobiodiversity management is people themselves. A livelihoods perspective facilitates a more thorough analysis of different social groups, including the distribution of benefits and access to resources from a gender perspective. Adoption of a livelihoods perspective will, therefore, facilitate identification of the multiple functions and purposes agrobiodiversity plays. Be it for different social groups and different environments, it will place the food security of poor people at the centre of the discussion.

Holistic

From a livelihoods perspective, agrobiodiversity management is not seen as a separate activity that aims to conserve individual species, varieties or breeds. Rather, it is seen to be part of the day-to-day livelihood strategies around the world. Farmers do not maintain agrobiodiversity for the mere purpose of conservation. They apply a more integrated and holistic perspective to the use of species, varieties and breeds within their agricultural system. Agrobiodiversity is managed by farmers, for a wide range of reasons, and the success of conservation and improvement depends on the benefits people obtain.

Dynamic

The use and management of agrobiodiversity is dynamic. Different components of agrobiodiversity are used by different people at various times and places, thus contributing to the development of complex livelihood strategies. Understanding how this use differs according to wealth, gender, age and ecological situation is essential to the understanding of agrobiodiversity's contribution to the livelihoods of different members in a community.

Building on Strength and Assets

If we take a livelihoods perspective it means we focus on livelihoods' existing strengths and assets, rather than on weaknesses and needs. From a livelihoods perspective, local knowledge and genetic resources are considered important assets. The knowledge held by farmers, for example, on their local plant and livestock species is a crucial component of species selection, conservation and improvement.

Local plants and animals form part of a complex agroecosystem; farmers have built up a significant stock of knowledge on how these have to be managed under specific conditions.

Macro-micro Linkages

Research and development activities tend to focus on either the macro or micro level. Applying a livelihoods perspective, it is important to link these levels for the successful management of agrobiodiversity. As many factors related to the loss of agrobiodiversity are linked to the macro level. Factors contributing to the loss of agrobiodiversity include globalisation of markets, funding strategies and the setting of priorities for research and development and access rights to genetic resources. On the other hand, the micro level is relevant to the consideration of agrobiodiversity as a valuable asset managed by a variety of people.

Sustainability

The livelihoods approach emphasises the importance of building on existing strengths and capacities. Key aspects are the empowerment of local people through information sharing and capacity building. In addition, the negotiation of Farmers' Rights and the equitable sharing of these benefits will contribute to livelihood sustainability.

Overall, the livelihoods perspective is concerned first and foremost with people. An accurate and realistic understanding is sought of people's strengths (assets or capital endowments) and how they may convert these into positive livelihood outcomes. The approach is based on the belief that people require a range of assets to achieve positive livelihood outcomes. No single category of assets, on its own, is sufficient to yield the many and varied livelihood outcomes that people seek. This is particularly true for the poor whose access, to any given category of assets, tends to be very limited. They have to seek ways of nurturing and combining the assets they have in innovative ways to ensure survival.

The example from Kenya shows the complexity behind a simple activity such as bean growing. Women farmers try to achieve a range of different livelihood outcomes, by using a diversity of bean varieties. In this case, their bean varieties form a central asset in their livelihoods strategy. The land they use to plant these crops is another important asset, and so is their labour, which they use to manage these crops. The livelihood outcomes they achieve include food security, health issues, pest management strategies.

The livelihoods approach furthermore emphasises the relevance of the wider context in which people's livelihoods and their assets are embedded. This is very important to bear in mind, when agrobiodiversity and its potential contribution to people's livelihoods are discussed, people's vulnerability context, existing policies, institutions and processes need to be considered as well. We must consider the different livelihood strategies and outcomes that strongly determine how these assets can be used. The Fig. 9.1 is a schematic view of the sustainable livelihoods framework. The terms used in this framework will now be explained and presented in more detail.

The sustainable livelihoods framework presents the main factors affecting people's livelihoods, and typical relationships between these. The framework can be used in both planning new development activities and assessing the contribution to livelihood sustainability made by existing activities. In particular the framework:

1. Provides a checklist of important issues and sketches out the way these link to each other.
2. Draws attention to core influences and processes.
3. Emphasises the multiple interactions between the various factors affecting livelihoods.

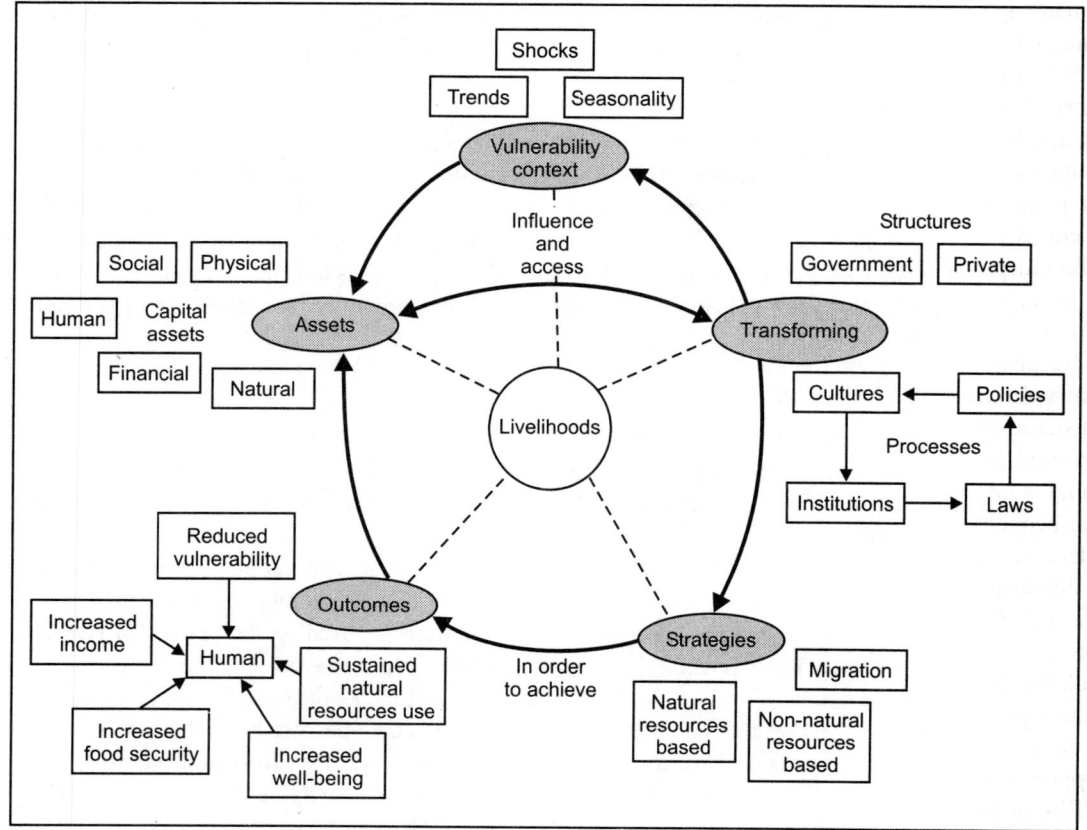

Fig. 9.1. Sustainable livelihoods framework.

The framework does not work in a linear manner and does not try to present a model of reality. Its aim is to help stakeholders, with their different perspectives, engage in structured and coherent debate of the many factors affecting livelihoods, their relative importance and the way in which they interact. In our case, the framework should help exploring linkages between agrobiodiversity, gender and local knowledge and to better understand their potential in contributing to improved livelihoods.

Livelihoods are shaped by a multitude of different forces and factors, which are themselves constantly changing. People-centred analysis is most likely to begin with the simultaneous investigation of people's assets, their objectives (the livelihoods outcomes they seek) and the livelihood strategies they adopt to achieve these objectives. Following terms used in the framework and their relevance will be explained.

Assets

Assets are what people use to gain a living. They are the core aspects of a livelihood. Assets can be classified into five types—human, social, natural, physical and financial. People will access assets in different ways, e.g. through private ownership or as customary rights for groups.

Human capital is the part of human resources that is determined by people's qualities, e.g. personalities, attitudes, aptitudes, skills, knowledge, also their physical, mental and spiritual health.

Human capital is the most important, not only for its intrinsic value, but because other capital assets cannot be used without it. Like social capital, described below, it can be difficult to define and measure.

Social capital is that part of human resources determined by the relationships people have with others. These relationships may be between family members, friends, workers, communities and organisations. They can be defined by their purpose and qualities such as trust, closeness, strength, flexibility. Social capital is important because of its intrinsic value. This is because it increases well-being, facilitates the generation of other capital and serves to generate the framework of society in general; with its cultural, religious, political and other norms of behaviour. With agrobiodiversity, we could think of the linkages between generations that facilitate the flow of information and knowledge. Or, we could think of seed exchange strategies between households, as part of a safety-net, in case of crop losses, etc.

Natural capital is made up of the natural resources used by people: air, land, soil, minerals, water, plant and animal life. They provide goods and services, either without people's influence (forest wildlife, soil stabilisation) or with their active intervention (farm crops, tree plantations). Natural capital can be measured in terms of quantity and quality (acreage, head of cattle, diversity and fertility). Natural capital is important for its general environmental benefits, and because it is the essential basis of many rural economies (in providing food, building material, fodder). This is probably the easiest asset to understand, because agrobiodiversity, as such, forms a natural capital.

Physical capital is derived from the resources created by people. These include buildings, roads, transport, drinking-water, electricity, communication systems and equipment and machinery that produce more capital. Physical capital is made up of producer goods and services and consumer goods that are available for people to use. Physical capital is important, because it directly meets the needs of people through provision of access to other capital via transport or infrastructure. A relevant example related to the management of agrobiodiversity is the availability of storage facilities to keep seeds from one cropping cycle to the next.

Financial capital is a specific and important part of created resources. It comprises the finance available to people in the form of wages, savings, supplies of credit, remittances or pensions. It is often, by definition, poor people's most limiting asset. Although it may be the most important, as it can be used to purchase other types of capital, and can have an influence, good and bad, over other people. With regard to agrobiodiversity, financial assets may be important in that they prevent people from having to eat or sell all their crops and seeds or slaughter all their livestock.

Balance

The relative amount of assets possessed or available to an individual, will vary depending on gender, location and other factors. The pentagon diagram representing assets can be redrawn, as shown in the example, to visualise the relative amount of each capital that is available to be accessed by an individual or community. It is important to know how this access and availability varies over time.

Vulnerability Context

The extent, to which people's assets can be built up, balanced; and how they contribute towards their livelihoods, depends on a range of external factors that change people's abilities to gain a living. Some of these factors will be beyond their control and may exert a negative influence. This aspect of livelihoods can be called the vulnerability context. This context must be understood, as far as possible, so as to design ways to mitigate the effects. There are three main types of change:

Trends: These are gradual and are relatively predictable. Changes may relate to population, resources, economy, governance or technology. They can have a positive effect, although here we focus on negative effects. Examples are:

1. Gradual degradation of natural resource quality. The processes of desertification can lead to the loss of valuable plant and animal species.
2. Excessive population increase because of migration, which can lead to increased pressure on local resources resulting in unsustainable use and depletion.
3. Inappropriate developments in technology may displace local crop or livestock species or varieties.
4. Undesired changes in political representation might lead to political systems that exploit local natural resources.
5. General economic stagnation may lead to increased poverty, and result in the unsustainable management of local resources. This could, for instance, lead to the depletion of certain plant genetic resources.

Shocks: Some external changes can be sudden and unpredictable. They may be related to health, nature, economy or relations. Generally, they are far more problematic. Examples are:

1. Climatic extremes (drought, flood, earthquake), which could wipe out existing plant or animal resources.
2. Civil disturbance (revolution) could affect social structures. May result in the interruption of knowledge transfers for the management of animal or plant genetic resources.
3. Outbreaks of disease, e.g. HIV/AIDS could lead to changes in labour resources for agricultural activities. Certain crops might be abandoned along with the related knowledge of their management.

Seasonality: Many changes are determined by the seasonal effects of crop production, access and living conditions. Although short-term, enduring for a season, they can be critical for poor people who have a subsistence livelihood. Examples are changes in:

1. Prices: Could make production of certain products, and their related plant resources, too expensive and therefore unattractive. In turn, this may lead to their abandonment.
2. Employment opportunities: Could change the availability of labour resources, for agricultural production in important seasons, leading to the loss of some agricultural practices and crops.

Policies, Institutions and Processes (PIPs)

In addition to the factors that determine the vulnerability context, there is a range of policies, institutions and processes designed to influence people and the way they make a living. If designed well, these influences on society should be positive. However, depending on their original purpose, some people may be affected negatively.

Policies, institutions and processes, within the livelihoods framework, are the institutions, organisations, policies and legislation that shape livelihoods. Their importance cannot be over-emphasised. They operate at all levels, from the household to the international arena. They function in all spheres, from the most private to the most public. They effectively determine:

1. Access to various types of capital, to livelihood strategies, and to decision-making bodies and sources of influence.
2. Terms of exchange between different types of capital.
3. Returns, economic and otherwise, to any given livelihood strategy.

In addition, they directly impact people's feelings of inclusion and well-being. Because culture is included in this area, PIPs account for other unexplained differences in the way things are done in different societies.

Examples of PIPs include:

1. Policies: On plant genetic resource use and biodiversity management.
2. Legislation: On patenting of plant genetic resources, property rights.
3. Taxes, incentives, etc. Incentives for growing cash crops or improved varieties that could replace local varieties.
4. Institutions: Extension or research institutions that promote external innovations, and represent the interest of prosperous farmers who depend less on agrobiodiversity.
5. Cultures: Concerning gender relationships, which may affect access and decision-making on crop and livestock selection and management.

Livelihood Strategies

To sum up the features of livelihoods: people use *assets* to make a living. They cope as best they can with factors beyond their control that make their livelihoods vulnerable. They are affected by existing policies, institutions and processes, which they can partly influence themselves. There are three main types of strategies, which can be combined in multiple ways:

1. Natural resource based: The majority of rural dwellers will plan on ways to make a living, based directly on the natural resources around them, e.g. subsistence farmers, fishers, hunter/ gatherers, plantation managers.
2. Non-natural resource based: Some rural dwellers, and most urban-based people, will opt to make a living based on created resources ranging from begging, service jobs, drivers, government jobs to shop-keeping.
3. Migration: If there are no appropriate opportunities for people to make a living, then a third option may be to migrate away from the area to a place where they can make a living. Examples vary from nomadic tribes to the expatriate academic. This migration can be seasonal or permanent.

Recent studies have drawn attention to the enormous diversity of livelihood strategies at every level—within geographic areas, across sectors, within households and over time. This is not a question of people moving from one form of employment or 'own-account' activity (farming, fishing), to another. Rather it is a dynamic process in which people combine their activities to meet their various needs at different times. A common manifestation of this, at the household level, is 'straddling', whereby different members of the household live and work in different places temporarily, e.g. seasonal migration or permanently.

Livelihood Outcomes

The aim of these livelihood strategies is to meet people's needs, as efficiently and effectively as possible. These needs can be expressed as desired livelihood outcomes of a chosen livelihood strategy. When considering 'poor' people, there are five basic outcomes that will usually be most important to them. The priority given to each will depend on the individual's perception of his or her circumstances. They are as follows:

1. Increased food security: A basic requirement for any livelihood is to achieve food security. It is not enough to have adequate food for part of the year and insufficient in another. There must be a secure supply all year round.
2. Increased well-being: An increased feeling of physical, mental and spiritual well-being is an important and basic need. To a certain extent, it is dependant on other needs being met.

3. Reduced vulnerability: As far as possible, a chosen livelihood should help reduce the effect of the various factors that make life more vulnerable, e.g. drought, conflict.

4. Increased income: Clearly, most poor people will want their income increased to an adequate level, and to have the maximum flexibility in meeting their needs.

5. Sustainable natural resource use: Since many livelihoods of the rural poor depend on access to natural resources, it is important that their strategies lead to more sustainable use of these resources.

Key Points

1. The sustainable livelihoods framework presents the main factors affecting people's livelihoods and the typical relationships that exist between these features.

2. The entry point to agrobiodiversity management is people themselves.

3. Agrobiodiversity management is not a separate activity that aims to conserve individual species, varieties or breeds. Rather, it is seen as part of the day-to-day livelihood strategies of people throughout the world.

4. Taking a livelihoods perspective means focusing on existing strengths and livelihoods assets, rather than on weaknesses and needs.

5. It is important to link macro and micro levels for the successful management of agrobiodiversity.

6. The use and management of agrobiodiversity is dynamic. Different components of agrobiodiversity are used by different people at different times and in different places, contributing to the development of complex livelihood strategies.

7. The livelihood approach emphasises the relevance of the wider context in which people's livelihoods and their assets are embedded.

8. The empowerment of local people, through information sharing and capacity building, are key aspects of a livelihoods approach.

LINKAGES BETWEEN AGROBIODIVERSITY, LOCAL KNOWLEDGE AND GENDER FROM A LIVELIOODS PERSPECTIVE

In fact agrobiodiversity can be considered an important natural capital or asset, for poor people's livelihoods, having the potential of contributing to food security and income generation. Human capital—such as local knowledge is considered to be a livelihood asset that can contribute to different livelihood strategies. Gender roles and relations form part of the policies, institutions and processes influencing the probability that people will use their assets to achieve their desired livelihood outcomes.

The challenge, faced by us and the research and development community, is to understand the linkages and complexities between these different livelihood components. Only then can we achieve the sustainable management of agrobiodiversity and can we contribute to the improvement of livelihoods, economic development as well as the maintenance of genetic diversity and associated local knowledge.

There is sufficient evidence, from past and current experiences, that these linkages and the way they function, result in positive or negative livelihood outcomes.

Relationships Between Assets

Assets combine in a multitude of different ways to generate positive livelihood outcomes. Two types of relationships are particularly important:

Sequencing: Do those who escape poverty start with a particular combination of assets? Is access to one type of asset or a recognisable subset of assets, either necessary or sufficient to escape poverty?

This is an important question to consider, in terms of the conservation efforts employed to maintain agrobiodiversity. Is it enough to have access to a wide range of diversity? Or, do people need other types of assets to make effective use of agrobiodiversity? The short case study from Cameroon and Uganda (Table 9.2) shows that the availability of a market structure is crucial to the successful selling of products. Usually, the livelihoods of poor people are quite complex and draw on very different resources for their survival. Therefore, it seems unlikely that only one type of asset will be sufficient to make a living. Moreover, increasing evidence suggests that access to information, knowledge and market infrastructure are important factors governing the successful management of agrobiodiversity.

Table 9.2. Indigenous vegetables in Cameroon and Uganda.

In Cameroon and Uganda indigenous vegetables play an important role, in both income generation and subsistence production. Indigenous vegetables offer a significant opportunity for the poorest people to earn a living, as producers and/or traders, without requiring a large capital investment. These vegetables are an important commodity in poor households. This is because their prices are relatively affordable, compared with other food items. Arguably, the indigenous vegetable market is one of the few opportunities for poor, unemployed women to earn a living. Despite the growth in exotic vegetable production, indigenous vegetables remain popular, especially in rural areas, where they are often considered to be more tasty and nutritious than exotic vegetables. Indigenous vegetables often have a ceremonial role, and are an essential ingredient in traditional dishes.

Substitution: Can one type of capital be substituted for others? For example, can increased human capital compensate for a lack of financial capital in any given circumstance?

Existing research and development results show that poor people especially depend on natural capital. The possibility of their replacing the loss of diversity with other types of assets is extremely limited. However, this question cannot be answered in general terms and depends very much on individual or case specific conditions. For example, if there are alternative employment possibilities outside the agricultural sector, people having the relevant skills could move away from agriculture to other sectors.

Relationships with Other Framework Components

Relationships within the livelihood framework are highly complex. Understanding them is a major challenge, and a core step in the process of livelihoods analysis, leading to actions to eliminate poverty.

Assets and the vulnerability context: assets are both destroyed and created, as a result of the trends, shocks and seasonality of the *vulnerability context* (Fig. 9.1). For example, the sudden disappearance of formal seed distribution systems in a given area could cause people to return to local crop varieties and seed systems, which would enhance diversity. Or a natural or human-induced disaster could lead to the loss of local seeds in a region.

Assets and policies, institutions and processes (PIPs): Policies, institutions and processes have a profound influence on access to assets. They:

1. Create assets: Government policy to invest in basic infrastructure, physical capital or technology generation, yielding human capital or the existence of local institutions that reinforce social capital. For instance, these could be important for the maintenance of local seed systems or livestock management practices.
2. Determine access: Ownership rights, institutions regulating access to common resources. This is extremely relevant with respect to agrobiodiversity for intellectual property rights, patents, etc.
3. Influence rates of asset accumulation: Policies affecting returns to different livelihood strategies, taxation, etc. With respect to agrobiodiversity management one could think about incentive structures to enhance various systems.

However, this is not a simple one-way relationship. Individuals and groups themselves influence policies, institutions and processes. Generally speaking, the greater people's asset endowment, the more influence they can exert. Hence, one way to achieve empowerment may be to support people in building up their assets.

Assets and livelihood strategies: People with more assets tend to have a greater range of options. They also have the ability to switch between multiple strategies to secure their livelihoods. When looking at available assets and livelihood strategies, there is an important gender dimension. As men and women have different livelihood strategies, they manage agrobiodiversity in different ways.

Assets and livelihood outcomes: Poverty analyses have shown that people's ability to escape from poverty is critically dependent upon their access to assets. Different assets are required to achieve different livelihood outcomes. For example, some people may consider a minimum level of social capital essential to the achievement of a sense of well-being. Or, in a remote rural area, people may feel they require a certain level of access to natural capital to provide security.

The following short example illustrates most of the issues mentioned above. It shows how a natural asset (indigenous vegetables) is used to contribute to various desired livelihood outcomes. It also illustrates that the existence of certain infrastructure (markets) is required to successfully carry out a particular livelihood strategy (in this case the marketing of these vegetables). Furthermore, it shows that trends, such as the increasing production of exotic vegetables, do not necessarily negatively affect this livelihood strategy.

Linkages Between Policies, Institutions and Processes Within the Framework

The influence of PIPs extends throughout the framework:

1. There is direct feedback to the vulnerability context. PIPs affect trends both directly, policies for agricultural research and technology development/economic trends, and indirectly, health policy/population trends. They can help cushion the impact of external shocks, policy on drought relief, food aid, etc. Other types of PIPs are also important, for example, well-functioning markets can help reduce the effects of seasonality by facilitating inter-area trade. In turn this could be an incentive for local farmers to maintain certain crop varieties, which would otherwise be replaced by marketable crops.

2. PIPs can restrict people's choice of livelihood strategies. Common examples are policies and regulations that affect the attractiveness of particular livelihood choices through their impact upon expected returns. For instance, establishment of quality norms of fruit and vegetables can cause the production of local varieties to be less attractive, as these may be less uniform than improved varieties.

3. There may also be a direct impact on livelihood outcomes. Responsive political structures that implement propoor policies, including the extension of social services into the areas in which the poor live, can significantly increase people's sense of well-being. They can promote awareness of rights and a sense of self-control. They can also help reduce vulnerability, through the provision of social safety nets. Relationships between various policies and the sustainability of resource use are complex and sometimes significant.

Key Points

1. Assets combine in a multitude of different ways to generate positive livelihood outcomes. Two types of relationship are particularly important: sequencing and substitution.

2. Livelihood assets are both destroyed and created as a result of the trends, shocks and seasonality of the vulnerability context.

3. Policies, institutions and processes have a profound influence on access to assets.
4. Those with more assets tend to have a greater range of options, and an ability to switch between multiple strategies to secure their livelihoods.
5. Men and women have different livelihood strategies, and therefore manage agrobiodiversity in different ways.
6. Poverty analyses have shown that people's ability to escape from poverty is critically dependent upon their access to assets. Different assets are required to achieve different livelihood outcomes.

FARMERS AND THE FUTURE OF AGROBIODIVERSITY

Agroecosystems and agrobiodiversity contribute to sustainable livelihood securities at the local, national and global levels. They provide a range of goods and services including food, fodder, climate change mitigation, biodiversity conservation and water quality options. Farmers and farming communities have a significant role to play in the preservation and conservation of these resources and ecosystems. The role of agriculture in the provision of ecosystem services depends, however, on the incentives available to it. At present incentives are designed to pay for the goods rather than the services provided by agricultural ecosystems. Payments for ecosystem services interventions often do not reflect correctly the social, environmental, economic and cultural aspects of the environmental services that farmers and farming communities deliver.

Arguments in favour of payments for ecosystem services in agroecosystems should be based not only on the current provisioning and regulatory services offered but also on anticipated future services. With growing reliance on increasing productivity systems in agricultural landscapes, food production and allied services agroecosystems offer the best option for provisioning services.

Some of the key issues or consideration by policy-makers to ensure the continued engagement of farmers in conservation and the use of agrobiodiversity include the following:

1. In spite of concerns relating to the depletion of agrobiodiversity, awareness of their potential uses is increasing. With growing pressure on land, demand for crops for activities such as biofuel production and bio-fortified food will increase the demand for agrobiodiversity. This pressure will affect the way in which farming, will be carried out in the future. It is crucial that farmers are encouraged to continue farming rather than moving to non-farming activities as the mainstay of their livelihoods. This requires some form of incentive system for farmers.

The importance of the services provided by agroecosystems including increasing the diversity of pollinator biodiversity and soil biodiversity, maintaining the natural enemy populations, providing of water and climate mitigation and adaptation options means that the conservation of agrobiodiversity is a necessary investment for countries and governments to make. Farmers and farming communities can benefit from the design of payment for ecosystem services mechanisms to encourage them to continue farming. In either case, the key challenge is to ensure farming remains attractive option. Failing crops owing climate variations, decreasing State support to fix fair prices for crops, disconnected markets influenced by middlemen and a lack of incentives mainstreamed into national procurement and export policies in agriculture all result in farmers being wary of continuing to work in the agricultural sector. Suitable policy and support packages are required to remove these negative forces that threaten farmers and farming communities.

2. Payment for ecosystem services in agricultural systems requires a careful assessment of the characteristics of services provided and the social and economic context within which such payment for ecosystem service schemes are discussed and designed. What should be paid for?

Who should it be paid for? How should be paid? These are all the key questions that require consideration by those people who are designing payment for ecosystem services interventions in agroecosystems.

3. Policy and regulatory aspects related to access to genetic resources, farmers' rights provisions and the safe use of genetically modified organisms are all relevant to farming practices. Farmers should be made aware of the debates on these issues and educated so that they can make informed decisions.

4. Payment for ecosystem services schemes are designed to encourage farmers to continue providing their service and should not be seen as a poverty reduction tool. This understanding is critical to delivering an appropriately-designed payment for ecosystem services mechanism for agrobiodiversity conservation. While agrobiodiversity is seen as a critical provisioning service, regulatory services such as the provision of water for irrigation and consumption through watershed management should also be considered in the design of payment for ecosystem services.

5. Direct drivers of agricultural production systems and their services include demand for food consumption, availability of crop diversity and their management, land use patterns, climate variability and change, energy provisions and availability of labour. Careful assessment of the links between these drivers and their individual and combined impact on agricultural production systems are critical to ensuring the development of suitable national agricultural and economic packages.

6. Access to credit, capital and assets often pose the greatest challenge to farmers who are engaged in subsistence farming. Most agrobiodiversity occurs in areas where subsistence farming is practiced owing to difficult growing conditions and the importance of farming as a low-risk option. Policy and development packages should consider appropriate interventions to deliver access to credit, capital and assets. Inappropriate credit structure, untimely capital support and unclear asset regimes are bound to create more problems rather than removing obstacles to enable farmers to continue farming in fragile environments and on small land holdings.

The agricultural sector is one of the largest contributors to greenhouse gas emissions, second only to the energy sector. Conversely, climate change affects agriculture throughout the world. According to the fourth assessment report of the Inter-governmental Panel on Climate Change, crop yield losses as a result of climate change will be more severe in the tropics than in temperate regions. Estimates indicate that between 75 million and 250 million people in Africa will be affected by water shortages caused by climate change. As in any situation of economic imbalance, the poor will be the most affected—losing livelihood opportunities and access to food and water. Many mitigation and adaptation measures are beyond the reach of countries with severe resource constraints.

Adapting to Climate Change

Adaptation to climate change should be considered from a contingency planning process perspective. Many least developed countries have had the opportunity to develop National Adaptation Plans of Action in the context of the United Nations Framework Convention on Climate Change but implementation of those programmes and strategic links to resourcing actions are often lacking. Adaptation in the agricultural sector can be seen in terms of both short-term and long-term actions. The provision of crop and livestock insurance, social safety nets, new irrigation schemes and local management strategies, as well as research and development of stress resistant crop varieties form the core of short-term responses. Long-term responses include redesigning irrigation systems, developing land management systems and raising finances to sustain adoption of those systems.

The Canadian agricultural sector has identified 96 distinct adaptation measures in agriculture including altering the topography of land; changing farming systems; using different crop varieties; making governmental and institutional changes and researching new technologies to take up the challenges posed to agriculture by climate change.

Agriculture and Mitigation

Livestock and crops emit carbon dioxide, methane and nitrous oxide making agriculture a major source of greenhouse gases. Some 80 per cent of these emissions come from developing countries. Agriculture is also a major cause of deforestation according to reports of the United Nations Framework, Convention on Climate Change. Nitrous oxide emissions from soils, because of the use of fertilisers and manures and methane from livestock production account for a third of non-carbon dioxide emissions. Land use change, compounded by agriculture, also reduces carbon sequestration.

Challenges

In light of the foregoing, the agricultural sector faces multiple challenges. While intensification and diversification of agriculture is key to securing food for local people, in the absence of clear understanding of their impacts on agriculture, they can be problematic. Though measures to reduce the use of fertilisers, to increase organic inputs and to deploy new varieties of crops are suggested as better agronomic practices, more clarity is required regarding their impacts on climate. For example, the selection of rice varieties that include wetland rice in sub-Saharan Africa can reduce deforestation as well as management costs and emissions. Agriculture could also benefit from emerging areas of climate change action. For example, it could profit from the benefits of land uses that sequester carbon, from the emerging markets for trading carbon emissions. Such activities offer higher returns than those arising from forest conversion to agricultural land. Post-2012 discussions under the Kyoto Protocol to the United Nations Framework Convention on Climate Change might consider exploring credits for the sequestration of carbon in soils through conservation tillage in agriculture as well as agroforestry in agricultural landscapes.

Livestock improvements brought about by more research on ruminant animals, storage and capture technologies for manure and conversion of emissions into biogas are additional contributions that agriculture can make towards mitigating climate change. National agricultural priority setting should consider climate change responses. While the biophysical impacts of climate change on agriculture and *vice versa* are better understood, the social and economic impacts have not been researched adequately in many developing countries. With increasing trade distortions and the changing prioritisation of agriculture in developed countries, developing countries affected by climate change should focus on developing suitable national, regional and global measures that will provide a safety net in the short-term, should productivity fail owing to climate variability and change. Institutional and human resource capacities supported by sustained funding options in the form of direct or indirect investments into adaptation to climate change in agriculture are essential. Mainstreaming climate change issues into national economic and development plans is critical to enabling countries tackle the impacts of climate change on agriculture and reducing the negative effects of agricultural practices on climate change.

Some key interventions

Some key answers may be derived from the observation that public expenditure decisions often focus on short-term payoffs that are politically correct. Long-term assessments of investments are often not considered. Increasing trade distortions and related national policies on the fixing of fair prices and the spreading of subsidies reduce incentives for farmers to continue working in agriculture, making public and private investments in agriculture unattractive.

The challenges to increasing investments and returns in agriculture and related conservation activities are based on micro-meso and macro-level actions. At the local level, there is a need to provide organised, sustained and relevant credit and technological systems for poor farmers to help increase productivity and reduce output costs. At the regional level, input supply and financial delivery systems need to be organised better and perhaps supported by suitable and trained institutional mechanisms. At the national level, there is a need for governments to provide adequate coordination (fair procurement prices, appropriate concessions, insurance mechanisms) and infrastructural services (storage, movement, processing, marketing) that are based on rational assessment of types of crops (such as cash crops, staples, etc.), agronomic conditions (rain-fed or irrigated) and future potential for private sector investment. This kind of sequencing that works in both directions would enhance investments in agriculture and conservation and also provide for better local economic empowerment.

Attracting investment from the private sector for agrobiodiversity conservation and agricultural development is a challenge, as the private sector traditionally has not been interested in crops without a good market value. Unfortunately for the agriculture of developing countries non-cash crops are a priority in ensuring local food security. One way to break this deadlock would be to develop and use innovative mechanisms such as tax concessions for private sector investment in research and development for non-cash crops. Complementary funding activities from the public sector and external donors that attract the private sector could increase research into staple crops.

Adequate protection measures for investments, through options such as suitable protection of intellectual property rights, breeders' rights or farmers' rights are another area that developing countries need to focus on to attract investment in agriculture.

In conclusion, it is important to note that attracting and sustaining investment into agriculture and agrobiodiversity conservation needs strategic long-term planning, understanding of local needs and the dynamics of agronomic practices and market potential, dialogue with the private sector and donors to encourage partnerships and enormous political will to move from securing the vote bank to securing sustainable livelihoods.

Agricultural Innovations and their Impact on Agrobiodiversity

Agriculture and agrobiodiversity sustain human life. Today, more than a decade after the 1996 World Food Summit, there are 820 million more hungry people around the globe than there were in 2006. If we are to keep the promise of the Summit, 31 million people must be removed from undernourishment every year until 2015. At present the number is climbing at a rate of about 4 million people per year. The proportion of people suffering from lack of food has, however, decreased as the global population has increased over the last decade.

In addition to the issue of absolute hunger, there is 'hidden hunger' that results from the lack of micronutrients available to people around the world, especially in sub-Sharan Africa. Agro, biodiversity offers the best solution to overcoming that 'hidden hunger'. Reduced yields of traditional crops, increasing reliance on farm-based subsidies and the decreasing market value of many traditional crops have taken their toll on agrobiodiversity at the rural and household levels. The decreasing availability of land, labour, water and energy mean that engaging in agriculture is increasingly challenging.

Research, institutional and policy settings for technological innovations in agriculture are, however, changing rapidly. Innovations now require plurality of systems and multiple sources. Linking technological progress with institutional and market changes is the need of the hour. Some key recent conclusions emanating from technological innovations include that: genetic improvements are successful, but not everywhere; there is a need for management and system technologies to complement genetic improvement; more investments are needed into research and development; the use of available technologies such as information and communication technology have still not permeated some parts of the world to improve the efficiency of agricultural systems; and innovative partnerships are key. Sharing of data and information and infrastructural developments and deployment will help to ensure better reach and impacts of available innovations.

Innovations and impacts

The characteristics of the various paradigms of agricultural innovation are changing rapidly, thereby having distinct effects, including, a move away from key innovators being scientists to potentially any of a range of actors, including farmers; the intended outcomes of interventions changing from mere technology transfer and uptake to enhanced capacities to innovate; and the changing role of policy from setting priorities and allocating resources to being a integral part of innovation capacity and strengthening the enabling environment.

Technical and technological innovations

Some regions such as sub-Sharan Africa continue to experience limited gains from the green revolution owing to the slow adoption of new and improved varieties by the farmers; agroecological heterogeneity; lack of infrastructure and lack of public policies that encourage better use of land and adoption of technologies; and poor market structures and related policies. Although the Consultative Group on International Agricultural Research, for example, invests some 35 per cent of its resources (twice the amount of investment in genetic improvement) in sustainable production systems the adoption and use of those systems remain limited, warranting an assessment of agricultural policies in these regions.

Public funding that contributes to about 94 per cent of current investment in agricultural research and development is scarcely able to match investments needs. There is limited private sector investment in research and development in agriculture in many developing countries, which is a cause for concern. Globally, research in agriculture is focusing more on maintaining yields than on improving them. In the absence of national policies on investment in agricultural research and development commercial interests may erode the thin base available to farmers in the form of agrobiodiversity.

Innovative ideas such as participatory plant breeding offer sustainable solutions to address the need for improved research and development in agriculture. Through participatory plant breeding, for example, farmers will be trained to be more efficient in the use of their varieties and to improve them to suit local agroclimatic conditions in addition to providing them with an opportunity to be mainstreamed into commercialising and protecting their varieties through mechanisms such as 'farmers' rights'.

Policy innovations

Recent studies have demonstrated that there is a particular trend with respect to the changing nature of food security and consequent policy reform. The general direction is towards greater openness and competition in national markets with respect to domestic and international trade. Countries, however, differ in their institutional and infrastructural set ups and their capacity to deal with these changes.

Policy reform should focus on developing and sustaining rural infrastructure, developing non-farming employment options, transitional compensatory measures and improving productivity and market access to rural people. Trade and related domestic policy reforms do not always run contrary to food security and the agricultural sector, although may appear to do so owing to the lack of subtlety in the design and sequencing of actions to implement the reforms. Countries require an innovative and responsive policy package that is designed to offset the negative impacts of liberalisation.

Biofuels and Agrobiodiversity

One of the great benefits of using biomass for energy production is its potential to reduce the green house gas emissions associated with fossil fuels. The associated risks, however, include potential effects on land use, agricultural production systems, habitats, biodiversity, land, air and water qualities. For this reason, discussions and activities related to bioenergy and biofuels are receiving attention at the environmental, economic and social levels especially with regard to the effects linked to agrobiodiversity, food security and the costs of food production.

The increase in the price of maize, for example, by 60 per cent between 2005 and 2007 in countries such as the United States of America is prompting economists and agriculturalists to assess the impacts of the global surge in biofuel production and distribution policies globally; with 240 kilograms of maize required to produce just 100 litres of ethanol, surely the choice is between producing more food or more fuel. That quantity of maize is equal to the food requirement of one person for one year!

According to a recent World Bank review, the potential environmental benefits of biofuels including their impacts on biodiversity, air, water and soil qualities cannot be generalised and need to be assessed on a case-by-case basis, evaluating cropping and land use patterns as well as the type of crop used for biofuel production.

Bioenergy and biofuel production planning and policies require the integration of relevant sectoral policies, such as energy, agriculture, environment, economics and rural development. According to the World Bank reviews, the reduction of greenhouse gases in the United States of America owing to the production of ethanol from maize will be in the range of 10 to 30 per cent, at best, while in Brazil the use of ethanol could reduce greenhouse gas emissions by 90 per cent. According to several studies, although biofuels have the potential to enhance national energy security, the benefits of this technology will remain limited for smallholder farmers. Countries will need sound schemes to ensure smallholder participation catering to local energy demands.

It is widely known that biofuel production in developed countries is backed by high protective tariffs in conjunction with large subsidies paid to biofuel producers. It is still debatable as to whether developing countries can afford such policies in addition to supporting policies that may have to deal with increased grain prices because of the diversion of grain to biofuel production. Challenges will remain until we are able to rely on first generation technologies for biofuel production. Second generation technologies such as production of fuels from microbes, may offer some respite, but not in the immediate future.

Impacts of biofuels on agrobiodiversity

According to a number of recent reports, biofuels have the potential to contribute to an increase food grain prices, increased competition for land and water, as well as deforestation, and destruction of agrobiodiversity. Socially and environmentally relevant strategies and policies on biofuels are therefore needed for developing countries. Significant international activity is underway to determine sustainability criteria, including through the Global Bioenergy Partnership and the Roundtable on Sustainable Biofuels. Tools developed by the Food and Agricultural Organisation of the United Nations led consortium such as the 'analytical framework' are expected to provide guidance for countries on dealing with the bioenergy industry to respond to energy security in a manner that does not jeopardise the food security of poor farmers.

Biofuels are currently produced using conventional food crops such as maize, wheat, sugar cane, oil palm and oilseed rape. The use of these crops for biofuels that too in large scale, will in near future create a direct competition between biofuels and food security as well as eroding agrobiodiversity. Research has shown that perennial rhizomatous grasses such as Arundo donax (native to Asia) and Phalaris arundinacea (reed canary grass native to Europe, Asia and North America) that are currently regarded as biofuel species are in fact invasives. Species such as Miscanthus that are regarded as potential sources of biofuel are again reportedly invasives. The following policy and research needs may assist in avoiding these difficulties:

Policy and research needs

Existing policy frameworks focus on supply targets rather than on delivery mechanisms and the impacts of production strategies on environmental, social and economic benefits. National strategies on biofuels should consider the cost of technologies, their deployment and use, the short, medium and long-term impacts of their use on farmers and biodiversity and assess the economic viability of biofuel technology.

Developing a coherent policy that balances feedstock supply against other uses of land and water would reduce the impacts of biofuel production on agrobiodiversity conservation.

Adequate investments and fiscal and regulatory incentives to develop further technologies that minimise the effects of biofuel feedstock on food crops are needed.

Continuing the discussion on national food, energy and environmental security in an integrated fashion to assess impacts would improve the development of research and policy options that are balanced towards future food and security needs.

Investing in second generation technologies such as the broadening range of feedstock, including dedicated energy crops, perennial grasses, forestry species that combine the properties of high carbon to nitrogen ratios, higher yields of biomass or oils, easily processable lignocellulosic material, use of organisms such as those from the marine environment (micro-algae) and direct production of hydrocarbons from plants or microbial systems are critical to reduce the potential impacts of biofuels on agrobiodiversity. Care should be taken, however, that such technologies are managed properly so as not to endanger agriculture and biodiversity.

Governance and Agriculture

Agriculture is one of the most promising instruments for reducing poverty and securing local livelihoods. One of the critical conditions required of the agricultural sector is to ensure that good governance structures and related policies are in place at all levels. Studies have shown that agriculture based economies around the world are to be found mostly in developing countries and countries with economies in transition. It is in these countries, however, that scores for governance are at their lowest and often negative. Good governance has a number of dimensions: political stability; the rule of law; voice and accountability; effective governments; regulatory quality and control of corruption, among others. In spite of the recommendations of the 1982 World Development Report that governments should focus on governance to improve agriculture, very little has been achieved in that regard to date.

Weak governance in agriculture led to major problems during the 1980s and 1990s when strong State interventions were undermined by structural adjustment programs, emphasising the role of markets.

Lack of macroeconomic policies and unstable political situations are pre-conditions to governance problems. Policy biases, underinvestment, mis-investment and lack of capacities underpin weak governance in agriculture. To address these issues, it is important to focus on the following key challenges.

Invigorate the roles of the public sector, private sector and civil society with guidance and support

The time has come for the public sector to think and act differently in relation to agriculture in all countries. The private sector and its role in dealing with agricultural governance need to be strengthened. Getting market dynamics right, improving macroeconomic policies and ensuring that State policies are strengthened to compliment the needs of civil society are critical to ensuring that partnerships are stronger and more responsive to the emerging challenges of national and global agricultural scenarios. Strong political will and support are essential.

Governance reform can be approached from the demand or the supply side

Reforms that improve accountability, transparency, voice and impact are needed to promote a sound political economy that favours good governance. It is important to understand the characteristics of agrarian communities to enable the demand-side approach to agriculture to be better addressed. Similarly, reforms to improve public sector capacity, efficiency and delivery must take into account problems affecting the performance of agriculture ministries, agencies and extension services. This focus will enable the supply side to be more responsive. Introduction of *e*-governance programs in India and ISO 9000 certification management schemes in EI Salvador and Mexico, for example, improved the supply side of the equation considerably by making agricultural administration more accountable. Similarly, the involvement of non-governmental organisations in decision-making processes and the implementation of reforms have provided positive experiences to dealing with the demand side of the issue in countries such as Ethiopia and Senegal. Decentralisation is seen as an effective answer to some of the questions related to better governance in agriculture.

Better implementation of 'Paris' agenda on aid effectiveness

Along with national and local partners, donors also have a role to play in improving governance in agriculture. Donors need to coordinate their efforts to provide support and mainstream their aid policies to ensure that they respond to national needs than the needs of their own capitals. There is a need for replication of ideas such as the Global Donor Platform for Rural Development, TerrAfrica or the Neuchatel Initiative, which provides an informal platform for bilateral and multilateral donors to develop common views and guidelines for in-country donor coordination.

Global action for agriculture

Managing global common challenges such as climate change; regional challenges such as pandemic plant and animal diseases and invasive species; conducting research and development into 'orphan crops' that are important for national and local food security (e.g. cassava) and reducing transaction costs through standards and rules all need support at the global level. In addition to reforms in national and regional level governance, reforms in the global governance of agriculture are much needed. From the United Nations reforms to changing agribusiness and increasing the influence of non-governmental organisations in setting the global agenda on governance all will have to consider their role in ensuring that good governance structures are adopted and provided for in developing countries. Training and human resource development to address agricultural governance, as in other areas of focus, needs special attention. Ethics, equity and justice are the three pillars to ensuring that global action for better governance makes sense at the local level in addition to making a difference to the livelihoods of millions of people in developing and least developed countries. Unless effective governance systems are put in place, reforms in agriculture and their impact on achieving food security and poverty reduction will remain ineffective.

Chapter 10

Impact of Genetically Engineered and Genetically Modified Crops on Biodiversity

INTRODUCTION

The potential impact of genetically engineered (GE) crops on biodiversity has been a topic of interest both in general as well as specifically in the context of the convention on biological diversity. Overall, the chapter discusses currently commercialised GE crops which have reduced the impacts of agriculture on biodiversity, through enhanced adoption of conservation tillage practices, reduction of insecticide use, and use of more environmentally benign herbicides. Increasing yields also alleviate pressure to convert additional land into agricultural use. The potential impact of genetically modified (GM) crops on biodiversity has been a topic of general interest as well as specifically in the context of the Convention on Biological Diversity. Agricultural biodiversity has been defined at levels from genes to ecosystems that are involved or impacted by agricultural production. After fifteen years of commercial cultivation, a substantial body of literature now exists addressing the potential impacts of GM crops on the environment.

GENETICALLY ENGINEERED (GE) CROPS

Crop Diversity

Crop genetic diversity is considered a source of continuing advances in yield, pest resistance and quality improvement. It is widely accepted that greater varietal and species diversity would enable agricultural systems to maintain productivity over a wide range of conditions. With the introduction of GE crops, concern has been raised that crop genetic diversity will decrease because breeding programs will concentrate on a smaller number of high value cultivars. Three studies have analysed the impact of the introduction of GE crops on within-crop genetic diversity. Studies of genetic diversity in cotton and soyabean in the US both concluded that the introduction of GE varieties was found to have little or no impact on diversity. In contrast, the introduction of *Bt* cotton in India initially resulted in a reduction in on-farm varietal diversity due to the introduction of the technology in only a small number of varieties, which has since been offset by more *Bt* varieties becoming available over time. From a broader perspective, GE crops may actually increase crop diversity by enhancing underutilised alternative crops, making them more suitable for widespread domestication.

Farm-scale Diversity

Plants have a major influence on soil communities of micro- and other organisms that are fundamental to many functions of soil systems, such as nitrogen cycling, decomposition of wastes and mobilisation of nutrients. The potential impact of *Bt* crops on soil organisms is well studied. A comprehensive review of the available literature, by Icoz and Stotzky, on the effects of *Bt* crops on soil ecosystems included the results of 70 scientific articles. The review found that, in general, few or no toxic effects of *Cry* proteins on woodlice, collembolans, mites, earthworms, nematodes, protozoa and the activity of various enzymes in soil have been reported. Although some effects, ranging from no effect to minor and significant effects, of *Bt* plants on microbial communities in soil have been reported, they were mostly the result of differences in geography, temperature, plant variety and soil type and, in general, were transient and not related to the presence of the *Cry* proteins. Studies published since the Icoz and Stotzky review have reached similar conclusions, including novel studies on snails.

Crop production practices also have significant effects on the composition of weed communities. Changes in the kinds of weeds that are important locally are termed weed shifts. Such shifts are particularly relevant for managing weeds in herbicide tolerant crop systems in which tillage practices and herbicide use both play major roles in shaping the weed community. There are reports in the literature of fourteen weed species or groups of closely related species that have increased in abundance in glyphosate resistant (GR) crops. At the same time, in a survey of corn, soyabean, and cotton growers in six states, between 36 per cent and 70 per cent of growers indicated that weed pressure had declined after implementing rotations using GR crops.

The use of herbicides can also result in changes to weed communities through the development of herbicide tolerant weed populations. Globally, GR weeds have been confirmed for 21 weeds in 15 countries. Most of these cases have been reported where GR crops are commonly grown. The development of weeds resistant to glyphosate will likely require modification to weed control programs where practices in addition to applying glyphosate are needed to control the resistant populations.

Landscape-scale Diversity

The most direct negative impact of agriculture on biodiversity is due to the considerable loss of natural habitats, which is caused by the conversion of natural ecosystems into agricultural land. Increases in crop yields allow less land to be dedicated to agriculture than would otherwise be necessary. A large and growing body of literature has shown that the adoption of GE crops has increased yields, particularly in developing countries. A review of the results of global farmer surveys found that the average yield increases for developing countries range from 16 per cent for insect-resistant corn to 30 per cent for insect-resistant cotton, with an 85 per cent yield increase observed in a single study on herbicide-tolerant corn. On average, developed-country farmers report yield increases that range from no change for herbicide-tolerant cotton to a 7 per cent increase for herbicide-tolerant soyabean and insect-resistant cotton. Researchers have estimated the benefit of these yield improvements on reducing conversion of land into agricultural use. They estimate that 2.64 million hectares of land would probably be brought into grain and oilseed production if biotech traits were no longer used.

The most direct landscape-level effects of growing *Bt* crops would be expected for target pest species for which the crop is a primary food source and that are mobile across the landscape. Area-wide pest suppression not only reduces losses to adopters of the technology, but may also benefit non-adopters and growers of other crops by reducing crops losses and/or the need to use pest control measures such as insecticides. Several studies have investigated the impact of the introduction of *Bt* corn and cotton on regional outbreaks of pest populations, reporting evidence of regional pest suppression in *Bt* corn and cotton in various areas of the US and in *Bt* cotton growing regions of China. The effects of GE crops on above-ground non-target invertebrates have been the subject of a large number of laboratory and field studies. By the end of 2008, over 360 original research papers had been published on non-target effects of *Bt* crops. A comprehensive review of the literature by Naranjo included 135 laboratory-based studies on nine *Bt* crops from 17 countries and 63 field-based studies on five *Bt* crops from 13 countries, which were analysed using meta-analysis techniques. In general, laboratory studies identified greater levels of hazard than field studies, at least partially explained by differences in organisms studied, and frequently higher protein exposure in lab studies compared to exposure levels in the field. Field studies demonstrated few harmful non-target effects, with the non-target effects of insecticides being much greater than *Bt* crops. More recent literature on the non-target impacts of *Bt* crops are largely consistent with Naranjo's conclusions.

Studies on the non-target impacts of herbicide tolerant crops, such as the UK Farm Scale Evaluations (FSE), have found that the effects on various groups of arthropods followed the effects on the abundance of their resources. Where weed control was more effective, the reduction in weeds and weed seeds led to decreases in insects that live in or on weeds, and *vice versa*. Other studies on the non-target impacts of herbicide tolerant crops, conducted for HT soyabean and corn in the US and HT canola in Canada, have reached similar conclusions.

The bird survey results of the FSE were in accord with differences in food availability found in the studies. Specifically, a greater abundance of granivores was found on conventional than genetically engineered herbicide tolerant sugar beet, as well as on genetically engineered herbicide tolerant maize after application of herbicides to the GE HT field. No differences were detected in spring oilseed rape. In the subsequent winter season, granivores were more abundant in fields where conventional sugar beet had been grown than on GE HT fields. Several bird species were more abundant on maize stubbles following GE HT treatment.

Indirect Indicators

The introduction of herbicide tolerant crops has been associated with the increased adoption of conservation tillage practices, which decreases run-off, increases water infiltration, and reduces erosion. Trends in the adoption of conservation tillage have been studied in the US and Argentina, the largest growers of herbicide tolerant crops. While conservation tillage was already being adopted by some growers prior to the introduction of GE herbicide tolerant crops in both countries, studies have shown a positive two-way causal relationship between the adoption of conservation tillage and the adoption of GE herbicide tolerant crops.

The pest management traits that are embodied in currently commercialised GE crops have led to changes in the use of pesticides that may have impacts on biodiversity. If the planting of GE pest-resistant crop varieties eliminates the need for broad-spectrum insecticidal control of primary pests, naturally occurring control agents are more likely to suppress secondary pest populations, maintaining a diversity and abundance of prey for birds, rodents and amphibians. In addition to the studies on the non-target impacts of GE crops compared to conventional practices, many studies have quantified changes in pesticide use since the introduction of GE crops. Reductions ranging from 14 per cent to 75 per cent of total active ingredient have been reported for *Bt* crops compared to conventional crops in Argentina, Australia, China, India and the US.

Fewer surveys have captured changes in herbicide use in GE herbicide tolerant crops, perhaps because the impact of GE herbicide tolerant crops has largely been a substitution between herbicides that are applied at different rates, and therefore, changes in the amount of herbicide used is a poor indicator of environmental impact. Several studies have been done to apply environmental indicators to observed changes in pesticide use related to the adoption of both insect resistant and herbicide tolerant crops, which all show a reduction in the environmental impact of pesticides used on GE crops.

Thus the knowledge gained over the past 15 years that GE crops have been grown commercially indicates that the impacts on biodiversity are positive on balance. By increasing yields, decreasing insecticide use, increasing use of more environmentally friendly herbicides, and facilitating adoption of conservation tillage, GE crops have contributed to increasing agricultural sustainability.

Previous reviews have also reached the general conclusion that GE crops have had little to no negative impact on the environment. Most recently, the US National Research Council released a comprehensive assessment of the effect of GE crop adoption on farm sustainability in the US that concluded, 'generally, GE crops have had fewer adverse effects on the environment than non-GE crops produced conventionally'.

GE crops can continue to decrease pressure on biodiversity as global agricultural systems expand to feed a world population that is expected to continue to increase for the next 30 to 40 years. Due to higher income elasticities of demand and population growth, these pressures will be greater in developing countries. Both current and pipeline technology hold great potential in this regard. The potential of currently commercialised GE crops to increase yields, decrease pesticide use, and facilitate the adoption of conservation tillage has yet to be realised, as there continue to be countries where there is a good technological fit, but they have not yet approved these technologies for commercialisation.

In addition to the potential benefits of expanded adoption of current technology, several pipeline technologies offer additional promise of alleviating the impacts of agriculture on biodiversity. Continued yield improvements in crops such as rice and wheat are expected with insect resistant and herbicide tolerant traits that are already commercialised in other crops.

Technologies such as drought tolerance and salinity tolerance would alleviate the pressure to convert high biodiversity areas into agricultural use by enabling crop production on suboptimal soils. Drought tolerance technology, which allows crops to withstand prolonged periods of low soil moisture, is anticipated to be commercialised within five years. The technology has particular relevance for areas like sub-Saharan Africa, where drought is a common occurrence and access to irrigation is limited. Salt tolerance addresses the increasing problem of saltwater encroachment on freshwater resources.

Nitrogen use efficiency technology is also under development, which can reduce run-off of nitrogen fertiliser into surface waters. The technology promises to decrease the use of fertilisers while maintaining yields, or increase yields achievable with reduced fertiliser rates where access to fertiliser inputs is limited. The technology is slated to be commercialised within the next 10 years.

GENETICALLY MODIFIED (GM) CROPS

This review takes a biodiversity lens to this literature, considering the impacts at three levels: the crop, farm and landscape scales. Within that framework, this review covers potential impacts of the introduction of genetically engineered crops on: crop diversity, non-target soil organisms, weeds, land use, non-target above-ground organisms and area-wide pest suppression. The emphasis of the review is on peer-reviewed literature that presents direct measures of impacts on biodiversity. In addition, possible impacts of changes in management practices such as tillage and pesticide use are also discussed to complement the literature on direct measures. The focus of the review is on technologies that have been commercialised somewhere in the world, while results may emanate from non-adopting countries and regions. Overall, the review finds that currently commercialised GM crops have reduced the impacts of agriculture on biodiversity, through enhanced adoption of conservation tillage practices, reduction of insecticide use and use of more environmentally benign herbicides and increasing yields to alleviate pressure to convert additional land into agricultural use.

After fifteen years of commercial cultivation, a substantial body of literature now exists addressing the potential impacts of GM crops on the environment. This review takes a biodiversity lens to this literature, considering the impacts at three levels: the crop, farm and landscape scales. Within that framework, this review covers potential impacts of the introduction of genetically engineered crops on: crop diversity, biodiversity of wild relatives, non-target soil organisms, weeds, land use, non-target above-ground organisms and area-wide pest suppression.

The emphasis of the review is on peer-reviewed literature that presents direct measures of impacts on biodiversity. In addition, possible impacts of changes in management practices such as tillage and pesticide use are also discussed to complement the literature on direct measures. The focus of the review is on technologies that have been commercialised somewhere in the world, while results may emanate from non-adopting countries and regions. The most direct negative impact of agriculture on biodiversity is due to the conversion of natural ecosystems into agricultural land. In that context, the potential impacts of GM crops are most appropriately considered in relation to prevailing modern agricultural practices. The categories of potential impacts of GM crops are similar to those of non-GM crops.

Crop Diversity

Modern agriculture is the result of a long process of plant domestication to create new and better agricultural produce for society. Conventional breeding has focused on improving economic efficiency, and as such has narrowed the number and genetic basis of current crops. It has been estimated that 7000 plant species have been used for human consumption, but that just four crops (wheat, maize, rice and potato) provide one-half of the total world food production and 15 crops contribute two-thirds. Crop genetic diversity is considered a source of continuing advances in yield, pest resistance and quality improvement. It is widely accepted that greater varietal and species diversity would enable agricultural systems to maintain productivity over a wide range of conditions. Particularly in light of climate change, maintaining and enhancing the diversity of crop genetic resources is of increasing importance to ensure the resilience of food crop production. A meta-analysis of studies on genetic diversity trends in crop cultivars released in the last century found no clear general trends in diversity. While a significant reduction of 6 per cent in diversity in the 1980s as compared with the diversity in the 1970s was observed, diversity in released varieties appears to have increased after the 1980s and 1990s.

With the introduction of GM crops, concern has been raised that crop genetic diversity will decrease because breeding programs will concentrate on a smaller number of high value cultivars. Three studies have analysed the impact of the introduction of GM crops on within-crop genetic diversity. A study of field genetic uniformity, a measure of genetic relatedness, in US cotton comparing 2000 to 2007, a year in which 72 per cent of cotton acreage was planted to GM varieties, found a 28 per cent reduction in uniformity across the US. Similarly, a study of 312 glyphosate tolerant and conventional released cultivars or advanced breeding lines analysed the coefficient of parentage, which measures the average degree of relationship among a population among other indicators of diversity within a population. The introduction of glyphosate tolerant varieties was found to have had little impact on diversity due to its incorporation into many breeding programs. In contrast, the introduction of Bt cotton in India initially resulted in a reduction in on-farm varietal diversity due to the introduction of the technology in only a small number of varieties, which has been offset by more Bt varieties becoming available over time. From a broader perspective, GM crops may actually increase crop diversity by enhancing underutilised alternative crops, making them more suitable for widespread domestication. Transgenic approaches are being used to improve so-called orphan crops, such as sweet potato. Crop diversity may also be impacted by gene flow between crops and wild relatives if the gene flow reduces genetic diversity available for crop improvement.

Farm-Scale Diversity

For the purposes of this review, impacts at the farm scale are considered to encompass any impacts on organisms that live primarily within the boundaries of the farm, including soil-organisms and weeds.

Non-target soil organisms: Plants have a major influence on communities of micro- and other organisms in soil which are fundamental to many functions of soil systems, such as nitrogen cycling, decomposition of wastes and mobilisation of nutrients. The type and amount of nutrients released will affect both the numbers of organisms and their diversity.

The potential impact of *Bt* crops on soil organisms is well studied. A comprehensive review of the available literature on the effects of *Bt* crops on soil ecosystems included the results of 70 scientific articles. The review found that, in general, few or no toxic effects of Cry proteins on woodlice, collembolans, mites, earthworms, nematodes, protozoa and the activity of various enzymes in soil have been reported. Although some effects, ranging from no effect to minor and significant effects, of *Bt* plants on microbial communities in soil have been reported, they were mostly the result of differences in geography, temperature, plant variety and soil type and, in general, were transient and not related to the presence of the Cry proteins.

The review found that the respiration of soils cultivated with *Bt* maize or amended with biomass of *Bt* maize and other *Bt* crops was generally lower than from soils cultivated with or amended with biomass of the respective non-*Bt* isolines, which may have been a result of differences in chemical composition between *Bt* plants and their near-isogenic counterparts. Studies have shown differences in the persistence of Cry proteins in soil, which appear to be the result primarily of differences in microbial activity, which in turn is dependent on soil type, season, crop species, crop management practices and other environmental factors that vary with location and climate zones.

Studies published since the Icoz and Stotzky review have reached similar conclusions (Table 10.1). Notably, two recent studies have investigated the potential impacts of *Bt* corn on snails, which had not been previously studied. The first study, using purified protein found no negative effect of the *Bt* toxin on the snail *H. aspersa* during the observed life stages. Subsequent work using plant material and soil from fields where *Bt* corn had been grown in a no-choice feeding experiment showed reduced growth at long exposure times, which was considered a worst case scenario.

Weeds: Crop production practices have significant effects on the composition of weed communities. Changes in the kinds of weeds that are important locally are termed weed shifts. Such shifts are particularly relevant for managing weeds in herbicide tolerant crop systems, in which tillage practices and herbicide use both play major roles in shaping the weed community. There are reports in the literature of fourteen weed species or groups of closely related species that have increased in abundance in glyphosate resistant crops. At the same time, in a survey of corn, soyabean and cotton growers in six states, between 36 and 70 per cent of growers indicated that weed pressure had declined after implementing rotations using glyphosate resistant crops. The potential impact of herbicide tolerant crops and their management systems on weed biodiversity was studied as part of the Farm Scale Evaluations (FSE's), supported by the

Table 10.1. Recent results of studies on impact of *Bt* crops on soil organisms.

Country	Organism	Species	Location	Experimental variable	Crop	Event (protein)	Comparison type	Effect
Germany	Enchytraeids	*Enchytraeus albidus*	Laboratory	Survival, reproduction	Corn	*Bt* 11(Cry 1Ab) MON88017 (Cry 3Bb1)	No-choice diet of *Bt* or non-*Bt* near isoline	No significant differences for Cry 3Bb1; significantly higher survival and significantly lower reproduction for Cry 1Ab, likely to be caused by differences in plant components
US	Soil microbes		Field	Microbial community function by quantification of extracellular enzymes	Corn	MON863 (Cry 3Bb1)	*Bt*, non-*Bt* near isoline with insecticide, untreated non-*Bt* near isoline	No appearance of adverse effects on saprophytic microbial communities of soil and decaying roots or on decomposition
Portugal	Soil microbes		Field	Number of culturable aerobic bacteria, activity of dehydrogenase and nitrogenase enzyme and ATP content	Corn	Event 176 (Cry 1Ab), MON810 (Cry 1Ab)	*Bt* and non-*Bt* near isoline	The presence of *Bt* Maize did not cause, in a general way, changes in the microbial populations of the soil or in the activity of the microbial community
US	Earthworms	*Aporrectodea caliginosa, Aporrectodea trapozoides, Aporrectodea tuberculata, Lumbricus terrestris*	Field	Biomass of juveniles and adults	Corn	*Bt*11(Cry 1Ab), MON810 (Cry 1Ab), MON863 (Cry 3Bb1)	MON810 and non-*Bt* near isoline, MON863 and non-*Bt* near isoline with and without insecticide seed treatment	No significant differences in biomass of juveniles and adults

(Contd...)

Country	Organism	Species	Location	Experimental variable	Crop	Event (protein)	Comparison type	Effect
China	Earthworms	*Eisenia fetida*	Laboratory	Acute toxicity weight, SOD activity, growth and reproduction	Cotton	GK19 (Cry 1Ac)	No choice diet of *Bt* or non-*Bt* parent line, insecticide treated soil and sterile manure controls	No significant acute toxicity; average weight, numbers of cocoons and new offspring not significantly different
US	Soil microbes		Field	Number of culturable bacteria, carbon substrate utilisation, total soil DNA	Corn	MON810 (Cry 1Ab)	*Bt* and non-*Bt* near isoline	Altered functional activity substrate metabolism and structure of microbial communities attributed to higher lignin content of *Bt* variety
Switzerland	Soil meso- and macrofauna	Collembola, Acari and Clitellata	Field	Number of extracted organisms	Corn	MON810 (Cry 1Ab), Bt 11 (Cry 1Ab) MON88017 (Cry 3Bb1)	*Bt* and non-*Bt* near isoline	Corn varieties had no impact on the soil fauna community
Denmark	Snails	*Cantareus aspersus*	Laboratory	Survival, growth and egg hatchability	Corn	MON810 (Cry 1Ab)	No choice *Bt* and non-*Bt*, with and without soil from *Bt* fields	25% lower growth coefficient for snails fed *Bt* corn leaves and exposed to soil from *Bt* fields; significant difference after first period of reproduction in body mass between snails fed *Bt* corn and exposed to soil from *Bt* fields and

(Contd...)

Country	Organism	Species	Location	Experimental variable	Crop	Event (protein)	Comparison type	Effect
								unexposed snails, which was no longer apparent at the end of exposure; hatchability of eggs from unexposed snails similar in *Bt* and non-*Bt* soil
Denmark	Snails	*Helix aspersa*	Laboratory	Egg hatchability, juvenile growth and survival		Cry 1Ab	Purified protein alone and in soil	No negative effect during the observed life stages
US	Surface dwelling collembola	32 species	Field	Abundance and diversity	Corn	MON810 (Cry 1Ab)	*Bt* and non-*Bt* near isoline	No negative effects

United Kingdom government. The trials were undertaken on over 60 fields of sugar beet, maize and oilseed rape in the UK, to allow the comparison of large-scale management systems of conventional and genetically engineered herbicide tolerant crops. Effects on weeds and associated arthropods and on detrital food webs were evaluated and presented in a series of papers published in the *Transactions of the Royal Society of London*. These studies showed that for genetically engineered herbicide tolerant sugar beet and oilseed rape fewer weeds and weed seeds, whereas for genetically engineered herbicide tolerant corn, an increase in dicot weeds and weed seed was observed. The results on associated arthropods, detrital food webs and birds are discussed further in the relevant sections below.

The use of herbicides can also result in changes to weed communities through the development of herbicide tolerant weed populations. The first confirmed report of a weed population expressing tolerance to an herbicide was in 1964, where field bindweed in Kansas was found to be resistant to 2,4-D. The US has the highest number of herbicide resistant weeds, with over 130 herbicide resistant weeds confirmed in the United States. The first confirmed case of glyphosate resistant (GR) weeds was in Australia in 1996, prior to the commercialisation of GR crops. The first confirmed case of glyphosate resistance in an area growing GR crops was in horseweed in Delaware in 2000. Globally, GR weeds have been confirmed for 21 weeds in 15 countries. Most of these cases have been reported where GR crops are commonly grown. However, GR weeds have also been reported in California in almonds and roadsides, orchards in Oregon and nurseries in Michigan, none of which are related to GR crops. The development of weeds resistant to glyphosate will likely require modification to weed control programs where practices in addition to applying glyphosate are needed to control the resistant populations.

Landscape-Scale Diversity

For the purposes of this review, potential impacts at the landscape scale include land use, area-wide pest suppression and nontarget above-ground invertebrates.

Land use: The most direct negative impact of agriculture on biodiversity is due to the considerable loss of natural habitats, which is caused by the conversion of natural ecosystems into agricultural land. Increases in crop yields allow less land to be dedicated to agriculture than would otherwise be necessary. For example, it was estimated that if wheat yields in India had stagnated at 1961–1966 levels, farmers there would have had to cultivate an area almost three times greater to produce the same amount of wheat that was harvested in 1990. A more recent analysis compared actual agricultural production between 1961 and 2005 with hypothetical scenarios where increases in food production were realised by expanding farmland instead of increasing yields, finding that between 864 and 1,514 million hectares would have had to be converted to agricultural production, depending on the living standard assumed.

A large and growing body of literature has shown that the adoption of GM crops has increased yields. A recent review of results from 49 peer-reviewed publications reporting on farmer surveys from 12 countries that compare yields of adopters and non-adopters of currently commercialised GM crops showed increased yields for adopters. Of 168 results comparing yields of GM and conventional crops, 124 show positive results for adopters compared to non-adopters, 32 indicate no difference and 13 are negative. Yield increases were greatest for developing country farmers. The average yield increases for developing countries range from 16 per cent for insect-resistant corn to 30 per cent for insect-resistant cotton, with an 85 per cent yield increase observed in a single study on herbicide-tolerant corn (Table 10.2). On average, developed-country farmers report yield increases that range from no change for herbicide-tolerant cotton to a 7 per cent increase for herbicide-tolerant soyabean and insect resistant cotton.

Table 10.2. Average percentage changes in yield by technology for developed and developing countries [(GM-conventional)/conventional].

Technology	Change in yield	# of results	Min.	Max.	Std. Err.
Developed countries	6%	59	–12%	26%	1.0%
HT cotton	0%	6	–12%	17%	3.8%
HT soyabean	7%	14	0%	20%	1.7%
HT/IR cotton	3%	2	–3%	9%	5.8%
IR corn	4%	13	–3%	13%	1.6%
IR cotton	7%	24	–8%	26%	1.9%
Developing countries	29%	107	–25%	150%	2.9%
HT corn	85%	1			
HT soyabean	21%	3	0%	35%	11%
IR corn	16%	12	0%	38%	4%
IR corn (white)	22%	9	0%	62%	6.9%
IR cotton	30%	82	–25%	150%	3.5%

Averages calculated across surveys, geographies, years and methodologies. A two-tailed t-test shows a significant difference between the average yields of developed and developing countries ($t = 7.48$, df = 134, $p < 0.0005$).

Researchers have estimated the benefit of these yield improvements on reducing conversion of land into agricultural use. Brookes modelled the impact of the introduction of GE corn, soyabean and canola on global production and prices, accounting for market effects on the planting decisions of farmers in adopting and non-adopting countries. Their analysis assumed yield impacts of between 0 (US) and 31 per cent (Romania) for soyabean, between 5 (US) and 24 per cent (Philippines) for corn and 3.7 (Canada) to 6 per cent (US) for canola. They estimate that 2.64 million hectares of land would probably be brought into grain and oilseed production if biotech traits were no longer used.

A potential impact of increased productivity related to adoption of GM crops is an increase in crop acreage, by farmers deciding to increase acreage planted to GM crops at the expenses of other farming activities or through expansion into natural areas. Both effects have been observed in Brazil and Argentina, where the introduction of GM crops combined with an enabling policy environment and facilitated adoption of no-till and doublecropping to cause the expansion of soyabean acreage into areas previously planted to other crops or used as pasture, as well as into some natural areas. To the extent that soya was planted into degraded pastures, it may be seen as an environmentally friendly way of expanding arable land.

Area-wide pest suppression: The most direct landscape-level effects of growing *Bt* crops would be expected for target pest species for which the crop is a primary food source and that are mobile across the landscape. Observations of target pest populations over time reveal a high level of variability that may be driven by *Bt* crops as well as other factors such as weather, local and distant cropping patterns, other crop management practices and pest population dynamics. Area-wide pest suppression not only reduces losses to adopters of the technology, but may also benefit non-adopters and growers of other crops by reducing crops losses and/or the need to use pest control measures such as insecticides.

Several studies have investigated the impact of the introduction of *Bt* corn and cotton on regional outbreaks of pest populations. Evidence of regional suppression of the target pests *Ostrinia nubilalis* and *Helicoverpa zea* in corn was gathered from an area of Maryland where *Bt* corn adoption was over

60 per cent. Moth trap records from 35 years were used to capture variability in pest populations both before and after the introduction of *Bt* corn. Moth activity was 63 per cent and 48 per cent lower than the long-term average for *O. nubilalis* and *H. zea*, respectively, declines that are believed to have led to pest management benefits in other host crops, such as soyabean and vegetables.

Populations of *O. nubilalis* are also observed to have declined in Midwestern US maize growing areas. Also using long-term data, on larval and moth flight, researchers have found significantly different per capita population growth rates in areas with different levels of adoption. The analysis found that the majority of benefits of *Bt* corn adoption has accrued to non-adopters.

In an-investigation of the impact of *Bt* cotton on populations of the target pests *H. zea* and *Heliothis virescens* in Washington County, Mississippi, data from adult pheromone trap captures from 1986 through 2008 were analysed. Despite yearly fluctuations, adult populations of both species were found to have declined annually since 1997. Declines in adult populations of *H. virescens* were dramatic, particularly over the years from 2000 to 2008, and may have been due to wide-scale plantings of *Bt* cotton among other factors.

In a ten year study across 15 regions of Arizona, researchers concluded that *Bt* cotton suppressed a major pest, *Pectinophora gossypiella*, independent of the effects of weather and regional variation. Population densities were found to have declined only in regions where *Bt* cotton was abundant.

In the cotton growing area of the Imperial Valley, California, gossyplure-baited trap catch data for 1989 to 2003 were analysed, which covered periods when different area-wide control strategies were used to control pink bollworm. Catches were significantly lower in 1998 to 2008 than in 1995 to 1997, except in 1999, although high populations in 1995 to 1997 may have been related to moth migrations from the large cotton acreages grown in the Mexicali Valley, which borders the Imperial Valley.

A study of the population dynamics of cotton bollworm from 1992 to 2007 covered six provinces in northern China, a major growing area of cotton and other crops which are also hosts to bollworm, such as corn, peanuts, soyabeans and vegetables. The analysis indicated that a significant decrease in regional outbreaks in multiple crops was associated with the planting of *Bt* cotton, which suggests a reduced need for insecticide sprays in general.

Non-target above-ground invertebrates

Insect resistant crops: The effects of GM crops on above-ground non-target invertebrates have been the subject of a large number of laboratory and field studies. By the end of 2008, over 360 original research papers had been published on non-target effects of *Bt* crops. Several reviews have summarised the literature. The overall conclusion of the reviews is that studies of the potential impact of *Bt* crops on non-target herbivores and beneficials have not detected significant adverse effects, and no evidence of landscape-level effects. Laboratory and glass house studies have revealed effects on natural enemies only when *Bt*-susceptible, sublethally damaged herbivores were used as prey or host, with no indication of direct toxic effects. Field studies have confirmed that the abundance and activity of parasitoids and predators are similar in *Bt* and non-*Bt* crops. The indirect impacts of *Bt* crops on beneficials, due to multitrophic exposure, loss of prey or reduction of prey quality, are considered to be negligible compared with the direct effects of agricultural practices.

The first published quantitative review analysed the results of 45 laboratory studies of the impact of insect resistant crops (including *Bt* and proteinase inhibitors) on 32 species of natural enemies, finding that 30 per cent of studies for predators and nearly 40 per cent of studies for parasitoids reported significant negative effects on multiple life history characteristics. This review was later updated with subsequent published literature to include a total of 80 studies on 48 species, finding 21.2 per cent negative results. The Lovei analysis has been criticised for using multiple nonindependent measures of life history and behavioural traits, which can inflate the purported effects in the data, lack of consideration of prey/host mediated effects in tri-trophic studies, inclusion of studies with irrelevant or unrealistic experimental designs and generalisation across *Bt*, proteinase inhibitors and lectins.

Another series of quantitative reviews has been conducted based on a common dataset, originally compiled by Marvier which covers impacts of *Bt* crops on invertebrates. Using meta-analysis, the results of 42 field-based studies of *Bt* corn and cotton were analysed, showing that the abundance of non-target invertebrates was higher in *Bt* crops compared to non-*Bt* crops that had been treated with insecticides, although abundance was slightly lower compared to non-sprayed non-*Bt* crops.

A later study used a modified version of the Marvier database, examining the results of field-based studies for corn, cotton and potato by functional guild using meta-analysis. Predators were found to be less abundant in *Bt* cotton compared to unsprayed non-*Bt* controls, and fewer specialist parasitoids of the target pest occurred in *Bt* corn compared to unsprayed non-*Bt* controls, though no significant reduction was detected for other parasitoids.

The abundance of predators and herbivores was higher in *Bt* crops compared to sprayed non-*Bt* controls, the difference affected by the type of insecticide used. However, omnivores and detritivores were more abundant in insecticide treated non-*Bt* crops. The study found no uniform effects of *Bt* crops on non-target arthropods by functional guild, and that the impact of insecticide use was much greater than *Bt* crops.

A separate study analysed the results of 25 laboratory studies that specifically looked at the impacts of *Bt* crops on honey bees, also using meta-analysis. *Bt* proteins in lepidopteran or coleopteran resistant crops were found to have had no effect on the survival of larvae or adults.

Most recently, the Marvier database was updated with literature published subsequent to the compilation of the original database. The updated database included 135 laboratory-based studies on nine *Bt* crops from 17 countries and 63 field-based studies on five *Bt* crops from 13 countries, which were analysed using meta-analysis techniques.

Laboratory studies were found to have identified negative effects of *Bt* crops when organisms were exposed directly to *Bt* proteins, some of which were expected because of the relatedness of the non-target pest to the groups targeted by the *Bt* crop. In tri-trophic studies, the quality of prey or hosts was associated with negative effects, whereas no negative effects were found when unaffected, high quality prey or hosts were used. In general, laboratory studies identified greater levels of hazard than field studies, at least partially explained by differences in organisms studied, and frequently higher protein exposure in lab studies compared to exposure levels in the field. Field studies demonstrated few harmful non-target effects, with the non-target effects of insecticides being much greater than *Bt* crops. More recent literature on the non-target impacts of *Bt* crops are largely consistent with Naranjo's conclusions (Table 10.3).

A modelling study of the potential impact of *Bt* corn on non-target butterfly and moth species in four European countries estimated mortality both within the field and at the field margin at varying distances from the crop edge. That study estimated low environmental impacts in all regions, with a calculated mortality rate of less than one individual in every 1572 for the butterflies and one in 392 for the moth in the worst case scenarios.

Table 10.3. Recent results of research on potential impact of *Bt* crops on non-target above-ground organisms.

Country	Organism	Species	Location	Experimental variable	Crop	Event (protein)	Comparison type	Effect
US	Caddisflies, crane fly larva, aquatic isopod	*Lepidostoma* spp., *Pycnopsyche* cf. *scabripennis* (Rambur), *Tipula* (*Nippotipula*) cf. *abdominalis* (Say), *caecidotia communis* (Williams)	Laboratory	Head width, dry mass, survival	Corn	MON810 (Cry 1Ab) MON810/MON863 (Cry 1Ab/Cry 3Bb1)	*Bt* and non-*Bt* near isoline, single trait and stacked	No bioactivity of Cry 1Ab protein in senesced corn tissue after 2 weeks of environmental exposure; growth of trichopterans not negatively affected, reduced growth rates of tipulid crane fly, reduced growth and survivorship of isopod or *Bt* near-isoline but not on stacked near-isoline
Poland	Bird cherry-oat aphid, green lacewing	*Rhopalosiphum padi* L., *Chrysoperla carnea* Steph.	Greenhouse/ laboratory	Population, feeding, moulting data	Corn	MON810 (Cry 1Ab)	*Bt* and non-*Bt* near isoline	No significant differences
South Africa	*Lepidoptera*	15 species	Field	Infestation, damage	Corn	MON810 (Cry 1Ab)	*Bt* and non-*Bt* near isoline	Incidence of infestation and infestation levels generally lower in *Bt* fields than non-*Bt* fields
India	Sucking insects, foliage feeder and predators	*Amrasca biguttula biguttula*, *Bemisia tabaci*, *Aphis gossypi*, *Thrips tabaci*, *Myllocerus undecimpustulatus*, *Chrysoperla* spp., *Orius* spp.,	Field	Densities, time of first appearance	Cotton	Bollgard (Cry 1Ac), Bollgard II (Cry 1Ac/ Cry 2Ab)	*Bt* and non-*Bt* cultivars with and without insecticide	Similar densities and no difference in time of first appearance of non-target insects

(Contd...)

Country	Organism	Species	Location	Experimental variable	Crop	Event (protein)	Comparison type	Effect
		Coccinella spp. *Brumus* spp., *Vespa* spp., *Lycosa* spp., *Aramews* spp.						
Hungary	Rove beetles	21 species	Field	Abundance, species richness, diversity and similarity	Corn	Mon810 (Cry 1Ab)	*Bt* and non-*Bt* near isoline	No influence on overall community structure; no significant differences for non-aphidophagous predators and parasitoids; significantly and marginally significantly higher abundances for predators with aphids in their diet in isogenic maize in 2 of years
France	Moth	*Cotesia marginiventris*	Laboratory	Rates of parasitism, host suitability, developmental periods, offspring longevity, size and sex ratio	Corn	Mon 810 (Cry 1Ab)	No choice purified protein in artificial diet; no choice plant tissue	No effect of purified protein at evaluated concentrations; exposure via *Bt* maize tissue affected developmental time, adult size and fecundity
Switzerland	Spider mite, ladybird beetle	*Tetranychus urticae*, *Stethorus punctillum*	Laboratory	Development, adult survival and reproduction	Corn	MON88017 (Cry 3Bb1)	No choice *Bt* and non-*Bt* near isoline	No differences for *T. urticae*; female *S. punctillum* fed *Bt* corn had shorter preoviposition period, increased fecundity

(Contd...)

Impact of Genetically Engineered and Genetically Modified Crops on Biodiversity 245

Country	Organism	Species	Location	Experimental variable	Crop	Event (protein)	Comparison type	Effect
								and increased fertility, otherwise no differences for *S. punctillum*
Spain	Leafworm, ground beetle	*Spodoptera littoralis*, *Poecilus cupreus*	Laboratory	Development mortality, adult weight	Corn	Event 176 (Cry 1Ab)	No choice *Bt* and non-*Bt* near isoline	No differences in developmental time of larvae and pupae, adult weight or larval and pupal mortality of *P. cupreus*
China	Beet armyworm	*Spodoptera exigua*	Laboratory	Development, food utilisation and population performance	Cotton	NC33B (Cry 1Ac)	No choice *Bt* and non-*Bt* near isoline	Longer larval life-span and lower pupal weight, higher survival rate and adult fecundity, significantly lower consumption, frass and relative growth rate observed in three successive generations
Germany	Spiders, carabid beetles	50 species of spiders, 57 species of carabid beetles	Field	Activity abundance	Corn	MON810 (Cry 1Ab)	*Bt* and non-*Bt* near isoline with and without insecticide	Significantly different activity abundances observed for spiders and carabid beetles in one of three years
Switzerland	Ladybird beetle	*Adalia bipunctate*	Laboratory	Larval/pupal mortality, development time, overall body mass accumulation		Cry 1Ab, Cry 3Bb1	No choice diet with solutions including either purified protein solution or solution with empty vector cassette	Significantly higher mortality for Cry 1Ab at all concentrations; marginally significantly higher mortality for Cry 3Bb1 at intermediate concentration; no

(Contd...)

Country	Organism	Species	Location	Experimental variable	Crop	Event (protein)	Comparison type	Effect
								differences in development time and body mass of newly emerged adults
Germany	Plant bugs	Six taxa, of which *Trigonotylus Caelestialium* was most abundant	Field	Density	Corn	MON88017 (Cry 3Bb1)	*Bt* and non-*Bt* near isoline	No evidence of negative impact on most abundant plant bug
US	Carabid beetles	15 species	Field	Abundance and diversity	Corn	MON810 (Cry 1Ab)	*Bt* and non-*Bt* near isoline	No negative effects
Switzerland	Aphids	*Aphis gossypii*	Laboratory	Nymphal developmental time, number of nymphs produced, reproductive rate, total fecundity per female, adult longevity, total longevity, instrinsic rate of population increase	Cotton	BG-1 (Cry 1Ac)	*Bt* and non-*Bt* near isoline	No effect
US	Arthropods	16 arthropod taxa	Field	Abundance	Corn	DAS-01507 (Cry 1F)	*Bt* and non-*Bt* near isoline	No significant impact on community abundance or abundance of individual taxa
Canada	Ground beetles	39 species	Field	Abundance	Corn	MON810 (Cry 1Ac)	*Bt* and non-*Bt* near isoline with and without insecticide	No effect on total beetle abundance or species richness; negative effect for 3 of 39 species by year combinatio:.; examined

(Contd...)

Country	Organism	Species	Location	Experimental variable	Crop	Event (protein)	Comparison type	Effect
India	Ladybird beetle	*Cheilomenes sexmaculatus*	Laboratory	Larval and pupal periods, larval survival, weights of male and female larvae, adult emergence, weights of male and female adults		Cry 1Ab, Cry 1Ac	No choice purified protein or aphids fed purified protein	Reduced larval survival and adult emergence from direct exposure, attributed to long-term exposure; no effect of indirect exposure through aphids
Germany	Planthoppers and leafhoppers	5 taxa, *Zyginidia scutellaris* most prevalent	Field	Abundance	Corn	MON810 (Cry 1Ac)	*Bt* and non-*Bt* near isoline with and without insecticide	No consistent differences for *Z. scutellaris* between *Bt* and untreated non-*Bt*; decrease in *Z. scutellaris* for treated non-*Bt* compared to untreated non-*Bt*
US	Midge	*Chironomus dilutus*	Laboratory	Mortality and dry weight	Corn	MON863 (Cry 3Bb1)	*Bt* and non-*Bt* near isoline	Significant decrease in survival at intermediate concentration tested; no effect on growth of surviving larvae
Switzerland	Green lacewing	*Chrysoperla carnea*	Laboratory	Larval development and survival, pupal development time and survival and adult dry weight		Cry 1Ac and Cry 1Ab	Purified protein	No effect

(Contd...)

Country	Organism	Species	Location	Experimental variable	Crop	Event (protein)	Comparison type	Effect
Spain	Ladybird beetle	*Stethorus punctillum*	Laboratory	Survival and development	Corn	*Bt 176* (Cry 1Ab) MON810 (Cry 1Ab)	*Bt* and non-*Bt* near isoline	No effect on survival or development time to adulthood or fecundity
US	Green lacewing, flower bug	*Chrysoperla carnea, Orius tristicolor*	Field	Abundance	Cotton	Bollgard (Cry 1Ac)	*Bt* and non-*Bt*, with and without insecticide	No significant difference between *Bt* and unsprayed non-*Bt*; significantly lower abundance for insecticide-treated fields in one of two years
India	Predatory spiders, ladybird beetles, green lacewing, ichneumonid parasitoid	*Clubiona* sp., *Neoscona* sp., *Cheilomenes sexmaculatus, Chrysoperla carnea, Campoletis chlorideae*	Field	Abundance	Cotton	Bollgard (Cry 1Ac)	*Bt* and non-*Bt*, with and without insecticide	No effect on predatory spiders, ladybird beetle or green lacewing; significant reduction in cocoon formation and adult emergence of *C. chlorideae*
UK	Aphids, parasitic wasps	*Rhopalosiphum maidis, Cotesia marginiventris*	Laboratory	Performance	Corn	*Bt 11* (Cry 1Ac), MON810 (Cry 1Ac), Event 176 (Cry 1Ac)	*Bt* and non-*Bt* near isolines	No difference in mean relative growth rates in individual comparisons; significantly more nymphs on transgenic lines than respective near isolines; differences might be partially explained by higher amino acid levels in *Bt* lines

In a study of the landscape level effects of *Bt* cotton, population densities of ants and beetles in non-cultivated sites near agricultural fields were monitored for up to 5–6 years. The sequence of crops planted in neighbouring fields, crops diversity and abundance were more frequently associated with greater insect density than characteristics of the non-cultivated sites. Results suggest that the farming of *Bt* cotton in neighboring fields frequently resulted in positive short- and long-term landscape effects on ants and beetles in non-cultivated sites, while *Bt* cotton planted farther away had less frequent negative short-term impacts.

In a study of the impact of reduced use of broad spectrum insecticides in *Bt* cotton on secondary insect pests in China, mirid populations were found to be higher in *Bt* cotton fields compared to conventional cotton sprayed with conventional insecticides as well as in fruit crops in regions with higher proportions of *Bt* cotton. In another analysis of the observed increase in secondary pests in *Bt* cotton in China, the rise and fall of mirid populations was found to be largely related to local temperature and rainfall.

Herbicide tolerant crops: As described above, the UK Farm Scale Evaluations (FSEs) evaluated the effects of conventional and herbicide tolerant weed management systems on weeds and associated arthropods and on detrital food webs. These studies showed that for genetically engineered herbicide tolerant sugar beet and oilseed rape, fewer weeds and weed seeds, and fewer insects from species that live in or on weeds, were observed. Increased numbers of detritivores were found in HT sugar beet and maize, with a corresponding increase in predators in sugar beet. Whereas for genetically engineered herbicide tolerant corn, an increase in dicot weeds and weed seed was observed, with a corresponding increase in seed-eating beetles.

The Farm-Scale Evaluations have been criticised for leading to inappropriate conclusions about the environmental impacts of genetically engineered crops. The FSE was not designed to test the direct effect of the genetically engineered trait itself, but rather the broader herbicide and crop management practices. Glyphosate and glufosinate as used in conjunction with genetically engineered herbicide tolerant sugar beet and oilseed rape provided highly effective weed control, which resulted in fewer weeds and weed seeds. Not surprisingly, the effects on various groups of arthropods followed the effects on the abundance of their resources. It has been noted that the introduction of other technologies that increase control of weeds is routine and occurs without public debate.

Other studies on the non-target impacts of herbicide tolerant crops have come to similar conclusions. The effects of genetically engineered herbicide tolerant soyabeans and corresponding weed management strategies on various soyabean insects were the focus of a study from 1996 to 1998 in Iowa. Weed management systems that allowed more weed escapes typically had higher insect population densities. Springtail numbers were similar to or higher in herbicide tolerant crops than those in conventional plots.

The potential impact of GM HT canola on bees was studied in a two-year field trial and one year semi-field trial in Canada. No differences in larval survival, adult recovery and pupal weight were detected in a comparison of colonies in conventional and GM canola fields. A significantly higher amount of hemolymph protein was found in one of two years in newly emerged bees from GM fields compared to conventional fields. Authors conclude that the results suggest that transgenic canola pollen does not have adverse effects on honey bee development.

In a study of the potential impact of GM HT soyabean weed management on soyabean insect pest populations in South Georgia in 2007 and 2008, few differences in pest population densities were observed. Observed differences were associated with soyabean maturity grouping on some dates, and were not associated with the GM HT trait.

A two-year study of the impact of the deployment of herbicide tolerant corn on corn arthropods compared weeds, insect herbivores and their natural enemies in plots treated either with glyphosate or conventional pre-emergence herbicides. Despite significant differences in weed abundance, there were very few significant differences in the arthropod groups monitored. These results contrast with prior findings of the same study comparing plots treated with glyphosate to plots with no herbicide treatment.

Given the flexibility of herbicide treatment offered by the herbicides used in herbicide tolerant crops, some have proposed novel weed management systems that crops can be managed for enhanced weed and insect biomass without compromising yields in order to increase food and shelter for farmland birds and other wildlife.

Stacked insect resistant/herbicide tolerant crops: Only one study was located that addressed the potential impacts of stacked crops on non-target above-ground invertebrates. In a 2-year farm-scale evaluation of 81 commercial fields in Arizona, insecticides used on conventional cotton was related to reduced diversity of non-target insects. However, the effects of cultivation of cotton, whether transgenic or not, were found to result in similar effects on biodiversity compared with diversity in adjacent noncultivated sites.

Birds: The authors of the FSE reports suggest that the observed decreases in weed seeds and insects might reduce the number of birds that feed on these insects and seeds, though the results of a bird survey that was conducted on a sub-set of fields used in the FSE were not published until 2007. The bird survey results were in accord with differences in food availability found in the FSE. Specifically, a greater abundance of granivores was found on conventional than genetically engineered herbicide tolerant sugar beet, as well as on genetically engineered herbicide tolerant maize after application of herbicides to the GM HT field. No differences were detected in spring oilseed rape. In the subsequent winter season, granivores were more abundant in the fields where conventional sugar beet had been grown than on the GM HT fields. Several bird species were found to be more abundant on maize stubbles following GM HT treatment.

Indirect Indicators

Changes in tillage practices: The introduction of herbicide tolerant crops has been associated with the increased adoption of conservation tillage practices, which decreases run-off, increases water infiltration and reduces erosion. Trends in the adoption of conservation tillage have been studied in the US and Argentina, the largest growers of herbicide tolerant crops.

In the US, soyabean growers were already adopting conservation tillage practices prior to the introduction of glyphosate tolerant soyabean, using other post-emergence selective herbicides which became available in the 1980s and 1990s. Already by 1989, 30 per cent of US soyabean acreage was under conservation tillage. Herbicide-tolerant crops facilitated adoption by making it easier and less risky to adopt conservation tillage and no-till. Between 1996 and 2008, adoption increased from 51 to 63 per cent of planted acres. In particular, the adoption of no-till in full-season soyabean, which leaves the most crop residue on the soil surface, is estimated to have increased from 27 per cent in 1995 to 39 per cent of planted acreage according to the latest surveys by the Conservation Tillage Information Center.

The results of the CTIC are reinforced by two additional independent surveys. A survey of corn, cotton and soyabean growers in six states (Illinois, Indiana, Iowa, Mississippi, Nebraska and North Carolina) found that the percentage of growers using no-till and reduced-till systems was increased as a result of the adoption of glyphosate-resistant (GR) crops. Tillage intensity declined more in continuous

GR cotton and GR soyabean (45 and 23 per cent, respectively) than in rotations that included GR corn or non-GR crops. A survey of 610 soyabean growers across 19 states found that growers of GR soyabeans made 25 per cent fewer tillage passes than growers of conventional soyabean.

The introduction of glyphosate-tolerant soyabeans is also cited as a contributing factor in the rapid increase of no-till in Argentina, where adoption of no-till increased from about 1/3 of soyabean acreage in 1996 to over 80 per cent in 2008. Other factors that also contributed to the expansion of no-till in soyabean are: favourable macroeconomic policies, reduction in price of herbicides and continued research and promotion efforts. A 2001 survey of 59 soyabean growers in Argentina found that the number of tillage operations was 58 per cent lower on glyphosate-tolerant acreage than on conventional soyabean fields.

Whether the introduction of herbicide tolerant crops has caused an increase in the adoption of conservation tillage or *vice versa*, has been the subject of several studies. In an analysis of farmer decision-making using national survey data for 1997, researchers found that conservation tillage adoption led to adoption of glyphosate tolerant soyabeans, but that glyphosate tolerant soyabean adoption did not lead to increased adoption of conservation tillage. However, more recent surveys have shown a positive two-way relationship. Using data from 1998 to 2007, a simultaneous relationship was found between the adoption of conservation tillage and the adoption of herbicide tolerant cotton in Tennessee. A broader study of herbicide tolerant cotton producers used state level data from 1997 to 2007 also found a simultaneous relationship between adoption of herbicide tolerant varieties and conservation tillage.

Changes in pesticide use: The pest management traits that are embodied in currently commercialised GM crops have led to changes in the use of pesticides that may have impacts on biodiversity. If the planting of GM pest-resistant crop varieties eliminates the need for broad-spectrum insecticidal control of primary pests, naturally occurring control agents are more likely to suppress secondary pest populations, maintaining a diversity and abundance of prey for birds, rodents and amphibians. In addition to the studies on the non-target impacts of GM crops compared to conventional practices, many studies have quantified changes in pesticide use since the introduction of GM crops. In a review of farmer surveys that report changes in yields and production practices, 45 results show decreases in the amount of insecticide and/or number of insecticide applications used on *Bt* crops compared to conventional crops in Argentina, Australia, China, India and the US. The reductions range from 14 to 75 per cent in terms of amount of active ingredient and 14 to 76 per cent for number of applications. A small sample survey in South Africa observed a reduction in the number of insecticide sprays in one of two years studied and an insignificant difference in the other year. There are no results indicating an increase in insecticide use for adopters of GM insect resistant crops.

Fewer surveys have captured changes in herbicide use in GM herbicide tolerant crops, perhaps because the impact of GM herbicide tolerant crops has largely been a substitution between herbicides that are applied at different rates, and therefore, changes in the amount of herbicide used is a poor indicator of environmental impact. Indeed, studies show both increases and decreases in the total amount of herbicide active ingredient applied per acre. Several studies have been done to apply environmental indicators to observed changes in pesticide use related to the adoption of both insect resistant and herbicide tolerant crops, which all show a reduction in the environmental impact of pesticides used on GM crops (Table 10.4). Thus the knowledge gained over the past 15 years that GM crops have been grown commercially indicates that the impacts on biodiversity are positive on balance. By increasing yields, decreasing insecticide use, increasing the use of more environmentally friendly herbicides and facilitating the adoption of conservation tillage, GM crops have already contributed to increasing agricultural sustainability.

Table 10.4. Summary of results of studies that quantify environmental impacts of pesticide use changes.

Country	Technology	Data source	Crop year	Indicator	Result
Argentina	*Bt* cotton	Farmer survey	2001	Amount of insecticide used (kg/ha) by toxicity class	Reductions of 47% and 78% for toxicity classes 1 and 2; no significant chance for less toxic classes 3 and 4
Argentina	HT soyabean	Farmer survey	2001	Amount of herbicide use (kg/ha) by toxicity class	Reductions of 83% and 100% for toxicity classes 2 and 3; increase of 248% in the use of less toxic class 4
US	HT cotton	Farmer survey	2000	Pesticide leaching potential (PLP/ha)	25% lower for herbicides used on HT cotton
US	HT/*Bt* cotton	Farmer survey	2000	Pesticide leaching potential (PLP/ha)	42% lower for herbicides used on HT/IR cotton; 5% lower for insecticides used on HT/IR cotton
South Africa	*Bt* cotton	Farmer survey	1998/99–2000/01	Biocide index	Reductions between 40% and 62% for total insecticide use
US	HT soyabean	USDA NASS chemical usage	1994–96 compared to 2006	Kovach's environmental impact	Reduction of 12% for 2006 compared to 1994–96
Canada	HT canola	Farmer survey	1995 compared to 2000	Kovach's environmental impact	Reduction of 36.8%
Belgium	HT corn	Current herbicide regime compared to possible HT regime	Ex ante	Pesticide occupational and environmental risk (POCER) indicator	Reduction of 1/6 when glyphosate or glufosinate used alone
US	HT soyabean	Expert opinion on common HT and conventional herbicide regimes	2004	Kovach's environmental impact	Reduction of 59%
US	HT canola	Expert opinion on common HT and conventional herbicide regimes	2004	Kovach's environmental impact	Reduction of 42%
US	HT cotton	Expert opinion on common HT and conventional herbicide regimes	2004	Kovach's environmental impact	Reduction of 42%

(Contd...)

Country	Technology	Data source	Crop year	Indicator	Result
US	HT corn	Expert opinion on common HT and conventional herbicide regimes	2004	Kovach's environmental impact	Reduction of 39%
Global	HT soyabean, HT corn, HT cotton, HT canola, IR corn, IR cotton	Several	1996–2007	Kovach's environmental impact	Aggregate reduction of 15.4%
Australia	Bt cotton	Farmer survey	2002/03–2003/04	Kovach's environmental impact	Reduction of 64%
US	HT soyabean	Farmer survey of weed pressure combined with herbicide selector software	Not indicated	LD$_{50}$ doses for rats	Reduction of 40% of total doses

Overall, the review finds that currently commercialised GM crops have reduced the impacts of agriculture on biodiversity, through enhanced adoption of conservation tillage practices, reduction of insecticide use and use of more environmentally benign herbicides and increasing yields to alleviate pressure to convert additional land into agricultural use.

The key findings of the review are: The impact of the introduction of GM crops on crop diversity has not been thoroughly studied. However, the small number of studies that have been done (cotton in the US and India, soyabeans in the US) find that the introduction of GM crops has not decreased crop diversity.

The potential impact of *Bt* crops on soil organisms is well studied. Few or no effects on soil organisms have been reported. Some reported differences, particularly in soil microbial communities, may have been due to differences in geography, temperature, plant variety and soil type.

Changes in weed community composition and abundance have been reported, due to changes in herbicide regimes and tillage practices associated with herbicide tolerant crops.

GM crops have increased yields, and therefore may have already alleviated pressure to convert natural habitat into agricultural use.

Pest populations have declined in some areas with high adoption rates of *Bt* crops, which benefits growers of other host crops and reduces the need for insecticide use.

The potential non-target impacts of 'insect resistant *Bt* crops on above-ground invertebrates have been extensively studied, with several hundred studies comparing effects of *Bt* and non-*Bt* crops. Several comprehensive reviews of the literature have been published, concluding that effects on natural enemies were observed only when *Bt*-susceptible, sublethally damaged herbivores were used as prey or host, with no indication of direct toxic effects. Field studies have confirmed that the abundance and activity of parasitoids and predators are similar in *Bt* and non-*Bt* crops.

Potential non-target impacts of herbicide tolerant crops on above-ground invertebrates have been the subject of several studies, mostly notably, the Farm Scale Evaluation in the UK. Changes in abundance of invertebrates and birds followed changes in weed control efficacy.

The introduction of herbicide tolerant crops has facilitated adoption of conservation tillage, which is expected to decrease erosion, increase water infiltration and decrease pesticide-runoff.

Adopters of GM crops have reduced insecticide use and switched to more environmentally friendly herbicides.

GM crops can continue to decrease the pressure on biodiversity as global agricultural systems expand to feed a world population that is expected to continue to increase for the next 30 to 40 years. Due to higher income elasticities of demand and population growth, these pressures will be greater in developing countries. Both current and pipeline technology hold great potential in this regard. The potential of currently commercialised GM crops to increase yields, decrease pesticide use and facilitate the adoption of conservation tillage has yet to be realised, as there continue to be countries where there is a good technological fit, but have not approved these technologies for commercialisation.

In addition to the potential benefits of expanded adoption of current technology, several pipeline technologies offer additional promise of alleviating the impacts of agriculture on biodiversity.

Continued yield improvements in crops such as rice and wheat are expected with insect resistant and herbicide tolerant traits that are already commercialised in other crops. *Bt* eggplant, which is expected to increase yields and decrease insecticide use significantly, is currently under consideration by Indian regulators. Technologies such as drought tolerance and salinity tolerance would alleviate the pressure to convert high biodiversity areas into agricultural use by enabling crop production on suboptimal soils.

Drought tolerance technology, which allows crops to withstand prolonged periods of low soil moisture, are anticipated to be commercialised within 5 years. The technology has particular relevance for areas like sub-Saharan Africa, where drought is a common occurrence and access to irrigation is limited. Salt tolerance addresses the increasing problem of saltwater encroachment on freshwater resources.

Nitrogen use efficiency technology is also under development, which can reduce run-off of nitrogen fertiliser into surface waters. The technology promises to decrease the use of fertilisers while maintaining yields or increase yields achievable with reduced fertiliser rates where access to fertiliser inputs is limited. The technology is slated to be commercialised within the next 10 years.

Chapter 11

Agrobiodiversity Informatics with Special Reference to Spices

INTRODUCTION

Biodiversity stands for all living things on earth. It refers to the range of variations among a set of entities and is commonly used to describe variety and variability of living organisms in terms of genetic diversity (heritable variations within populations), species diversity (species richness in a habitat) and ecological diversity (biophysical diversity). India, because of its unique biogeographic location embraces three major biological realms, viz. Indo-Malayan, Eurasian and Afro-tropical. It is notable in its species-richness and endemism, and is ranked tenth amongst the biodiversity-rich countries. India is one of the world's 12 Vavilovian centres of origin and diversification of cultivated plants (with 167 species of agrihorticultural crops and 320 species of their relatives known to have originated here). Conservative estimates of species richness show that around 1,27,000 species (plants, animals and micro-organisms) have been so far reported from India and 4,00,000 species are yet to be explored. Biodiversity also occurs at a genetic level. In other words, even within a single species there is variation. This is known as genetic biodiversity. Different populations of the same species can differ in their genetic profiles. It is this genetic biodiversity which facilitates the breeding of different varieties of plants and animals. Diversity in genes provides the basis for continued survival of the species in view of the changing environment.

IMPORTANCE OF BIODIVERSITY

Biodiversity is part of our daily lives and livelihood and constitutes the resources upon which families, communities, nations and future generations depend. The well-being and prosperity of earth's ecological balance as well as human society directly depend on the extent and status of biological diversity. Plant and animal diversity ensures a constant and varied source of food, medicine and raw material for human populations. Much of traditional medicine is based around plant extracts found in nature. Modern researchers are increasingly looking towards biological resources to find treatments and cures for illnesses. In agriculture, biodiversity provides us with a varied food supply, which is needed for balanced human nutrition. It is important to conserve all the variability in the species gene pool including wild relative, land races and cultivars which are highly vulnerable and prone for extinction. Loss in diversity is a direct threat to the ecological security of the country and the livelihood security of millions of people. As human populations and their demands on the natural world grow, our accumulated knowledge about biodiversity and the environment will become ever more important in the effort to develop a sustainable world. Beside the profound ethical and aesthetic implications, it is clear that the loss of biodiversity has serious economic and social costs.

The biological diversity we see today is the result of billions of years of evolution. While we all welcome our modern thriving economy, we must acknowledge the pressure that increased transportation, construction, inappropriate agricultural practices, poorly managed afforestation and climate change have put on biodiversity. Other factors such as overexploitation and climate change have also had a significant impact on biodiversity. Anthropogenic activities coupled with the burgeoning human population, have led to the grim biodiversity scenario; numerous important plant and animal species are on the verge of extinction, while others are threatened or vulnerable. Biological diversity must be treated more seriously as a global resource, to be indexed, used, and above all, preserved.

BIODIVERSITY INFORMATICS

Biodiversity Informatics is the application of informatics techniques to biodiversity information for improved management, presentation, discovery, exploration and analysis. It typically builds on a foundation of taxonomic, biogeographic or ecological information stored in digital form, which, with the application of modern computer techniques, can yield new ways to view and analyse existing information, as well as predictive models for information that does not yet exist. Biodiversity informatics is a relatively young discipline (the term was coined in or around 1992) but has hundreds of practitioners worldwide, including the numerous individuals involved with the design and construction of taxonomic databases. The term 'biodiversity informatics' is generally used in the broad sense to apply to computerised handling of any biodiversity information; the somewhat broader term 'bioinformatics' is often used synonymously with the computerised handling of data in the specialised area of molecular biology.

Biodiversity informatics has been defined as 'the creation, integration, analysis, and understanding of information regarding biological diversity', and 'the field that brings information science and technologies to bear on the data and information generated by the study of organisms, their genes, and their interactions'. Broadly speaking, it seeks to draw upon and integrate information held in various taxonomic databases and other digital sources to answer biodiversity questions at scales ranging from global to local. Such questions might range from 'How many described species exist in the world?' (answer: still not known for certain, as all the relevant data are not currently compiled in any coherent manner) to 'Predict the effects of a global temperature rise of X°C on the geographic range of species Y', a question which involves not only biodiversity in the basic sense but related domains of ecology, geographic distributions of environmental parameters, global climate models and more. In addition to handling formally named taxa, biodiversity informatics may also have to cope with managing information from unnamed taxa such as that produced by environmental sampling and sequencing of mixed-field samples. The term biodiversity informatics is also used to cover the computational problems specific to the names of biological entities, such as the development of algorithms to cope with variant representations of identifiers such as species names and authorities, and the multiple classification schemes within which these entities may reside according to the preferences of different workers in the field, as well as the syntax and semantics by which the content in taxonomic databases can be made machine queryable and interoperable for biodiversity informatics purposes.

Biodiversity informatics can be considered to have commenced with the construction of the first computerised taxonomic databases in the early 1970s, and progressed through subsequent developing of distributed search tools towards the late 1990s including the Species Analyst from Kansas University, the North American Biodiversity Information Network NABIN, CONABIO in Mexico, and others, the establishment of the Global Biodiversity Information Facility in 2001, and the parallel development of a variety of niche modelling and other tools to operate on digitised biodiversity data from the mid-980s onwards. In September 2000, the US journal science devoted a special issue to 'Bioinformatics for biodiversity', the journal 'biodiversity informatics' commenced publication in 2004, and several international conferences through the 2000s have brought together biodiversity informatics practitioners, most recently the London e-Biosphere conference in June 2009.

Current Biodiversity Informatics Issues

Global list of all species

One major issue for biodiversity informatics at a global scale is the present absence of a machine queryable (or even non-digital) master list of currently recognised species of the world, although this is an aim of the catalogue of life project which has been quoted as aiming to achieve this goal (for extant species only) by 2012; in its 2009 Annual Checklist edition a total of 1.16 million valid species names and 0.76 million synonyms were included, out of an estimated target 1.8 million extant described species. A similar effort for fossil taxa, the Paleobiology Database documents some 1,00,000+ names for fossil species, out of an unknown total number.

Problems with genus and species scientific names as unique and persistent identifiers

Application of the Linnaean system of binomial nomenclature for species, and uninomials for genera and higher ranks, has led to many advantages but also problems with homonyms (the same name being used for multiple taxa, either inadvertently or legitimately across multiple kingdoms), synonyms (multiple names for the same taxon), as well as variant representations of the same name due to orthographic differences, minor spelling errors, variation in the manner of citation of author names and dates, and more. In addition, names can change through time on account of changing taxonomic opinions (for example, the correct generic placement of a species or the elevation of a subspecies to species rank or *vice versa*), and also the circumscription of a taxon can change according to different authors' taxonomic concepts. One proposed solution to this problem is the usage of life science identifiers (LSIDs) for machine-machine communication purposes, although there are both proponents and opponents of this approach.

Achieving a consensus classification of organisms

Organisms can be classified in a multitude of ways, which can create design problems for Biodiversity Informatics systems aimed at incorporating either a single or multiple classification to suit the needs of users or to guide them towards a single 'preferred' system. Whether a single consensus classification system can ever be achieved is probably an open question, however in an attempt to provide at least a degree of consensus, the catalogue of life project has recently released a document that attempts to list some of the issues in this area, and may lead to a more coherent classification that can be promoted via that project's future products at least.

Mobilising Primary Biodiversity Information

'Primary' biodiversity information can be considered the basic data on the occurrence and diversity of species (or indeed, any recognizable taxa), commonly in association with information regarding their distribution in either space, time or both. Such information may be in the form of retained specimens and associated information, for example as assembled in the natural history collections of museums and herbaria or as observational records, for example either from formal faunal or floristic surveys undertaken by professional biologists and students or as amateur and other planned or unplanned observations including those increasingly coming under the scope of citizen science. Providing online, coherent digital access to this vast collection of disparate primary data is a core biodiversity informatics function that is at the heart of regional and global biodiversity data networks, examples of the latter including OBIS and GBIF.

As a secondary source of biodiversity data, relevant scientific literature can be parsed either by humans or (potentially) by specialised information retrieval algorithms to extract the relevant primary biodiversity information that is reported therein, sometimes in aggregated/summary form but frequently as primary observations in narrative or tabular form. Elements of such activity (such as extracting key taxonomic identifiers, keywording/index terms, etc.) have been practiced for many years at a higher level by selected academic databases and search engines. However, for the maximum biodiversity Informatics value, the actual primary occurrence data should ideally be retrieved and then made available in a standardised form or forms; for example both the Plazi and INOTAXA projects are transforming taxonomic literature into XML formats that can then be read by client applications, the former using Taxon X-XML and the latter using the taXMLit format. The biodiversity heritage library is also making significant progress in its aim to digitise substantial portions of the out-of-copyright taxonomic literature, which is then subjected to OCR (optical character recognition) so as to be amenable to further processing using biodiversity informatics tools.

Biodiversity Informatics Standards and Protocols

In common with other data-related disciplines, biodiversity informatics benefits from the adoption of appropriate standards and protocols in order to support machine-machine transmission and interoperability of information within its particular domain. Examples of relevant standards include the Darwin Core XML schema for specimen- and observation-based biodiversity data developed from 1998 onwards, plus extensions of the same, Taxonomic Concept Transfer Schema, plus standards for Structured Descriptive Data and Access to biological collection data (ABCD); while data retrieval and transfer protocols include DiGIR (now mostly superseded) and TAPIR (TDWG access protocol for information retrieval). Many of these standards and protocols are currently maintained, and their development overseen, by the taxonomic databases working group (TDWG).

Current Biodiversity Informatics Activities

At the recent, large scale *e*-Biosphere conference in the UK, contributions (e.g. as posters) were grouped into the following themes, which is indicative of a broad range of current Biodiversity Informatics activities and how they might be categorised:

1. Application: Conservation/agriculture/fisheries/industry/forestry.
2. Application: Invasive alien species.
3. Application: Systematic and evolutionary biology.
4. Application: Taxonomy and identification systems.
5. New tools, services and standards for data management and access.
    (a) New modelling tools.
    (b) New tools for data integration.

(c) New approaches to biodiversity infrastructure.

(d) New approaches to species identification.

(e) New approaches to mapping biodiversity.

6. National and regional biodiversity databases and networks.

A post-conference workshop of key persons with current significant biodiversity informatics roles also resulted in a workshop resolution that stressed, among other aspects, the need to create durable, global registries for the resources that are basic to biodiversity informatics (e.g. repositories, collections); complete the construction of a solid taxonomic infrastructure; and create ontologies for biodiversity data.

WORLD BIODIVERSITY DATABASE CD-ROM SERIES

The World Biodiversity Database CD-ROM Series was set up by ETI based on an initiative of the United Nations Educational, Scientific and Cultural Organisation (UNESCO) to provide scientists, students, and other interested parties all over the world with easily accessible high-quality taxonomic and ecologoigcal information on the world's biological diversity. This series of electronic publications provides the user with a unique digital library of basic information on species and higher taxa.

BIODIVERSITY DATABASE INITIATIVES IN INDIA

The development of biodiversity informatics resulted in the establishment of a number of biodiversity databases and information resources (Table 11.1). India has developed a number of excellent biodiversity databases such as the Flora of Karnataka, Traditional Knowledge Digital Library or the National Register of Green Grassroots Innovations and Traditional Knowledge. There exist therefore rich possibilities of building upon country's biodiversity resources and associated knowledge; to promote biodiversity-based enterprises in the modern, as well as traditional sectors; to develop biotechnology industries at the cutting edge of new technologies as well as to encourage local level value addition to biodiversity resources.

Table 11.1. Web resources on biodiversity informatics with special reference to plants.

Database name	Description about the database (organism)	URL
Bioversity International	Bioversity is the world's largest international research organisation dedicated solely to the conservation and use of agricultural biodiversity.	http://www.bioversityinternational.org/
Botanical Survey of India	Plant database—three million accessioned specimens of plants	http://envfor.nic.in/bsi/
Plants of India	Plants—plant family and genus information database	http://www.ecoinfoindia.org/pdb_query.php
FRLHT's Encyclopedia of Indian Medicinal Plants	Indian Medicinal Plants	http://www.frlht.org.in/meta/
Sahyadri: Western ghats Biodiversity Information System	Western ghats biodiversity information	http://wgbis.ces.iisc.ernet.in/biodiversity/
Botanic Gardens Conservation International (BGCI)	Plant diversity	http://www.bgci.org/worldwide/home/

(Contd...)

Database name	Description about the database (organism)	URL
ETI's World Biodiversity Database	Taxonomic database and information system, which converse a wide variety of organisms. Also has 20 species specific databases.	http://www.eti.uva.nl/tools/wbd.php
Flora of NW Europe	Vascular plants (pteridophytes, gymnosperms and angiosperms)	http://nlbif.eti.uva.nl/bis/flora.php
Ethnoforestry	Indigenous knowledge about forestry	http://www.iifm.ac.in/databank/ef/ethnoforestry.html
Sal Borer Problem in India	Shorea *robusta*-timber species	http://www.iifm.ac.in/databank/problems/salborer.html
Catalogue of the Benthic Marine Algae of the Indian Ocean	All published records of species and infraspecific taxa of benthic marine algae from the Indian Ocean.	http://ucjeps.berkeley.edu/rlmoe/tioc/ioctoc.html
CITES-Species Database	Species database, currently holds 7 million records of trade in wildlife and 50,000 scientific names of taxa.	http://www.cites.org/eng/resources/species.html
Integrated Taxonomic Information System (ITIS)	Taxonomic information on plants, animals, fungi, and microbes of North America and the world.	http://www.itis.gov/
Species 2000	Species database	http://www.species2000.org/
International Plant Name Index (IPNI)	Vascular plants	http://www.ipni.org/
Internet Directory of Botany	Plant names	http://www.botany.net/IDB/
BioCISE	European biological collections	http://www.bgbm.org/BioCise/
Global Biodiversity information facility	Biodiversity information	http://www.gbif.org/
BioCASE	Transnational network of biological collections of all kinds.	http://www.biocase.org/
BioLib (Biological Library)	International encyclopedia of plants, fungi and animals.	http://www.biolib.cz/en/main/
Barley DB	Barley germplasm	http://www.shigen.nig.ac.jp/barley/
Conservation Commons	Open access system to data, information, expertise and knowledge related to the conservation of biodiversity.	http://biodiversity.org/
Biodiversity Information System	Indian biodiversity information	http://www.bisindia.org/
Biodiversity Informatics Facility at the American Museum of Natural History's Center for Biodiversity and Conservation	Biodiversity and conservation information	http://biodiversityinformatics.amnh.org/

(Contd...)

Database name	Description about the database (organism)	URL
Chinese Biodiversity Information System (CBIS)	Chinese biodiversity information	http://cbis.brim.ac.cn/cbise/index.html
Biodiversity Information System(Plants of Western Ghats)	Information about 5000 species of flowering plants	http://ces.iisc.ernet.in/hpg/cesmg/pew/bis.html
The Biotica Information System (BIOTICA)	The system designed to handle curatorial, nomenclatural, geographical, bibliographical and ecological data.	http://www.conabio.gob.mx/informacion/biotica_ingles/doctos/acerca_biotica.html
Biodiversity, Department of the Environment, Water, Heritage and the Arts	Biodiversity portal of Australia	http://www.environment.gov.au/biodiversity/index.html
European Natural History Specimen Information Network (ENHSIN)	Specimen databases and resource sharing	http://www.nhm.ac.uk/research-curation/projects/ENHSIN/index.html
FloraBase	Floras information	http://florabase.calm.wa.gov.au/
GrainGenes	Triticeae and Avena	http://wheat.pw.usda.gov/GG2/index.shtml
International Legume Database and Information Service (ILDIS)	Fabaceae (Leguminosae)	http://www.ildis.org/
International Organisation for Plant Information (IOPI)	Plant taxonomic information	http://plantnet.rbgsyd.nsw.gov.au/iopi/iopihome.htm
Centre for Plant Biodiversity Research and Australian National Herbarium	Botanical names and biodiversity information	http://www.anbg.gov.au/cpbr/index.html
KOMUGI (Integrated Wheat Science Database)	Wheat science	http://www.shigen.nig.ac.jp/wheat/komugi/top/top.jsp
National Plant Germplasm System	Genetic diversity in plants	www.ars-grin.gov/npgs
NatureServe	Biodiversity and conservation information	http://www.natureserve.org/
OryzaBase	Rice Science	http://www.shigen.nig.ac.jp/rice/oryzabase/top/top.jsp
IABIN	The Inter-American biodiversity information network	http://www.iabin-us.org/
The System-wide Information Network for Genetic Resources (SINGER)	Germplasm information exchange network of the Consultative Group on International Agricultural Research (CGIAR) and its partners	http://www.singer.cgiar.org/
The *Arabidopsis* Information Resource (TAIR)	Genetic and molecular biology data for the model higher plant *Arabidopsis thaliana*.	http://www.arabidopsis.org/index.jsp
The Tree of Life (TOL)	Diversity of organisms on earth, their evolutionary history (phylogeny) and characteristics.	http://www.tolweb.org/tree/

The Indian Institute of Spices Research, Calicut has played a phenomenal role in collecting and conserving the genetic resources of spices, which include cultivated, wild, hybrids and several endangered species. The national repository of spice germplasm maintained in *ex situ* and *in situ* conservatories are enriched regularly by undertaking collection surveys in primary and secondary centres of origin. Some of the valuable collections in the germplasm are endangered species like *Piper barberi* and *P. arboretum*. *P. silentvalleyensis*, *P. sugandhi* and *P. nigrum var. hirtellosum* are three new taxa identified and reported. *Vanilla anadamanica*; *P. colubrinum*, a source of resistance against *Phytophthora*, *pollu* beetle and *Radopholus similis*; multibranch types and natural *katte* resistant lines of cardamom; king cloves; putative wild types of ginger and high curcumin types of turmeric were also collected and conserved. The institute also conserves diverse strains of micro-organisms associated with spices. Several information resources were developed pertaining to these germplasm collections and micro-organisms associated with spices (Table 11.2).

Table 11. 2. Biodiversity databases developed at IISR, Calicut

Biodiversity database/tools name	Crop/organism	Availability
Spice Genes I	Black pepper	Online and CD-ROM
Spice Genes II	Curcuma species	CD-ROM
Spice Genes III	Nutmeg	CD-ROM
Piper base	*Piper* species	CD-ROM
PhyDish	*Phytophthora* spp.	Online
PIR databases	*Phytophthora* spp.	Online
PLASBID	Plant associated bacteria	Online

Black Pepper

Black pepper (*Piper nigrum* L.) is one of the most important spices and India (Western Ghats) is the centre of origin and diversity for *Piper nigrum*. India leads in area and/or production of this commodity. The Indian Institute of Spices Research, Kozhikode is the National Repository for black pepper germplasm and it has a collection of 1075 wild accessions and 1272 cultivar accessions in the gene bank. This comprises of about nineteen indigenous species and six exotic species besides more than 80 local cultivars. A portion of this germplasm is also maintained in the *in vitro* repository in the Institute. Wide variability occurs in the genus *Piper* with regard to morphological and reproductive characters like leaf size, shape, pubescence, nature of spike (size, colour, erect or pendent), fruit size, etc. The germplasm at IISR, Kozhikode is characterised for about 50 characters based on IPGRI descriptor. The data on these characters are recorded and a database has been prepared using MS-Access–2000 program. This database 'Spice Genes I' is used for entering and retrieving the data. This is useful for researchers and all those who are involved or interested in *Piper*. A unique spike proliferating black pepper accession collected from a farmer's nursery is being maintained in the pepper repository. The database is available on CD-ROM as well as online. Another database, PiperBase on *Piper* species in India is also available on CD-ROM from the Institute. It includes botany, taxonomy, agronomy, biochemistry and medicinal properties of various *Piper* species present in India.

Curcuma Species

The genus *Curcuma* belonging to the family *Zingiberaceae* has a widespread occurrence in the tropics of Asia to Africa and Australia. Apart from *Curcuma longa* or *C. domestica*, the common culinary turmeric, there are several other species of *Curcuma* which are mainly used as colouring agents, for production of arrowroot and for medicinal purposes. Species such as *C. aromatica* (Kasturi manjal), *C. caesia* (black turmeric) *C. amada, C. zedoaria, C. purpurescens, C. mangga, C. heyneana, C. xanthorrhiza, C. aeruginosa, C. phaeocaulis* and *C. petiolata* are also cultivated in different places and regions. The genus Curcuma consists of about 117 species whereas from India around 40 species are reported. *Curcuma* species differ in floral characters, aerial morphology and underground rhizome features besides the chemical traits. Spice Genes II, a database on *Curcuma* species has been prepared using Visual Basic and MS Access. The database is available on CD-ROM.

Nutmeg

Nutmeg belong to the family Myristicaceae with about 18 genera and 300 species of which genus *Myristica* is the most primitive. At present Myristicaceae is considered as a member of Magnolides or its taxonomic equivalents. *Myristica fragrans* Houtt is the cultivated nutmeg tree and gives two spices — nutmeg and mace. The major nutmeg growing areas are Indonesia and Grenada and to a limited extent in Sri Lanka, India, China, Malaysia, Western Sumatra, Zanzibar, Mauritius and the Solomon Islands. Though *M. fragrans* is indigenous to the Banda islands in the Moluccas, many related species of *Myristica* are distributed from India and South East Asia to North America and the Pacific Islands. Sinclair listed a total of 72 species distributed in these areas. Nutmeg was introduced to Southern India in 18th century where it was naturalised. Nutmeg is dioecious with male and female flowers occurring on different trees. About 12 species of *Myristica* occurs in Indo Malayan region. A collection of 465 accessions of *Myristica* including cultivated types and related species are conserved at IISR gene bank. The present day nutmeg population in India has evolved from a few sources trees introduced originally. The biodiversity information on *Myristica* germplasm accessions and related species has been organised as searchable database called Spice Genes III using MS Access and Visual Basic.

Phytophthora spp.

Apart from these biodiversity databases, the institute is having germplasm of *Phytophthora*, an important pathogen of horticultural crops. *Phytophthora* root rot is a serious, widespread and difficult to control fungal disease affecting a wide range of plants across the world. *Phytophthora* species are mostly pathogens of dicotyledons, and are relatively host-specific parasites. Many species of *Phytophthora* are plant pathogens of considerable economic importance. *Phytophthora infestans* was the infective agent of the potato blight that caused the Great Irish Famine. The foot rot of black pepper is caused by *P. capsici*. The Institute has developed two databases, which deals with *Phytophthora* information namely PIR and PhyDisH. PIR database contains information on 64 species and an expert identification tool based on morphological characteristics of each species. Extensive information about literature collection and molecular information details are also included. The other database PhyDisH (*Phytophthora* Diseases of Horticulture Crops) deals with the major diseases caused by *Phytophthora* spp. on horticultural crops in India. One can find different species of *Phytophthora* affecting different crops and also their management methods. Information on 447 *Phytophthora* isolates belonging to nine species obtained from 30 different hosts that are maintained in the National Repository of *Phytophthora* (NARPH) is also made available in PhyDisH. It also integrates a literature database which contains abstracts of publications on *Phytophthora* from Pubmed database. This database is developed primarily from the information generated through a National Network Project on Phytophthora Diseases of Horticultural

Crops (PHYTONET) funded by the Indian Council of Agricultural Research (ICAR), New Delhi. Both these databases are available online.

PLASBID

Bacteria associated with plants have been observed frequently to form assemblages referred to as aggregates, microcolonies, symplasmata or biofilms on leaves and on root surfaces and within intercellular spaces of plant tissues. Bacteria associated with plants are diverse in their ability to affect plant health, their genotypic and phenotypic characteristics and their phylogeny. A database named PLASBID (plant associated bacteria identification system) has been developed by consolidating information available on bacteria reported to be associated with plants. The core information made available is nucleotide sequences belonging to 16S rRNA, primer information, G + C content, NCBI link and their taxonomy. It has a unique sequence similarity search facility using NCBI BLAST to find the similar sequence in the database with respect to query sequence. The sequence provided by the user is searched against PLASBID 16S rRNA nucleotide database and identifies the species based on their sequence similarity. Available primer information also has been compiled. Another feature of the database is Restriction Analyzer, which is designed to provide the information regarding restriction enzymes and their cutting sites based on the sequence entered by the user.

Chapter 12

Dry Land Biodiversity

INTRODUCTION

Drylands have an immense scientific, economic and social value. They are the habitat and source of livelihood for about one quarter of the earth's population. It is estimated that these ecosystems cover one-third of the earth total land surface and about half of this area is in economically productive use as range- or agricultural land.

Dryland ecosystems contain a variety of native animal, plant and microbial species that have developed special strategies to cope with the low and sporadic rainfall, and extreme variability in temperatures that prevail in these ecosystems. Such adaptive traits have global importance, especially in the context of predicted climate change.

Dryland pastoralists and farmers have developed efficient pastoral and mixed cropping systems adapted to the difficult conditions of drylands. These systems have sustained the livelihoods of generations of dryland people. Furthermore, dryland pastoralists and farmers have successfully created and maintained high levels of agrobiodiversity of crops and livestock breeds. Yet, global awareness about the great value of drylands remains frustratingly low. Compared to tropical rainforests, for example, the wealth of dryland biodiversity and indigenous knowledge is less well documented, and has received much less support and advocacy in conservation media.

Although remnants of healthy dryland biodiversity and indigenous knowledge still exist at various locations, drylands face increasing threats of further degradation. Already, as sources quoted by Subsidiary Body on Scientific, Technical and Technological Advice (SBSTTA) report, it is estimated that 60 per cent of drylands is already degraded resulting in an estimated annual economic loss of US$ 42 billion world-wide. Thus, continued degradation of drylands is a major threat to the ecological functions of drylands and to the species and genes living in these ecosystems and thus to human welfare.

Drylands need urgent attention. Efforts in the past to address dryland issues have achieved much less than expected. Thus new paradigms are needed to go beyond the status quo with imagination and courage. The purpose of this chapter is to: (i) draw greater attention to the potentials of drylands and (ii) suggest options for better conservation and sustainable use of dryland biodiversity.

DEFINITION AND EXTENT OF DRYLANDS

The term drylands is used in this chapter to cover hyper-arid, arid, semi-arid and dry subhumid ecosystems. Aridity zones as widely used in the scientific literature are based on the ratio P/PET (where P is the area's mean annual precipitation and PET is the mean potential evapotranspiration). This ratio is referred

266

to as aridity index and is used to classify drylands as hyper-arid (ratio less than 0.05), arid (0.05 to 0.20), semi-arid (0.20 to 0.50) and dry subhumid areas (0.50 to 0.65).

Drylands as defined above are equivalent to what the CBD's Subsidiary Body for Scientific, Technical and Technological Advice refers to as 'drylands, arid, semi-arid, savannah, and grassland and Mediterranean ecosystems'. Dryland ecosystems cover extensive land areas stretching across more than one-third of the earth's land surface. They are found on all continents in both the northern and southern hemispheres, and are home to about one quarter of the earth's population. They cover a variety of terrestrial biomes which are extremely heterogeneous with wide variations in topography, climatic, geological and biological conditions. The main dryland ecosystem-types have been described in some details including geographic distribution, land area, vegetation type and various aspects of biodiversity.

Despite differences between various dryland ecosystem-types caused by differences in levels of aridity, topographic elevation, geological and biological conditions, etc. these ecosystems have in common a unifying characteristic: precipitation is low and extremely variable. Recurrent droughts that may persist for several consecutive years are the rule, not the exception. Furthermore, and particularly in the more arid areas, diurnal temperature variability is high, thus required special adaptations from all species.

DRYLAND BIODIVERSITY STATUS AND TRENDS

The United Nations Convention on Biological Diversity defines biological diversity as 'the variability among living organisms from all sources including, *inter alia*, terrestrial, marine and other aquatic ecosystems and the ecological complexes of which they are part; this includes diversity within species, between species and of ecosystems'. This section reviews the status of dryland biodiversity and assesses its future trends.

Driving Forces of Biodiversification in Drylands

As indicated earlier, dryland ecosystems cover a variety of terrestrial biomes (i.e. arid steppe, grasslands at various altitudes and latitudes; tropical and sub-tropical savannahs; dry forest ecosystems and coastal areas) which are extremely heterogeneous. A major driving force of biological diversification in dryland environment is relative aridity. Within each aridity zone, significant variations between sites are introduced by topography and geology as mentioned earlier, and also by variations in the most limiting factors, i.e. water and soil nutrients. The resulting patchwork of habitats determines the distribution of living organisms.

In addition to habitat differentiation, other driving forces of biodiversification in drylands include the seasonal pattern of rainfall, fires and herbivore pressure. Types and intensities of these environmental stresses combine to determine the main selection pressures in drylands: low and highly variable rainfall in time and space; recurrent but unpredictable droughts that may persist for several consecutive years; high temperatures; inherently low soil fertility; high incidence of salinity; prevailing herbivore pressure; and fires. These stresses have selected for a large diversity of adaptive traits.

The seasonal pattern of rainfall, for example, has selected for plant and animal species and micro-organisms able to develop rapidly and complete their *life-cycle* in a very short period of time. Adaptation to drought has selected for a wide range of strategies. Some species have mechanisms to escape drought whereas other species have appropriate organs to resist drought. Plant species, for example, have large below ground systems to store water and nutrient or corky bark to insulate living cells from desiccation and fire burning. Another powerful selection agent is human population. Local people have developed complex pastoral and cropping systems for which they have selected and maintained the biological diversity they value most in domestic livestock and crops.

Status of Dryland Biodiversity

Dryland ecosystems are unique. One can site such examples as the Mediterranean systems (e.g. the distinctive sclerophyllous vegetation of the Mediterranean Basin, drylands of Southern Australia and California, Chile, Cape Floral Kingdom of South Africa, and shrublands of Australia); the cold deserts of Mongolia and Chile; the Sahara and Sahel of Africa; the arctic circle drylands; and the high altitude drylands of Iran and Afghanistan; just to name a few.

Total number of named species in the world is approximately 1.4 million or more. How many are from drylands, however, is not well documented. Furthermore, the degree of threat and extinction is not well-known. Specifically, it is not possible to ascertain the correlation between the rate of degradation of drylands (estimated as 60 per cent) and the rate of extinction of species because of the lack of data on endemic species distribution.

In a very challenging paper titled 'There is more to biodiversity than the tropical rainforests', Readford pointed out that much of the publicity concerning threats to global ecosystems and biodiversity has centred on tropical rainforests. While the authors concede that there is no doubt of the importance of moist tropical forests in terms of biodiversity, they also voiced concern that the almost exclusive attention to rainforests may act as blinders, limiting the vision of major conservation stakeholders: donor agencies, governmental bodies, etc. This would in turn lead to the neglect of other ecosystems. One such neglected ecosystem is the world dryland.

The sub-group on biodiversity of the CBD's international panel of experts also noted that the scientific community and international agencies that spearheaded efforts to raise awareness about rainforest biodiversity are conspicuously absent from the dryland debate. Low awareness about the importance of drylands *vis-à-vis* biodiversity and livelihoods is one of the reasons why this ecosystem has received inadequate attention.

Dryland wildlife is best known through its endangered species, the symbols of which include the rhino, the elephant and the impressive large herds of herbivores in eastern and southern Africa. These have become important sources of income through conservation campaigns, game hunting and ecotourism. However, the popular scientific media has also popularised the image of drylands as harsh environments where only scorpions, snakes and specially adapted creatures (including man) can survive. In many ways, this dramatisation has been useful in getting extra attention for drylands, but it has also helped to prolong the myth that drylands must be 'tamed', i.e. they have little intrinsic value in and of themselves.

Protected areas have long been a major pillar of biodiversity conservation strategies. In Africa, for example, about 8.5 per cent of the total land area of the continent is designated as protected area. Drylands have a slightly higher share than forests. About 16 per cent of Africa's population live within 20 km of designated protected areas and population growth in these buffer zones has been found to be higher than elsewhere. This is indicative of the importance of these sites for local people's livelihoods, and a clear warning of the weakness of the 'pure conservation' approach.

Special Features of Dryland Biodiversity

Dryland biodiversity has distinguishable features that are often overlooked. These include heterogeneity, remarkable diversity of micro-organisms, presence of wild relatives of globally important domesticated species, and traditionally adapted land use systems.

Diversity of habitats: Biodiversity in drylands varies among ecosystems, but also among habitats. Species and management systems alike, have adapted to the heterogeneity and special driving forces.

Of particular importance are the natural and human-induced habitats, such as those described in Table 12.1.

Table 12.1. Diverse habitats in drylands.

Wetlands, oases and protected areas constitute islands of enhanced biodiversity in drylands. These are often the life lines and biodiversity hot spots of drylands.

Ponds, lakes and rivers are major poles of socioeconomic activities with significant effects on biodiversity. Global significance of these habitats is increasingly realised. For example, migratory birds depend on these sites, such as the Djoudj national park in Senegal for survival. Another global dimension of these habitats is that degradation of the basins of transboundary rivers has serious repercussions for a huge population covering several countries.

Oases are small islands of greenery in the dry landscape. They are sites of intensive and highly productive systems in drylands. Accessibility to water makes these sites the engines of desert life. Biodiversity per unit land area is probably very high, although not well documented.

Groves: A grove is a forest patch. It may be a remnant of the original vegetation or not. Many groves are worshiped as shrines. Sacred groves and taboo species of animals and plants are part of a community's cultural heritage of natural resource management and biodiversity conservation. Sacred groves, like oases, are biodiversity micro 'hot spots'. The persistence of shrines and groves in many dryland rural communities despite degradation of the surrounding environment are testimony to their importance for these communities.

The many diverse habitats are important factors of intra-specific variation in drylands. Thus, although species richness is higher in tropical forests than in drylands, within–species diversity is probably much higher in drylands than in forest ecosystems, because of the isolation of populations. But there is a lack of reliable data to document this diversity.

Wild relatives of domesticated species: Historically, drylands have been the living basis for mankind. The first humans originated in the savannah grasslands of eastern and southern Africa. The origins of many of the earth's most important food crops are found in drylands. For example, maize, beans, tomato and potatoes originate from the drylands of Mexico, Peru, Bolivia and Chile. Millet and sorghum, and various species of wheat and rice come from the African drylands. The Mediterranean basin has given the world date palm and olive trees. And drylands continue to provide new food, as traditional food products are increasingly becoming commercialised globally in the age of health-consciousness (e.g. wild millet, wild rice, etc.).

Micro-organisms: The world of micro-organisms is notoriously poorly documented. The level of biodiversity is unknown. The importance of micro-organisms, however, is well appreciated in food and pharmaceutical circles. Their ecological role is also crucial. In the particularly difficult environment of drylands, micro-organisms play a major role in key ecological processes that sustain the functioning of these ecosystems. In response to the inherently low soil moisture and nutrients (especially nitrogen and phosphorus) in dryland soils, there exist a great diversity of symbiotic associations between plants and micro-organisms which efficiently: (i) fix atmospheric nitrogen through legume-rhizobia associations and (ii) extract phosphorus, various micro-nutrients, and water under very low moisture conditions, through mycorrhizal associations. Also, some free-living organisms contribute significantly to the nitrogen balance of dryland ecosystems. Blue green algae, for example, have been shown to fix nitrogen in amounts large enough to partially offset nitrogen losses through burning. These are major assets of drylands with global significance.

Agrobiodiversity in drylands: Agricultural biodiversity is a vital subset of biodiversity. It is the result of the careful selection and inventive development of farmers whose food and livelihood security depends on the sustained management of this biodiversity. Since the dawn of agriculture some

12,000 years ago, humans have selectively used and bred certain species of plants to provide food and other goods. These varieties of plants and breeds of animals, the species, and the agroecosystems that support them comprise agrobiodiversity.

Pastoralism: Dryland herders have engineered efficient pastoral systems and promoted livestock diversity

Progress in range science and better appreciation for indigenous knowledge have increased our awareness of the resilience of rangelands and the reversibility of the alleged degradation of rangeland ecosystems. In fact, due to strong seasonal dynamics, the risk of overgrazing is limited to a short period in time. The two basic properties of dryland pastoral ecosystems, instability and resilience, support the continued practice of transhumance and of nomadism. Both methods of utilising grazing land resources are based on mobility and maximal dispersion during the growing season.

The new non-equilibrium ecological theory undermines earlier approaches to range ecology, and represents an alternative theory of the functioning of pastoral ecosystems. This approach stresses the need for flexibility and mobility in dryland opportunistic grazing strategies, principally due to the strong rainfall fluctuations found in these arid environments. There is now a greater appreciation of the efficiency of traditional pastoral systems based on mobility and the exploitation of extensive resources.

In Australia, for example, rangelands have been shown to be remarkably resilient. Although overgrazing leads to reduction of vegetation cover in the shrublands, the system is known to bounce back to earlier vegetation type and species composition when favourable conditions return, i.e. reduced grazing pressure and better rainfall. It is only where extensive commercial ranching is practised that rangeland degradation becomes severe.

In the Sahel, Denève reports that comparison of productivity between livestock in this ecosystem and ranches in Texas and Australia shows that animal productivity per surface unit is 1.5 to 10 times higher in the Sahel than in modern ranches. The author also challenges a rooted misconception about the role of livestock in desertification, indicating that when everything has been grazed, animals have to leave or die, but this does not mean that vegetation has disappeared. Indeed, like in the case of Australian rangelands, the capacity of the highly resilient vegetation of the Sahel to bounce back when good rains come after a drought period is impressive.

Contrary to the widely held view that herders in their ignorance are raising their animals in an irrational way and destroying the environment, traditional herders have accumulated invaluable indigenous knowledge and skills for optimum use of their marginal land.

Dryland systems have in fact adapted to herbivory. Evidence shows that dryland vegetation can degrade if grazing is reduced or prohibited. Most drylands are grazing-dependent systems. But drylands are now seriously threatened and in some places have broken down, due to commercial (continuous, non-mobile) ranching in the case of Australia or pressures to provide land for agriculture, human settlements, protected areas, etc. in the Sahel. The breakdown of the system has serious impact on the diversity of living organisms but also on the wealth of indigenous knowledge accumulated by traditional herders.

Indeed, as indicated above, indigenous pastoralists have acquired extensive knowledge of species, habitats and key ecological processes in grazing lands and have developed efficient management skills for these systems. In addition, they have contributed enormously to the promotion and conservation of great diversity in domestic livestock as illustrated. All these assets may be lost if nothing is done to alleviate the root causes of ecosystem degradation.

Agriculture: Dryland Farmers Promote Agrobiodiversity in farming systems and crop genetic resources

Management of crop-biodiversity by local farmers in drylands dates back to the dawn of agriculture. There is general agreement that many cultivated plants originated from drylands, including species of sorghum, millet in Africa, beans, potatoes, tomatoes from Latin America, etc. Management strategies developed over several millennia have generated a vast array of farming systems and crop genetic resources.

Diversity of farming systems

The last three decades of farming systems research have shown the tremendous diversity and vitality of many traditional cropping systems in drylands, as elsewhere. We now have a better appreciation of why farmers continue to nurture biodiversity despite pressures to convert to mechanised mono-cropping (Table 12.2). The reasons have to do with risk management, balancing long-term ecological sustainability vs. short-term gains, and multiple uses and products rather than specialisation in productivity.

Table 12.2. Diversity in farming systems.

In the drylands of India and Pakistan, for example, farmers still maintain many of their traditions of nurturing biodiversity of wild and cultivated food crops and medicinal plants, despite introduction of monocropping by the Green Revolution. In addition, the traditional reverence of Indian farmers for multipurpose trees such as *Prosopis cineraria* in croplands enhances agrobiodiversity.

In a study of indigenous practices in farming systems and crop planting methods in eastern Kenya, Mathenge described no less than 10 distinct farming systems with 6 different types for what outsiders would only refer to as one 'slash and burn' system. The crop planting strategy also was found to be very diverse, with 6 different planting methods of mixed seed cropping in which seeds of more than one species are planted in one hole. One of the most commonly practised method is to plant millet, sorghum and cowpea in one hole.

Another major farming strategy in drylands is the deliberate preservation of valued trees and shrubs in crop fields, a traditional agroforestry system known as parklands. In West Africa, for example, *Faidherbia* (Acacia) *albida* parklands in semi-arid and sub-humid areas are known to have sustained continuous cropping for generations without fallow periods. The system is similar to the *Prosopis cineraria*-based system in India.

Although biodiversity may induce reduced crop yields in many cases, through crop competition, farmers consider that the overall benefits of the biodiversity rich system override the shortcomings. During years of drought, annual crops may fail completely. Farmers then rely heavily on the products of trees and shrubs for survival. Even during years of good harvest, people depend on tree products for vitamins, minerals, medicine, etc. Also, trees and herbaceous hedges on farm contribute to increased carbon storage in biomass above and below ground, thus reducing emissions of carbon dioxide to the atmosphere. This runs counter to the widely-held view that conversion of natural vegetation to agriculture reduces biodiversity and carbon storage. Agroforestry systems in drylands need better documentation to shed light on this apparent paradox of agroforestry parklands that increase biodiversity and carbon stocks on-farm. This is clearly a dryland asset with global significance.

Diversity of crop genetic resources

Promotion of crop genetic diversity is part of farmer's coping strategies for mitigating weather unpredictability; it also reduces the so-called 'hunger period' by spreading availability of food products over time. For example, in mixed farming, green leaves from cowpeas may be harvested as early as 21 days after sowing, whereas early green maize harvest is done at 60 days and late material at

120 days. Potato farming communities in Cusco, Peru, are reported to be managing up to 150 varieties. This diversity is an important element of food security in rural areas.

Rubyogo investigated farmers' crop variety ranking criteria in Kenya and reports the use of the following criteria:

1. Early maturity (drought escaping).
2. Drought tolerant.
3. Stable and if possible high yield.
4. Pest/disease and weed tolerance.
5. Socioeconomic criteria, e.g. variety for market production or household consumption.

Farmers maintain different varieties of maize, sorghum and other crops for each of these objectives. Evidence from other dryland ecosystems types support these findings that farmers value having agricultural biodiversity in their farming systems. This has been the experience with millet farmers in Rajasthan in India, with sorghum farmers in Tharaka, Kenya, with sorghum and millet farmers in Zimbabwe, and with potato farmers in Peru.

Green pharmacy: Traditional healers are knowledgeable in medicinal plant diversity

Old civilisations in drylands, like everywhere else, care for medicinal plants because they have built up generations of tried and tested curative methods and products. In India, for example, Shankar reported that 4671 plant species are used in folk medicine. Although not unique to drylands, it is a remarkable fact that the use of medicinal plants is a living tradition of dryland rural people. In addition to the 'professional' healers, countless of millions of women and elders have invaluable knowledge of herbal home-remedies and food and nutrition. Much of this knowledge has begun to be documented, preserved and fed into *sui generis* systems of IPR around the world. More investigation and analysis would be necessary, for example, in correlating dryland epidemiology with pharmacopae, developing benefit sharing regimes, and enhancing effectiveness of remedies for countless of rural populations without adequate health coverage.

Threats to Dryland Ecosystems and Species Diversity

Extinction is a feature of all biological systems, but extinction of global biodiversity is now proceeding at an alarmingly high rate. Before human life on earth, the speed of extinction was about one species per year. Today, the rate is estimated to be 1000 to 10,000 times this natural rate. According to IUCN's 2002 'red list', about 34 per cent of all fishes, 25 per cent of mammals, 25 per cent of amphibians, 20 per cent of reptiles and 11 per cent of birds were threatened with extinction at that time.

It is generally agreed that the following five causes are responsible for the loss of biodiversity:

1. Fragmentation, degradation or outright loss of habitats.
2. Overexploitation.
3. Pollution.
4. Introduction of non-native (alien or exotic) species.
5. Climate change.

The specific contribution of each of these causes to dryland ecosystem degradation and biodiversity loss varies with regions.

In Australia, 2/3 of the continent consists of rangelands, of which 60 per cent are used for commercial extensive ranching worth an annual revenue of 2 billions Australian dollars. It is estimated that over the last century, 12 per cent of Australian dryland mammals have gone extinct and 8 per cent of bird species are under threat. Most of these losses occur in areas used for large scale commercial ranching which, therefore, is the main cause of ecosystem degradation and biodiversity loss in Australian drylands.

In the Indian and Pakistani drylands, the main threats are: (i) overcutting of trees and shrubs for fuel and building material, (ii) large scale irrigation and expansion of agricultural fields and (iii) urbanisation, mining and industrialisation. Wildlife and natural vegetation are thought to be severely affected, but tree-based and other diverse production systems contribute to diversification of agricultural production and enhancement of agrobiodiversity.

Latin American drylands experience overgrazing by goats and/or overexploitation of natural vegetation for fuelwood, but these do not seem to cause irreversible degradation. Major disturbances are caused by mining operations, irrigation and urbanisation. Increasing conversion of natural vegetation into commercial monoculture also cause serious damage to ecosystem integrity. Traditional farming systems, however, still retain high levels of agrobiodiversity.

In the African drylands, the Convention to Combat Desertification recognises that poverty-induced overexploitation of the land are a major cause of environmental degradation. This occurs through encroachment of agriculture on grazing lands (marginal to cropping but high quality for wild or domestic animal grazing), and overcutting of natural vegetation for fuelwood. These problems are compounded by rapid urbanisation, with its concomitant exponential increase in demand for charcoal, other wood products, construction gravel and soil, and other natural resources, not to mention negative impacts from lack of waste management. All of these factors have caused severe disruption of the traditional pastoral and rainfed cropping systems. Mobility, which is key to Sahelian pastoral systems of transhumance and nomadism, as well as wildlife, is becoming increasingly difficult because of fragmentation or outright destruction of grazing lands by agriculture and human settlements.

The status and trend in dryland biodiversity can be summarised as follows:

1. The status of biodiversity loss in wild species is poorly documented; it may be that endemism is still healthy in pockets and hot spots, but the viability of these populations is not known.
2. Most likely, genetic diversity has decreased as specific populations have been wiped out.
3. Species diversity in traditional production systems is relatively well retained in farmers' crop fields and family herds.

Degradation of habitats due to changes in land use is the immediate most serious threat to dryland biodiversity. This first and foremost affects wild biodiversity, but as populations increase, and the urban and industrial sectors are not able to absorb this increase, there will be pressure on agricultural land, leading to agro-habitat degradation as well.

New and powerful factors cause ecosystem degradation that are far beyond the coping capacity of the legendary resilience of dryland systems: commercial ranching and monoculture, mining and industrialisation, urbanisation and other forms of human settlements, widescale irrigation schemes, strictly protected areas, etc.

The root causes of these factors are population increase, decades of market distortions and disincentives that encourage natural resource exploitation and mono-cultures, at the expense of dryland adapted sustainable development. This has been compounded by the 'benign neglect' of drylands. Loss of biodiversity caused by land use systems is further exacerbated by climatic factors at both local and global scales.

BIODIVERSITY, DESERTIFICATION AND CLIMATE INTERACTIONS IN DRYLANDS

The physical processes of land degradation, biodiversity evolution or extinction, and climate change are intimately inter-twined, especially in drylands. Land degradation reduces natural vegetation cover, and affects productivity of crops, livestock and wildlife. Soil micro-organisms are also affected through soil erosion. The loss of biodiversity likewise undermines the environmental health of drylands and makes them more prone to further degradation. The vicious cycle fuels increased soil erosion, which causes increase in sedimentation of rivers and lakes. This contributes to the degradation of international waters and affects biodiversity in rivers, lakes and coastal ecosystems.

Desertification is also related to climate in many ways. Degradation of vegetation cover decreases carbon sequestration capacity of drylands, thus increasing emissions of carbon dioxide into the atmosphere. But carbon storage capacity of drylands is poorly documented. Given the increasing recognition that many dryland plant species develop extensive below ground biomass, current estimates of carbon (C) sequestration in drylands are probably vastly underestimated. In the Sahel, for example, tree below ground biomass has been shown to be as high as the above ground biomass, with roots extending to 70 m away from the trunk or as deep as 30 m. This is yet another manifestation of dryland adaptation. Another link between desertification and climate is through the effect of dryland dust on atmospheric composition. Arid lands are significant contributors of dust. Reduction of vegetation cover caused by land degradation increases these effects. The exact pathways in which increase in atmospheric dust can modify climate us still hotly debated, but a recent study reported by National Oceanic and Atmospheric Administration (NOAA) has provided the first substantiated evidence that atmospheric dust can affect both regional and global climates. Periodic burning of savannah landscape has also been shown to have implications for atmospheric chemistry. Thus ecological processes in drylands influence local and global climate. Climate change in turn affects drylands biodiversity by influencing species distribution range, water supplies, heat extremes, the humidity and temperature of soils and thus the albedo. The predicted global climate warming resulting from the build-up of greenhouse gases in the atmosphere is expected to have profound impacts on global biodiversity at levels that may compromise the sustainability of human development on the planet. Climate warming will cause, *inter alia*, higher evaporation rates and lower rainfall both of which are major determinants of dryland ecological processes. Simulation models of climate change predict shifts in species distribution and reduced productivity in drylands. Each one-degree rise in temperature is expected to displace the adaptation of terrestrial species some 125 km towards the poles or 150 metres in altitude. Approximately 30 per cent of the earth's vegetation could experience a shift as a result of climate change. Millet crop yields in Africa, for example are expected to drop by 6–8 per cent. The yield decrease may even be as high as 11 per cent and 38 per cent in some localities more severely affected. The maize crop yield in Asia and Latin America may shrink by between 10 and 65 per cent. The expected impact on wild biodiversity is much less known or analysed.

Simulation models by Sala and Chapin to assess biodiversity change over the next 100 years predict that dryland biomes such as savannahs, grasslands and Mediterranean ecosystems will be among the biomes experiencing the largest biodiversity change, and will be affected significantly by a combination of land use change and climate change. The strong linkages between desertification, biodiversity and climate are a clear invitation to CCD, CBD and FCCC for more effective collaboration. Increased efforts and resources must be used to promote stronger synergies. Such collaboration needs to be effective all the way from the level of local communities up to national, regional and international levels.

CONSERVATION AND SUSTAINABLE USE OF DRYLAND BIODIVERSITY: A CALL FOR ACTION

Past conservation efforts have focused mostly on: (i) *ex situ* conservation of major food crop genetic resources and (ii) *in situ* conservation of natural systems in protected areas. Tangible results have been achieved in both cases, but serious limits and conspicuous outright failures have also been recorded.

Building on lessons learnt from past experiences, this section argues that dryland conservation strategies must focus on *people*, the end-users of genetic resources, whether for present or future generations. In this context, and in line with the precepts of the convention on biodiversity, successful management of drylands will depend on our collective ability to formulate and implement appropriate policies and design/conduct proper field activities. The main objective will be to maintain and restore dryland ecosystems through conservation, the sustainable use of biodiversity and the fair and equitable sharing of benefits. Within this general framework, however, program and projects on the ground will have to be specific to ecosystem-types since ecological processes and root causes of land degradation vary between regions, as shown above. Very often the root causes can be traced to a history of well intentioned, yet misguided interventions. Enabling environments may have deteriorated to the point where conservation and sustainable use of biodiversity can only be part of integrated strategies to promote sustainable economic growth and social development.

Lessons Learnt From Past and Ongoing Efforts

Ex situ conservation

Various national, regional and international institutions, program and projects have participated in wide-range collections and conservation in genebanks of tropical food crop genetic resources, including dryland species (Table 12.3).

Table 12.3. Sources for *in situ* conservation of dryland crops.

Centre name and location	Crop
CIMMYT, Mexico	Wheat, maize, triticale
CIP, Perou	Potato, sweet potato, Andean roots and tubers
ICARDA, Syria	Lentil, faba bean, chickpea, barley and wheat
ICRISAT, India	Sorghum, millet, chickpea, pigeon pea and groundnut
ILRI, Nairobi (ex ILCA, Addis)	Grasses, legume and browse species

Although this represents an asset of immense value, the limits of *ex situ* conservation are increasingly recognised. Several weaknesses have been identified, including the fact that gene banks cannot evolve or 'store' the farmers' knowledge and experimentation that creates and maintains agricultural biodiversity. This problem is not specific to drylands but it is particularly significant in drylands. Migration of people caused by land degradation or displacement/resettlement projects seriously undermine the conservation of indigenous knowledge. The limits of *ex situ* conservation and the need to complement it with *in situ* conservation have been discussed in detail throughout the Convention on biological diversity.

In situ conservation

In the past, *in situ* conservation strategies have tended to focus mostly on parks and protected areas. There is increasing recognition, however, that it is no longer sufficient to protect isolated fragments of land and water, given the critical role of biodiversity in maintaining human livelihoods, and *vice-versa*.

Current approaches to both *ex situ* and *in situ* conservation have a common serious constraint that limit their chances of success and impact. The problem concerns the fair and equitable sharing of benefits arising from conservation and more generally, the issue of incentives for conservation. Passionate and emotional debates are ongoing on the Trade Related Aspects of Intellectual Property Rights (TRIPs).

Many stakeholders question the ability of TRIPs in its current approach to settle the issue of fair and equitable sharing of conservation benefits. As early as a decade ago, Altieri wondered, « How common is our common future » given the divide between North and South on the issue. More recently, Plahe referred to TRIPs as a 'Licence to loot'. The problem is not specific to drylands, but it has far-reaching implications on the future of global conservation. Finding the right approach that will ensure perceived fairness and equity in the sharing of conservation benefits between local communities, nations of origin and private multinationals remains a major stumbling block.

Framework for Action

Building on strengths

In a radical departure from the pessimistic view about drylands, the proposed framework for action builds on positive and strong points that emerged over the last decade on the potentials of drylands, namely that:

1. Drylands (rangelands and agricultural lands) are highly resilient and dryland people have developed indigenous knowledge and know-how to manage these lands productively and sustainably.
2. Dryland Farmers and pastoralists value biodiversity in the management of their crop fields and family herds and actively create and maintain high diversity in traditional production systems.
3. There is evidence that participatory involvement of local communities in identification of conservation priorities and sharing of benefits enhances the chances of success. Barrow have shown in Kenya that densities of herbivores increased in ranches where communities benefited from conservation, whereas densities decreased by 40–80 per cent where there was no such organised and perceived equitable sharing of benefits.

The results also highlight the need for participatory involvement of civil societies in community-based conservation efforts. This requires the forging of partnership among a wide range of grassroot stakeholders: local NGOs, farmer organisations, women groups, individual conservation champions and/or opinion leaders, etc.

1. The no-use preservation model of *in situ* conservation does not work in drylands of developing countries where there is strong dependence of people on natural resources for survival. By breaking new and important grounds which reconcile the need for conservation with the concern for development, CBD provides a framework which puts conservation on a more favourable ground for adoption by dryland people.
2. Globally, there are stronger legally-binding instruments for conservation today than there were before, at both national and international levels, prior to the Rio Conventions. CCD specifically addresses desertification issues and therefore focuses on drylands. CBD's article 20, alinea 7, stipulates that 'consideration should also be given to the special situation of developing countries, including those that are most environmentally vulnerable, such as those with arid and semi-arid zones, coastal and mountainous areas'.

Pillars for a plan of action

Based on the main results on dryland biodiversity reviewed in this section, and on the opportunities for dryland conservation as presented above, the current renewed interest for drylands should be seized to put the Global Dryland Partnership on a secure footing in order to address the enabling environment:

Documentation of status and trends: There is urgent need to launch a concerted global initiative to ensure in depth documentation of dryland biodiversity and maintain a credible database. This can be done through desk reviews but will have to be complemented with field studies where required. This in turn will enhance the scope for advocacy. The vicious cycle is that low awareness about dryland potentials has a negative feedback on resource mobilisation. The proposed initiative will break this vicious cycle.

Capacity building. These include: incentive measures; policy reform; letting dryland voices be heard; resource mobilisation; relationship with other international Conventions. All of these will require a concerted effort at capacity building at all levels.

, *Awareness and exchanges between regions*: Research and training; sharing of information; technical and scientific co-operation.

CONCLUSION

This chapter has highlighted the importance of drylands: extensive land areas which are habitat and source of livelihood for about one quarter of the earth's population. Drylands contain highly resilient species adapted to the seasonal pattern of rainfall and recurrent droughts that prevail in these ecosystems. These attributes have global significance in the context of predicted global warming.

Drylands have been neglected in both conservation and sustainable use efforts. It is therefore difficult to provide a definitive picture of biodiversity status and trends. We may conjecture that biodiversity in drylands survives in pockets and hot spots or in transhumant and nomadic areas not affected by habitat conversion or in traditional farming systems. But, following the precautionary principle, and until such time as additional data proves otherwise, it is safe to say that habitat degradation is an imminent and immediate problem affecting biodiversity loss in drylands. It is important that biodiversity in drylands be addressed in two parallel fronts: addressing potential and actual biodiversity loss, through documentation, advocacy, capacity building and improvement of the enabling environment, and highlighting and encouraging instances where biodiversity is healthy and managed sustainably.

Chapter 13

Biodiversity in Grasslands

INTRODUCTION

Loss of biodiversity is a pressing problem for the biosphere. Current estimations for one of the attributes of biodiversity—species richness—indicate that extinction rates are higher than in the recent past and are still increasing. The environmental drivers of species extinction rates are also changing other structural and functional attributes of biodiversity. The main causes of biodiversity loss are global; land use and land cover change has been proposed as having the biggest impact. The grassland biome, which includes a wide range of ecosystem types, from humid prairies to arid shrub–grass steppes, has been subjected to particularly intense pressure for the production of food and fibres, so current extinction rates of grassland species are expected to remain high or even increase. The main aim of this chapter is to consider current processes and future scenarios for grassland biodiversity. As one considers the future, it is useful to address three questions regarding biodiversity loss. First, what are the effects of the main drivers of biodiversity loss, and what are the main trends in the biodiversity of grasslands? Second, how will biodiversity continue to change after extinctions take place? Third, how can we manage grassland biodiversity in the transition to the new scenario? Currently, there is a lot more information available to answer the first question than there is information to address the other two. Nevertheless, it is necessary to address the second and third questions in order to develop proactive strategies for managing the transition to new scenarios.

On the verge of the sixth major extinction event in the geological history of the earth, this chapter reviews both current and probable near-future trends in grassland biodiversity. Concern about biodiversity derives from two certainties. First, that changes in major drivers of ecosystem structure and functioning are global and affect all biomes. Second, that these changes will affect humanity through the erosion of the life-support systems of the earth. In other words, there will be a reduction in ecosystem capability to provide natural resources and to process wastes (i.e. ecosystem services). Grasslands have been and still are central to the production of food and fibres for human use.

The main land uses in grasslands—agronomy and animal husbandry—involve four main activities:
1. Extraction of resources (e.g. mineral nutrients).
2. Changes in energy and material transfers (e.g. individual plant growth and nutrient mineralisation).
3. Changes in species composition (e.g. addition of crops and weeds) and substances (i.e. agrochemicals, principally pesticides and fertilisers).
4. Changes in the disturbance regime (e.g. cultivation, grazing).

These activities introduce major changes in the structure and functioning of ecosystems. However, superimposed on land use changes, grasslands are experiencing changes in other drivers, such as climate, atmospheric composition, and non-planned species exchanges, and these changes combine to threaten ecosystem integrity on a global scale.

Biodiversity has been identified as one of the best descriptors of ecosystem condition. Biodiversity is the sum of total biotic variation, from gene to landscape, but researchers have primarily focused on species diversity. This has resulted in the use of biodiversity and species richness as synonyms and a consequent need to relate species richness with ecosystem functioning. Over the past decades, ecologists have studied the determinants of species diversity and the connection with ecosystem functioning. The issues are far from completely resolved, since many processes at different temporal and spatial scales contribute to or reduce species diversity and change ecosystem functioning. Nevertheless, our current understanding confirms that human appropriation of earth's ecosystems promotes the extinction of species at higher rates than in the past, in addition to extensive changes in ecosystem functioning.

Management for biodiversity conservation requires the integration of social and ecological systems into regional systems. Management of regional systems requires understanding of the main interactions between social systems and ecosystems. Scientists working in biophysical and social science, land managers, stakeholders and decision-makers should reach a common understanding in order to address the major causes of biodiversity loss, and to plan for the future use of natural resources. An integrated view of the social-ecological system is one of the main components of the new generation of projects studying biocomplexity. Biocomplexity theory deals with the study of complex systems and proposes complementing the study of parts of systems with the study of systems as a whole.

This chapter summarises the main scenarios foreseen for the coming decades concerning the changes in biodiversity in grasslands. Written for grassland managers, the chapter starts by introducing a framework for discussing biodiversity. This framework allows the reader to escape from the usual trap of confounding biodiversity with species richness. Subsequently it addresses the question: 'What are the scenarios of grassland biodiversity over the next few decades?' To answer this, the main drivers of changes in biodiversity and their impact on biodiversity are analysed. The chapter then addresses the question: 'What will happen after extinction occurs?' In particular, 'What types of species will dominate landscapes?' and 'What will be the main pattern of evolution for the remaining species?' In the final section, the emerging concept of biocomplexity is discussed, and the main challenges that grassland ecologists will face during the next decades are proposed.

BIODIVERSITY AS A HIERARCHY

Biodiversity is, in some way, a fuzzy term, one that includes the sum of all biotic variation at three levels: intraspecific genetic variation; species diversity; and ecosystem diversity. Being such a multidimensional system attribute, biodiversity cannot be reduced to a single number, such as species richness, so there is the need for a series of indicators that allows us to identify and measure the different attributes of biodiversity. Noss and others proposed a hierarchical framework to study the attributes of biodiversity. They proposed three complementary sets of attributes of biodiversity: compositional; structural; and functional (Table 13.1). The compositional attributes refer to the identity and variety of the elements included in the target system. They range from intraspecific genetic diversity (i.e. alleles), to landscape types (i.e. proportion of habitats). The structural attributes refer to the physical organisation of the elements. They range from genetic structure (i.e. effective population size) to landscape structure (i.e. patchiness and connectivity among patches). The functional attributes include the ecological and evolutionary processes that organise ecosystems. They include gene flow, matter and energy exchanges, and disturbance regimes at the landscape level. In summary, biodiversity is a hierarchical concept that can be evaluated based on three complementary sets of attributes. Each set of attributes includes a variety of indices that can be measured.

Table 13.1. Biodiversity set of attributes according to Noss. The three sets of attributes can be studied at four levels, from landscapes to genes. Levels are nested in a hierarchy and attributes are related by their effects on each other.

Biodiversity sets of attributes		
Compositional	*Structural*	*Functional*
Landscape types	Landscape patterns	Landscape processes (disturbance regime, land use trends)
Communities	Habitat structure	Interspecific interactions, ecosystem processes
Species, populations	Population structure (size distribution, sex ratio, spatial)	Demographic processes
Genes	Genetic structure	Genetic processes

Why is the Noss framework relevant to biodiversity assessment and monitoring in grasslands? There are two issues worth considering. First, the framework helps us to move away from using just a single attribute, such as composition, and a single indicator: diversity or richness. Considering the structural and functional attributes of biodiversity helps us to include essential aspects for understanding and monitoring ecosystem functioning.

For example, at the population level, an important indicator of the population structure is spatial organisation of genotypes. Spatial organisation also affects other indicators, such as interspecific interactions (e.g. competition or facilitation) among plant populations. Another benefit of this framework is that it promotes the use of hierarchy theory. In other words, it promotes consideration of the levels above and below a target level of inquiry, from temporal and spatial perspectives. For example, fully understanding the effect of grazing on an individual species depends not only on knowledge of the interactions of that species with other species in the plant community (i.e. a functional attribute above the species level), but also awareness of the variability of ecotype diversity (i.e. a compositional attribute in the hierarchy below the species level).

GLOBAL CHANGE AND BIODIVERSITY SCENARIOS IN GRASSLANDS

Grasslands, in common with other major biomes, are experiencing the effects of major global changes. There are various possible scenarios of change in biodiversity for the next century, according to the report by Sala and others. The report discusses the sensitivity of biomes to the different global changes. They screened five drivers of change: land use; climate; nitrogen deposition; biotic exchanges; and atmospheric CO_2. A projected scenario of biodiversity results from multiplying the relative importance of the change in the drivers, by the impact on biodiversity of those changes.

Grasslands show intermediate effects from most of the drivers, except for land use. Changing land use will have a major impact on biodiversity, since it reduces habitat availability. Climate changes are expected to have less impact on grasslands—which are located mostly at temperate and intermediate latitudes—compared with higher latitude biomes, such as tundra. Nitrogen deposition will certainly have a major impact in N-limited systems. In general, grasslands are not the most N-limited biome, so the impact of this driver was assumed to be intermediate. Biotic introductions are also intermediate, because while mild habitat conditions facilitate establishment of invaders, existing levels of species diversity constrain resource availability.

Finally, atmospheric changes in CO_2 will have a relatively high impact on grasslands because of the frequent mixed composition in terms of C_3 and C_4 species. In general, however, it is considered that these drivers have only an intermediate impact on biodiversity.

Since none of these drivers acts alone, the report proposes that it is possible to consider three types of interaction: additive; response to only the driver with the maximum effect; and multiplicative. After considering these interactions, grasslands were ranked in the upper half of the biomes studied, in terms of altered biodiversity. This means that grasslands are ecosystems where some of the largest changes in biodiversity are expected to take place. Other systems, such as Mediterranean scrub or savannahs, usually linked to grasslands under the category of 'rangelands', ranked close to grasslands when the additive or the multiplicative model of interaction among drivers was assumed. The following use Noss' framework, presented in the previous section, to discuss the consequences of changes in the drivers that Sala and others identified as those most important for grassland biodiversity, namely land use, climate, nitrogen deposition, biotic exchanges and atmospheric CO_2.

LAND USE EFFECTS ON BIODIVERSITY

Grassland 'land use' includes all human activities involving grasslands. These activities range from grazing and management of domestic animals, to land conversion to crops, forest or urbanisation. This section discusses the effect on biodiversity of grazing and land conversion to crops and forests. Since grazing and cropping in grasslands have been extensively discussed elsewhere, the objective is to focus on Noss' framework in order to organise and review the information that already has been presented. The objective is therefore twofold. First, to review the ways in which land use affects grassland biodiversity, and, second, to use the hierarchical and multiperspective model proposed by Noss.

Grazing by Domestic Herbivores

Domestic herbivores are selective grazers and promote changes in plant community composition (i.e. the composition attributes of biodiversity). This was soon noted by rangeland ecologists. Dykterhuis proposed a theory of rangeland management based on this knowledge. According to this theory, by changing stocking rates it was possible to manage plant community composition. This statement proved, however, not to be valid for all conditions and ecosystems. New knowledge enriched the model of rangeland management. For example, it is widely accepted that transition from one community composition to another may not be linear. In other words, changes in plant community composition may not be gradual, but triggered as disturbance reaches a certain threshold. The currently accepted model of grazing effects on plant dynamics indicates that a given ecosystem includes several states (i.e. different species compositions).

The transition from one state to another is not solely the result of stocking rates. For example, variability between years in climate and in disturbance regime also play major roles in determining the occurrence of change. In addition to community composition, other attributes of biodiversity change as a result of grazing. In arid and semi-arid ecosystems, vegetation is frequently organised in a dual-phase mosaic composed of patches with high vegetation cover (or plant density) dispersed in a matrix of low plant cover (Fig. 13.1). This horizontal organisation of plant canopy was reported in an early study, but its origin is not fully comprehended. A growing number of studies enrich our knowledge of the processes that maintain this particular community pattern. More importantly, ecologists are using the concept of a dual-phase mosaic to create and build communities with a particular structure [Fig. 13.1(a)]. This structural dimension in plant community biodiversity is gaining recognition as an issue needing consideration.

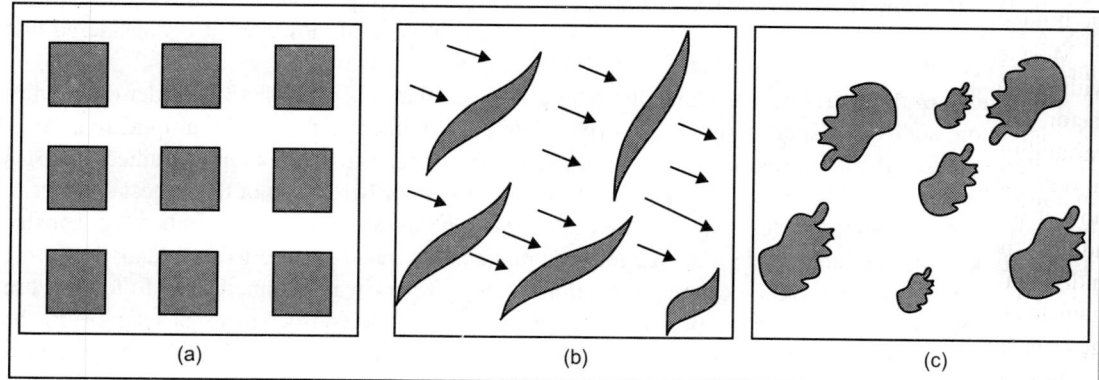

(a) (b) (c)

Fig. 13.1. Structural biodiversity in vegetation mosaics at the landscape and community levels. (a) A schematic example of a vegetation mosaic with a regular pattern in agroecosystems. (b) A banded pattern, such as those described in semi-arid landscapes with a gradual slope, found in Australia, Africa or North America. Arrows indicate the direction of water flow in the inter-band areas. Water infiltrates in the bands. (c) The mosaic is a patch pattern as described in semi-arid ecosystems of South and North America.

Many ecosystem and population processes are influenced by the spatial organisation of the vegetation mosaic. For example, in banded vegetation in Mexico and Australia, water availability and hydrological balance are controlled by the vegetation pattern [Fig. 13.1(b)]. The matrix functions as a source area that transfers water to the vegetation patches, allowing greater primary production or a particular flora. The effects of grazing on spatial structure are not clear. Since grazing generates a pattern in vegetation, the effect of grazing depends on the vegetation pattern being analysed (banded or spotted mosaic). In some cases, such as in the banded vegetation in Australia, it has been suggested that overgrazing disrupts the mosaic. A recent review of the effects of grazing on spatial structure of plant communities found that currently there is no strong evidence that grazing disrupts vegetation patterns. In summary, a spatially-explicit view of community and ecosystems allows new perspectives for processes (i.e. dispersal, population dynamics, and plant–plant and plant–animal interactions) and functions (i.e. soil organic matter dynamics, and water and nutrient circulation). Whereas structural diversity has been recognised at the landscape level, structural diversity has been ignored at the community level, even though long proposed.

From the compositional perspective, grazing has a major impact not only on species composition (as already discussed), but also at the population level. Population genetic diversity and structure of preferred species can change as grazing drives plant density down. Rangeland theory assumes that, with proper management, species reduced by overgrazing can recover if the physical environment has not changed drastically. In other words, there are no demographic constraints to recovery for the few remaining plants. From the genetic point of view, however, the situation may be different. As the size of a population decreases, genetic variability is reduced. Genetic drift, the loss of genetic diversity that occurs in small size populations, contributes to the process. Grazing can exert strong selection pressure, promoting traits that increase tolerance or avoidance of herbivory. In small populations, genetic drift is a common process, and erodes genetic variability in a random fashion, without increase in fitness. Low genetic diversity levels may therefore constrain population sustainability by reducing the capacity of the population to cope with environmental variability. Essentially, adequate management can promote an increase in population numbers, but not necessarily an increase in genetic variability, and therefore the limited genetic diversity of the recovering population can restrain population viability.

Most reviews of biodiversity mention genetic diversity as one of the most pressing problems associated with human-induced global changes. However, for grasslands, there is not enough information on this factor. More information is needed in order to understand the relationship between grazing and genetic variability. More information is needed on the structural and functional attributes of genetic diversity. In other words, it is necessary not only to know how much variability exists in a population, but also how it is spatially distributed (i.e. structural attributes of genetic diversity). Also, it is necessary to address the genetic processes, such as migration of genes. As plant density decreases, distances between individuals increase. Two processes maintain the unity of this fragmented population: movement of gametes (i.e. pollen) and seed dispersal. All the information must be spatially explicit since grazing is heterogeneous in space at different scales.

In summary, since the emergence of new theories for range management, most efforts have been aimed at evaluating changes in species abundance and diversity, and at estimating the effect of these changes on ecosystem processes, such as primary production and decomposition. These changes increase along productivity gradients from arid to subhumid grasslands. The next efforts to understand the effects of grazing on grasslands need to be oriented towards understanding of the structural attributes of biodiversity, from population to community. There is evidence that the spatial structure of populations and communities control many processes. Gathering spatially explicit data will promote our understanding of the effect of a spatially heterogeneous process such as grazing. Currently, there is information on the three sets of attributes of biodiversity at regional and landscape scales.

Replacement of Grassland by Other Land Cover Types

Grassland biodiversity is negatively affected by two sources of land cover change: cropping and forestry. These activities differ in their impacts on ecosystem structure and function. Cropping changes above- and below-ground plant biomass in a sudden fashion, but it also promotes large changes in soil organic matter. In contrast, forestry changes land cover more slowly, from a grass-dominated to a tree-dominated system. A forest-dominated system changes the light environment, but can also change soil nutrient dynamics.

Cropping has affected the three sets of biodiversity attributes (Table 13.1). For example, in the Rolling Pampa of Argentina, one of the most productive grasslands in temperate South America, the structural and compositional attributes of the region have changed because of land-cover changes. The original grassland during the nineteenth century was turned into a mosaic of crops, cultivated pastures and semi-natural grasslands. The total cultivated area at the start of the twentieth century was 6 million hectares, but by 1984 it had increased to 26 million hectares. The spatial structure of the mosaic changes the inherent patterns of the landscape, which in turn changes the attributes of biodiversity.

Land use regimes in the pampas changed the compositional attributes of biodiversity according to the particular management application at the paddock level. Therefore diversity of the landscape changed as new crops (e.g. soyabean) or management (e.g. no-tillage cropping) were added or with changes in the relative area devoted to different crops in the mosaic.

Associated with the change in the structure of the vegetation mosaic, several community processes (functional attributes) changed as well. For example, many species became locally extinct, while others colonised and invaded the landscape. The original number of plant species in well-drained soils has been estimated to have been around 222; in 2006, in maize crops, the number of species had decreased to 99. Of the 99 species, only 54 species had been present in the original grassland, the remaining 45 species were exotic. At the same time that such changes in community composition and processes occur, other ecosystem processes change too.

For example, agricultural practices foster changes in decomposition rates, exportation of nutrients, soil structure degradation and erosion, and seasonality in production. Noss proposed the inclusion of all these processes as functional attributes of biodiversity.

As new or old species increase in abundance in the different areas of the agricultural mosaic, new interactions develop within the biotic community. For example, in the pampas agroecosystems there has been an increase in species that produce secondary products such as terpenoids, thyophenes and alkaloids. It is suggested that this increase is associated with soil degradation that promotes stress-tolerant species in particular locations within the mosaic. Further changes in the rates of processes may develop from change in community composition. For example, pest outbreaks may be negatively affected since, from the perspective of the pest, the landscape has become fragmented and diverse from a structural point of view.

As cropping fragments a landscape, other human activities generate different impacts on species population dynamics. Fences, roads, railways and rights of way under telephone and electricity transmission lines are not cropped. These areas develop semi-natural vegetation and form a network of corridors that facilitates organism dispersal throughout the region. In this sense, analyses of the effects of land use require the assessment of the spatial structure of the landscape and the interactions of species with the landscape mosaic. In perspective, data from the pampas agricultural mosaic indicate that land use will lead to more changes than just a reduction in the number of species. Indeed, the structural attributes of biodiversity change dramatically in grasslands that are transformed to crop production. At the same time, the functional attributes of biodiversity will also change. Furthermore, these changes occur not only at the scale of individual fields or paddocks but also at the landscape and regional scale.

Conversion to forest has been the other major land cover change experienced by grasslands. This change occurs directly as a result of land management decisions or as an indirect effect of other decisions. Many countries, such as Argentina, are promoting forestation as a way to ensure meeting requirements for cellulose fibres. In general, forestations are monospecific. Many of these forestations are financed by credit from international agencies. In the case of Argentina, the World Bank has financed a forestation plan that has been successful in capturing the attention of landowners. From 1992 to 2006, the surface area converted to forest in the pampas grassland was 46,870 ha. Forestation clearly changes the structural attributes of biodiversity.

Unplanned conversion of grassland to forest has been described in many ecosystems distributed throughout the Americas, Australia and Africa. Rates of change are high. Data from the tall-grass prairie in North America indicate that 40 years are enough to develop a closed canopy forest once there are some focal plants established. Many processes contribute to tree invasion and establishment in a grassland ecosystem. Seed availability is affected by changes in dispersal corridors. Seedling establishment and survivorship is influenced by changes in the disturbance regime. In the pampas, the human-developed network of corridors host up to 40 woody species, mostly trees. Either wind or animals, including domestic herbivores, and birds disperse the successful tree invaders. At the same time, changes in disturbance regime might be the principal cause of successful establishment and growth of seedlings. Briggs, Hoch and Johnson pointed out that grazers reduce fuel load and reduce fire frequency, promoting *Juniperus virginiana* success in the tall-grass prairie.

Data from the pampas indicate that there are species-specific traits that promote success of trees in grasslands. For example, in the pampas, competition from tussock grasses was intense enough to preclude establishment of two out of four tree species tested.

Effects of both planned and unplanned forestation on the functional attributes of grassland biodiversity are similar. Both biodiversity and herbaceous productivity decrease. Briggs, Hoch and Johnson report a decrease in richness (10 m² plots) from around 30 species to less than five species. At the same time, invasion susceptibility of the new forest community also changes.

Mazia and others indicates that under forest canopies none of the seedlings survived of the four tree species tested. In other words, this indicates that forests were less susceptible to invasion than grasslands. Herbaceous biomass decreased from 384 g/m² in annually burned grassland to 0.2 g/m² underneath the *Juniperus* forest. In the long-term, forestation promotes other changes as well. In the pampas, mineral soils in 50-year-old forests had lower pH (4.6 to 5.6), 40 per cent lower exchangeable Ca, and three times the Na level compared with soils in neighbouring grasslands. Data indicate that these changes result from recycling and redistribution of elements rather than from leaching facilitated by organic acids produced by trees.

NITROGEN INPUTS AND ATMOSPHERIC CO$_2$ INCREASES

N and atmospheric CO$_2$, two limiting plant resources, are globally increasing as a consequence of human activities and are therefore treated together here. Nitrogen inputs in ecosystems over approximately the last 50 years have been roughly double natural N inputs. This widespread change has been particularly intense in agricultural lands, mostly located in grassland regions. Fertilisation changes the functional attributes of biodiversity, changes the rate of processes of such denitrification or leaching, and changes rates of extinction and colonisation in plant communities. Management of nitrogen cycling includes the addition of fertilisers, land rotations that include nitrogen-fixing crops, and tillage to increase mobilisation of N stored in soil organic matter. N fertiliser use in croplands has increased constantly since 1940. Increments in N input occurred first in the developed countries, and subsequently in developing countries. Most of the information on the effects of N addition refers to changes in community composition and ecosystem functioning.

Nitrogen addition experiments in grasslands indicated that one of the consequences is a reduction in plant diversity. For example, the Rothamsted Park Grass Experiment, one of the longest and oldest experiments, indicates that N addition promotes the dominance of a few species and the suppression of many other species. Species loss was five times greater in the treatment with the highest N addition, compared with no N addition. Similar results were found in North American grasslands. In Europe, agriculture fertilisation and N deposition on heathland communities have promoted a replacement of the original shrub community with one dominated by grass species.

In general, N addition increases the dominance of fast-growing species with high shoot–root ratio. This trait enables those species to successfully compete for light, which is usually limiting in productive habitats. This contrasts with the natural situation in grasslands, where dominant species usually exhibit traits suitable to dealing with soil limitations. Additionally, use of N fertilisation promotes acidification of soil.

Atmospheric CO_2 enrichment is also expected to effect grassland composition and functioning. The importance of CO_2 enrichment derives from its effects on plant carbon gain and water status. Absorption of CO_2 through stomata is associated with plant water losses. Water use efficiency (i.e. grams of CO_2 absorbed per gram of water lost), increases as CO_2 increases. Early predictions of the impact of CO_2 enrichment on grasslands indicated that plant community composition would change. Essentially, it was proposed that C_3 grass species would outcompete C_4 grasses. However, a meta-analysis of experimental data indicated that C_3 and C_4 species responded similarly to CO_2 enrichment. C_3 species increased total biomass by 44 per cent and C_4 by 33 per cent.

Results from a Mediterranean annual grassland showed that the dominant grass (*Avena barabata*) increased seed production and plant survival, resulting in an overall increase in plant density. These changes occurred not only because of higher CO_2 concentration, but changes were also found in soil water dynamics associated with higher water use efficiency. Plant transpiration was less under elevated CO_2 than under ambient concentrations. Lower transpiration at the beginning of the growing season also increased soil water content.

In temperate grasslands in Canada, Potvin and Vasseur compared community dynamics of ambient and CO_2-enriched plots. After three years, they found that ambient plots had lower species richness and more dominance than enriched plots. In enriched plots, early successional species (*Plantago major*) responded to CO_2, reducing the increase in later successional species (*Agropyron repens*). In general, these authors found that dicotyledenous species advanced while grasses retreated. Their results parallel responses measured in a tallgrass prairie in central North America and in a Swiss pasture. In contrast, Paruelo refers to a study in a tallgrass prairie that found no changes in the C_3–C_4 balance after eight years of enrichment. Other disturbances, such as fire and grazing, had a greater effect on species composition.

In synthesis, increments in N and CO_2 availability might promote changes both in the compositional attributes of biodiversity (genotypes in a population or species in a community) and in the functional attributes of biodiversity (transpiration, carbon gain). In essence, change in the compositional attributes parallels change in the functional attributes of biodiversity.

BIOTIC EXCHANGE

Biotic exchange among ecosystems has been profoundly affected by human activities worldwide. Human activities facilitate both species migration and community invasiveness. Species migration can increase by exchanges of species, both accidentally and deliberate. Community susceptibility to invasion can increase because of changes in the disturbance regime or in other environmental drivers. In the case of temperate grasslands, species invasions have been particularly conspicuous.

Biotic exchange greatly increases genotypic and species homogenisation at various levels from patches to regions. Species extirpation and invasions increase the abundance of a few, very successful invaders. Rusch and Oesterheld found that grazing promoted the introduction of many exotic dicotyledenous forb species, increasing species diversity. Species invading grassland ecosystems include various functional groups of plants. For example, in the pampas in South America, Chaneton found that most of the cool and warm season invasive species were forbs. In North American grasslands, the invasion by the annual grass *Bromus tectorum* is one of best case studies. Invasion of *Bromus* took place at the end of the nineteenth century in the Great Basin region, displacing both shrubs and perennial grasses. Plant cover reduction by heavy grazing prompted and facilitated invasion. Dry biomass of *Bromus tectorum* increased fire frequency, promoting the disappearance of perennial and dominant grasses and shrubs. The site became dominated by annual grasses and shrubs (*Gutierrezia sarothrae* and *Chrysothamnus* spp.). *Bromus tectorum* competes with established perennial grasses for water, but the critical element in its success is the increased fire frequency: from every 80 years to every 4 years. Perennial grasses went locally extinct, as they could not cope with the new fire regime. D' Antonio and Vitousek reviewed information on the ecosystem effects of biological invasion by exotic grasses and found that this pattern of invasion and change of fire regime seems to be common. According to them, biological invasions have caused more species extinction than climate change or atmospheric composition change. They stress the fact that biological invasions change not only the community diversity (compositional attributes of biodiversity) but also the functioning of ecosystems (functional attributes of biodiversity). Other ecosystems processes such as energy fluxes and material cycles are also affected by invasions.

AFTER EXTINCTION

Previous sections analysed scenarios for different attributes of biodiversity. They also presented examples in which changes in the drivers promoted changes in the various attributes of biodiversity. This section abandons the use of the three sets of attributes of biodiversity proposed by Noss and focuses on species diversity alone. It moves on one step and assumes that diversity reduction has effectively occurred. According to current estimations, the extinction rate is higher than in the recent past, and it is increasing. What will occur after extinction? Will extinction alter the evolutionary processes by which diversity is generated? Will any particular types of species dominate future ecosystems? Will biodiversity eventually recover? What will be the recovery rates? These are relevant questions, if we hope to develop proactive management strategies for the 'new world'. Sponsored by the National Academy of Science of the United States of America, a colloquium discussed these issues (*The Future of Evolution* [contributed papers can be downloaded free from the *Proceedings of the National Academy of Science* Internet site]). Experts in different fields evaluated the information recovered from, for example, the geological record. They also evaluated current knowledge on species requirements and species strategies in relation to environmental templates in new ecosystems. The direction of evolution is not predictable, but knowing what happened in the past may indicate how evolution might proceed.

Various patterns of evolution might occur after extinctions, since major disruptions are taking place in very diverse biomes, such as tropical forests, grasslands, wetlands, coral reefs and estuaries. Several ideas summarise the current view of the future. As extinction proceeds, an outburst of speciation may occur because of a large number of vacant niches and the fragmentation of populations. A few ecological types may possibly account for most of the speciation because of the predominance of human-dominated ecosystems throughout the biosphere.

For example, it is possible that opportunistic and generalist species may proliferate. Disruption in particular ecosystems, such as tropical forests, may affect evolutionary patterns in other ecosystems. It has been proposed that tropical forests are the engines for new species, and their disappearance may hamper future evolution in neighbouring grasslands. As we evaluate future evolution, it is necessary to consider not only environmental changes but also biotic changes. For example, extinction will affect the 'biodisparity' within the biota. In other words, extinction can preferentially affect species with particular morphology and physiology. In this sense, future patterns of evolution will develop from an incomplete biota. Because of anthropogenic effects it is probable that higher taxa and guilds are candidates for extinction, and for this reason, re-diversification might be limited.

Recovery rates (i.e. the velocity of rebuilding species diversity) after extinction are difficult to estimate from the geological record. These records are also not completely appropriate for this endeavour.

Perturbations that caused extinction events were probably intense, but short-lived. In contrast, we are currently facing global changes that may alter the biosphere and the major drivers of extinction over longer time periods. Jablonski indicates that recovery would be slow in terms of a human time scale. However, many aspects of recovery, such as the spatial context (e.g. source–sink dynamics) and the reorganisation of ecosystems after extinction, increase the complexity and decrease the predictability of the recovery process.

More geographical analyses of the fossil record may shed light on the issue. Erwin found no clear relationship between the magnitude of extinction and the rate of recovery. Theoretical modelling confirmed that empty niches should refill with new species rapidly, but empirical studies showed complex dynamics. Erwin submits that, since extinction may disrupt ecosystem structure and function, it is necessary to accept that recovery will involve the concurrent redevelopment of species diversity and ecosystem processes. In this sense, some studies reveal a time lag in recovery, as well as the reappearance of groups that disappeared again subsequently in the recovery process.

Other evidence indicates that rates of diversification after extinction may not be slow. Cowling and Pressey present the case of the Cape Floristic Region in South Africa, where species diversification occurred rapidly after climate change in the late Pliocene. During that time, the climate became Mediterranean, and fires became frequent. In the Cape Floristic Region, diversification did not occur randomly among ecological groups. Cowling and Pressey indicate that, for example, species from the woody groups are 'low [stature], fire killed (i.e. non-sprouting) shrubs with poorly dispersed seeds, small and short-lived seed banks, and insect-pollinated flowers'. According to them, all these characteristics lead to rapid generation turnover, which promotes diversification. A short-lived seed bank favours non-overlapping generations and promotes the expression of novel characteristics in each generation. Fire induces population fragmentation and local population extinction, and short dispersal distances induce isolation and diversification of populations in different habitats.

Tilman and Lehman discuss the particular case of plant species extinction and the subsequent diversification in the grasslands biome. They use a set of well-established theoretical ideas to address two issues. First, what will be the expected changes in the factors that constrain plant population growth, and what will be the consequences of the predicted changes? Second, assuming a certain type of new environment, what will be the probable types or guilds of plant species evolving from the remaining guilds? Plants respond to major constraints by adjusting allocation of resources to roots, stems, leaves and seeds (recruitment of new individuals). Most grasslands or woody–grass ecosystems are strongly constrained by soil resources (e.g. nitrogen, water). The plant types adapted to these constraints allocate an important amount of growth to roots. The consequence is that these species guilds are poor competitors for light or poor recruiters or dispersers or a combination. As discussed above, two of the major environmental changes are an increase in nitrogen availability and environmental fragmentation. In grasslands, most species are not suited to successfully compete in this new environment. They either lack the ability to compete for light or lack dispersal capacity to overcome human-created barriers or fragmentation. Few species will succeed in this environment. These species are usually classified as weedy because of their ability to respond rapidly to resource availability, their poor competitive ability, and their good dispersal ability.

To address the second issue, Tilman and Lehman use a simple model that considers a trade-off between competitive ability and dispersal ability. They assume that after extinction a single type of a plant species remains. This type, as already discussed, is a good disperser but a poor competitor. Their model projects that, after enough generations, two types of species will dominate: one will be a very poor competitor but a good disperser, whereas the other will be a good competitor, but a poor disperser. After a very long time (0.5 million years) the model predicts the evolution of more than 20 species types, scattered along a competition-dispersal gradient.

The model does not consider other factors that can alter these results, such as periodic invasions of species that can displace the species evolving *in situ*. As was reviewed above, biotic exchange would be a predominant process in the future. There is evidence that, in the past, invasive species altered the evolution pathway of native species. Alteration of evolution derives from many possible processes, including competitive exclusion, niche displacement, hybridisation, introgression, predation and extinction itself. It is also necessary to say that invaders contribute to biodiversity and also evolve in response to the physical and biotic environment.

It is clear that estimating post-extinction patterns and rates of evolution of species diversity (compositional attribute of biodiversity) is a hard task and our predictive capabilities are still poor. Trying to include in the picture the other attributes of biodiversity (i.e. structural and functional) is a remaining task. A complementary strategy would be to monitor changes in the three attribute sets as a way to gain an understanding of the entire process. As several participants in the colloquium agreed, a proactive response to the extinction process is necessary.

Monitoring and adaptive management are complementary tasks. Adaptive management is not an easy task due to the many issues at stake (social, economic, cultural and technical), and also due to our currently poor forecasting power and because of the science-policy gap.

MANAGEMENT OF BIOCOMPLEXITY—THE NEXT FRONTIER

Human pressure on the biosphere will increase because of both population growth and increasing per capita consumption. This pressure will not only be intense but also widespread in grasslands, as crop and animal husbandry occurs mostly in this biome. As discussed above, any project oriented to managing the biodiversity transition must include, among other drivers of global change, human-driven land use change as a major component of the scenario. Since the social system affects and is affected by ecosystem structure and functioning, there is increasing interest in studying the interactions between social and natural systems.

Social systems and ecological systems have interacted at least since the beginning of human existence. In this sense, the study of ancient human impact on ecosystems could be used as the prologue to the design of sound management strategies. Redman integrates archaeological and historical records in a model of the interactions between natural and social systems. He identifies three different phases in human systems: expansion, intensification and abandonment. In this conceptual model, human society is initially constrained by factors from the social system and the natural systems. Humans promote changes in land use and land cover, acting on landscape organisation and biological and biogeochemical processes. As a consequence, human society is released from the constraints and initiates a phase of human expansion.

Paralleling changes in the natural system, the human system also becomes more complex as new social groups develop (e.g. administrators, religious organisations and military). More resources are needed to support the expansion and the new functions in the society. There is also the need to accumulate reserves to meet society needs in bad years.

All these changes result in intensification of natural resource exploitation. Pressures on ecosystems move them to a structure and function beyond their resilience, leading to the next phase: collapse and abandonment. One important feature of this model is that the social group that makes the decisions about natural resource management is not the same group that is involved in working the land, i.e. decision-makers have no direct contact with natural resources.

Based on the present review, grasslands, as socio-ecological systems, should be observed from three different perspectives: biophysical, productive-technical and socioeconomic (Fig. 13.2).

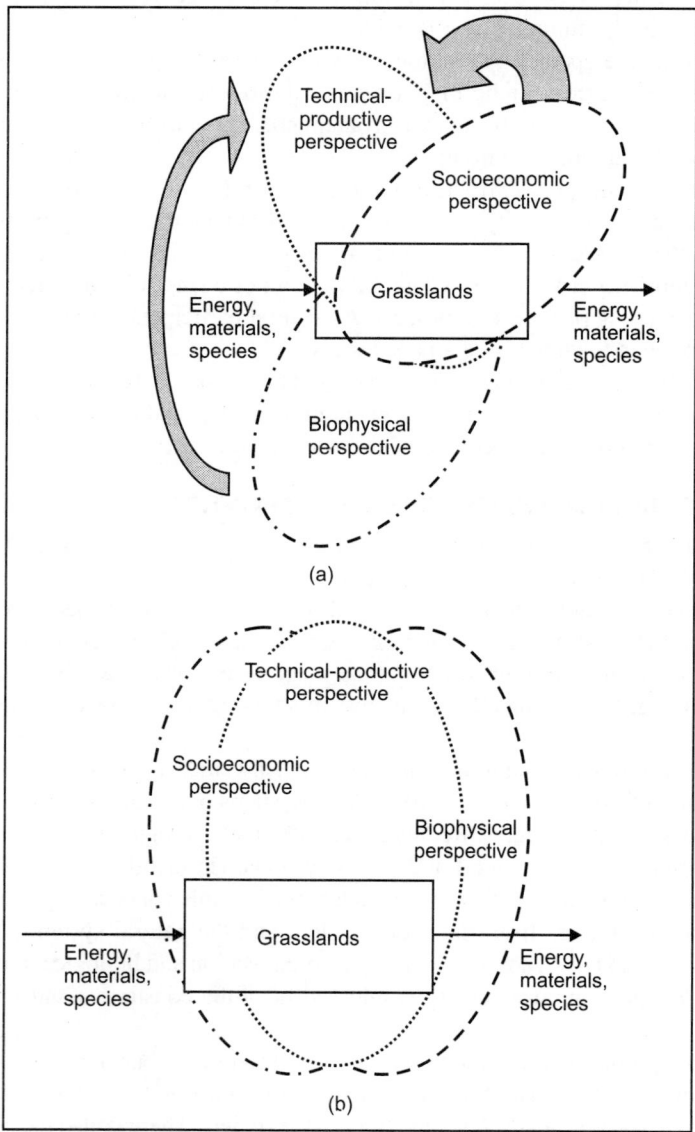

Fig. 13.2. Schematic representation of the three perspectives from which grasslands can be analysed. (a) Currently the biophysical perspective has little influence on the technical-productive and socioeconomic perspectives, being the least important determinant of the way that grasslands are utilised. (b) Represents a complementary evaluation of grassland management.

The biophysical perspective is the one with which most ecologists are familiar. Biodiversity is studied to understand its relationship with the physical and biotic environment. The productive-technical perspective refers to all the management technology and tools developed to modify grassland structure and function. It includes, for example, the selection of species genotypes, control of natural flora or fauna, and modification of soil conditions through tillage, fertilisation or irrigation. The socioeconomic perspective includes the social, political, economic, cultural and religious or spiritual aspects of grasslands. These three perspectives should be complementary, although not always true in practice. The biophysical perspective, in theory, may influence the productive-technical perspective, since the biological and physical components constrain what can be done and achieved. In practice, productive-technical decisions are mostly associated with the socioeconomic perspective.

For example, commercial strategy is a rather large determinant in management decisions. On many occasions, it goes against what the biophysical perspective indicates (sometimes even against what the technical-productive perspective indicates). For example, the development of a new herbicide is an effective weed control, but in time (i.e. several generations of the weed species or several cultivation cycles) the weed community changes its composition or the weed population develops resistance, thus reducing the herbicide's value for crop production.

Delaying the appearance of herbicide resistance includes the cost associated with incomplete or imperfect weed control, but it results in a long-term benefit. Incomplete weed control permits survival of susceptible genotypes and delays the dominance of resistance in the weed population. Another example, at a global scale, is the failure to fully implement the Kyoto protocol on CO_2 emissions. Here, policy decisions may hamper efforts to adjust technical-productive procedures that reflect the biophysical perspective.

Conserving and managing biodiversity both need the integration of all three perspectives in a complementary way [Fig. 13.2(b)]. This is not an easy task and requires a new framework, rooted in the understanding that a collaboration of social and biophysical sciences is necessary. That framework also needs to incorporate the idea that socio-ecological systems are complex systems. A new area of interest is the study of biocomplexity. According to the US National Science Foundation:

'Biocomplexity refers to phenomena that arise as a result of dynamic interactions that occur within biological systems, including humans, and between these systems and the physical environment. From individual cells to ecosystems, these systems exhibit properties that depend not only on the individual actions of their components, but also on the interactions among these components and between these components and the environment.'

The complexity of these systems derives not necessarily from them having many components. On the contrary, the main reason is that their behaviour cannot be predicted from the knowledge of their parts and feedback among the system components. Complexity derives from different system properties, such as variability in their components, context-dependent processes, non-linear processes, time-delay responses and chaotic behaviour. Both social and ecological systems have these characteristics. For example, communities and populations are composed of different species and genotypes, respectively. Usually, biological responses have short-time lags, if we consider morphological, physiological or behavioural responses. However, time lags become long if we consider numerical responses, such as birth and mortality rates. Also, many population and ecosystem processes are non-linear.

The study of socio-ecological systems requires a new approach that includes not only the study of the parts but also the study of the whole. It also requires a major collaborative effort of social and biophysical sciences. Complexity theory can bridge the gap between sciences, since both social and biophysical systems are complex. It also requires that biophysical sciences are considered as part of society and therefore objects of study themselves.

Another issue that hampers the integration of the three perspectives of grasslands systems is that the power of prediction of biophysical models decreases as spatial and temporal scales increase from the local paddock to the global, and from days to decades, respectively. Uncertainty in forecasts is at the root of the science-policy gap.

Uncertainty and complexity are part of the normal lives of scientists, whereas 'society and policy-makers seek certainty and deterministic solutions'. In this scenario of uncertainty, it is difficult to apply a standard decision-making process in social-ecological systems. An alternative approach is a resilience analysis and management that seeks to define a desirable configuration of systems that reinforces system resilience, which is the capacity to cope with future shocks. Resilience results from three characteristics: the amount of change that the system can undergo without altering its configuration; the capability of self-organisation; and the capacity to learn and adapt.

In this framework, system stakeholders are key players in identifying both threats to the system and management practices that maintain and promote resilience. Their experiences are used to define thresholds in system attributes (i.e. desirable structure and functions, such as debt–income ratio). For example, perception of the desertification process by stakeholders may be very vague, but income reduction is readily perceived.

Connecting with some key features of an ecosystem, e.g. woodiness or presence of weed seeds in wool, may be easier than comprehending any indicator of biodiversity, such as the ones presented here. Participation of stakeholders in the analysis fosters a change in perception of uncertainty from ignorance to knowledge. This is a key change in order to reduce the science-policy gap.

CONCLUSION

Future scenarios of biodiversity indicate that grasslands, as with other biomes, will suffer from biodiversity loss or erosion. The main causes of biodiversity losses are global changes; however, land use (a regional-scale phenomenon) might have the biggest impact. As we move into this new scenario with diminished biodiversity, it is useful to address three aspects of the biodiversity erosion. First, 'What are the main trends in the biodiversity of grasslands, and what are the effects of the main drivers of biodiversity loss?' Second, 'What are the changes expected in the biodiversity of grasslands after extinctions have occurred.' Third, 'How can we manage the biodiversity of grasslands in the transition to the new scenario?'

Biodiversity includes many levels of organisation, and many temporal and spatial scales. It is a useful and necessary concept to address the main characteristics of ecosystems. In time, this concept has become vague, as scientists commonly use 'biodiversity' as a synonym for species richness. Noss proposed a framework that allows us to monitor and manage biodiversity. This framework for biodiversity has three sets of attributes: compositional, structural and functional. Adopting this three-attribute-set framework allows us to address the many changes that are occurring at different spatio-temporal scales and in biological hierarchies. It also ensures that we include the spatial structure of ecological systems in our analyses. Our current understanding points to the fact that spatial structure is a key determinant of many processes, across populations, communities and ecosystems. Many examples show that land cover and land use changes are eroding all sets of attributes, not just the compositional.

It is necessary to begin to understand the future patterns of evolution and change of ecological systems as many species disappear, landscapes becomes fragmented, communities are invaded, and nutrient cycles are modified. A simulation exercise, based on the compositional attributes alone, allows a first approximating of the evolutionary paths. In the case of plants, it seems that weedy species will predominate. However, in time it is possible that new species will evolve from this group.

Since grasslands will continue to be heavily used ecosystems, it is necessary to implement new, sound management practices to address biodiversity. Humankind is a component of ecosocial systems, and therefore management needs to be based on a complementary use of biophysical, technical and socioeconomic perspectives. In order to address a complementary view of grasslands, more interactions between social and ecological scientists are necessary. Complex system theory, and the biocomplexity framework in particular, provides a template for studying these interactions. Accepting that uncertainty is a key attribute of both social and ecological systems is the next frontier in studies of biocomplexity and biodiversity.

Chapter 14

Effects of Pesticides on Biodiversity

INTRODUCTION

Various determinants play an important role in agriculture with regard to biological diversity, as shown in Table 14.1. It is often difficult to separate the individual aspects and their effects on biological diversity from one another. The choice of crop, the type of crop rotation and the use of fertilisers and pesticides co-determine which plants and animals survive and establish themselves on agricultural land, which are displaced or harmed, and whether natural regulatory systems are disrupted or supported. Pesticides allow the type of agriculture that contributes to the loss of biological diversity. Special attention must therefore be paid to coherency between the protection of biodiversity and pesticide legislation.

Table 14.1. Influences of agriculture on biodiversity.

Determining factors	Characteristics
Cultivation	
Methods of working the soil, fertilising, harvest techniques	Tillage or non-tillage practices, mulch sowing, direct sowing, with or without pesticide use; fertiliser type (chemical, mineral, organic); harvesting with or without pesticide use, with or without, e.g. straw removal (type, intensity, frequency, time).
Cultivated crops	
Crop rotation, choice of type and variety	Number of rotating crops; alteration between summer/winter crops, leaf/stem crops. Variety: traditional, regional or hybrid types.
Structure	
Size and form of cropland, contact with landscape components	Presence of free strips on the borders of arable land, hedge banks, hedges, trees, groups of trees and bushes, waters, etc. as space for exchange, retreats, food sources and habitats, and breeding grounds for living creatures.
Genetic engineering	
Herbicide resistant plants, insecticide plants, terminator technology	Active substance gap (risk of resistances), species crossing (uncontrollable transfer of manipulated genetic information to formerly unaltered cultivated and wild plants).

The use of pesticides affects biological diversity in three ways. Firstly, pesticides enable certain methods of cultivation that would otherwise be almost impossible without their use, such as monocultures, tight crop rotation or the cultivation of crops that are hardly adapted to the site. Secondly, pesticides can damage organisms directly, for example aquatic coenosis, if they enter surface water. Thirdly, they indirectly affect habitats, as the use of pesticides deteriorates the quantity and quality of food sources for wild animals, or in the long-term, causes a shift in the composition of species in ecosystems.

Pesticides which are used for preventing or destroying pest is having more negative impact on our ecological system when compared to its desired action. Pesticides are carried by wind to other areas and make them contaminate. Pesticides are also causing water pollution and some pesticides are persistent organic pollutants which contribute to soil contamination.

The amount of pesticide that migrates from the intended application area is influenced by the particular chemical's properties: its propensity for binding to soil, its vapour pressure, its water solubility, and its resistance to being broken down over time. Some pesticides contribute to global warming and the depletion of the ozone layer.

WATER

Pesticides were found to pollute every source of water including wells. Pesticide residues have also been found in rain- and groundwater. Pesticide impacts on aquatic systems are often studied using a hydrology transport model to study movement and fate of chemicals in rivers and streams. Studies by the UK government showed that pesticide concentrations exceeded those allowable for drinking water in some samples of river water and groundwater.

The main routes through which pesticides reach the water are:
1. It may drift outside of the intended area when it is sprayed.
2. It may percolate, or leach, through the soil.
3. It may be carried to the water as runoff.
4. It may be spilled accidentally or through neglect.

They may also be carried to water by eroding soil. Factors that affect a pesticide's ability to contaminate water include its water solubility, the distance from an application site to a body of water, weather, soil type, presence of a growing crop, and the method used to apply the chemical. Maximum limits of allowable concentrations for individual pesticides in public bodies of water are set by the Environmental Protection Agency in the US.

SOIL

Many of the chemicals used in pesticides are persistent soil contaminants, whose impact may endure for decades and adversely affect soil conservation. The use of pesticides decreases the general biodiversity in the soil. Not using the chemicals results in higher soil quality verified needed, with the additional effect that more organic matter in the soil allows for higher water retention. This helps increase yields for farms in drought years, when organic farms have had yields 20–40 per cent higher than their conventional counterparts. A smaller content of organic matter in the soil increases the amount of pesticide that will leave the area of application, because organic matter binds to and helps break down pesticides.

AIR

Pesticides can contribute to air pollution. Pesticide drift occurs when pesticides suspended in the air as particles are carried by wind to other areas, potentially contaminating them. Volatile pesticides applied to crops will volatilise and are blown by winds to nearby areas posing a threat to wildlife. Sprayed pesticides or particles from pesticides applied as dusts may travel on the wind to other areas, or pesticides may adhere to particles that blow in the wind, such as dust particles. Compared to aerial spraying ground spraying produces less pesticide drift. Farmers can employ a buffer zone around their crop, consisting of empty land or non-crop plants such as evergreen trees to serve as windbreaks and absorb the pesticides, preventing drift into other areas.

EFFECTS ON BIOTA

Plants

Nitrogen fixation, which is required for the growth of higher plants, is hindered by pesticides in soil. The insecticides DDT, methyl parathion, and especially pentachlorophenol have been shown to interfere with legume-rhizobium chemical signalling. Reduction of this symbiotic chemical signalling results in reduced nitrogen fixation and thus reduces crop yields. Root nodule formation in these plants saves the world economy $10 billion in synthetic nitrogen fertiliser every year.

Pesticides can kill bees and are strongly implicated in pollinator decline, the loss of species that pollinate plants, including through the mechanism of Colony Collapse Disorder, in which worker bees from a beehive or Western honey bee colony abruptly disappear. Application of pesticides to crops that are in bloom can kill honeybees, which act as pollinators. The USDA and USFWS estimate that US farmers lose at least $200 million a year from reduced crop pollination because pesticides applied to fields eliminate about a fifth of honeybee colonies in the US and harm an additional 15 per cent.

Animals

Pesticides inflict extremely widespread damage to biota, and many countries have acted to discourage pesticide usage through their Biodiversity Action Plans. Animals may be poisoned by pesticide residues that remain on food after spraying, for example when wild animals enter sprayed fields or nearby areas shortly after spraying.

Widespread application of pesticides can eliminate food sources that certain types of animals need, causing the animals to relocate, change their diet, or starve. Poisoning from pesticides can travel up the food chain; for example, birds can be harmed when they eat insects and worms that have consumed pesticides. Some pesticides can cause bioaccumulation, or build up to toxic levels in the bodies of organisms that consume them over time, a phenomenon that impacts species high on the food chain especially hard.

Pesticides can affect animal reproduction directly, as evidenced by the deleterious effect of the persistent organochlorine insecticides on reproduction in raptors and other birds. Eggshell thinning due to the uptake of organochlorine insecticides that affect calcium (Ca) metabolism has been observed in predacious birds. Fish-eating birds are more severely affected than terrestrial predatory birds, because the fish-eating birds acquire more pesticides via their food chain than the other predators.

Pesticides can also affect reproduction in the invertebrates; for example, sublethal doses of DDT, dieldrin, and parathion increased egg production by the Colorado potato beetle by 50, 33 and 65 per cent, respectively, after two weeks. The herbicide 2,4,5-T was found to reduce the reproduction of soil-inhabiting Collembola. Populations of invertebrates with high rates of increase can recover stable populations much more rapidly than those of bird and mammal populations.

Human

Pesticides can enter the human body through inhalation of aerosols, dust and vapour that contain pesticides; through oral exposure by consuming food and water; and through dermal exposure by direct contact of pesticides with skin. Pesticides are sprayed onto food, especially fruits and vegetables, they secrete into soils and groundwater which can end up in drinking water, and pesticide spray can drift and pollute the air.

There is increasing anxiety about the importance of small residues of pesticides, often suspected of being carcinogens or disrupting endocrine activities, in drinking water and food. In spite of stringent regulations by international and national regulatory agencies, reports of pesticide residues in human foods, both imported and home-produced, are numerous.

Over the last fifty years many human illnesses and deaths have occurred as a result of exposure to pesticides, with up to 20,000 deaths reported annually. Some of these are suicides, but most involve some form of accidental exposure to pesticides, particularly among farmers and spray operators in developing countries, who are careless in handling pesticides or wear insufficient protective clothing and equipment. Moreover, there have been major accidents involving pesticides that have led to the death or illness of many thousands. One instance occurred in Bhopal, India, where more than 5000 deaths resulted from exposure to accidental emissions of methyl isocyanate from a pesticide factory.

The effects of pesticides on human health are more harmful based on the toxicity of the chemical and the length and magnitude of exposure. Farm workers and their families experience the greatest exposure to agricultural pesticides through direct contact with the chemicals. But every human contains a percentage of pesticides found in fat samples in their body. Children are most susceptible and sensitive to pesticides due to their small size and underdevelopment. The chemicals can bioaccumulation in the body over time.

Exposure to pesticides can range from mild skin irritation to birth defects, tumours, genetic changes, blood and nerve disorders, endocrine disruption, and even coma or death.

Aquatic Life

A major environmental impact has been the widespread mortality of fish and marine invertebrates due to the contamination of aquatic systems by pesticides. This has resulted from the agricultural contamination of waterways through fallout, drainage or runoff erosion, and from the discharge of industrial effluents containing pesticides into waterways. Historically, most of the fish in Europe's Rhine River were killed by the discharge of pesticides, and at one time fish populations in the Great Lakes became very low due to pesticide contamination.

Fish and other aquatic biota may be harmed by pesticide-contaminated water. Pesticide surface runoff into rivers and streams can be highly lethal to aquatic life, sometimes killing all the fish in a particular stream. Application of herbicides to bodies of water can cause fish kills when the dead plants rot and use up the water's oxygen, suffocating the fish. Some herbicides, such as copper sulphite, that are applied to water to kill plants are toxic to fish and other water animals at concentrations similar to those used to kill the plants. Repeated exposure to sublethal doses of some pesticides can cause physiological and behavioural changes in fish that reduce populations, such as abandonment of nests and broods, decreased immunity to disease, and increased failure to avoid predators.

Application of herbicides to bodies of water can kill off plants on which fish depend for their habitat. Pesticides can accumulate in bodies of water to levels that kill off zooplankton, the main source of food for young fish. Pesticides can kill off the insects on which some fish feed, causing the fish to travel farther in search of food and exposing them to greater risk from predators. The faster a given pesticide breaks down in the environment, the less threat it poses to aquatic life. Insecticides are more toxic to aquatic life than herbicides and fungicides.

Birds

Pesticides had created striking effects on birds, those in the higher trophic levels of food chains, such as bald eagles, hawks and owls. These birds are often rare endangered, and susceptible to pesticide residues such as those occurring from the bioconcentration of organochlorine insecticides through terrestrial food chains. Pesticides will also kill grain- and plant-feeding birds, and the elimination of many rare species of ducks and geese has been reported. Populations of insect-eating birds such as partridges, grouse and pheasants have decreased due to the loss of their insect food in agricultural fields through the use of insecticides.

Bees are extremely important in the pollination of crops and wild plants, and although pesticides are screened for toxicity to bees, and the use of pesticides toxic to bees is permitted only under stringent conditions, many bees are killed by pesticides, resulting in the considerably reduced yield of crops dependent on bee pollination.

Bald eagles are common examples of nontarget organisms that are impacted by pesticide use. Rachel Carson's landmark book *Silent Spring* dealt with the of loss of bird species due to bioaccumulation of pesticides in their tissues. There is evidence that birds are continuing to be harmed by pesticide use. In the farmland of Britain, populations of ten different species of birds have declined by 10 million breeding individuals between 1989 and 2007, a phenomenon thought to have resulted from loss of plant and invertebrate species on which the birds feed. Throughout Europe, 116 species of birds are now threatened. Reductions in bird populations have been found to be associated with times and areas in which pesticides are used. In another example, some types of fungicides used in peanut farming are only slightly toxic to birds and mammals, but may kill off earthworms, which can in turn reduce populations of the birds and mammals that feed on them.

Some pesticides come in granular form, and birds and other wildlife may eat the granules, mistaking them for grains of food. A few granules of a pesticide is enough to kill a small bird.

The herbicide parquet, when sprayed onto bird eggs, causes growth abnormalities in embryos and reduces the number of chicks that hatch successfully, but most herbicides do not directly cause much harm to birds. Herbicides may endanger bird populations by reducing their habitat.

THREATENING REPORTS ON HAZARDOUS EFFECTS OF PESTICIDES

Endosulfan is a harmful insecticide, it causes several health hazards in human beings, endosulfan was aerial sprayed on cashew plantations in India especially in northern parts of Kerala for more than 20 years, the terrain was unsuitable for aerial spraying considering the relatively high rainfall and its geological structure, unsual diseases and even deaths were observed in and around the region.

Endosulfan is a chlorinated hydrocarbon insecticide of the cyclodiene subgroup which act as a contact poison in wide variety of insects and mice is primarily used on food crops like tea, fruits, vegetable and grains. Exposure to endosulfan will result from ingestion of contaminated food. It does not easily dissolve in water and transport is likely occur if it attached to soil particles in surface runoff, endosulfan residues have been found in numerous food products at very low concentration.

Endosulfan is rapidly degraded and eliminated in mammals with very little absorption in gastrointestinal tract. In these areas, where aerial spraying was done lot of children who have exposed are considered to be living martyr.

Studies consistently show that endosulfan is highly poisonous and easily causes death and severe acute and chronic toxicity to various organ system including mental impairment, neurologic disturbances, immunotoxicity and reproductive toxicity and most of the new born were physically handicapped and showing epilepsy.

Classified by the US environmental protection agency as highly hazardous endosulfan was at the centre of controversy in the Philippinese in 1990's.

ALTERNATIVE METHODS FOR ELIMINATING PESTICIDES

Diversified Planting

A common practice among home gardeners is to plant a single crop in a straight row. This encourages pest infestation because it facilitates easy travel of an insect or disease from one host plant to another. By intermingling different types of plants and by not planting in straight rows, an insect is forced to search for a new host plant thus exposing itself to predators. Also, this approach corresponds well with companion planting.

Low Toxicity Pesticides

Formulated, biodegradable pest-control substances are commercially available. Although these products are pesticides, they have low toxicity to mammals and do not last long in the environment. The local County Extension Service can provide information on these and other pesticide products.

Many alternatives are available to reduce the effects pesticides have on the environment. There are a variety of alternative pesticides such as manually removing weeds and pests from plants, applying heat, covering weeds with plastic, and placing traps and lures to catch or move pests. Pests can be prevented by removing pest breeding sites, maintaining healthy soils which breed healthy plants that are resistant to pests, planting native species that are naturally more resistant to native pests, and use biocontrol agents such as birds and other pest eating organisms.

CASE STUDY: PESTICIDES AND LOSS OF BIODIVERSITY IN EUROPE

Biodiversity Loss and the Use of Pesticides

Pesticides are a major factor affecting biological diversity, along with habitat loss and climate change. They can have toxic effects in the short-term in directly exposed organisms or long-term effects by causing changes in habitat and the food chain.

Charles Darwin and Alfred Wallace were among the first scientists to recognise the importance of biodiversity for ecosystems. They suggested that a diverse mixture of crop plants ought to be more productive than a monoculture. Though there are exceptions, recent studies confirm the idea that an intact, diverse community generally performs better than one which has lost species. Ecosystem stability (resilience to disturbance) seems to arise from groups of connected species being able to interact in more varied positive and complimentary ways. Biological diversity manifests itself at different levels. It includes the diversity of ecosystems, species, populations and individuals. In an ecosystem, interdependent populations of various species deliver 'services' such as the supply of food and soil resources or the retention and cycling of nutrients, water and energy. Although it seems that the average species loss can affect the functioning of a wide variety of organisms and ecosystems, the magnitude of effect depends on which particular species is becoming extinct.

Communities of different animal and plant species perform vital functions within ecosystems. In general, communities which have higher diversity tend to be more stable.

Why conserve threatened species

Rachel Carson provided clear evidence of the far-reaching environmental impact of pesticides in her pioneering work 50 years ago. In 'Silent spring' she showed that organochlorines, a large group of insecticides, accumulated in wildlife and the food chain. This had a devastating effect on many species.

Only a decade after the 'green revolution' began it became obvious that large-scale spraying of pesticides was causing serious damage. In 1963, Rachel Carson emphasised human dependence on an intact environment: 'But man is a part of nature, and his war against nature is inevitably a war against himself.' Human well-being depends on the services delivered by intact ecosystems. While biodiversity loss is in itself a cause for concern, biodiversity conservation also aims to sustain humanity. People's livelihoods ultimately depend on biological resources. Thus lacking progress towards the target of the Convention on Biological Diversity, 'to achieve by 2013 a significant reduction of the current rate of biodiversity loss' could undermine achievement of the Millenium Development Goals and poverty reduction in the long-term. The 2013 target has inspired action but will not be fully attained. Biodiversity loss and degradation of ecosystems have increasingly dangerous consequences for people, and may threaten some societies' survival.

When EU cereal yield was doubled it resulted in the loss of half the plant species and one-third of carabid beetles and farmland bird species. Of the components of agricultural intensification, pesticide use, especially insecticides and fungicides, had the most consistently negative effects on species diversity, and insecticides also reduced the potential for biological pest control. In the EU, up to 80 per cent of protected habitat types and 50 per cent of species of conservation interest now have an unfavourable conservation status. Much greater effort is needed to reverse the decline in threatened species or habitats on a larger scale. A 'business-as-usual' scenario would mean that the current decline of biodiversity will continue and even accelerate, and by 2050 a further 11 per cent of natural areas which existed in 2000 will be lost, while 40 per cent of land currently under low-impact agriculture could be converted to intensive agricultural use. Human survival is inextricably linked to the survival of numerous other species on which intact ecosystems depend.

Increasing pressure of agriculture on habitats and biodiversity

The use of pesticides (particularly herbicides) and synthetic fertilisers has increased dramatically over the past 60 years. In industrialised countries, farming practices have fundamentally changed. In the UK and many other places, mixed agriculture has been lost and farms have become increasingly specialised. Arable farming (field crops) and pastoral grassland are now largely separated as traditional crop rotation has been abandoned. In the British lowlands, field sizes have risen and field margins have shrunk. Harvesting has become more efficient and hedgerows have been lost. There has been a marked population decline of many species living on farmland.

Worldwide humans are estimated to use about 20 per cent of the net primary production (plant organic matter produced in photosynthesis). In South America and Africa, humans use about 6 per cent and 12 per cent, respectively, of the regional net primary production, while the fraction consumed by humans is 72 per cent in western Europe and 80 per cent in south central Asia. The proportion of plant organic matter consumed by humans varies enormously between regions. For example, consumption of regionally-produced plant matter per inhabitant is nearly five times higher in North America than in south central Asia. Changes of habitat and biodiversity have been due both to changing climate and people's increasing use of plant and animal resources.

1. Heavy pesticide input has been a key feature of agricultural intensification. This is closely linked to changes in farming practices and habitat destruction or loss.
2. Between 1990 and 2006, the total area treated with pesticides increased by 30 per cent in the UK, and the herbicide-treated area increased by 38 per cent.
3. In farmland habitats, population declines have occurred in about half of plants, a third of insects and four-fifths of bird species.

Impact of pesticides on wildlife populations and species diversity

Many pesticides are toxic to beneficial insects, birds, mammals, amphibians or fish. Wildlife poisoning depends a pesticide's toxicity and other properties (e.g. water-soluble pesticides may pollute surface waters), the quantity applied, frequency, timing and method of spraying (e.g. fine spray is prone to drift), weather, vegetation structure, and soil type. Insecticides, rodenticides, fungicides (for seed treatment) and the more toxic herbicides threaten exposed wildlife. Over the past 40 years, the use of highly toxic carbamate and organophosphate has strongly increased. In the South Europe, organochlorines such as endosulfan, highly persistent in the environment, are still used on a large scale. With habitat change, pesticide poisoning can cause major population decline which may threaten rare species.

Agricultural pesticides can reduce the abundance of weeds and insects which are important food sources for many species. Herbicides can change habitats by altering vegetation structure, ultimately leading to population decline. Fungicide use has also allowed farmers to stop growing 'break crops' like grass or roots. This has led to the decline of some arable weeds. In Canada, losses among 62 imperilled species were significantly more closely related to rates of pesticide use than to agricultural area in a region. Species loss was highest in areas with intensive agriculture (aerial spraying). Smith concluded that either pesticides or other features of intensive agriculture linked to pesticide use in Canada, played a major part in the decline of imperilled species.

1. Pesticides affect wildlife directly and indirectly via food sources and habitats.
2. Wildlife poisoning by highly toxic insecticides, rodenticides, fungicides (on treated seed) and toxic herbicides can cause major population decline.
3. Pesticides accumulating in the food chain, particularly those which cause endocrine disruption, pose a long-term risk to mammals, birds, amphibians, and fish.
4. Broad-spectrum insecticides and herbicides reduce food sources for birds and mammals. This can produce a substantial decline in rare species populations.
5. By changing vegetation structure, herbicides can render habitats unsuitable for certain species. This threatens insects, farmland birds and mammals.

Bird Species Decline Owing to Pesticides

In Western Europe the number of farmland birds is now just half that of 1980, even among formerly abundant species. While average populations of all common and forest birds declined by about 10 per cent in Europe between 1980 and 2006, populations of farmland birds have fallen by 48 per cent. This figure is based on surveys in 21 EU countries. Forest birds have declined less than have specialist birds living on farmland. A recent survey found that in the USA one in three bird species is endangered, threatened or of conservation concern. Forty per cent of grassland and arid-land birds are affected by population decline. Populations of raptors and other birds recovered after DDT and other toxic pesticides were banned in Europe. In North America between 1980 and 1999, populations of grassland species declined more than species living in shrubland. In 78 per cent of species there was at least one association between population trend and change in agricultural land use, and for most species this factor accounted for 25–30 per cent of the variation in trend among states.

In Europe, the population decline among farmland birds was far greater in countries with more intensive agriculture, and in a statistical analysis 'cereal yield' explained over 30 per cent of the trend in population change. Smith predicted that introducing EU agricultural policy into accession countries will result in a major decline in key bird populations. This occurred in the German state of Saxony-Anhalt. After 1990, farming in this region shifted from rotational cultivation (e.g. root-crops) to oilseed rape and winter cereals, which led to a reduction in grassland area and increased insecticide and herbicide use. In the same period, red kite (*Milvus milvus*) numbers fell by 50 per cent from over 40 nesting pairs to about 20 pairs per 100 km^2.

Important Bird Areas (IBAs) include agricultural areas with important bird populations. Although IBAs are appointed as priority conservation sites, they have no official protected status. Agricultural expansion and intensification threaten half of IBAs in Africa and one-third in Europe. It is estimated that worldwide bird populations have declined by 20 per cent to 25 per cent since pre-agricultural times. Altogether, 1211 bird species (12 per cent of the total) are considered globally threatened, and 86 per cent of these are threatened by habitat destruction or degradation. For 187 globally threatened bird species, the primary pressure is chemical pollution, including fertilisers, pesticides and heavy metals entering surface water and the terrestrial environment.

Bird poisonings caused by pesticides

In the UK, the volume of seeds eaten by many bird species is large enough to pose a potential risk if the seeds are treated with one of the more toxic fungicides. Organophosphate insecticides, including disulfoton, fenthion and parathion are highly toxic to birds. These have frequently poisonised raptors foraging in fields. Field studies have led to the conclusion that given the usual amounts of insecticide used, 'direct mortality of exposed birds is both inevitable and relatively frequent with a large number of insecticides currently registered'. In the USA, some 50 pesticides have killed songbirds, gamebirds, raptors, seabirds and shorebirds. In a small area of the Argentine pampas, monocrotophos, an organophosphate, has killed 6000 Swainson's hawks. Worldwide, over 1,00,000 bird deaths caused by this chemical have been documented.

The number of birds found killed by pesticides, in the UK was at least 60 in 2006, and 55 in 2007. Pesticides were investigated as a possible cause of death in another 90 cases. The affected species included buzzard, red kite, raven, crow, peregrine falcon, golden eagle, gull, barn owl, tawny owl, magpie, pheasant, rook, marsh harrier, dove, jackdaw and chaffinch. The following pesticides were identified as a cause of fatal bird poisoning: carbamates (aldicarb, bendiocarb, carbofuran), organophosphates (chlorpyrifos, diazinon, isofenphos, malathion, mevinphos, phorate), anticoagulant rodenticides (bromadiolone, brodifacoum, difenacoum) and alphachloralose.

In 2005, from 20 dead barn owls and ten kestrels that contained one or more anticoagulant rodenticides, six barn owls and five kestrels had residues in the potentially lethal range. It was concluded that rodenticides may have contributed to the death of one barn owl and two kestrels, based on the circumstances of death and examination of carcasses. Residues in five of 23 red kites found dead would be potentially lethal to barn owls, while 17 of these had residues of at least one rodenticide, and ten had residues from two or three rodenticides. From dead tawny owls collected under the Predatory Bird Monitoring Scheme, 20 per cent (and 33 per cent of owl livers) contained residues of one or more rodenticides.

Negative impacts of pesticides on food sources of birds

Herbicides and avermectin residues (used as worming livestock agents) affect birds indirectly by reducing food abundance. Lower availability of key invertebrates and seed food for farmland birds in northern Europe was likely due to insecticides and herbicides, intensification and specialisation of farmland, loss of field margins and ploughing. Insecticides generally had a negative effect on yellowhammer when spraying occurred during the breeding season. Spraying at this time may cause more damage than repeated use throughout the year. Spraying insecticides within 20 days of hatching led to smaller brood size of yellowhammer, lower mean weight of skylark chicks and lower survival of corn bunting chicks. More frequent spraying of insecticides, herbicides or fungicides was linked to a considerably smaller abundance of food invertebrates. This resulted in lower breeding success of corn buntings and may have contributed to their decline. In Sussex, herbicides were a major cause of the decline of grey partridge populations by removing weeds which are important insect hosts.

Pesticide use trends (measured by the percentage of treated area) was linked to periods of rapid bird decline. Bird species at risk from indirect effects caused by pesticides in the UK include grey partridge, corn bunting, yellowhammer, red-backed shrike, skylark, tree sparrow and yellow wagtail. The main causes of farmland bird decline have been: (i) pesticides and weed-control with herbicides, particularly, (ii) change from spring-sown to autumn-sown cereals, (iii) drainage and intensified management of grassland and (iv) increased cattle or sheep density. Sublethal effects on the nervous system can cause changes in behaviour. In an orchard, parent birds made fewer feeding trips after azinphos-methyl, an organophosphate, had been sprayed.

1. Bird populations are directly affected by poisoning from organophosphate or carbamate insecticides and anticoagulant rodenticides. Sublethal poisoning of birds by organophosphates can lead to detrimental changes in behaviour.
2. Broad-spectrum herbicides threaten rare and endangered bird species by reducing the abundance of weeds (eaten by birds) and insects hosted by weeds. Insecticides reduce the number of insects which are important food sources for birds.

Risk to Mammals of Hazardous Pesticides

Pesticides and other chemicals have caused population declines in Britain's wild mammals. Mostly bats and rodents (and 38 per cent of species) were affected. Certain pesticides can gradually accumulate in the food chain. This is of concern to vertebrates, particularly species in higher orders and top predators such as mammals or raptors. Anticoagulant rodenticides are highly toxic and some can bioaccumulate. Non-target predatory mammals (e.g. dogs and foxes) and raptors frequently suffer 'secondary poisoning' by eating rats or mice which are poisoned by rodenticides. In France, foxes were poisoned by residues of bromadiolone in prey tissue. In the UK, following rat control with rodenticides, local wood mice, bank vole and field vole populations has declined significantly.

At least 25–35 per cent of small mammal predators (polecats, stoats, and weasels) sampled had been exposed to rodenticides and this may be an underestimate. However, it is not well-known how frequently rodenticides cause secondary mammal poisoning, and what the impact on their populations might be.

Herbicide use can affect mammals such as the common shrew, wood mouse and badger by removing plant food sources and changing the microclimate. Hares prefer a more diverse habitat. They are thus likely to benefit from increased fallow land. On organic farms, foraging activity by bats was significantly higher than on conventional farms, which may be due to a larger abundance of prey insects. Less intensive farming systems may help to reverse bat decline.

1. Anticoagulant rodenticides often indirectly poison predatory mammals and raptors.
2. Herbicides can cause changes in vegetation and habitat which threaten mammals, while insecticides may reduce the availability of important food insects (Fig. 14.1).

Fig. 14.1. Britain's wild mammal.

Impact of Pesticides on Butterflies, Bees and Natural Enemies

Broad-spectrum insecticides (e.g. carbamates, organophosphates and pyrethroids) can cause population declines of beneficial insects such as bees, spiders or beetles. Many of these species play an important role in the food web or as natural enemies of pest insects. Since 1970, insect numbers in cereal fields in Sussex have dropped by half. Numbers of bugs, spiders and beetles were considerably higher in untreated fields. On British organic farms, numbers and species richness of butterflies was greater than on conventional farms. The number of carabid beetles and spiders was usually higher on organic farms. Conventional management practice appeared to affect natural enemies far more than other insects or target pests. Moths were considerably more abundant on organic farms and species richness was higher. In arable fields, insecticide use was an important factor influencing communities of epigeic spiders. On sites with an increased pesticide input, communities of bugs, wild bees and spiders were more uniform, indicating less exchange between communities in areas with intensive agriculture.

Bees perform essential pollination. Honey bees are under pressure from parasitic mites, viral diseases, habitat loss and pesticides. Intensified agricultural practices, habitat loss and agrochemicals are considered to be among the chief environmental threats to Europe's honey and wild bees. Agricultural policy must reduce these pressures to ensure adequate pollinator populations. On organic farms in the USA, near natural habitat, diverse native wild bee communities provided full pollination services, while diversity and numbers of native bees were greatly reduced on other farms. In the UK, of the 95 incidents of bee poisoning (where the cause could be identified) between 1995 and 2001, organophosphates caused 42 per cent, carbamates 29 per cent and pyrethroids 14 per cent of cases. In the last decade in the UK, insecticides which poisoned bee colonies included bendiocarb (a carbamate) and three pyrethroids: cypermethrin, deltamethrin and permethrin. Synergistic effects between pyrethroids and EBI fungicides (imidazole or triazole fungicides) can increase the risk to honeybees.

Clothianidin, and to a lesser extent, imidacloprid are highly toxic to bumble bees and other wild bees. These two neonicotinoid insecticides are used to treat corn and sunflower seeds. In 2008, clothianidin caused many bee poisonings and colony deaths in southern Germany. The product has since been withdrawn. When imidacloprid-treated seed is grown, a large enough amount can enter the environment to poison bees. Residues of imidacloprid in maize pollen grown from treated seed can be a high risk to bees owing to sublethal effects. Even at low doses of imidacloprid, bees' foraging behaviour was negatively affected. Exposure to low doses of imidacloprid over a longer period led to reduced learning capacity among bees. In alfalfa, imidacloprid affected the number and species diversity in communities of arthropods (natural enemies such as spiders) more strongly than among target pest insects. Imidacloprid has been banned in France. Field margins without use of pesticides (herbicides in particular) had a positive effect on the number of Lepidoptera (such as moths or butterflies), bugs and staphylinid beetles at the edges of arable fields. In organic plots, average numbers of spiders and carabid or staphylinid beetles were almost twice as high as those in conventional plots.

1. Pesticides which are highly toxic to bees, bumblebees and other beneficial insects: carbamates (e.g. aldicarb, benomyl, carbofuran, methiocarb), organophosphates (e.g. chlorpyrifos, diazinon, dimethoate, fenitrothion), pyrethroids (e.g. cyfluthrin, cyhalothrin), and neonicotinoids (imidacloprid, thiamethoxam, clothianidin).

2. Recently, clothianidin used in seed treatments have caused widespread bee poisoning. Imidacloprid residues in plants can negatively alter bee behaviour.

Pesticides Affecting Amphibians and Aquatic Species

One-third of 6000 amphibian species worldwide are threatened. Besides habitat loss, overexploitation or introduced species, amphibians are affected by the pollution of surface waters with fertilisers and pesticides from agriculture. In the USA, spray drift of hexazinone, a triazine herbicide, was considered 'likely to adversely affect' the endangered red-legged California frog and its habitat. Atrazine is moderately toxic to some fish species. It can indirectly affect aquatic ecosystems by damaging aquatic plants. A review concluded that further study is needed on the potential hormonal effects of atrazine on frogs or fish. In Europe, the authorisation for atrazine has been withdrawn due to health and environmental risks. Urea herbicides such as isoproturon and diuron often contaminate rivers, lakes and groundwater. Most breakdown products of diuron were more toxic to cellular micro-organisms than the parent compound. Fungicides based on copper are highly toxic to aquatic organisms. In fish and some other aquatic organisms, the risk of copper accumulating may be high. The EU aims eventually to eliminate copper in organic vineyards and apple orchards. A major study investigating amphibian communities in the USA found that, among other factors, agricultural fields near surface water and pesticides (at sufficiently high concentrations to affect insects or plants) will harm amphibian species richness. In particular, if water contains herbicides at concentrations which substantially reduce populations of aquatic plants this is likely to be associated with low numbers of amphibians relative to predator populations, and increased numbers of trematode parasites in amphibians. Parasitic nematodes were more abundant in agricultural wetlands during the growing season. Agricultural activity may intensify the infection of frogs by harmful nematodes. Atrazine suppressed the immune system of tiger salamanders by causing reduced numbers of white blood cells. The rate of infection by pathogenic viruses was higher in salamanders which were exposed to atrazine. In field studies, atrazine affected the immune system of tadpoles of northern leopard frogs, a declining species. Atrazine and phosphate fertiliser were the main factors linked to numbers of larval trematodes in frogs. In California, tadpoles of the Pacific treefrog in areas with a poor population status had reduced levels of the enzyme cholinesterase, indicating exposure to organophosphates and/or carbamates. Endosulfan was highly toxic to the declining yellow-legged frog. The insecticides chlorpyrifos and endosulfan have the potential to cause serious damage to amphibians at concentrations occurring in the environment under normal conditions of use. In laboratory tests, the survival of juvenile Great Plains toads and New Mexico spadefoot toads was reduced after exposure to certain formulations of the herbicides glufosinate and glyphosate.

Circumstances for pesticide use which present a high risk to communities of aquatic species include spray drift for insecticides and run-off from fields for herbicides. However, in a study of the risks of 261 pesticides to aquatic ecosystems in field ditches, about 95 per cent of the predicted risk was caused by only seven pesticides. More selective pesticides (with no or minimal impact on non-target organisms) should clearly be preferred. Surface water is frequently contaminated with insecticides through normal use at levels above those known to affect fish and aquatic invertebrates such as daphnia or shrimp. For example, this was observed for levels of azinphos-methyl, chlorpyrifos and endosulfan in the aquatic environment. Similarly, chlorpyrifos and endosulfan were rated as 'chemicals of potential ecological concern'. An assessment concluded that adverse effects of endosulfan on fish and invertebrates are a concern when this insecticide is used near aquatic ecosystems. In field tests, the insecticide carbaryl affected the composition of an aquatic community of amphibians and insects by changing colonisation of pools and numbers of eggs laid.

1. Insecticides and herbicides in surface waters (from spray-drift or run-off) can alter the species composition of aquatic communities and affect fish and invertebrates.

2. Insecticides (organophosphates, carbamates) have toxic effects on the nervous systems of amphibians which may alter their behaviour. Herbicides (e.g. atrazine) can impair the immune system of frog tadpoles, which can make amphibians more susceptible to harmful parasitic nematodes. Indirect effects can be fatal.

3. Urea herbicides such as diuron frequently contaminate surface and groundwater. Copper-based fungicides are highly toxic to fish and have a potential to accumulate.

Effect of Pesticides on Plant Communities

In recent decades, the use of herbicides has dramatically increased. Today, some non-crop plants (or 'weeds') are threatened with extinction in Britain. Although the total volume of herbicides applied in the UK decreased slightly between 1990 and 2006, the herbicide-treated area increased by 38 per cent. Diversity of wild plants in agricultural fields and field margins is declining, especially in infertile grassland and hedge bottoms. A slight increase in plant diversity in arable fields in 1998 may have been due to the introduction of set-aside. The number of plants providing food for butterfly caterpillars decreased in Britain between 1998 and 2007. Field margins created within agri-environment schemes supported a higher number of plant species than crop areas, but plant cover and species richness are still low (on average 11 species per plot and 21 per cent cover) when compared to other habitats (e.g. horticultural land) and set-aside. By providing an unsprayed field margin at least three metres wide, the diversity and number of arable plants and insects hosted by them increased substantially. Over five years, harmful weed cover did not increase in field margins.

On European farms using IPM methods (and an average of half the herbicides), the seed bank of weeds in soil doubled in autumn-sown crops. This was considered acceptable, while in spring-sown crops the seed bank increased more than threefold for certain weeds. In lowland areas in England, species diversity and abundance of plants, birds, bats, invertebrates and plants were typically higher on organic farms than conventional ones. Positive effects were strongest for plants. It was estimated that organic fields held up to twice as many plant species and, on average, a weed cover twice as large. Flora in Britain is changing as arable plants such as the corn marigold introduced long ago are decreasing. This could be due to intensive agriculture and a decline in mixed farming.

Some herbicides are highly toxic to plants at very low doses, e.g. sulphonylureas, sulphonamides and imidazolinones. Tribenuron-methyl affected growth of algae and activity of microalgae at very low concentrations. In a study on the effects of sulphonylurea herbicides on phytoplankton it was concluded that these herbicides present a potential hazard to aquatic systems even at low environmental concentrations. Sulphonylureas have replaced other herbicides which are more toxic to animals. In potatoes, sulphometuron-methyl caused major yield losses even when used at rates below the recommended dose. Experts have warned that the widespread use of sulphonylureas 'could have a devastating impact on the productivity of nontarget crops and the makeup of natural plant communities and wildlife food chains'. Hexazinone is a persistent triazine with high leachability. In the USA, at all application rates the EPA's levels of concern for aquatic and terrestrial nontarget plants were exceeded. Aquatic ecosystems within or next to hexazinone-treated areas could be altered by the effects on aquatic plants. Other triazines affect aquatic plants similarly, e.g. terbuthylazine and atrazine. In field tests, the herbicide glyphosate altered the composition of freshwater microbial communities by decreasing the abundance of microbial phytoplankton and increasing cyanobacteria.

1. Many plants which were previously common on British farmland are declining owing to the abandonment of mixed farming and increased herbicide use.
2. Large-scale use of sulphonylurea herbicides, and presumably also sulfonamides and imidazolinones, poses a risk to nontarget plants, algae, and ecosystems.
3. Triazine herbicides may present a risk to nontarget and aquatic plants.

Are Pesticides Diminishing Soil Fertility

What is soil fertility? A fertile soil provides the nutrients needed to promote growth of plants, is a habitat of an active and diverse community of organisms, and exhibits a structure which is characteristic of the location, and which enables a continuous decomposition of organic residues.

In South Africa, the feeding activity of soil organisms was higher in soil from organic vineyards than from conventionally treated sites. The number of earthworms was 1.3–3.2 times higher in organic compared to conventional plots, and the length of plant roots colonised by mycorrhizae was 40 per cent higher in organic than in conventional systems. Triclopyr, a herbicide, caused a major reduction in the growth of mycorrhizae at levated soil levels.

The sulphonylurea herbicides metsulfuron and (to a lesser extent) chlorsulfuron caused a reduction in the growth of Pseudomonas soil bacteria. In laboratory tests, a combination of two sulphonylurea herbicides, bensulfuron-methyl (B) and metsulfuron-methyl, caused a considerable reduction in soil microbial biomass over the first 15 days. In bacterial communities in soil, bromoxynile (a nitrile herbicide) caused major changes in species composition and diversity. Bromoxynile inhibited the growth of bacteria capable of degrading chemicals in soil. Also captan (a fungicide) and the herbicide glyphosate caused a shift among species in bacterial communities in soil. Certain organophosphate insecticides (e.g.

dimethoate) can decrease the activity and biomass of soil micro-organisms, while others (such as fosthiazate) may actually result in an increase in microbial biomass. How pesticides affect long-term soil fertility is not well understood as this depends on many factors.

1. Pesticides affect earthworms, symbiotic mycorrhizae and other organisms in soil.
2. Composition and activity of bacterial communities can be changed by pesticides.

Policies and Methods for Biodiversity Conservation

In the EU, national policies set targets for biodiversity conservation. The convention on biological diversity provides national strategies and action plans for conserving species at national level. These include the establishment of national targets. For example, the UK biodiversity action plan (BAP) currently lists 1150 species and 65 habitats with a priority for conservation. In 2002, of 78 farmland priority species, 39 per cent were declining, 21 per cent had unknown or unclear status, 18 per cent were stable, 15 per cent on the increase, and 7 per cent had been lost. From the total area of one million hectares of nationally-important wildlife sites (Sites of Special Scientific Interest) in the UK in 2008, about 3,80,000 hectares or 38 per cent, were in an unfavourable condition owing mainly to agriculture. Only 47 per cent of important wildlife sites on farmland were in a favourable condition. One of the BAP's targets is to reverse the decline in farmland birds in Britain by 2020. In winter, farmland bird density is much higher on stubble (rotational set-aside) than on cereal fields. However, EU policy recently changed and set-aside is no longer compulsory. Ornithologists have warned that this could have serious negative impacts on farmland biodiversity across the EU.

Maintaining an appropriate population of weed species to support farmland wildlife is a challenge. It may be achieved by providing conservation headlands, by developing much more selective herbicides, and through their selective use. In England between 1978 and 1990, plant diversity on arable land was declining. Between 1998 and 2007, plant diversity in main plots increased by 36 per cent. This was due to increases in the area of set-aside or fallow land, driven by agri-environment schemes. On plots with reduced herbicide input, farmland birds used winter cereal stubble more often than on conventional plots. To reverse the decline of birds, farming needs to change substantially and incorporate appropriate practices. Species diversity is usually higher in non-crop areas than in grassland or fields. Unsprayed headlands of fields contain rare weed species and the highest diversity of invertebrates. As a feeding area for birds, the headland is most important. Weeds on the ground provide refuge and host many food insects. Using more selective herbicides in winter cereals could benefit farmland bird species which feed their chicks weed seeds e.g. linnets or finches. In the EU on arable or mixed farms which use integrated management practices, on average the use of herbicides was reduced by 43 per cent, use of insecticides or molluscicides by 55 per cent, and fungicide use was 50 per cent lower when compared to conventional farms. On farms using IPM, the number of arthropods (such as beetles, spiders, springtails or sawflies), plants and earthworms increased significantly. Similar positive effects were observed for soil organisms, birds and mammals such as wood mice.

National systems for pesticide authorisation aim to limit of the harm pesticides inflict on nontarget species. But measures for reducing risks from pesticides are still being developed. Regulatory controls alone will not eliminate the impact on nontarget species. Additional initiatives are needed which mitigate the effects of pesticides on biodiversity. The EU's Sixth Environment Action Program identified biodiversity conservation as a high priority. Areas protected under the Birds and Habitats Directives are connected in the 'Natura 2000' network. The proposed strategy on sustainable pesticide use in the EU aims to minimise risks to health and the environment from pesticides. Member states must eliminate or reduce the use of pesticides as far as possible in Natura 2000 sites, and promote farming with low pesticide input, particularly integrated pest management (IPM), and establish the necessary conditions for implementing IPM techniques.

One of the leading organisations in the development of IPM standards is the International Organisation for Biological and Integrated Control of Noxious Animals and Plants (IOBC). Its principles for integrated production emphasise the importance of biodiversity.

According to the IOBC, integrated production is a farming system which produces high quality food and other products by using natural resources and regulating mechanisms to replace polluting input and to secure sustainable farming.

Agri-environment schemes in the EU provide payments to farmers taking measures to preserve the environment and countryside. But spending on these measures are to date marginal. Farmers who practise less intensive farming and who conserve nature need to be rewarded. The UN Convention on Biological Diversity requires that countries develop national strategies and action plans for conserving biodiversity, and define national or sub-national targets and indicators. The number and quality of conservation targets vary strongly between countries. Incentive schemes should be continuously evaluated and adapted. To assess the effectiveness of measures for conserving threatened species such as farmland birds, quantitative and measurable goals are needed, plus monitoring of species. Extensive farming may need to stretch over larger connected areas to achieve substantial gains. Organic farms, together with agri-environment schemes to some degree, have positive effects on the diversity of plants and beetles in the EU, while bird species were not significantly more diverse. This may be due to widespread chemical pollution. So a shift towards farming with minimal pesticide use over large areas is urgently needed.

Corn, sugar cane and palm oil are increasingly being used to produce biofuels. These crop plants are linked to a high input of pesticide and fertiliser. Their use as biofuels is threatening biodiversity. Corn-based bioethanol is the worst among the alternatives currently available, so alternatives to corn as a biofuel source must be urgently pursued. In the EU, to produce energy from plant biomass in an environmentally-friendly way, the proportion of environmentally-orientated farming would need to increase to about 30 per cent of the Utilised Agricultural Area in most Member States by 2030, except for the most densely-populated countries.

Need for a Biodiversity Rescue Plan

The UN Convention on Biological Diversity requires the EU's 27-Member States to develop national policies to set biodiversity conservation targets. Not all Member States are equally ambitious, meaning that the 2010 objectives to halt further biodiversity loss need a new quantitative rescue plan for 2020, setting clear quantitative and qualitative targets, timetables and requiring ambitious monitoring. They also need to ensure coherence and better targeting, on these and a number of other EU policies (for sensitive 'Natura 2000' areas and water), the establishment of new EU policies (on soil and bio-waste). However, the success of the biodiversity rescue plan will also to a large extent depend on the EU's implementation of the new 'Regulation on the Placing of Plant Protection Products on the Market', as well as on how seriously member states implement the new framework directive on the sustainable use of pesticides. An important tool would be for member states to use this new opportunity to set dependency/use pesticide reduction targets and clear timetables.

A biodiversity rescue plan also needs to be accompanied by further reform of the EU's Common Agricultural Policy (CAP), departing from the current model where farmers receive income support for upkeep of their land into a model where farmers receive funding to provide public benefits, which includes paying farmers to use sustainable agricultural practices based on prevention first, also called integrated production, whereby the more farmers provide environmental and health services, the greater the public funding they receive. In the International Year of Biodiversity, we should fight together for reform of the CAP to encourage better agricultural practices. We should start by encouraging more mixed agriculture, crop rotation and pastoral grassland and lower field size. Even more so, we should encourage the development of practices such as bigger field margins and the re-establishment of hedgerows. We should put prevention first, in a dynamic system, encouraging front-runners who are willing to make environmental improvements, and incorporate a policy of making truly integrated agricultural production the basis of the post-2013 CAP.

Such an approach would be a step in the right direction in reversing the decline of birds, bees, bats, arthropods and earthworms, which thrive best in association with organic farming. It is also the best way to re-establish communities of different animal and plant species which perform vital functions within ecosystems, bringing higher diversity which tends to be more stable, and as a result will also help ensure greater long-term food security.

Chapter 15

Salt Marsh

INTRODUCTION

A salt marsh is an environment in the upper coastal intertidal zone between land and salt water or brackish water, it is dominated by dense stands of halophytic (salt-tolerant) plants such as herbs, grasses or low shrubs. These plants are terrestrial in origin and are essential to the stability of the salt marsh in trapping and binding sediments. Salt marshes play a large role in the aquatic food web and the exporting of nutrients to coastal waters. They also provide support to terrestrial animals such as migrating birds as well as providing coastal protection.

BASIC INFORMATION ON SALT MARSH

Salt marshes occur on low-energy coasts in temperate and high-latitudes. These typically include sheltered environments such as embankments, estuaries and the leeward side of barrier islands and spits. In the tropics and sub-tropics they are replaced by mangroves; an area that differs to a salt marsh in that instead of herbaceous plants, they are dominated by salt-tolerant trees. Most salt marshes have a low topography with low elevations but a vast wide area, making them largely popular for human populations.

Salt marshes are located among different landforms based on their physical and geomorphological settings. Such marsh landforms include deltaic marshes, estuarine, back-barrier, open coast, embayments and drowned-valley marshes. Deltaic marshes are associated with large rivers where many occur in Southern Europe such as the Camargue in the Rhone delta or the Ebro delta. They are also highly extensive within the rivers of the Mississippi Delta. In New Zealand, most salt marshes occur at the head of estuaries in areas where there is little wave action and high sedimentation. Such marshes are located in Awhitu Regional Park in Auckland, the Manawatu Estuary and the Avon-Heathcote Estuary in Christchurch. Back-barrier marshes are sensitive to the reshaping of barriers in the landward side of which they have been formed.

They are common along much of the eastern coast of the United States and the Frisian Islands. Large, shallow coastal embayments can hold salt marshes with examples including Morecambe Bay and Portsmouth in Britain and the Bay of Fundy in North America. Salt marshes are sometime included in lagoons, and the difference is not very marked: e.g. a major part of the Venetian Lagoon is made up of these sorts of animals and/or living organisms belonging to this ecosystem (Fig. 15.1).

Fig. 15.1. An estuarine salt marsh along the Heathcote River, Christchurch, New Zealand.

TIDAL FLOODING AND VEGETATION ZONATION

Coastal salt marshes can be distinguished from terrestrial habitats by the daily tidal flow that occurs and continuously floods the area. It is an important process in delivering sediments, nutrients and plant water supply to the marsh. At higher elevations in the upper marsh zone, there is much less tidal inflow, resulting in lower salinity levels. Soil salinity in the lower marsh zone is fairly constant due to everyday annual tidal flow. However, in the upper marsh, variability in salinity is shown as a result of less frequent flooding and climate variations. Rainfall can reduce salinity and evapotranspiration can increase levels during dry periods. As a result, there are microhabitats populated by different species of flora and fauna dependent on their physiological abilities. The flora of a salt marsh is differentiated into levels according to the plants' individual tolerance of salinity and water table levels. Vegetation found at the water must be able to survive high salt concentrations, periodical submersion, and a certain amount of water movement, while plants further inland in the marsh can sometimes experience dry, low-nutrient conditions. It has been found that the upper marsh zones limit species through competition and the lack of habitat protection, while lower marsh zones are determined through the ability of plants to tolerate physiological stresses such as salinity, water submergence and low oxygen levels.

The New England salt marsh is subject to strong tidal influences and shows distinct patterns of zonation. In low marsh areas with high tidal flooding, a monoculture of the smooth cordgrass, *Spartina alterniflora* dominate, then heading landwards, zones of the salt hay, *Spartina patens*, black rush, *Juncus gerardii* and the shrub *Iva frutescens* are seen respectively. These species all have different tolerances that make the different zones along the marsh best suited for each individual.

Plant species diversity is relatively low, since the flora must be tolerant of salt, complete or partial submersion, and anoxic mud substrate. The most common salt marsh plants are glassworts (*Salicornia* spp.) and the cordgrass (*Spartina* spp.), which have worldwide distribution. They are often the first plants to take hold in a mudflat and begin its ecological succession into a salt marsh. Their shoots lift the main flow of the tide above the mud surface while their roots spread into the substrate and stabilise the sticky mud and carry oxygen into it so that other plants can establish themselves as well. Plants such as sea lavenders (*Limonium* spp.), plantains (*Plantago* spp.), and varied sedges and rushes grow once the mud has been vegetated by the pioneer species.

Salt marshes are quite photosynthetically active and are extremely productive habitats. They serve as depositories for a large amount of organic matter and are full of decomposition, which feeds a broad food chain of organisms from bacteria to mammals. Many of the halophytic plants such as cordgrass are not grazed at all by higher animals but die off and decompose to become food for micro-organisms, which in turn become food for fish and birds.

HUMAN IMPACTS

The coast is a highly attractive natural feature to humans through its beauty, resources and accessibility. As of 2008, over half of the world's population was estimated to being living within 60 km of the coastal shoreline, making our coastlines highly vulnerable to human impacts from daily activities that put pressure on these surrounding natural environments. In the past, salt marshes were perceived as coastal 'wastelands,' causing considerable loss and change of these ecosystems through land reclamation for agriculture, urban development, salt production and recreation. The indirect effects of human activities such as nitrogen loading also play a major role in the salt marsh area.

Land Reclamation

Reclamation of land for agriculture by converting marshland to upland was historically a common practice. Dikes were often built to allow for this shift in land change and to provide flood protection further inland. For centuries, livestock such as sheep and cattle grazed on the highly fertile salt marsh land. Land reclamation for agriculture has resulted in many changes such as shifts in vegetation structure, sedimentation, salinity, water flow, biodiversity loss and high nutrient inputs. There have been many attempts made to eradicate these problems for example, in New Zealand, the cordgrass Spartina anglica was introduced from England into the Manawatu River mouth in 1913 to try and reclaim the estuary land for farming. A shift in structure from bare tidal flat to pastureland resulted from increased sedimentation and the cordgrass extended out into other estuaries around New Zealand. Native plants and animals struggled to survive as non-natives out competed them. Efforts are now being made to remove these cordgrass species, as the damages are slowly being recognised.

Nitrogen Loading

Following the period of marshland conversion for agriculture, these areas were later transformed into urban, residential and industrial land. Major cities such as Boston, San Francisco, Amsterdam, Rotterdam, Venice and Tokyo have all expanded out into areas of salt marsh. The remaining marshes surrounding these urban areas are under immense pressure from the human population as human-induced nitrogen enrichment enters these habitats.

Nitrogen loading through human-use indirectly affects salt marshes causing shifts in vegetation structure and the invasion of non-native species. Human impacts such as sewage, urban run-off, agricultural and industrial wastes are running into the marshes from nearby sources. Salt marshes are nitrogen limited and with an increasing level of nutrients entering the system from anthropogenic effects, the plant species associated with salt marshes are being restructured through change in competition. For example, the New England salt marsh is experiencing a shift in vegetation structure where *S. alterniflora* is spreading from the lower marsh where it predominately resides up into the upper marsh zone.

Additionally, in the same marshes, the reed *Phragmites australis* has been invading the area expanding to lower marshes and becoming a dominant species. *P. australis* is an aggressive halophyte that can invade disturbed areas in large numbers outcompeting native plants. This loss in biodiversity is not only seen in flora assemblages but also in many animals such as insects and birds as their habitat and food resources are altered.

Mosquito Control

Earlier in the 20th century, it was believed that draining the salt marshes would help mitigate mosquito populations. In many locations, particularly in the northeastern United States, residents and local and state agencies dug straight-lined ditches deep into the marsh flats. The end result, however, was a depletion of killifish habitat. The killifish is a mosquito predator, so the loss of habitat actually led to higher mosquito populations, and adversely affected wading birds that preyed on the killifish. These ditches can still be seen, though with the realisation of the impacts on the marshes' health, some places have begun efforts to refill the ditches.

RESTORATION AND MANAGEMENT

The perception of bay salt marshes as a coastal 'wasteland' has since changed, acknowledging that they are one of the most biologically productive habitats on earth, rivalling tropical rainforests. Salt marshes are ecologically important providing habitats for native migratory fish and acting as sheltered feeding and nursery grounds. They are now protected by legislation in many countries to look after these ecologically important habitats. In the United States and Europe, they are now accorded to a high level of protection by the Clean Water Act and the Habitats Directive respectively. With the impacts of this habitat and its importance now realised, a growing interest in restoring salt marshes, through managed retreat or the reclamation of land has been established. However, many Asian countries such as China are still to recognise the value of marshlands. With their ever-growing populations and intense development along the coast, the value of salt marshes tends to be ignored and the land continues to be reclaimed, Fig. 15.2 shows Glasswort (*Salicornia* spp.) a species endemic to the high marsh zone.

Fig. 15.2. Glasswort (*Salicornia* spp.) a species endemic to the high marsh zone.

Bakker suggests two options available for restoring salt marshes. The first is to abandon all human interference and leave the salt marsh to complete its natural development. These types of restoration projects are often unsuccessful as vegetation tends to struggle to revert back to its original structure and the natural tidal cycles are shifted due to land changes. The second option suggested by Bakker is to restore the destroyed habitat into its natural state either at the original site or as a replacement at a different site. Under natural conditions, recovery can take 2–10 years or even longer depending on the nature and degree of the disturbance and the relative maturity of the marsh involved. Marshes in their pioneer stages of development will recover more rapidly than mature marshes as they are often first to colonise the land. It is important to note, that restoration can often be sped up through the replanting of native vegetation.

This last approach is often the most practised and generally more successful than allowing the area to naturally recover on its own. The salt marshes in the state of Connecticut in the United States have long been an area lost to fill and dredging. As of 1969, the Tidal Wetland Act was introduced that seized this practice, but despite the introduction of the act, the system was still degrading due to alterations in tidal flow. One area in Connecticut is the marshes on Barn Island. These marshes were diked then impounded with salt and brackish marsh during 1946–1966. As a result the marsh shifted to a freshwater state and became dominated by the invasive species *P. australis*, *Typha angustifolia* and *T. latifolia* that have little ecological connection to the area. By 1980, a restoration program was put in place that has now been running for over 20 years. This program has aimed to reconnect the marshes by returning tidal flow along with the ecological functions and characteristics of the marshes back to their original state. In the case of Barn Island, declines in the invasive species have initiated, re-establishing the tidal-marsh vegetation along with animal species such as fish and insects. This example highlights that considerable time and effort is needed to effectively restore salt marsh systems. Times in marsh recovery can depend on the development stage of the marsh; type and extent of the disturbance; geographical location; and the environmental and physiological stress factors to the marsh-associated flora and fauna.

Although much effort has gone into restoring salt marshes worldwide, further research is needed. There are many setbacks and problems associated with marsh restoration that requires careful long-term monitoring. Information on all components of the salt marsh ecosystem should be understood and monitored from sedimentation, nutrient, and tidal influences, to behaviour patterns and tolerances of both flora and fauna species. Once we have a better understanding of these processes and not just locally, but over a global scale, we can then suggest more sound and practical management and restoration efforts that can be used to preserve our valuable marshes and put them back to their original state.

While humans are situated along coastlines, there will always be the possibility of human-induced disturbances despite the number of restoration efforts we plan to implement. Dredging, pipelines for offshore petroleum resources, highway construction, accidental toxic spills or just plain carelessness are examples that will for some time now and into the future be the major influences of salt marsh degradation.

In addition to restoring and managing salt marsh systems based on scientific principles, the opportunity should be taken to educate public audiences of their importance biologically and their purpose as serving as a natural buffer for flood protection. Because salt marshes are often located next to urban areas, they are likely to receive more visitors than remote wetlands. By physically seeing the marsh, people are more likely to take notice and be more aware of the environment around them. An example of public involvement occurred at the Famosa Slough wetland in San Diego, where a 'friends' group worked for over a decade in trying to prevent the area from being developed. Eventually, the 5 hectare site was bought by the City and the group worked together to restore the area. The project involved removing invasive species and replanting with natives, along with public talks to other locals, frequent bird walks and clean-up events.

RESTORATION OF THE MARSH

Restoration of the marsh will restore normal function for all of the functions assessed. In addition, restoration will reverse the trend toward marsh surface subsidence. Research indicates that back-barrier marshes in Maine have the highest sediment accumulation rates of the four types of marshes studied. Four-fifths of the back-barrier marshes they studied were accumulating sediment at a rate that outpaced current sea level rise. Below is a discussion of these assessments for each of the seven functions evaluated.

Fish and Shellfish Habitat

Fish habitat under existing conditions

The present 48 inch culvert (the trunk) provides access for some fish species. However, the small size of this culvert severely restricts the total area of fully functioning salt marsh to an estimated 30 acres. This reduction in total area of salt marsh from its historical extent of nearly 200 acres reduces two critical salt marsh habitats for fish, tidal creeks and rivers and salt marsh pannes. Tidal creek habitat is presently restricted by tidal flow restrictions at Little Boar's Head, Appledore Road and Huckleberry Lane. In addition, there is up to 2.2 feet of sediment in the Little River Channel and the 'Great Ditch'. Several factors have likely contributed to this sedimentation. First, the restrictions to tidal flow greatly reduce the flushing action that would normally occur with each tidal cycle. Second, the origin of much of the sediment is probably 'rotten' peat which has sloughed from the degraded portions of the marsh and tidal creeks.

Roman discussed water quality problems that can affect fish habitat in salt marshes degraded by restrictions to tidal flow. These problems are primarily due to changes in hydrology, which lead to an anaerobic decomposition of marsh peat. This in turn causes marsh subsidence and an increase in organic matter to tidal streams. The decay of additional organic matter in the streams reduces dissolved oxygen levels. The oxidation of salt marsh peat under these conditions may also result in the release of acidic leachate caused by the oxidation of pyrites.

A site visit by Doug Grout of NH Fish and Game, indicated the following about current conditions in the marsh:

1. Winter flounder are unlikely due to the small culvert. They may be more likely with a larger culvert.
2. River herring are not likely in the river due to the small size and long length of the trunk (existing 48 inch culvert).
3. A fish kill of freshwater fish such as bass and catfish occurred during the October 2007 storm as these fish were washed down the river to the North Hampton Beach.
4. The weir across the Little River at the upstream end of the marsh is an impediment to fish passage except during very high tides.
5. There is presently very little cobble bottom for spawning in the river (approximately 150 feet) near the trunk.
6. Salinities on the upstream of the trunk were measured at 13 ppt (low tide) and 30 ppt (high tide).

Diers sampled fish populations June 30, 2004, at three stations on the Little River using a seine. The results were as follows:

Station 1: Located near the trunk - Stickleback, banded killifish, mummichog, sand shrimp and green crab.

Station 2: Located at the intersection of Little River and the 'great ditch'—mummichog, banded killifish, stickleback, American eel, and green crab.

Station 3: Located just below the weir—stickleback, mummichog, banded killifish, sand shrimp, American eel and green crab.

Fish habitat in restored marsh

Channel dredging/excavation will be on one side only except in rare cases where this is impractical. This will protect existing overhanging channel banks that provide shade and contribute to the diversity of fish habitat. Enlarging the opening to a more natural size for a salt marsh of this size would greatly increase the total habitat for anadromous, catadromous and amphidromous fish. One of the most important functions of salt marshes is to provide food chain support for the coastal fisheries in the Gulf of Maine. Much of the nutrients exported to coastal fisheries is in the form of forage fish such as Atlantic silversides (*Menidia menidia*) and mummichogs (*Fundulus heteroclitus*). These amphidromous fish move between salt and brackish water feeding on detritus and invertebrates.

Little indicates that Striped Bass (*Morone saxatilis*) once spawned in the larger tidal rivers in New Hampshire including the Piscataqua. They are also known to have moved in and out of smaller tidal streams around the Gulf of Maine with the tide. Many populations were eliminated in the 17th and 18th centuries by overfishing. Early fishermen placed nets across tidal creeks effectively trapping large portions of a population on the outgoing tide. While it is not known if striped bass once used the Little River, it is certainly possible that they did. This important game fish might utilise a restored marsh at least for feeding.

Another imporant habitat function is to serve as nurseries/refuges for important fish such as winter flounder (*Pseudopleuronectes americanus*) and alewife (*alosa pseudoharengus*).

Specific habitat types that would be increased include:

Tidal Creeks and Rivers: Removal of sediment from the 'great ditch' and Little River Channel will improve this habitat type. The sediment to be removed is primarily degraded peat and other sediments deposited due to the present restriction of tidal flow. The restriction to tidal flow has contributed to sedimentation in two ways. First, much of the peat in the upper portion of the marsh has lost its integrity due to the replacement of spartina by phragmites, narrow leaf cattail and other invasive species. Second, rotten peat and other sediments have accumulated in the channel and 'great ditch' because there is presently inadequate tidal flow to keep the channels flushed. This habitat will also improve because of an increase in cobble beds for spawning.

These beds are formed by the inwash of rocks from the beach. Now these beds are limited about 150 feet upstream of the trunk. Increasing the culvert size will allow greater amounts of bed material to be washed into the Little River within the marsh.

Salt Marsh Pannes: At the present time pannes in most of the marsh, especially above Appledore Road, are in a highly degraded condition due to lack of tidal flow. Peat degradation and the replacement of spartina and other typical salt marsh plants with invasive plants have reduced their function as habitat for salt marsh fish to zero in many places.

Published literature on range and habits of fish indicates that restoration of the marsh could reasonably be expected to increase potential habitat for the fish listed below. This does not mean that any or all, of the listed fish will necessarily use the improved habitat or that the improved habitat is of high quality for all species. Grout for instance, believes that a smelt fishery is possible but not likely because of probable salinity ranges and lack of gravel for spawning. In addition, the actual use of the marsh by a particular species depends on nonhabitat factors including the actual presence or a given species along this section of the coast. The list then should be viewed as simply an indication of the presence of potential habitat in the restored marsh:

1. Rainbow Smelt *Osmerus mordax* (Mitchill)-Anadromous.
2. Alewife *Alosa pseudoharengus* (Wilson)-Anadromous.
3. Blueback Herring *Alosa Aestivalis* (Mitchill)-Anadromous.
4. Winter Flounder *Pseudopleuronectes americanus* (Walbaum)-Amphidromous.
5. Mummichog *Fundulus heroclitus* (Linnaeus)-Amphidromous.

6. Banded Killifish *Fundulus diaphanus* (Lesueur)-Amphidromous.
7. Threespine Stickleback *Gasterosteus aculeatus* (Linnaeus)-Amphidromous.
8. Fourspine Stickleback *Apeltes quadracus* (Mitchill)-Amphidromous.
9. Ninespine Stickleback *Pungitius pungitius* (Linnaeus)-Amphidromous.
10. American Eel *Anguilla rostrata* (Lesueur)-Catadromous.

Wildlife Habitat

Wildlife habitat under existing conditions

Existing wildlife habitat in the marsh is threatened by the invasion of phragmites, purple loosestrife and narrow leaf cattail, plants that generally lower the wildlife habitat value of tidally restricted salt marshes. This phenomenon has been widely observed in salt marshes with restricted tidal flow. These species, especially phragmites and purple loosestrife, tend to create a monoculture with low habitat value. Short learned from talks with knowledgeable residents, that purple loosestrife was first noticed on the marsh in the mid-1960's.

Smith, found a 70 per cent reduction in the area dominated by the salt marsh grass *Spartina patens*. This was most likely related to the greatly reduced soil salinities he found in most of the marsh. Only a small portion of the marsh had soil salinities above 10 parts per thousand (ppt) even at high tide. This was especially true for the portion of the marsh upstream of Appledore Road. Except for a narrow band along the river most of this area has a soil salinity of less than 5 ppt.

Salinity measurements on three replicate transects in areas being invaded by purple loosestrife indicate that this plant is rarely found in areas where soil pore water salinities exceed 6 ppt. The presence of soil salinities of 6 ppt in areas with Spartina patens is itself an indication of marsh degradation. *Spartina patens* tends to grow on the more saline portions of a salt marsh and soil pore salinities in the range of 18 ppt would be more usual. The soil salinity of her transects ranged from 3.3 ppt to 25.3 ppt. *S. patens* grew through this salinity range indicating that the area had once been functioning high marsh. Purple loosestrife, on the other hand, only grew at salinity range between 3.3 and 7.7 ppt.

Additional loss of salt marsh habitat has occurred by the invasion of shrubs and freshwater marsh plants around the edge of the marsh. While these plants have value for wildlife, they are nevertheless encroaching on areas of historic salt marsh, one of the rarest habitats in New Hampshire. The project will not eliminate the shrub habitat and other brackish and fresh habitats associated with the marsh. Rather these habitats will be restricted to the periphery of the marsh where they normally occur under natural conditions.

Wildlife habitat in a restored marsh

Sinicrope found that the reintroduction of tidal flow to a salt marsh generally increased salt marsh plants and decreased fresh and brackish species. Live coverage of *typha angustifolia* declined from 74–16 per cent and surviving stands were stressed. *Spartina alterniflora* expanded from 1– 45 per cent and high marsh species became established covering 20 per cent of the marsh. Live coverage of phragmites increased from 6–17 per cent primarily around the edge of the marsh. It most likely moved into areas from which typha had been eliminated by increased salinity. Nine per cent of phragmites encountered on transects was dead.

Salt marshes have been identified as having high habitat value by the New Hampshire Resources Protection Project. Obviously, increasing the area of sustainable functioning salt marsh associated with the Little River from approximately 30 acres to 150+ acres plus will significantly increase the amount of salt marsh habitat available along the New Hampshire seacoast.

Wildlife that will specifically benefit from the restoration of the marsh include:

1. Shore Birds (e.g. least sandpiper, willet, greater and lesser yellowlegs).
2. Waterfowl (e.g. black duck, mallard).
3. Song Birds (e.g. sharp-tailed sparrow, seaside sparrow).
4. Birds of Prey (e.g. harrier).
5. Wading Birds (e.g. green heron, great blue heron, snowy egret).
6. Small Mammals (e.g. muskrat).
7. Large Mammals (e.g. white-tailed deer).

Flood Flow Alteration

Flooding under present conditions

Under present conditions the restriction caused by the 48 inch culvert impounds storm water in the marsh causing flooding of some local residences. This flooding has been exacerbated by the development around the natural outlet of the marsh near the Hampton-North Hampton town line. The present level of development around the breach as well as the current litigation precludes permanently opening the breach for either tidal flow or outflow of storm water. It is likely that in a future without project, flood damage would continue.

Educational/Scientific Value

Educational potential under present conditions

Presently the marsh is in a seriously degraded condition. Its educational/scientific potential is limited to showing what a degraded salt marsh looks like. This condition is predicted to worsen due to a lag between the restriction of tidal flow and all of the degenerative effects of that restriction. Loss of peat integrity and subsidence are expected to continue. Also expected to continue and perhaps accelerate is the spread of invasive plants. Without restoration, the viable salt marsh will likely shrink to 5–10 acres near the trunk and a narrow fringe of low marsh along the river channel.

Little River has been the site of several scientific investigations by Richardson, Short, Waterman, Dzierzeski. The first study was done before the marsh had deteriorated to its present state. Its author advocates that the marsh be restored even if it means losing some of the existing beds of widgeon grass discussed below. The last two studies were done in the Little River Marsh specifically because tidal flow had been restricted. Their aim was to investigate negative impacts of tidal flow restriction on salt marshes. All of these studies contribute to the baseline understanding of the marsh and will provide needed data to understand the changes that will occur after restoration.

Beds of widgeon grass (*Ruppia maritima*) exist in pannes and the Little River channel upstream of Appledore Road. Richardson indicated that the primary habitat for this plant is areas of still or slow moving water such as deep pools and shallow pools (pannes). Ruppia also occurs in ditches and tidal creeks. Under present conditions most of the high marsh pools and pannes in this area that would support ruppia has been lost to invasive plants. This has greatly restricted potential habitat for this important plant. The Little River channel has become more pool like because of the restrictions to tidal flow and possibly better ruppia habitat than under natural conditions.

Educational/Scientific potential in a restored marsh

Restoration of the marsh will provide an opportunity to document the return to health of the second largest back barrier marsh on the New Hampshire Coast. This is in addition to the obvious educational and scientific value of having a large fully functioning salt marsh under conservation easement.

Some but certainly all or even most beds of ruppia south of Appledore Road may be lost due to the planned dredging of the river channel. These losses are expected due to the enlargement of the Appledore culvert and the resultant lessening of ponding in the river channel. The extent of this loss cannot be predicted precisely. For one thing, the pools in which the ruppia now grows have been in existence for well over 100 years. They existed even when there was an unrestricted opening at the breach. It is likely that most of the ponds visible on the 1877 Coastal Chart will remain even after restoration. In addition, the restoration of high marsh in this area will create opportunities for the development of new ponds and pannes. To minimise negative impacts ruppia beds will be avoided where possible. Sumps will be excavated in some ponds and pannes to prevent complete drying. Sills will be left where possible to retain some intertidal water in existing ruppia ponds. In addition new pannes and pools will be dug in the former high marsh to provide future habitat. Richardson believes that there is a good possibility that such areas will become populated with ruppia. The unavoidable loss of some ruppia beds must be weighed against the greatly increased habitat that will be provided by restoring the high marsh and its associated pannes and pools above Appledore as well as the rest of the marsh.

Production and Export

Production and export under present conditions

Nutrient export from the marsh is limited by the restriction to tidal flow. Nutrients leave a salt marsh as dissolved material, detritus and forage fish. Waterman studied the nutrient fluxes of three New Hampshire salt marshes, Little River, Bass Beach and Wallis Sands. He found that Bass Beach and Little River marshes, which had restricted tidal flow, had decreased exports of dissolved nutrients compared to the unrestricted Wallis Sands marsh. There was also greater removal of NH_4 and PO_4. The increased NH_4 retention in the marsh may be accelerating the growth of purple loosestrife. Dzierzeski found that purple loosestrife responded to increases in nitrogen fertilisation.

Production and export in a restored marsh

Increased tidal flow will probably increase the export of usable nutrients to Coastal fisheries. This increase will be due to the increase of dissolved nutrients, detritus and forage fish.

Uniqueness/Heritage

Uniqueness/heritage under present conditions

The Little River marsh is the second largest back barrier marsh in New Hampshire. It has a long history of human use, including strong local and state concern for its ecological health. Local citizens have known for several hundred years, at least, that the key to keeping the marsh healthy is to maintain tidal flow. Under present conditions, this unique and important marsh has deteriorated and will continue to deteriorate.

Uniqueness/heritage in restored marsh

The overall concept of this project is to reduce the most damaging human stressor to this marsh, namely restriction of tidal flow. This will restore this important native ecosystem. The aim of this project is not to return the marsh to its pristine state that is not realistic. Rather, the goal is to reestablish to the greatest practical degree the physical, chemical and biological conditions that existed before the marsh was stressed by human activity.

Visual Quality/Aesthetics

Visual quality/aesthetics under present conditions

Under present conditions, views of the marsh are marred by stands of phragmites and purple loosestrife. The open grassland look of a functioning salt marsh which contributes so much aesthetic appeal to the seacoast is disappearing.

The south of Appledore Road is almost completely overgrown by invasive plants. Local residents recognise that while purple loosestrife is attractive in bloom, its floral display is a clear warning of a degrading salt marsh. Long time residents have recounted impressions of the marsh in the 1940's when the area south of Appledore was predominantly salt marsh. At that time, residents could boat from a pool at the river mouth up to the weir. Subsequent installation of smaller culverts at Appledore and the breach stopped these scenic trips.

Visual quality/aesthetics in restored marsh

Over a long period, the marsh should recover most of its original vegetation. Areas of subsided high marsh may cycle through a phase of low marsh. The overall effect will be a return to the open vistas, which characterised the marsh in years past.

Thus under existing conditions of severely restricted tidal flow, the marsh has seriously deteriorated and will continue to do so until very little of the original marsh remains. This project represents an opportunity to restore this marsh to a reasonable level of ecological health.

Chapter 16

Biodiversity in Mangrove Ecosystems

INTRODUCTION

Mangroves are various kinds of trees up to medium height and shrubs that grow in saline coastal sediment habitats in the tropics and subtropics — mainly between latitudes 25°N and 25°S. The word is used in at least three senses: (i) most broadly to refer to the habitat and entire plant assemblage or mangal, for which the terms mangrove forest biome, mangrove swamp and mangrove forest are also used, (ii) to refer to all trees and large shrubs in the mangrove swamp and (iii) narrowly to refer to the mangrove family of plants, the Rhizophoraceae or even more specifically just to mangrove trees of the genus *Rhizophora*.

The mangrove biome or mangal, is a distinct saline woodland or shrubland habitat characterised by a depositional coastal environments, where fine sediments (often with high organic content) collect in areas protected from high-energy wave action. Mangroves dominate three quarters of tropical coastlines. The saline conditions tolerated by various mangrove species range from brackish water, through pure seawater (30 to 40 ppt), to water concentrated by evaporation to over twice the salinity of ocean seawater (up to 90 ppt).

FLORAL DIVERSITY: BACTERIA AND FUNGI

Micro-organisms can be defined as life forms that cannot be seen with the unaided eye. This broad definition encompasses an extensive and diverse assemblage of organisms, which exhibit widely different morphological, ecological, physiological characteristics. They are represented by groups like viruses, bacteria, fungi, diatoms, algae, protozoans, etc. Of these the first three groups require not only microscopic examination but also in many cases growth in pure culture is essential for identification and hence, they can be strictly treated as micro-organisms. As viruses are a cellular and obligatory parasites and they do not take part either in biogeochemical cycles or biological interactions usually the bacteria and fungi are considered important microbial groups.

The microbial diversity encompasses vast number of species and the number of species of these groups known to science is only a tip of the iceberg. The total number of known species of Bacteria (including Cyanobacteria and unculturables) in the world is only 4000 whereas, the estimated species number is 30,00,000. A meagre of 0.1 per cent alone has been described. In the case of fungi (including yeasts, lichenforming fungi, slime moulds and oomycetes) the described species were 70,000 but the estimated number is 15,00,000. Only 5 per cent of the species were described. Considerable difficulty arises in the estimation of those, which remain undescribed. This may be because they are unculturable by the conventional methods and lack of suitable methods of culture of all groups or because we have not explored enough.

There are evidences for the occurrences of large numbers of bacterial species, which are 'unculturable' forms. Further, the species number between different micro-organism group is complicated by variations in the concepts of species. The common practice of differentiating 'biological species' may not be applicable to micro-organisms. In microbiology especially in the case of bacteria the recognition of species or strain are based on the morphological, biochemical or molecular similarity.

MARINE MICROBIAL DIVERSITY

Nearly 75 per cent of the earth is covered with seawater. Microbial world in the marine ecosystem is complex and diverse in nature. The study of marine microbiology includes the forms that live from the coastal to offshore regions, from surface of the water to abyssal depths inclusive of trenches, specialised ecosystems like black smokers of hot thermal vents at depths, tropical coral reef ecosystems and specialised eco-systems like estuaries, lagoons, saltpans, backwaters and mangroves. They are nutritionally versatile, as that of terrestrial counterparts ranging from phototrophy to chemolithotrophy and chemoheterotrophy. In the marine environment seven types of habitats are available for marine micro-organisms. They are:

1. The air-water interface.
2. The water column (as free living micro-organisms).
3. The sediment-water interface.
4. The sediments.
5. Inorganic and organic particle surfaces.
6. External and internal surfaces of plants and animals.
7. Extreme environments.

Each of these habitats has complex internal structures. With regard to water, there are major differences between water masses in the photic zone, the upper mixed layer and the deep ocean. The available organic substrates, light, temperature and other variables are not the same with these biotopes. All these variables are important in controlling the microbial numbers, growth rates, and size or biomass and species diversity. In sediment-water interface since complex interactions of nutrients take place it provides different habitats for micro-organisms. In sediments, there are gradients in the concentration of oxygen (aerobic and anaerobic) and organic and inorganic nutrients; these are especially complex in the rhizosphere of seagrasses and mangroves. Suspended inorganic and organic particles (seston) with various size and composition provide a special niche for heterotrophic micro-organisms. Organisms of both plants and animals, (plankton to mammals) their surfaces (as epibiotic) and internal cavities (as endobiotic) provide different types of niches for growth of micro-organisms.

Many environments are generally considered extreme and are characterised by low species diversity. The classic extreme environments are either hot (submarine hot springs, black smokers) or salty (salt pans) and or with extreme pH (highly acidic and alkaline). Another characteristic extreme environment with low temperature (5°) and high pressure (<600 bar pressure) such as the bulk of the ocean contain wide biological diversity. Different micro-organisms including bacteria, fungi and yeast can grow very well. Since varied biotopes are available in the marine environment, a large number of microbial species of bacteria, actinomycetes and fungi are present in the marine environment. Still the known species of marine bacteria and filamentous fungi are numbered in hundred only.

MARINE BACTERIAL DIVERSITY

Definition of marine bacteria is based on their ability to grow only at certain seawater concentrations. Many marine bacteria require sodium, potassium and magnesium ions. Some of them also require chloride ions and ferric ions. Generally obligate marine bacteria are capable of growing exclusively in seawater. All bacteria that live in seawater are not true marine bacteria. Near coastal waters, 95 per cent of the bacterial population is salt tolerant forms; only 5 per cent is true marine forms. In the open ocean and in deep sea the true marine forms dominate.

The oceans contain 10^{22} litres of water and as there are 10^7 to 10^9 bacteria per litre, significant portion the world's bacteria (10^{30}) live in the water column of the oceans. Large proportions of the remaining bacteria live in the ocean sediments, which cover 70 per cent of the world's surface. The ultimate source of energy for them is sunlight, but light is available only at the surface waters, the majority of bacteria in water and sediments would be expected to experience various conditions for different periods. Most oceans are oligotrophic, i.e. the concentration and the rate of supply of available organic matter is generally low.

Bacterial forms are highly versatile in their nutritional requirement. It is very difficult to classify the bacteria and many systems have been adopted since their discovery. A classification system based on phylogenic similarity has been proposed and it has wide acceptance. These grouping split the previously unified prokaryotes into archaebacteria and the eubacteria. The bacteria are further categorised based on their mode of obtaining carbon and energy. They are classified as autotrophs and heterotrophs. Autotrophs, based on the source of energy used are further grouped into photoautotrophs (sunlight) and chemoautotrophs (oxidation of reduced inorganic substances). Chemoheterotrophic bacteria which represent large and heterogenous groups have been further divided according to morphological and physiological characteristics. Since the bacterial forms exhibit high degree of metabolic diversity even within taxonomic groups and some individual bacteria has the ability to utilise variety of nutritional products depending upon environmental conditions. Fenical and Jettsen classified marine bacteria into several groups, which are as given in Table 16.1.

Table 16.1. Groups of Marine bacteria.

Archaebacteria	Autotrophic eubacteria	Chemoheterotrophic eubacteria	Eukaryotes
Chemoautotrophs	Photoautotrophs	Gram-positive	Fungi
Methanogens	Anoxygenic photosynthesis	Endospore forming rods and cocci	Higher fungi
Thermoacidophiles	Purple and green photosynthetic bacteria (order Rhodospirillales)	Non-spore forming rods	Ascomycetes
Chemoheterotrophs	Oxygenic photosynthesis	Non-spore forming cocci (family Micrococceae)	Deuteromycetes
Halophiles	Cyanobacteria (order Cyanobacteriales)	Actinomycetes and related organisms	Basidiomycetes
	Prochlorophytes (order Prochlorales)		Lower fungi (class Phycomycetes)

(Contd...)

Archaebacteria	Autotrophic eubacteria	Chemohcterotrophic eubacteria	Eukaryotes
	Chemoautotrophs Nitrifying bacteria (family Nitrobacteriaceae)	Rods and Cocci Aerobic (family Pseudomonadaceae)	
	Colourless sulphur oxidising bacteria	Facultative (family Vibrionaceae)	
	Methane oxidising bacteria (family Methylococcaceae)	Anaerobic (sulphur reducing bacteria)	
		Gliding bacteria (order Cytophagales and Beggiatoales) Spirochaetes (order Spirochaetales) Spiral and curved bacteria (family Spirillaceae) Budding, and/or appendaged bacteria Mycoplasma (class Mollicutes)	

ARCHAEBACTERIA

The archaebacteria are a heterogenous group that includes both autotrophs and heterotrophs. They inhabit extreme environments. They differ from eubacteria in not having peptidoglycan as a cell wall structural polymer. There are three highly distinct groups the halophiles (salt lovers), the thermoacidophiles (temperature and acid loving), and the methanogens.

The halophilic archaebacteria are chemoheterotrophic but can also utilise sunlight, and they require at least 12–15 per cent NaCl to survive and grow well at concentrations upto saturation. They occur in saltpans and salt lakes and can easily be cultured on high salt media and their taxonomy and physiology are also described. The thermoacidic archaebacteria are able to grow at a low pH and at high temperatures. Some forms can live upto 90°C temperature and at a pH of less than 1, and these habitats are considered as the most extreme environments. The members of the genus *Sulfolobus* a choemoautotroph, oxidise sulphur compounds and can produce sulphuric acid, and are responsible for acidity of the habitat in which they live.

The methanogens are strict anaerobes and have the ability to produce methane by reducing simple organic materials. They are abundant in marine habitats, including hypersaline environment. The methanogen and other archaebacteria can produce heat stable lipids and enzymes to preserve membrane integrity and cell function and so they survive in extreme environments.

EUBACTERIA

Most prokaryotes, including the cyanobacteria are eubacteria. They are unicellular but some are filamentous and are diverse and versatile. Depending on environmental conditions they can utilise more than one nutritional type. The systematics of these bacteria are dealt in Bergey's manual of systematic Bacteriology.

The eubacteria are divided into three groups as photo and chemoautotrophic eubacteria. The photoautotrophic group is again divided based on the type of photosynthesis they perform. The forms that do not liberate oxygen during photosynthesis are represented by purple and green sulphur bacteria of the order Rhodospirillales. These forms possess bacteriochlorophylls which structurally differ from chlorophylls of cyanobacteria and so they can photosynthesis only under anaerobic conditions. Some bacterial forms can perform anoxygenic photosynthesis in the presence of oxygen; they are called facultative phototrophs. They are common in aerobic marine habitat and they constitute upto 6.3 per cent of bacterial population. These are include the marine genus *Erythrobacter* and certain methylotrophic species. The bacterial forms that perform oxygenic photosynthesis are represented by cyanobacteria (formerly known as blue green algae) and the prochlorales. The cyanobacteria are a diverse group represented by more than 1000 species of filamentous and unicellular phototrophs. The classification of this group is problematic, previously they have been placed under Cyanophyta and now have been described using the International Code of Nomenclature of Bacteria, under the order Cyanobacteriales. These forms are common in the marine habitat and occur in water and sediment layers where light is available. Also they occur as symbionts in the surface tissues of some sponges. Some cyanobacteria are also capable of performing anoxygenic photosynthesis. The prochlorales only one marine genus has been described *Prochloron*. It occurs as a symbiont in acidians and reported to be associated with other invertebrates. These bacteria have never been successfully cultured.

CHEMOAUTOTROPHIC EUBACTERIA

These bacteria are Gram-negative forms occur widely in marine environment and play an important role in the recycling of elements. There are three major groups and they oxidise different substrates. These groups are represented by the family Nitrobacteriaceae (Nitrifying bacteria), sulphur-oxidising bacteria, and family methylococcaceae (methane-oxidising bacteria).

The nitrifying bacteria include the genera *Nitrosococcus* (oxidise ammonia to nitrite) and *Nitrobacter* (nitrite to nitrate). They play a key role in nitrogen cycle. The sulphur oxidising bacteria has a diverse group including thermophilic archaebacteria e.g. *Sulfolobus* and certain gliding bacteria. A small unicellular species within Nitrobacteriaceae, such as *Thiobacillus* also can oxidise sulphur compounds. The sulphur-oxidising bacteria of the order Cytophagales show gliding motility. They occur abundantly in sulphide rich marine habitats, where they form mats. Besides, species of *Achromatium* a chemoautotroph grow by oxidation of reduced sulphur compounds. Also from free living forms, symbiotic sulphur oxidising bacteria have been found to be associated with hydrothermal vent-associated invertebrates; notably vestimentiferan worms.

The methanophiles are a diverse group, aerobic and found in upper sediment layers and utilise biologically produced methane rising from deeper anaerobic sediment.

CHEMOHETEROTROPHIC EUBACTERIA

These groups are extensively studied since they can easily be cultured. They are large and diverse group and have been subdivided based on differences in cell wall structure (Gram-positive V_s Gram-negative), morphology and affinity for oxygen.

Gram-positive Bacteria

Most of the sea bacteria belong to Gram-negative. Gram-positive bacteria are less than 10 per cent of the total bacterial population. Now evidences indicate that Gram-positive bacteria do occur at higher percentage in sediments. The actinomycetes (Order Actinomycetales) and related diverse group, are Gram-positive filamentous forms. The other Gram-positive genera like Arthrobacter and endospore producing forms *Bacillus* and *Clostridium* (family Bacillaceae) have also been isolated. Especially *Bacillus* species readily grow in medium containing nutrients. Besides spore formers, nonspore forming cocci (family Micrococceae) and rods are found in marine environs.

Gram-negative Bacteria

The Gram-negatives represent the largest and diverse group of marine chemoheterotrophic prokaryotes. These forms can be grouped on morphological characters as rods and cocci and their affinity for oxygen. Majority of bacteria belong to the families Pseudomonadaceae and Vibrionaceae. They can easily be isolated and cultured. They are rarely distinct morphologically but are biochemically and ecologically diverse and the genera include like *Photobacterium* and *Vibrio*. The bacteria in Pseudomonadaceae are collectively called as *Pseudomonas* and they belong to genus *Pseudomonas* but also include other genera like *Xanthomonas* and *Altetromonas*. Most marine bacteria are aerobic or facultatively anaerobic because large parts of the ocean are well oxygenated. Obligatory anaerobic forms do occur in marine environment. One such genus is *Desulfovibrio*, a sulphur-reducing group.

Many Gram-negative heterotrophic bacteria having distinct morphological features are usually associated with surfaces and not found in open ocean waters. The gliding bacteria in the orders of Cytophagales and Beggiatoales are filamentous in nature. Most of them are chemoheterotrophs. They frequently found attached to the surfaces and occur as mats. These orders include the aerobic genus *Leucothrix* and *Beggiatoa*, The *Myxobacteria* (order Myxococcales) are motile gliding forms and their occurrence in the marine environment is not well documented.

Spirochaetes belonging to the order Spirochaetales, are large and coiled bacteria occur as free living form in marine environment and as symbionts in crystalline style of certain mollusks. These Spirochaetes belonging to the genus *Cristispira*, has never been successfully cultured.

The spiral and curved bacteria of the Spirillaceae are Gram-negative rods can be distinguished from spirochaetes in that they move using flagella. They are common in marine environment. There also include the genus *Bdellovibrio*, a parasitic bacteria parasitise other bacterial forms.

In addition to Gram-negative bacteria, there is a group called the mollicutes the smallest forms characterised by lack of defined cell wall (formerly called as mycoplasmas) occur as parasites in invertebrates.

BACTERIAL LIFE IN EXTREME ENVIRONMENTS

Microbial life in extreme environments has been dealt in detail by Kushner. Many environments in the marine ecosystems are considered as extreme. In the extreme environments (high temperature, extreme pH and salinity) only a few types of organisms grow. So these environments are characterised by low species diversity. But in some extreme environments like cold, high-pressure areas such as the bulk of the oceans it is not true.

At high temperature environments like submarine hot springs ($>100°C$) and hydrothermal vents the 'black smokers' ($>300°C$) certain groups of bacteria grow very well. Some cyanobacteria and anoxygenic phototrophic bacteria can grow at $70°C$. 'Extreme thermophiles' with temperature optima of $>80°C$ are also reported. All are archaebacteria, e.g. *Thermoplasma* ($65°C$), *Methanobacterium* ($75°C$), *Methanothermus* ($97°C$), *Sulfolobus* ($87°C$), *Acidianus* ($95°C$), *Pyrococcus* ($103°C$) and *Pyrodictium* ($110°C$). The eubacterim *Thermotoga maritima* ($90°C$) is a Gram-negative thermophilic form.

Salt loving bacteria are called halophiles. Some of the extremely halophilic archaebacteria (species of *halobacterium* and *Halococcus*) were isolated from salted foods as contaminants of solar salt prepared from seawater. The other halophilic archaebacteria include *Haloferax*, *Nitrobacterium*, *Natronococcus* were isolated from salt lakes. Anaerobic archaebacteria methanogens are also found in hypersaline environments. *Halomethanococcs doii* is an extremely halophilic form.

Many acidic and alkaline environments are also well populated by micro-organisms. The acidophilic bacteria that include the species of *Acetobacter* and *Thiobacilli* can live in high acidic condition. The alkalophilic bacteria include several *Bacillus* species, salt tolerant and phototrophic bacteria of the genus *Ectothiorhodospira*, archaebacteria, such as *Natronobacterium* and *Natronococcus* species.

Living micro-organisms are found in the greatest depths of the sea, at over 1000 bar pressure. This revealed the presence of truly barophilic bacteria, strongly adapted to life under high pressure. Several hundred strains of the bacteria have been isolated, often isolated with animals or animal remains from the same depths.

Thus the above information indicates the existence of microbial life in quite unlikely conditions. Some micro-organisms can withstand more than one extreme environment. The study of different 'extremophiles' has indicated the possibilities of life on other planets also. Thus marine prokaryotes represent a large and heterogenous group of micro-organisms. They show tremendous diversity.

MARINE FUNGAL DIVERSITY

The distribution of fungi in the marine environment has not been studied well when compared with the studies on fungi in freshwater and terrestrial ecosystems. They are poorly represented in the sea, since the marine fungi account for only 5 per cent of the total fungal flora. Although fungi occur widely in marine environment on dead organic matter and as parasites of living organisms still their distribution has not fully evaluated.

True marine fungi are defined on the basis of their ability to grow and sporulate exclusively in seawater, whereas facultative marine fungi are forms from land, which are capable of growing in seawater. The first facultative marine fungus described was *Phaeosphaeria typharum* by and the first obligate marine fungus isolated from rhizome of a seagrass was *Posidonia oceanica*. Though the marine environment is more stable with regard to environmental variables, still the open ocean is more or less a 'fungal desert'. There is no proven evidence to indicate the presence of mycelial fungi growing planktonically in the sea. However, the unicellular fungi, yeasts are common in the water column but much lower in number than bacteria. They may be found attached to plants and pelagic animals. Fungi can be expected to survive better along the shore where organic substrates such as algae, marsh plants and mangrove plant litter are available in plenty for their growth.

The higher marine fungi occur as parasites on plants and animals and as symbionts in marine lichens and algae. They also live as saprophytes on dead organic materials of plants and animals origin.

The marine biflagellate fungi the thraustochytrids are common in the marine environs and can be isolated from variety of substrates.

Though extensive works have been carried out on fungi in wood submerged in the seas, dead tissues of seagrasses, mangrove leaves and seedlings still their distribution has not been adequately studied in marine eco-systems. A wide variety of fungi might be present, since niches suitable for the growth of different groups of fungi are plenty in the marine environment.

BACTERIAL FLORA IN MANGROVE ECOSYSTEMS

Mangrove ecosystems are rich in bacterial flora. Fertility of the mangrove waters results from the microbial decomposition of organic matter and recycling of nutrients. Among the microbes, the bacterial population in mangroves are many-fold greater than the fungi. The bacteria exist as symbionts with plants and animals, saprophytes on dead organic matter, and as parasites on living organisms. The bacteria performed varied activities in the mangrove ecosystems like photosynthesis, nitrogen fixation, methanogenesis, magnetic behaviour, human pathogens, production of antibiotics and enzymes (arylsulphatase, L-glutaminase, chitinase, L-asparaginase, cellulase, protease, phosphatase), etc. Most of the bacteriological studies in the mangrove ecosystems are quantitative and a serious effort is necessary to identify the bacteria. In this regard, isolation and identification of bacteria are detailed here.

SAMPLING METHODS

Water: Surface water samples can be hand collected directly from near the shore or from a boat using sterile bottle. For collecting water the bottle should be held near its base and plung it neck downward below the surface. The bottle should be turned until the neck points slightly upward, the mouth being directed toward the current. For collecting water from depth Zobell's microbiological water samplers must be used.

Sediment: Sediment samples near the shore can be collected using a sterile spatula and kept in sterile petri dishes. For collecting bottom sample at depth, Peterson's grab can be used. The central portion of the sediment collected by the grab should be removed by a sterile spatula and can be used.

Isolation Method

The samples collected for isolating marine bacteria should be examined as soon as possible after collection. If samples are to be transported from a long distance, samples must be maintained at 4° to 10°C using cold packs and then transported within three hours and the plating should be completed immediately after the arrival at the laboratory. In general, plating techniques are followed, using surface plating with aquatic samples and/or sediment samples. Water samples are filtered through sterile Whatman Number 1 filter paper to remove unwanted debris. The filterate is used directly as inoculum after serial dilution or filtered through the sterile membrane filter paper (0.22 µm) and the membrane filter paper is used as inoculum source for the isolation of aerobic heterotrophic bacteria. The serially diluted inoculum of 0.1 ml will be spread uniformly with sterile glass spreader on the Nutrient Agar (NA) Medium or the membrane filter is kept on the surface of NA medium.

These materials will be incubated at room temperature and observed from 1–5 days. Similarly the sediment samples are serially diluted in 50 per cent seawater and 0.1 ml of the dilutant will be used for inoculation. Colonies will be counted and total viable count will be made as no. of colonies appeared on the plates, multiplied by dilution factor and is expressed per ml of water sample or per gm dry weight of sediment sample. For isolation of specific groups of bacteria with respective selective medium and following the above mentioned procedure, the bacteria will be isolated. For culture of anaerobic bacteria, plates prepared as above will be incubated in an anaerobic chamber and then observed. For identification purpose, the colonies will be subcultured and checked for purity by streaking method.

Identification Method

For identification, a series of biochemical, physiological and serological tests have to be performed or using prescribed test kits, the identification will be confirmed.

Antibiotic assay: This can be used to identify strains within a given species, which is especially important in epidemiological studies. Among the variety of tests that are available, the disc method is probably the simplest to perform and interpret. Discs of filter paper impregnated with antibiotic solutions, are placed on an agar plate heavily and uniformly inoculated with an actively growing culture of the organism. The test organism grow in a lawn of confluent growths on the plate except for a clear zone around the antibiotic to which the organism is sensitive. Bacteria showing resistance to a given antibiotic show no such inhibition — they grow up to the very edge of the disc.

IMViC (Indole, Methylred, Voges Proskauer, Citrate test)

Indole test: The test requires additions to the growing culture a mixture containing five drops of 5 per cent solution of paradimethyl amino benzeldehyde in isoamyl alcohol to which HCl is slowly added. A deep red colour indicates indole.

Methyl red test: Inoculate a pure culture into 10 ml buffered glucose broth; Incubate for 5 days at 35°C; To 5 ml of five days culture, add 5 drops of methyl red indicator; A distinct red colour is positive and distinct yellow, negative. Orange colour is dubious may indicate mixed culture and should be repeated.

Voges-proskauer test: Use a pure culture to inoculate 10 ml of buffered glucose broth or 5 ml of salt peptone glucose broth or use the previously inoculated buffered glucose broth from the methyl–red test; Inoculate the inoculated salt peptone glucose broth or the buffered glucose broth at 35 ± 0.5°C for 48 hours; Add 0.6 ml naphthol solution and 0.2 ml KOH solution to 1 ml of the 48 hours salt peptone or buffered glucose broth culture in a separate clean test tube; Shake vigorously for 10 seconds and allow the mixture to stand for 2–4 hours; A pink to crimson colour is a positive test; Do not read after 4 hours. A negative test may develop a copper or faint brown colour.

Citrate test: Lightly inoculate a pure culture into a tube of Simmon's citrate agar, using needle to stab, then streak the medium; Be careful not to carry over any nutrient material; Inoculate 35°C for 48 hours; Examine agar tube for growth and colour change; A distinct Prussian blue colour in the presence of growth indicates a positive test; no colour change is negative test.

H_2S production: Inoculate pure culture by stabbing and streaking the triple sugar iron (TSI) slant; incubate at 35°C for 18–24 hours in a incubator. Read and record reactions; Black colour develop in the medium is positive for H_2S production.

Cytochrome oxidase test: Remove part of the culture from 18–24 hours old nutrient agar slants and rub on the surface of filter paper impregnated with a 1 per cent aqueous solution of NNN'N' Tetramethyl–para–phenylenediamine dihydrochloride; A purple colour with 15–30 seconds constitutes a positive oxidase test and colourless constitutes a negative oxidase test.

ONPG test (O-nitrophenyl-B-D-galacto-pyranoside): Emulsify a large loopful of growth from each culture in 0.25 ml of physiological saline in a 10 × 75 mm fermentation tube; Add one drop of toluence to each tube and shake well; Let tubes stand for 5 minutes in 35°C water bath; Add 0.25 ml of buffered 0.75 m ONPG solution to each tube and incubate again in 35°C water bath; Read tubes at ½, 1 and 24 hours. A positive result is development of yellow colour. Lactose fermenters have β-galactosidase.

Motility of bacteria: Bacterial motility or the lack of it is generally ascertained by microscopic examination of the culture in hanging drop or wet mount; or by the use of motility agar media. Motility of bacteria can be studied by 2 methods. *viz.*, Hanging drop technique and stabbing technique or motility agar method.

Hanging drop slide method: Transfer small drop of 18–24 hrs old nutrient broth culture on the centre of a clean cover glass with the help of a sterile inoculation needle. Apply vasline on the edge of the cover glass. Place a cavity slide with the cavity above the culture and press the slide with the cover slip and then invert the slide up so as the drop of culture remains suspended from the cover glass upon examination of the slide under a microscope the unidirectional movement of the motile bacteria can be seen clearly.

Motility agar method: The motility agar medium or semisolid medium consists of the following ingredients Peptone 3 g; Yeast extract 2 g; Glucose 20 g; NaCl 5 g; Distilled water 1000 ml; pH 7.5 ± 0.2; Agar 3 g. Using a sterile inoculation needle remove a small amount of 18–24 hrs old broth culture and stab the motility agar medium prepared in a test tube, in a straight line. After incubation period (24 hrs) examine the tubes to verify whether culture has grown only along the line of inoculation (non-motile) or spread throughout the medium (motile forms).

Gram staining: The gram staining is probably the most widely used in bacteriology and is valuable because it enables to differentiate between two bacterial cultures which are morphologically indistinguishable yet of different species. The stain divides bacteria into two large groups. Those that retain the primary stain crystal violet and iodine complex (CVI) throughout the staining procedure, termed Gram-positive and those that lose the CVI complex upon washing with Gram's alcohol but are stained the colour of counter stain (Safranin) termed Gram-negative. The staining procedure can be performed on a bacterial smear prepared on a glass slide. The smear can be prepared by spreading 18–24 hrs old bacterial culture on a clean glass slide and the smear fixed to the slide by gently warming the slide using a spirit lamp. Steps involved in the Gram stain procedure is given in the following Table 16.2.

Table 16.2. Steps involved in the Gram stain procedure.

		Results	
Steps	*Procedure*	*Gram +*	*Gram –*
Primary stain	Crystal violet upon the smear for 40–60 seconds	Stains purple	Stains purple
Mordant	Iodine for 30–40 seconds	Remains purple	Remains purple
Decolourisation	95% ethanol for 10–20 seconds or 2 to 3 dips	Remains purple	Becomes colourless
Counter stain	Safranin for 30 seconds	Remains purple	Stains pink

Fermentation of carbohydrates (Acid and gas production): It is not uncommon for two different bacterial cultures to be very similar in their morphologic and cultural characteristics, yet show striking differences in their ability to utilised various carbohydrates and in the end products produced from carbohydrate metabolism (fermentation). For example one culture may procedure only a single acid (homofermentation), while another may produce different acids or alcohols (heterofermentation or mixed acid fermentation) with or without production of gases such as CO_2, CH_4, H_2S and H_2. This fermentation technique serves as one of the methods of identification of bacteria.

The fermentation procedure can be performed using oxidation and fermentation. The basal medium can be prepared by using the following substrates. Substrate (Glucose, lactose, sucrose) 7 g; Beef extract 1 g; Peptone 10 g; NaCl 0.5 g; Phenol red 0.018 g; Distilled water 1000 ml; pH 7.0.

The test can be performed and the medium prepared in a test tube and containing a small vial (Durham tube) in an inverted position and sterilised. To this tube, 2 or 3 loopful of 18–24 hrs old broth culture are added and incubated for 24 to 48 hrs at 30°C and are observed for the colour change in the medium (acid production) and gas collected in the inverted Durham vial (gas production) and the results can be confirmed.

Flagellar staining (Leifson's Method): Of the three types of motility seen in bacteria (gliding, rotary and flagellar), the latter is most common and best understood. Bacterial flagella may be as long as 70 μm, but on the average it is in the 10–20 μm range. They are only 20 nm in diameter, which places them outside the visibility range of the bright field microscope. Visual studies of flagella prior to the advent of the electron microscope necessitated the use of flagellar stains. However, since more reliable information is attainable with electron microscopy and since flageller staining is not an easy technique, this procedure is not as frequently practiced today. Where electron microscopy is not available, however, Leifson's method works quite well. To make flagella visible, it is necessary to increase flagellar diameters by precipitating a coating of dye over their entire length. The method accomplishes this by using a single staining reagent that utilises pararosaniline as a staining agent and tannic acid as a mordant. Staining takes place within 5 to 15 minutes.

IDENTIFICATION USING DIAGNOSTIC KITS

For this purpose, a series of biochemical, physiological and serological tests have to be performed or using prescribed test kits, the culture identification will be confirmed.

IDENTIFICATION USING MICROBIAL IDENTIFICATION SYSTEM (MIS)

Micro-organisms can be identified with the help of MIS. It relies on qualitative and quantitative analysis of the fatty acid composition of organisms. The whole cell fatty acid extracted from the cells cultured for a controlled time and temperature and in selective medium are analysed in MIS. The fatty acid analysed by the MIS are compared with the fatty acid profile available in its library and thus the cultures are identified. For each library a standard medium and a set of environmental conditions are to be followed.

FUNGAL FLORA IN MANGROVE ECOSYSTEMS

The marine fungi play a significant role in degradation of litter and alive woody species of mangrove ecosystems. Since Barghoorn and Linder and till now, marine fungi have been extensively studied, however most attention has been devoted to the wood inhabiting fungi. These lignicolous fungi constitute over 50 per cent of 450 species of obligate marine fungi described so far. About 150 species are found exclusively on decaying mangrove wood, aerial roots and seedlings, and are categorised as 'Manglicolous fungi'. The easy availability of wood as a bait or as a naturally occurring substrates for marine fungi in the mangrove ecosystems is the reason for proliferation of such studies. However, studies on marine fungi in India are meagre, so far 72 species have been described. A key is provided here to identify the manglicolous fungi.

COLLECTION TECHNIQUES

The substrates have to be collected and packed in sterile polythene bags. They are directly examined under dissection microscope for the presence of ascocarps, basidiocarps, pycnidia or conidia. Such fruit bodies are transferred with a needle to a microscopic slide, torn apart in a drop of water to expose the spores and carefully squeezed under a cover glass. For identification of unknown species, the observation of ascospores, basidospores or condia usually suffices. In some cases, it is necessary to search for asci and sterile elements of the ascocarps, such as paraphyses and pseudoparaphyses.

Chapter 17

Marine Biodiversity: Threats and Conservation Needs

INTRODUCTION

Marine biodiversity is higher in benthic rather than pelagic systems, and in coasts rather than the open ocean since there is a greater range of habitats near the coast. The highest species diversity occurs in the Indonesian archipelago and decreases radially from there. The terrestrial pattern of increasing diversity from poles to tropics occurs from the Arctic to the tropics but does not seem to occur in the southern hemisphere where diversity is high at high latitudes. Losses of marine diversity are highest in coastal areas largely as a result of conflicting uses of coastal habitats. The best way to conserve marine diversity is to conserve habitat and landscape diversity in the coastal area. Marine protected areas are only a part of the conservation strategy needed. It is suggested that a framework for coastal conservation is integrated coastal area management where one of the primary goals is sustainable use of coastal biodiversity.

Although there are a number of general reviews of biodiversity, such as the global biodiversity Assessment and Huston's more theoretical approach, there is no concise synthesis of marine biodiversity in relation to conservation needs. Short general reviews cover coastal-zone biodiversity patterns, deep-sea benthic diversity, marine benthic biodiversity research, marine functional diversity, coral reefs, foraminifera, fish diversity in the Caribbean and whale and dolphin diversity.

Angel reviews possible causes for the patterns of the pelagic biodiversity in the ocean and Suchanek temperate coastal marine biodiversity showing that temperate systems are among the most productive and diverse. Coral reefs, with their associated flora and fauna, although highly diverse, are still relatively poorly described and their functioning is not well understood. However, not all coral reefs are highly diverse, inshore shallow habitats on the Pacific rim have physically tolerant species to elevated temperatures and surface irradiance and are threatened by exploitation, dredging and removal. Such low diversity areas are also in need of conservation. Rao has reviewed the threats to mangroves and states the objectives for their conservation as: maintenance of genetic resources, sustainable utilisation and conservation or recreation of suitable habitats.

BIODIVERSITY

Biological diversity is defined as: 'The variability among living organisms from all sources including, *inter alia*, terrestrial, marine and other aquatic ecosystems and the ecological complexes of which they are a part; this includes diversity within species and of ecosystems.'

Biological diversity is often written in shorthand as 'biodiversity', and here the two terms are taken to be synonymous.

Genetic Diversity

The most basic level of biological diversity is that found within a species and is known as genetic diversity. Genetic diversity encompasses the variation among individuals within a population in their genetic make-up and the genetic variation among populations. Each species consists of one or more populations of individuals. A population is usually defined as a group of individuals that can interbreed and, if sexually reproducing, can interchange genetic material. Different populations tend to diverge genetically due to their having limited genetic mixing or mutations, natural selection, genetic drift and the accumulation of selectively neutral mutations. Thus, there are genetic differences both among individuals and among populations. Populations with higher genetic diversity are more likely to have some individuals that can withstand environmental change and thereby pass on their genes to the next generation. On an evolutionary time scale, (over many generations) genetic diversity is higher in species which characterise unstable, stressed environments when compared with counterparts from more stable environments. However, on an ecological time scale (few generations), stress reduces genetic diversity. Gillespie and Guttman showed that long-term exposure to contaminants decreased genetic diversity and the remaining population was more vulnerable to extinction. Alberte has shown that stressed eelgrass has lower genetic diversity than non-stressed populations. Commercial fishing, concentrating on specific size ranges, has significantly altered of the genetic composition of populations.

In general, marine species have higher genetic diversity than freshwater and terrestrial species. In a comparative study of fish Ward showed that average heterozygosity was similar in marine and freshwater species subpopulations, but was considerably less in freshwater species. High genetic diversity is found in marine algae, and *Pinctada margaritifera* an exploited tropical bivalve. Elliott and Ward found that a minimum of only 200 migrants per year were enough to maintain the genetic diversity of the Orange Roughy (*Hoplostethus atlanticus*) which suggests that marine populations probably exchange between 10 and 100 times more migrants per generation than freshwater species. Not all marine populations have high numbers and Scudder argues that for marginal populations the best way to maintain genetic (and species) diversity is by 'marginal habitat conservation'. This is an alternative strategy to the conservation of high biodiversity 'hot spots' advocated by some.

Much work has been done on the genetics of species used in aquaculture: on clams; Manila clam; oysters; penaid shrimps; salmonids; and the Orange Roughy. Doyle have reviewed genetic aspects of aquaculture and conclude that current practices lead to reductions in genetic diversity and maintenance of many breeds and meta-populations of marine species is needed.

Grassle argues that a considerable proportion of the genetic diversity of the planet is probably found in deep-sea organisms and recommends genetic studies of hydrothermal vent fauna which are naturally tolerant of high concentrations of toxic elements produced by the vents.

Species Diversity

The most common usage of diversity is the number of species found in a given area, species diversity. Most ecologists would regard a community comprising of 50 individuals of species A and 50 of species B as more diverse than a community comprising 99 individuals of species A and 1 individual of B. Thus, in addition to the number of species in a given area diversity indices have been proposed that take into account the distribution of individuals among species.

The number of species currently described on Earth is between 1.4 and 1.7 million, but the Global Diversity Assessment suggests a conservative estimate of 1.75 million. However, this figure does not include microbial species. Little is known about microbial diversity in general. New genetic techniques will change this. For example, Giovannoni using ribosomal RNA techniques found a completely novel group of bacteria in the Sargasso Sea. On land there are more species known than in the sea. This is due largely to the extraordinary diversity of beetles (*Coleoptera*); 4,00,000 species are described. Recently, in a highly controversial paper, Grassle and Maciolek have suggested that there may be 10 million undescribed species in the deep-sea. Briggs and May disagree with the methods used and May suggests that a more realistic estimate may be around 5,00,000 undescribed deep-sea species. Nevertheless, even this lower figure would be a substantial increase in the approximate figure of 3,00,000 known marine species.

Over geological time there has been a large change in the ratios of orders of families to genera to species. A rapid increase occurred in higher taxa (orders and families) until the Ordivician when diversity levelled off. In the Permian, some 50 per cent of marine families became extinct. The number of species has increased enormously in recent geological time more than doubling compared to those present 100 million years ago.

Most of marine species diversity is benthic rather than pelagic. This is a consequence of the fact that the marine fauna originated in benthic sediments. The pelagic realm has an enormous volume compared with the inhabitable part of the benthic realm. Yet there are only 3500–4500 species of phytoplankton compared with the 2,50,000 species of flowering plants on land. Angel estimates that there are probably only 1200 oceanic fish species against 13,000 coastal species. In the pelagic realm, diversity is higher in coastal rather than oceanic areas and therefore, efforts should be concentrated in coastal areas.

Another highly important aspect of species diversity is endemism (that is the species occurring in a restricted locality). The Antarctic has a higher degree of endemism than the Arctic. In the Red Sea 90 per cent of some groups of fishes are endemic. Overall however, only 17 per cent of Red Sea fishes are endemic. In a survey of 799 pan-tropical fish species Roberts showed that 17 per cent occupied only one grid square (223 × 223 km). In the Indian Ocean, of the 482 coral species recorded, 27 per cent occur only at one site and of the 1200 species of echinoderms found at 16 sites 47 per cent occurred at only one site. High degrees of endemicity pose particularly severe problems for development of conservation strategies. Questions that need to be raised are: are all species equally important for conservation purposes? Do some endemic species play more significant roles than others in the structuring or functioning of the habitat concerned?

The urgency of the need for assessments of species diversity has led to the development of a number of new 'rapid assessment' techniques. Non-specialists have been trained in a few days to sort samples into taxonomic groups with a high degree of precision. While the identification of the actual species must be done later by specialists, these techniques allow rapid assessment of the species diversity of areas that have not been fully studied. These methods need to be further tested in tropical marine areas but show great promise.

Phyletic Diversity

In the marine domain there are more animal phyla than on land. Thirty-five phyla are marine and of these 14 are endemic whereas only 14 occur in freshwater, where none are endemic, 11 are terrestrial, with one phylum being endemic and 15 phyla are symbiotic with four being endemic. This figure includes the newly described phylum Cycliophora found in the gills of the Norway lobster. Thus phyletic diversity is highest in the sea. Of the 35 marine phyla only 11 are represented in the pelagic realm; most phyla occur in the benthos, which is the archetypal habitat. Despite the fact that there are some rare phyla containing only a few species, it is extremely unlikely that present environmental threats will lead to reduction of phyletic diversity.

Functional Diversity

Functional diversity is the range of functions that are performed by organisms in a system. The species within a habitat or community can be divided into different functional types such as feeding guilds or plant growth forms or into functionally similar taxa such as suspension feeders or deposit feeders. Functionally similar species may be from quite different taxonomic entities.

One of the major current topics of debate is that of functional redundancy where it is suggested that there are more species present in communities than are needed for efficient biogeochemical and trophic functions. Recent data, however, show that this is not the case and the higher the number of species in a community the greater the efficiency of biogeochemical processes. Such experiments, however, have not been done in the marine environment.

Steele defines functional diversity in a different and idiosyncratic way as 'the variety of different responses to environmental change, especially the diverse space and time scales with which organisms react to each other and to the environment'. Steele's main point is that marine organisms are closely linked to physical processes at decadal scales whereas on land undisturbed systems change at scales of centuries to millennia.

Community and Ecosystem Diversity

Biodiversity can also be considered at levels other than that of taxonomic organisation, for example at the level of the community and/or ecosystem. In fact, when biodiversity is measured quantitatively it is usually as the number of species or the value of a diversity index for a given community or area of habitat.

A great ecological debate started in the 1930s on whether or not species occurred in distinct groups which could be classified as communities. Today the generally accepted view is that species are distributed along environmental gradients in approximately lognormal abundance patterns. However, interactions between species (predator–prey, commensal, symbiotic and competitive) lead to there being co-occurring groups of species under given environmental conditions. Thus communities are convenient groupings of species which merge gradually into other groupings unless there are sharp boundaries in environmental conditions. Recently, another term has found favour, assemblage, which is a more neutral term and does not imply the tight inter-species organisation that is implied in the term 'community', with its anthropomorphic connotations. The diversity of a community (or assemblage) is often measured.

In the biodiversity convention an ecosystem is defined as 'A dynamic complex of plant, animal and micro-organism communities and their non-living environment interacting as a functional unit'.

Terms such as 'estuarine ecosystem' or 'coral reef ecosystem' are used commonly. Yet the boundaries of such systems are loosely defined and are especially difficult to demarcate in the sea since the fluxes of energy and material within and exported from a system are rarely known. It is perhaps significant that in the Research Agenda for Biodiversity no mention is made of ecosystem diversity. Huston in his book uses the terms 'community' and 'ecosystem' interchangeably, and a recent textbook on ecology states that 'Traditionally 'the ecosystem' comprises the biological community together with its physical environment. However, while the distinction between community and ecosystem may be helpful, in some ways the implication that communities and ecosystems can be studied as separate entities is wrong. No ecological system, whether individual, population or community, can be studied in isolation from the environment in which it exists. Thus we will not distinguish a separate ecosystem level of organisation.' Also 'knowledge of the role that communities play in biogeochemical cycling is essential if we are to understand and combat the effects of acid rain or increasing levels of atmospheric carbon dioxide or radioactivity'.

Habitat Diversity

The most frequently used quantitative measure of biodiversity is for a given area rather than for a given biological community. In ecological terms, physical areas and the biotic components they contain are termed habitats. Habitat diversity is a more useful term than that of ecosystem diversity since habitats are easy to envisage (e.g. a mangrove forest, a coral reef, an estuary). Furthermore, habitats often have clear boundaries. Habitats have been termed 'the template for ecology'.

There are strong relationships between sampling scale and the processes that influence diversity. At small scales all species are presumed to interact with each other and to be competing for similar limiting resources. Ecologists have called this within habitat (or alpha) diversity. At slightly larger scales habitat and/or community boundaries are crossed and sampling covers more than one habitat or community. This scale has been called between-habitat (or beta) diversity. At an even larger scale (regional scale) where evolutionary rather than ecological processes operate the pattern has been called gamma diversity or, more recently, 'landscape diversity'. Landscape diversity can be defined as the mosaic of habitats over larger scales often hundreds of kilometres. Franklin discusses landscape diversity in relation to biodiversity conservation. (Ray calls the marine equivalents seascapes). Much attention has been given to ways of conserving landscape diversity on land. Clearly, a given habitat can be maintained but landscape diversity can be reduced if the mosaic of habitats is altered. It is clearly important, therefore, to specify what scale (and hence type of diversity) is being studied.

In an important recent paper Tuomisto have shown from an analysis of satellite images followed by extensive ground truthing that beta diversity has been greatly underestimated in tropical rainforests. Since the between-habitat (beta) diversity has been underestimated then the landscape diversity will also be underestimated. The authors point out that the conservation value of different areas depends on a sound estimate of between-habitat and landscape diversity. This is a topic that will need thorough consideration and discussion in any future conservation strategy.

Within coastal areas there are a wide variety of habitats with known high species diversity such as sea grass beds, coastal sedimentary habitats, mangal and coral reefs. Ray and Gregg have analysed the coastal wetland areas of Virginia and the Carolinas, USA, and conclude that there are large differences in the proportions of salt and freshwater marshes, forest/scrub-shrub and tidal flat areas which lead to dfferences in biodiversity between the two areas. Ray classifies marine habitats into 20 categories as a basis for characterising coastal areas. Coral reefs are themselves highly variable with large differences between the reef flat, reef crest and reef slope both in coral and associated species and each component is probably best considered as between-habitat β diversity. Hard rocky surfaces have a rich encrusting flora and fauna, for example in clumps of mussels Suchanek found over 300 species in Washington, USA, and within kelp holdfasts (*Laminaria digitata*) in boreal areas. Moore found over 350 species on the species-poor East Coast of UK.

Yet it is not only the high diversity areas that are in need of conservation. It is often in low diversity areas that productivity is highest and humans exploit these systems (e.g. upwelling areas and estuaries) for food resources and other uses. Estuaries with low species numbers due to salinity stress are habitats that are under severe threats from urbanisation and industrialisation. Arctic marine systems have relatively low diversity and there are low diversity coral habitats that are subject to a variety of threats. Thus one cannot set priorities for marine diversity conservation based simply on habitats with high diversity.

Ray argues cogently that biodiversity assessments need to be made at the community, habitat and landscape levels if we are to predict changes over time. In a review (by WWF, IUCN and UNEP) of ways of conserving genetic diversity of freshwater fish it was recommended that the best way to conserve species diversity is to conserve habitats. Ogden and Ray give examples showing species that use a coral reef during the day and migrate to seagrass beds or mangroves at night. Often seagrass beds are an integral part of the coral reef system. Thus it is the mosaic of habitats, that must be protected if a complete protection of biodiversity is to be achieved. It is primarily the loss of habitats that leads to the loss of both genetic and species diversity.

CHARACTERISTIC PATTERNS OF MARINE BIODIVERSITY

Latitudinal Pattern of Diversity

In terrestrial systems species, genera and families increase in diversity from poles to tropics. A good example is that of tree species diversity where the highest diversity values are shown in tropical rain-forests. In the marine domain, there is an apparent increase in species diversity of hard substratum epifauna from the Arctic to tropics. The Arctic is much younger and has low biodiversity and low endemism compared with the older Antarctic. The longer period of geographic isolation of the Antarctic is also important for biodiversity generation. Production processes also differ and whereas the Arctic is dominated by many commercial fish species the Antarctic is characterised by invertebrates (krill and squid) which support birds and mammals and only a small fishery.

Stehli and Wells 1971 showed that bivalve molluscs at species, genus and family levels show increased diversity towards the tropics in the Indo-Pacific. Recent data from the deep sea purports to confirm this principle. However, in the latter data there is much scatter and if one removes the data from the Norwegian Sea, which has low diversity the trends are far less clear, if evident at all. The Norwegian Sea is a recently glaciated area and it is therefore not surprising that diversity is low. Not all groups show such trends. Seaweed (macroalgal) diversity is higher in temperate latitudes than the tropics and lowest at the poles.

In the Southern hemisphere, the pole-to-tropic gradient is far less clear since the Antarctic has high diversity for many taxa. Data from Australia show that in a coastal area 800 species have been recorded from just 10 m^2 of sediment in Bass Strait and 700 species occur in sediments of Port Phillip Bay. These values are as high as the highest values for soft sediments found anywhere.

In summary, it seems probable that there is a cline of increasing diversity from the Arctic to the tropics but the cline from the Antarctic to the tropics is far less well established if it occurs at all. There is clearly a need to better document diversity patterns in other areas of the Southern hemisphere such as the African and American continents.

Longitudinal Pattern of Tropical Diversity

Probably the most well-known diversity pattern in the marine domain is that of coral genera and species, which show highest values in the Indonesian archipelago and falling values radiating westwards across the Pacific Ocean. Across the Indian Ocean diversity decreases irregularly from the high diversity epicentre, dipping and then rising in the Red Sea and Africa in some groups and with lowest diversity in the Caribbean. Similar patterns have been shown for mangroves, and gastropod snails. It appears that the Indonesian archipelago is the 'epicentre' for evolution of marine tropical biodiversity. Using rRNA techniques Palumbi has recently shown that species have indeed radiated out into the Indo-Pacific region from the centre.

The reason for this high level of diversity in the Indo-Pacific region is thought not to be solely the result of a long period of evolutionary stability, but rather due to the fact that there is a large diversity of types of islands and archipelagos which differ in size, in their geological history and in distance from sources of colonising species. There have been periods of isolation over evolutionary time, which have given rise to allopatric speciation, (speciation caused by the erection of physical boundaries between populations) followed by periods of reunification which has given sympatric speciation (speciation within a population, usually caused by competitive interactions). Throughout geological time there have been massive extinctions followed by rapid evolution and speciation.

Other Marine Biodiversity Patterns

Another pattern that has received much attention is that in soft sediments with increasing diversity from shallow areas to the deep-sea. This has recently been confirmed by Grassle and Maciolek in a study along a transect of 176 km off the US East coast at depths of between 1500 m and 2100 m. A total of 798 species was found among 90,677 individuals from an area sampled of 21 m^2.

It has been assumed that the data presented by Sanders are representative of a general pattern of low species diversity in shallow coastal areas. Surprisingly no-one has questioned whether or not this is the case. This is all the more remarkable since there are very large numbers of studies done in coastal areas. Using data obtained from the Norwegian continental shelf in the North Sea, Gray found over a distance of 1200 km a total of 620 species from 39,582 individuals. These data together with those of Poore and Wilson raise the question of whether coastal biodiversity shows values as high as that of the deep-sea. More quantitative information from coastal areas is needed particularly from tropical coasts and from the Southern hemisphere.

THREATS TO MARINE BIODIVERSITY

With the exception of ocean dumping and UV-B radiation there are probably few human activities posing major threats to oceanic diversity. However, long-transported materials enter the open ocean system and there are concerns about effects of organochlorine compounds on planktonic and benthic systems. The oceanic system is open and continuous and it is unlikely that contaminants will lead to measurable effects on diversity, such as local or regional extinctions. Organisms that live near tectonically active zones where plates are diverging have high diversity and naturally high levels of heavy metals and derive their primary energy from chemosynthesis rather than from sedimenting products of photosynthesis.

Most of the threats to biodiversity are in the coastal zone and are a direct result of human population and demographic trends. The world population has more than doubled since World War II and is expected to increase from 5.5 billion in 1992 to 8.5 billion by 2025. More important however, are the demographic trends of increased population densities in coastal areas. It is estimated that 67 per cent of the global population lives on the coast or within 60 km of the coast and the percentage is increasing. Within 30 years this population will double. Furthermore, many of the largest cities in the world, where population growth rates are highest, are near the coast (e.g. Sao Paulo, Shanghai, Hong Kong, Manila, Jakarta). These burgeoning populations increase pressures on utilisation of resources in coastal areas and in addition lead to habitat degradation, fragmentation and destruction. This is a special problem in Indonesia where the highest marine diversity is found near to centres of high human population growth.

There are a number of recent reviews of threats to coastal systems. These threats are: habitat loss; global climate change; overexploitation and other effects of fishing; pollution (including direct and indirect effects of inorganic and organic chemicals; eutrophication and related problems such as pathogenic bacteria and algal toxins; radionuclides); species introductions/invasions; watershed alteration and physical alterations of coasts; tourism; marine litter; and the fact that humans have little perception of the oceans and their marine life. The threats are frequently interlinked. All the reviews agree that the most critical threat is habitat loss. This is echoed in the recent Global Biodiversity Assessment which states 'The most effective way to conserve biodiversity, by almost any reckoning is to prevent the conversion or degradation of habitat.'

Habitat Degradation, Fragmentation and Loss

Complete loss of habitat is the most serious threat to marine biodiversity, especially if contiguous but different habitats forming landscape diversity are lost. Southeast Asia contains 30 per cent of the world's coral reefs. Based on studies of coral cover, which is not a good indicator of reef condition, Wilkinson and Chou claim that 60 per cent are already destroyed or on the verge of destruction and make the prediction that unless drastic action is taken immediately most of the reefs will be eradicated during the next 40 years. The loss of the reefs is due to increased sedimentation, overexploitation by dynamite and chemical fishing and by sewage pollution. In an analysis of data on coral reefs in Japan Veron found that over 37 per cent of species are at some risk of regional extinction and 29 per cent at a substantial risk of extinction.

In Sri Lanka reef cover is declining by 10 per cent annually and in the Gulf of Thailand by 20 per cent annually. In the Philippines studies show that almost 70 per cent of 735 studied reefs are seriously damaged and in Eastern Indonesia 80 per cent of the reefs have been damaged by dynamite fishing. There is reason to believe that similar damage is occurring in East Africa and in the Caribbean. Recently the US State Department has launched an International Coral Reef Initiative which is endorsed by scientists, policy-makers, donor organisations and national representatives. This concludes that 'human activity is the primary agent of degradation' of reefs either from direct impacts or by inadequate planning and management of coastal land and upland activities. All these impacts are exacerbated by human population growth, increased pollution. Mangrove forest destruction is occurring on an equally alarming rate. Indonesia has by far the largest areas of mangroves (21,011 km^2) and 45 per cent have been lost and the rates of loss are increasing rapidly. Data from the World Resources Institute show losses of between 40 and 70 per cent in Africa, almost 70 per cent in Asia, 85 per cent in India and 87 per cent in Thailand. In both the Philippines and Ecuador over 70 per cent of the forest has been destroyed to make way for shrimp farms. The primary source of shrimp larvae to stock the farm is the mangrove forest and thus the long-term sustainability of farms is jeopardised by destruction of mangrove. Other problems such as soil erosion often accompany mangrove destruction.

While losses of coral reefs and mangrove habitats are probably the most significant in terms of losses of biodiversity it should not be forgotten that other critical coastal habitats are also disappearing. Wetland areas, estuaries and seagrass beds are known to be key nursery areas for coastal fisheries and yet are being destroyed rapidly without there being full ecological and economical appraisal of the consequences even in developed countries. Estuaries pose particular problems globally since there are often conflicting interests such as industrial development, shipping and associated harbour development, fishing, tourism and the needs for conservation. There are few published data on the loss of landscape diversity in the marine environment (e.g. the mosaic of wetland, estuary and sand and mud flats as a combined system). It is relatively straightforward to record and document habitat loss on land and in shallow and/or tidal areas using for example remote sensing and Geographical Information Systems (GIS). Regional scale assessments are urgently needed.

Biodiversity will also be lost if habitats become degraded so that species can no longer survive. Assessing the degree of degradation needs monitoring over space and time and this is a major task. GESAMP has recently produced a report on Biological Indicators and their Use in the Measurement of the Condition of the Marine Environment. This report describes the indicators that can be used to measure exposure to contaminants and their effects, sets out a tiered approach for a field assessment program and discusses sampling designs which are appropriate to the measurement of the condition of a given habitat or area and finally discusses the types of managerial action that are needed to complete an assessment.

Another severe problem is that, while habitats may be ostensibly maintained, they become divided into small fragments. There is a large ecological literature on these so-called 'habitat islands' with theories of maintenance and loss of diversity within such islands. Huston discusses this in a general context. Small 'habitat islands' that are remote from the main pool of species have higher rates of species extinctions and lower immigration rates than larger 'habitat islands' or 'habitat islands' that are nearer the main pool of species. Fragmentation of habitats is expected to lead to losses of species diversity. However, in marine coastal areas few studies have been done that quantify species loss with loss of a given area of habitat. Horn and Connell have shown that diversity is often higher in habitats that are subjected to some disturbance than in undisturbed habitats, 'the intermediate disturbance hypothesis'. This is due to the disturbance creating space for new species to colonise. The spatial and temporal scales of the disturbance determine whether or not diversity increases or decreases. The species within a given habitat are adapted to the natural disturbance scales and are not necessarily adapted to man-made disturbances so that one cannot assume that man-made disturbance will increase diversity.

One important aspect that also needs to be considered is habitat restoration. On land there is a long tradition of restoring habitats, such as mining waste tips. There are some examples of habitat restoration in the marine environment, such as the well-publicised clean-up of the River Thames in the UK where salmon can now be found in London. The developing science of restoration ecology should be a part of a strategy for conservation of coastal biodiversity.

Global Climate Change

Pernetta has reviewed the potential implications of climate change for a number of tropical areas. The most publicised consequence of global climate change is that of sea level rise with severe effects likely in the Maldives and Tuvalu which are only 2 m and 4.5 m respectively above sea level. Bangladesh is expected to lose 12 to 28 per cent of its total land area over the next century as a consequence of predicted sea-level rise. Coastal wetland habitats are likely to suffer since wetland subsistence and formation probably cannot occur at rates of sea level rise above 10 mm per year. Wetland areas are important not only for the species they contain, their function as nursery areas, but also for stabilising coastlines and for protection against hurricanes and storm surges.

The most significant effect of global climate change on coastal systems is, however, likely to be altered storm events and rainfall patterns. It is predicted that the return period of storms will alter so that the 100-year storm occurs every 10-years and the 10-year storm annually. Such events are likely to be highly significant for nutrient transport to the coasts, for mixing processes in coastal areas and for current and frontal systems. As yet, models available are not able to make sufficiently accurate predictions of likely consequences at regional levels mainly due to the lack of data. A Global Ocean Observing System (GOOS) has been proposed to redress this lack of data and its implementation is being planned by UNESCO-IOC, UNEP and WMO. There are component modules on the Health of the Oceans (HOTO) and on the coasts.

The warming of the coastal ocean is known to lead to severe effects on corals. In 1983, 1989 and 1990 the surface temperature of the Caribbean increased by 2°C from 28°–29° to 30°–31°C with massive bleaching followed by death of corals. The species that died were important in structuring the reef so that the consequences were severe and extended over wide areas. Similar events have been recorded in Panama and Indonesia but not with the widespread effects found in the Caribbean.

UV-B Radiation

As a consequence of ozone depletion (which is not related to climate change) UV-B radiation is increasing. There are few data that predict effects on marine systems. It has been suggested that there will be reduced productivity of phytoplankton in surface waters, which includes the open ocean. Effects on the symbiotic zoozanthellae in corals have been predicted by Gleason and Wellington but this is still controversial. There are also concerns about impacts on diatoms on sand and mud flats. More research is needed before reliable predictions can be made of effects on marine biodiversity.

Effects of Fishing and Other Forms of Over Exploitation

Despite the fact that most fisheries resources are now within the jurisdiction of coastal states nearly all the world's fish resources are overexploited. Between 1988 and 1990 the marine fish catch declined in nine key fishing areas and especially off Peru, pelagic fish off Japan, off the Northeast coast of the US and in European seas. The consequences of heavy fishing pressure on commercial species is that the size distribution changes and this leads to loss of genetic diversity, e.g. Orange Roughy.

In many areas of the Northwest Atlantic there have been dramatic changes in the composition of fish stocks as a consequence of fishing. Highly important commercial species have declined (e.g. herring and Arctic cod) and other less valuable species have increased, e.g. sandeels, and sharks. Several studies show that changes in fish species composition have dramatic effects on other species dependent on fish such as sea birds and mammals.

Exploitation of fish resources can lead to local or regional species extinctions. The Blue Walleye (*Stizostedlion vitreum glaucum*) was overfished in Lake Erie and became locally extinct. The Coelacanth (*Latimeria chalumnae*), which lives in caves in the Cormora islands, has a total world population of under 500 individuals and is being harvested accidentally as a bycatch of fishing for other species and is in real danger of becoming extinct. Local extinctions of fish can also occur where estuaries are made unfit for spawning.

Trawling for bottom-living fish species is having a major effect on the habitat for species other than target species. It has been estimated that all of the sea bed of the North Sea is trawled over at least twice per year and the gear is getting heavier over time. Trawls have destroyed long-lived species of molluscs and echinoderms in the North Sea. Since these species play important functional roles in biogeochemical cycling the consequences may be far-reaching. There are plans to designate trawl-free areas where by comparison with trawled areas effects of trawling can be assessed.

Fishing using explosives on coral reefs occurs globally in areas where reefs are not properly protected. The ensuing destruction of the reef habitat, which sustains not only the fish but all other species dependent on the reef, has catastrophic consequences for biodiversity. In the Phillipines in addition to dynamite fishing, and fishing for the aquarium industry there is a further serious problem that of the widespread and increasing use of cyanide to obtain live fish for restaurants. Although the fish recover when placed in clean water the cyanide has major effects on the reefs. It is not known what effects the loss of large numbers of reef fish will have on the reef system as a whole.

There are relatively few quantitative data on local species extinctions. A few known examples are the Red Coral (*Corallium* spp.) and Black Corals (*Anthiparia* spp.) which are heavily exploited for jewellery in the Mediterranean and throughout the tropics and are listed by IUCN on the Red Data list as species in danger of extinction, as are Triton's trumpet snail (*Caronia tritonis*) and the Knysna sea horse (*Hippocampus capensis*). Predatory gastropod snails are sought as souvenirs in many tropical areas and since they play key roles in controlling prey populations, their local extinction can lead to major changes in diversity (e.g. Paine's 1966 classic study on effects of removing keystone predators, but see Mills, for a critique of the keystone species idea). Many other species are heavily exploited and may be in danger but there is far too little information on which to make a proper evaluation. There is an urgent need for better information.

Marine mammal and sea turtle exploitation are well documented and will not be treated in detail here. The species that are in danger are listed in the appendices to the Convention on International Trade in Endangered Species of Wild Fauna and Flora (CITES).

Pollution and Marine Litter

The GESAMP State of the Marine Environment Report is still the most authoritative statement of the threats to marine life. The report emphasises that coastal areas are affected by man almost everywhere and stresses that habitat losses from a wide variety of causes if unchecked will lead to a global deterioration of the environment. There is little that has happened since 1990 to suggest that things have changed for the better. In recent years there has been a recognition that heavy metals seldom pose a threat to marine biodiversity, although there are local areas where high concentrations are still cause for concern, such as areas subjected to mining waste run-off and industrialised estuaries or fjords. There are major concerns about the long-term effects on marine populations of organic chemicals. PCBs and dioxins have been much in focus and there are recent concerns about the fact that many organic chemicals of quite different structures seem to mimic the effects of female oestrogenic hormones and have led to severe reproductive changes in terrestrial species. Clearly this is a topic where more research is needed before the threats to marine biodiversity can be quantified.

GESAMP states that eutrophication caused by excess nutrients and/or sewage discharged into coastal waters is an expanding problem and incidents are known from almost every coastal state. The initial effects are of altered species compositions both in the water columns and in benthic communities. This may lead to local changes in biodiversity. More severe effects due to low oxygen concentrations are mass mortalities. Other effects that have been linked to eutrophication are harmful algal blooms, but causal links to eutrophication are not yet proven. Nutrient abatement is recommended where eutrophication symptoms occur.

Ciguatera, a disease affecting the nervous and cardio-vascular systems is caused by eating tropical fish that have bioaccumulated toxins from natural algae. Where algal biomasses are significantly elevated, such as in nutrient/sewage enriched areas, the risks of ciguatera are high and it is a common problem in Asia and the Pacific and effects 50,000 people per year. Other toxins produced by algal blooms affect coastal aquaculture and occasionally human health in both developed and developing countries. Although oil is a highly visible pollutant and when spilled in large quantities can cause severe local affects it is not regarded as a significant pollutant on global scales.

Marine litter is an increasing problem for marine life and tourism. In the Mediterranean there are three main sources: litter from drainage sources on land; litter left on beaches; and litter discarded from ships, including discarded nets and other materials from fishing vessels. Almost 75 per cent of litter is plastic with Styrofoam, metal, glass and wood as the other major components. Turtles are particularly vulnerable to discarded litter. Of 51 carcasses stranded in Florida, 6 per cent were entangled in nets and over 50 per cent of Green Turtles (*Chelonia mydas*) had ingested debris which was thought to have been a major contributor to their deaths.

Species Introductions/Invasions

The ctenophore *Mnemiopsis leidyi* was imported from the US East coast to the Black Sea, probably in ballast water, and has led to a catastrophic alteration in the whole trophic web and contributed to a huge reduction in stocks of commercial fisheries. Other concerns covered by GESAMP are the transport of species of algae that may cause toxic blooms in new areas and other introductions which have led to dramatic effects at regional levels. Alterations in biodiversity are also highly likely although this is poorly documented.

Watershed Alteration and Physical Alterations of Coasts

Joint Group of Experts on the Scientific Aspects of Marine Environmental Protection (GESAMP) has reviewed how altered watershed use has led to significant changes in both nutrient and sediment transport to the coasts. Construction of dams for hydroelectricity generation or for irrigation purposes has led to dramatic reductions in sediment loads with severe consequences for coastal ecosystems. The Nile delta is sinking at the rate of tens of centimeters per year due to a combination of lack of sediment input and enhanced erosion and in addition nutrient loads have been so severely reduced that the fisheries have collapsed in much of the Eastern Mediterranean.

Deforestation and mining, often many hundreds of kilometres inland have led to large increases in sediment loads which have smothered coral reefs and other coastal habitats in the Philippines, Malaysia, Indonesia, Sri Lanka, Pacific Islands, the Gulf of Thailand, the Caribbean, Columbia, Costa Rica and Cuba. It is thought from remote-sensing that the sediment loads come principally from small streams, although quantitative data from the streams are lacking.

Tourism

There are greatly increasing stresses on coasts caused by tourism even in Antarctica and the Arctic. The most serious threats are those of habitat destruction. Mangroves are often removed, wetland areas filled in and estuaries reclaimed to make way for touristic complexes without there being any evaluation of the benefits of the intact systems. Once built the resort may lead to effects on adjacent habitats through sewage discharge and other threats and ultimately to the loss of habitats and their resources. Establishment of hotels on coral reefs is becoming popular and often leads to the destruction of the habitat that was the reason for the development in the first place. Coral reefs are vulnerable to trampling and in the Cayman Islands the one-day visit of a tourist ship to a coral reef led to 3000 m^2 of a previously intact reef being destroyed. What is needed is a better understanding by policy-makers and planners of the value and requirements for maintenance of the integrity of the natural habitat.

Human Perceptions of the Oceans

'Most people are familiar with terrestrial habitats and can relate to a walk in the woods. Few, however, have experienced the wonders of a coral reef except for occasionally viewing a Jacques Cousteau special. Whilst it is easy to capture images of rainforests being cut down and to collect data to quantify the magnitude of habitat destruction on land, it is more difficult to study and document coral reef processes and degradation.' This view is echoed by Suchanek who lists three reasons why marine conservation is less developed than that of the terrestrial environment. These are that the populations and communities are not normally visible; our knowledge of them is limited; and we are maintaining no ongoing monitoring.

Thus, apart from efforts devoted to protect marine mammals, turtles and sea birds, there is a very limited public response to the needs for marine biodiversity conservation compared with conservation of terrestrial habitat conservation. In North America it is only recently and after considerable efforts that the coastal zone has been highlighted as in need of conservation.

Summary of Threats

From this analysis it is clear that there are few threats to the open ocean and the threats are concentrated in coastal areas. Habitat destruction is particularly pervasive in tropical areas where mangroves, coral reefs and wetland areas are being destroyed at alarming rates. In temperate areas there are severe threats to wetland areas and estuaries and conflicts between industrial and tourist development and conservation are universal. The threats from commercial fishing on biodiversity of coastal areas has been neglected.

LEGAL FRAMEWORK OF BIODIVERSITY CONSERVATION

Apart from the Biodiversity Convention itself, the UN's Convention on the Law of the Sea, which came into force in November 1994, is of major significance in relation to biodiversity. IUCN has recently produced a comprehensive analysis of the Law of the Sea and other legal issues relating to marine conservation. UNCLOS establishes a comprehensive framework for use of the ocean and its resources. In addition to UNCLOS, Kimball lists other international agreements that relate to fishing and conservation of marine resources, such as conventions on whaling, marine mammal conservation, regional seas, Antarctic resources, transboundary fisheries (e.g. salmon and tuna), etc.

Other important conventions include:

1. The 1971 Convention on Wetlands of International Importance Especially as Waterfowl Habitat, Ramsar and Protocol (RAMSAR).
2. Convention Concerning the Protection of the World Cultural and Natural Heritage, Paris. (UNESCO)—this includes the Great Barrier Reef and the Galapagos Islands.
3. The 1973 Convention of International Trade in Endangered Species of Wild Fauna and Flora (CITES).
4. The 1979 Convention on the Conservation of Migratory Species of Wild Animal (CMS).
5. There are many regional conventions and agreements protecting given coastal areas and Kimball (*loc. cit.*) lists these.

Application of these conventions alone will not lead to protection of coastal biodiversity. Most problems lie at national and local community levels where there are conflicting uses of coastal areas.

HOW CAN MARINE BIODIVERSITY BEST BE CONSERVED

Beatley reviews briefly how biodiversity can be protected in coastal environments, but the review lacks detail and contains no clear conservation strategy. Norse has produced 'A strategy for building conservation into decision-making'. This covers the topic in a general way, but includes neither a strategy for conservation, nor an indication of the types of concrete action that are needed.

A number of national and regional assessments of biodiversity which suggest conservation needs have been made. The creation of marine protected areas is the general strategy adopted and the International Union for the Conservation of Nature (IUCN) has been heavily involved. It has been estimated that < 1 per cent of the coasts are covered by marine protected areas and often these are isolated habitats. If marine biodiversity is to be conserved better protection of the coasts outside marine protected areas is needed.

Habitats themselves occur as a mosaic of interconnected units thus the mosaic of habitats, the landscape, must be considered. Perrings states 'Understanding and managing the habitats, as well as the landscape matrix of ecosystems—including greenways and corridors to counteract habitat fragmentation—is therefore likely to be more effective than focusing on species and populations alone, and it has been argued that in order to sustain biodiversity over multiple human generations biodiversity policy should in fact be set at the landscape level.'

The economic value of coastal habitats is often not estimated. Barbier describes examples of the indirect benefits of wetlands which are often not taken into account, such as storm protection and groundwater recharge of floodplains. There are many similar 'ecological services' that are provided by coastal habitats that need full economic appraisal. Barbier have given an overview of the economics of biodiversity conservation. Hodgson and Dixon evaluated the advantages of logging versus tourism on the coast of Palawan, Philippines. They found that there were economic benefits of maintaining the forest and concentrating on tourism, an option that the authorities had not considered.

In a comprehensive study Ruitenbeek has analysed the competing options for exploiting the mangrove forests in Bintuni Bay, Indonesia. His analysis shows that the preferred economic option for sustainable use of the forest is to selectively cut 25 per cent of the harvestable mangrove as this option will allow alternative uses of the coast for among others offshore shrimp production, as well as maintaining biodiversity. He emphasises that an important aspect is how this work relates to policy, planning and decision-making processes. From inception through field work to analysis and input to the decision process took just 6 months, a time frame that fits well within a single government administration.

The greatest levels of marine biodiversity are found in tropical countries which are developing. Being poorer than their developed country counterparts in general they have less facilities, equipment, trained staff and resources available to devote to marine biodiversity conservation. In addition it is natural that their priorities focus more on food production and development than on conserving biodiversity. There is a need to explore the economic and other practical benefits of conservation of biodiversity, so that policy decisions are made in the full knowledge of the benefits that can be gained from biodiversity conservation. A strategy should be made partly for protection of biodiversity, but also to ensure sustainable use of coastal habitat resources which includes biodiversity. Sustainable use of these resources will require that all stakeholders are involved in the assessments and the decision-making processes that follow. In addition to natural scientists users of the habitats, managers, planners, economists and policy-makers must be included if marine biodiversity is to be conserved. A framework for integration of this type is that of Integrated Coastal Management (ICM).

Chapter 18

Microbial Diversity and Soil Functions

INTRODUCTION

Soil is a complex and dynamic biological system, and still it is difficult to determine the composition of microbial communities in soil. We are also limited in the determination of microbially mediated reactions because present assays for determining the overall rate of entire metabolic processes (such as respiration) or specific enzyme activities (such as urease, protease and phosphomonoesterase activity) do not allow any identification of the microbial species directly involved in the measured processes. The central problem posed by the link between microbial diversity and soil function is to understand the relations between genetic diversity and community structure and between community structure and function. A better understanding of the relations between microbial diversity and soil functions requires not only the use of more accurate assays for taxonomically and functionally characterising DNA and RNA extracted from soil, but also high-resolution techniques with which to detect inactive and active microbial cells in the soil matrix.

Soil seems to be characterised by a redundancy of functions; for example, no relationship has been shown to exist between microbial diversity and decomposition of organic matter. Generally, a reduction in any group of species has little effect on overall processes in soil because other micro-organisms can take on its function.

The determination of the composition of microbial communities in soil is not necessary for a better quantification of nutrient transformations. The holistic approach, based on the division of the systems in pools and the measurement of fluxes linking these pools, is the most efficient. The determination of microbial C, N, P and S contents by fumigation techniques has allowed a better quantification of nutrient dynamics in soil. However, further advances require determining new pools, such as active microbial biomass, also with molecular techniques. Recently investigators have separated ^{13}C- and ^{12}C-DNA, both extracted from soil treated with a ^{13}C source, by density-gradient centrifugation. This technique should allow us to calculate the active microbial C pool by multiplying the ratio between labelled and total DNA by the microbial biomass C content of soil. In addition, the taxonomic and functional characterisation of ^{13}C-DNA allows us to understand more precisely the changes in the composition of microbial communities affected by the C-substrate added to soil.

At present there is a particular interest in the relation between biodiversity, simply defined as the number of species present in the system, and function in the soil. This is part of a more general concern to conserve biodiversity and its role in maintaining a functional biosphere. The tacit assumptions in many current studies are that: (i) by characterising diversity one will be able to understand and manipulate the working of ecosystems and (ii) the ability of an ecosystem to withstand serious disturbances may depend in part on the diversity of the system. The importance of biodiversity in the functionality of ecosystems was stressed by Agenda 21, a document from the United Nations Conference on Environment and Development, prepared in Rio de Janeiro in 1992. The document promoted scientific and international cooperation for a better understanding of the importance of biodiversity and its functions in ecosystems. There is now a growing body of experimental evidence that most organisms are functionally redundant and that the functional characteristics of component species are atleast as important as the number of species *per se* for maintaining essential processes. We believe that at least some minimum number of species is essential for ecosystem functioning under steady conditions and that a large number of species is probably essential for maintaining stable processes in changing environments, the so-called 'insurance hypothesis'. However, our theories on terrestrial ecosystems have been developed from above-ground observations, whereas comparatively few studies have been made in soil. The links between biodiversity and soil functioning are therefore poorly understood.

Soil is fundamental and irreplaceable; it governs plant productivity of terrestrial ecosystems and it maintains biogeochemical cycles because micro-organisms in the soil degrade, sooner or later, virtually all organic compounds including persistent xenobiotics and naturally occurring polyphenolic compounds. The living population inhabiting soil includes macrofauna, mesofauna, microfauna and microflora. In this brief review we focus on the relationship between microbial diversity and soil functionality, by considering that 80–90 per cent of the processes in soil are reactions mediated by microbes. Indeed, bacteria and fungi are highly versatile; they can carry out almost all known biological reactions. To provide a comprehensive view of the complex relations between microbial diversity and soil functionality we consider:

1. The complexity of soil as a biological system.
2. The problems in measuring microbial diversity and microbial functions in soil and the meaning of these measurements.
3. Current ideas concerning the link between microbial diversity and soil functions.
4. Instances when measurements of microbial diversity are unnecessary for a better understanding of soil functionality.
5. The research needed for a better evaluation and manipulation of microbial diversity and soil functionality.

SOIL AS A MICROHABITAT

Progress in testing contemporary ecological hypotheses in soil has been limited by the difficulty of accurately measuring species richness and evenness and by the lack of information about the species really involved in the measured microbial activities. Soil is a complex microhabitat (Fig. 18.1) for the following distinctive properties.

The microbial population in soil is very diverse. Torsvik calculated the presence of about 6000 different bacterial genomes per gram of soil by taking the genome size of *Escherichia coli* as a unit. Microbial biomass is large: in a temperate grassland soil the bacterial and fungal biomass amounted to 1–2 and 2–5 t ha^{-1}, respectively.

Soil is a structured, heterogeneous and discontinuous system, generally poor in nutrients and energy sources (in comparison with the concentrations optimal for nutrient microbial growth *in vitro*), with micro-organisms living in discrete microhabitats. The chemical, physical and biological characteristics of these microhabitats differ in both time and space. Scales of the habitats depend mainly on the size of the organism: a few μm for bacteria; less than 100 μm for fungi; between 100 μm and 2 mm for Acari and Collembola; between 2 and 20 mm for Isopoda. Even if the available space is extensive in soil, the biological space, that is, the space occupied by living micro-organisms, represents a small proportion, generally less than 5 per cent of the overall available space. Another peculiarity is the presence of 'hot spots', zones of increased biological activity, such as aggregates with different physicochemical properties from the bulk of the soil, zones with accumulated particulate organic matter or animal manures, and the rhizosphere. Indeed, only a few microhabitats have the right set of conditions to allow microbial life. Several environmental factors, such as carbon and energy sources, mineral nutrients, growth factors, ionic composition, available water, temperature, pressure, air composition, electromagnetic radiation, pH, oxidation–reduction potential, surfaces, spatial relationships, genetics of the micro-organisms and interaction between micro-organisms, can affect the ecology, activity and population dynamics of micro-organisms in soil. These environmental factors can change markedly, and so microhabitats in soil are dynamic systems.

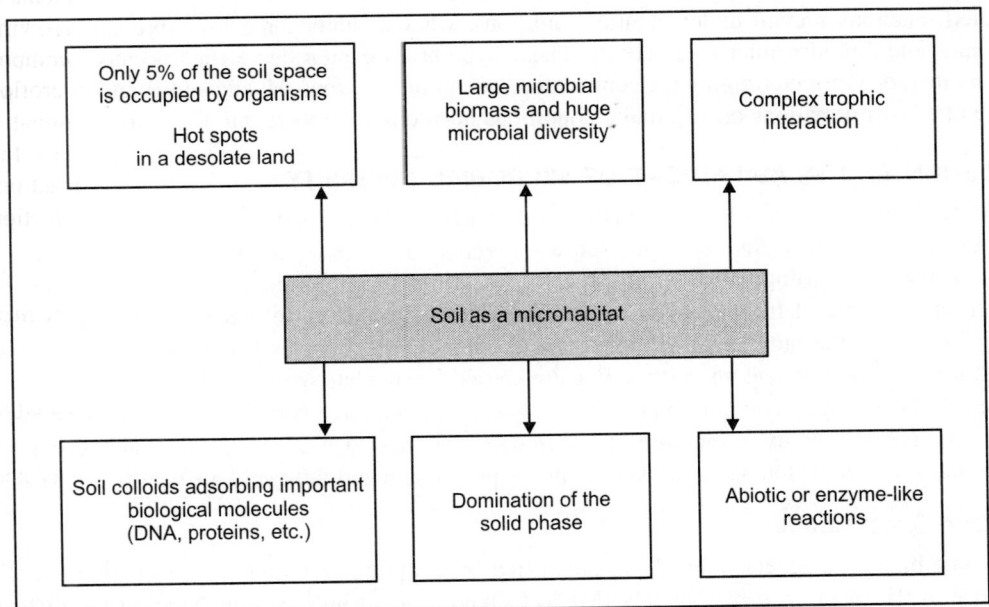

Fig. 18.1. Schematic representation of the main characteristics of soil as a microhabitat.

According to Hattori, almost 80–90 per cent of the micro-organisms inhabiting soil are on solid surfaces. Mechanisms by which these micro-organisms interact with soil surfaces have been much studied. Chen and Huang & Bollag showed that some bacterial cells produce extracellular polysaccharides interacting with clay particles and that these clay–polysaccharide complexes can persist even after the death of the microbes. The use of traditional and more recent (e.g. confocal laser scanning, use of micro-organisms with reporter genes) electron microscopy techniques with staining procedures has allowed us to locate the microbial groups, and inorganic and organic colloids in the soil matrix. All micro-organisms are aquatic and they live free or attached to surfaces, in water films surrounding solid particles, and inside aggregates.

Another distinctive characteristic of soil as a microhabitat is the property of the solid phase to adsorb important biological molecules such as proteins and nucleic acids. In this way some extracellular enzymes adsorbed by clay minerals or entrapped by humic molecules can maintain their activity, being protected against proteolysis, and thermal and pH denaturation. Deoxyribonucleic acid (DNA) molecules adsorbed or bound to humic molecules, clay and sand particles are protected against degradation by nucleases, but they can still transform competent bacterial cells, a process through which one of the two strands of extracellular DNA is taken up by the bacterial cell and inserted into its chromosomal DNA. In such a way the competent bacterial cell acquires all or part of the genes associated with the extracellular DNA.

The adsorption of organic compounds by soil colloids retards their microbial degradation; the location of potential substrates inside pores or microaggregates reduces their accessibility to soil micro-organisms.

The surfaces of soil mineral components can themselves catalyse many reactions. Clay minerals and Mn(III and IV) and Fe(III) oxides catalyse electron transfer reactions, such as the oxidation of phenols and polyphenols with formation of humic substances. Other abiotic reactions catalysed by soil minerals include deamination, polymerisation, polycondensation and ring cleavage. We think that microbe-mediated reactions prevail under natural conditions, whereas abiotic reactions prevail under harsh conditions and diminish microbial activity. These hypotheses cannot be verified because we have no accurate methods for determining the contribution of abiotic reactions. Huang and Ruggiero suggested that abiotic transformations prevail under conditions hostile to microbial activity in soil.

DEFINITION AND MEASUREMENT OF MICROBIAL DIVERSITY

Microbial diversity is a general term used to include genetic diversity, that is, the amount and distribution of genetic information, within microbial species; diversity of bacterial and fungal species in microbial communities; and ecological diversity, that is, variation in community structure, complexity of interactions, number of trophic levels, and number of guilds. Here we consider microbial diversity simply to include the number of different fungal and bacterial species (richness) and their relative abundance (evenness) in soil microflora. Equations used to calculate species richness and evenness and diversity indices, which combine both richness and evenness, have been discussed by Kennedy and Smith. Microbial diversity is measured by various techniques such as traditional plate counting and direct counts as well as the newer molecular-based procedures and fatty acid analysis.

Plate and Direct Counts

In the past microbial diversity has been quantified by counting techniques such as the plate count technique or the Most Probable Number (MPN) technique. Thus investigators have followed changes in the number of specific taxonomic or functional groups by plating on agar media, and have assessed precisely the culturable diversity by isolating colonies from these media followed by identification by a variety of typing methods. They have assessed the gross diversity of culturable micro-organisms by plotting the different colonies identified on the medium against the incubation time. However, the approach of determining microbial diversity from the number of isolates has become less popular, not only because a limited number of micro-organisms can be cultured but also because the procedures are laborious. As a consequence, an increasing number of molecular techniques are now popular because they do not rely on isolation-cultivation micro-organisms. It is well established that plate counts estimate only 1–10 per cent of the overall soil microflora. The discrepancy is essentially due to the interdependency of different organisms upon each other (for example, the endosymbiotic bacteria in specific worms and molluscs), to the inability to create in pure culture the environmental conditions faced by micro-organisms in the soil environment, and to the fact that some microbial species are cultivable only under certain physiological conditions. Generally the choice of the growing medium markedly affects the colony formed. Small bacteria cells (dwarf cells or ultramicrobacteria) are not culturable, that is, they cannot form colonies in agar. By considering that larger cells are supposed to account for about 80 per cent of the bacterial volume, Bakken hypothesised that the culturable bacteria have an ecological significance in soil more important than that which appears from their small numbers.

Direct counting by fluorescence microscopy can give 100–1000 times more than the numbers obtained by plate counting. Several stains specific to proteins or nucleic acids have been used; they include fluorescein isothiocyanate (FITC), acridine orange (AO), ethidium bromide (EB) and europium chelate with a fluorescent brightener differential stain (DFS) for bacteria. Phenol aniline blue (PAB) has been used to stain hyphae in agar films and on membrane filters, whereas metabolically active hyphae have been counted after their staining with fluorescent diacetate (FDA). Stains for active bacterial cells include FDA or redox probes such as 2-(p-iodophenyl)-3-(p-nitrophenyl)-5-phenyl tetrazolium chloride (INT) or 5-cyano-2,3-ditoyl tetrazolium chloride (CTC). Bloem improved the direct counting method with a video camera on an epifluorescence microscope. However, this procedure does not allow counting specific microbial species, and some of the stains used do not discriminate between living and dead microbial cells.

Molecular Techniques

The molecular techniques generally involve extraction of nucleic acid, directly or indirectly, from soil. They are independent of culture, and according to their sensitivity can detect species, genera, families or even higher taxonomic groups. Low-resolution methods include the analysis of base distribution (mole percentage guanine + cytosine) of DNA and the determination of rates at which denatured single-stranded DNA reassociates when temperature is lowered to approximately 25°C below the DNA melting point. These techniques give an overall indication of microbial diversity and can be used to monitor overall changes in the composition of a microbial community's composition after stress or changes in management.

High-resolution analyses allow the detection of microbial strains at the species and subspecies level. They usually give 'fingerprints' of non-coding DNA regions or involve the sequencing of both coding and non-coding regions. These techniques include *rep*-polymerase chain reaction (PCR) amplification of sequences between repetitive elements, ribosomal inter space analysis (RISA), which is based on the length polymorphism of the spacer region between 16S and 23S rRNA genes, and random amplified polymorphic DNA (RAPD), which does not require a preliminary knowledge of the genome. In RISA, PCR products are separated by gel electrophoresis, and the separated bands can be sequenced. A limit of this technique is the number of spacer sequences in the database. Recently, Borneman monitored micro-organisms responding to nutrient addition to soil with the thymidine analogue bromodeoxyuridine. Bromodeoxyuridine-labelled DNA extracted from soil DNA by immunocapture was subjected to RISA analysis.

The latest developments include the creation of bacterial artificial chromosomes (BAC) and the application of DNA microarrays. The former technique clones high-molecular-weight soil DNA (metagenome) which can be analysed at a phylogenetic and functional level. Microarrays of DNA are powerful for rapidly characterising the composition and functions of microbial communities because a single array can contain 500–1000 different DNA spots with the possibility of a very broad hybridisation with a broad identification capacity.

Most of the molecular methods have intermediate resolution, because they allow the detection of microbial groups rather than microbial species. Usually these techniques give fingerprints that allow multiple sample analysis to study spatial and temporal variation in the composition of soil microflora. Profiles of multiple replicates can be run next to each other, and thus allow easy detection of differences in band profiles. These techniques are characterised by 16S-rRNA PCR. Bacterial ribosomal DNA contains both conserved and variable regions; as sequencing of these genes has been carried out frequently in microbial ecology, there is a considerable database of known sequences. A successful analysis depends on the efficient DNA extraction from soil followed by a purification step.

The DNA can be extracted from soil by direct methods of *in situ* lysis or by indirect methods of initial cell extraction before lysis. In both cases, the methods used often include various combinations of bead beating, detergents, enzymatic lysis, and solvent extraction to obtain crude preparation of nucleic acids. Care is needed in interpreting the composition of microbial communities by molecular techniques because the method of extraction can influence patterns obtained by amplified ribosomal DNA restriction analysis (ARDRA) or RISA. Usually, an efficient extracting solution also solubilises humic molecules, which inhibit the PCR and thus it is followed by long procedures of purification. Another problem when extracting DNA from soil is the lysis of bacterial cells. Many Gram-positive bacteria require strong procedures to be lysed. However, strong extractants should be avoided because they degrade DNA molecules to fragments below 1 kb, and short DNA fragments may lead to the generation of chimeric 16S rRNA after amplification.

Among the intermediate-resolution techniques, denaturing gradient gel electrophoresis (DGGE) or temperature gradient gel electrophoresis (TGGE) are the most used for characterising bacterial communities in environments such as hydrothermal vents, hot springs, activated sludge, phyllosphere, biodegraded wall paintings, and soil. They are rapid and simple. The electrophoretic mobility of DNA fragments (obtained after PCR amplification) in polyacrylamide gels within a linear denaturant gradient (due to chemicals, DGGE or temperature, TGGE) depends on the base composition of the DNA. These techniques allow us to differentiate two molecules differing at the level of a single base. In addition to the simplicity and rapidity, another advantage of DGGE or TGGE is that the identity of bands can be investigated by hybridisation with specific probes or by extraction and sequencing.

However, we emphasise that all techniques based on PCR, such as DGGE and TGGE, have several drawbacks. Amplification of PCR can be inhibited if contaminants are not removed by the purification process, and the preferential or selective amplification in the presence of DNA from mixed communities can occur. Another bias of these techniques is the production of chimeric or heteroduplex DNA molecules. Many of these problems can be avoided if (i) all DNA molecules are equally accessible to primer hybridisation and form primer–template hybrids with equal efficiencies, (ii) in all templates there is the same extension efficiency of DNA polymerase and (iii) all templates are equally affected by the exhaustion of substrates.

In addition to the drawbacks of each PCR-based technique, DGGE and TGGE fingerprints present other biases (Table 18.1). Only the dominant populations are revealed, and bands from more than one species may be hidden behind a single band, resulting in an underestimation of the bacterial diversity. Finally, the same isolate can have different bands because multiple copies of operon in a single species are there. In addition, we must keep in mind that the fingerprinting depends on the primer used. Heuer reported that 14 different regions (A, B, C, D, E, F, V1–V3 and V5–V9) of the 16S rDNA have been used to generate fingerprints of bacterial communities. Usually V6 is the best one for optimal top-to-bottom analysis of bacterial communities of soil. According to Gomes, one can distinguish the different phylogenetic groups of bacteria with the following primers: F984GC, F27, R1378 and R1494 (bacteria), F243HGC (Actinomycetales), F203α (α-proteobacteria) and F948β (β-proteobacteria).

Table 18.1. Disadvantages of denaturing gradient gel electrophoresis (DGGE) or temperature gradient gel electrophoresis (TGGE).

Problem	Cause
Different fingerprints can be generated from the same DNA mixture	The available primers cover different regions of the 16S rDNA.
One single band might not correspond to a single bacterial species	Bands from more than one species might be hidden behind one band.
	DNA fragments of different species might have similar electrophoretic mobility.
	A single species might have different operons coding for 16S rDNA with sequence heterogeneity.
Dominant population representing at least 1% of the total community can be detected	This is the sensitivity limit of the technique in the fingerprint.

The assessment of fungal diversity in soil by molecular techniques has not been as successful as the characterisation of bacterial diversity because the concentration of fungal DNA is much less than that of bacterial DNA. Another problem has been the use of specific primers without co-amplification of DNA from other eukaryotic organisms such as plants, algae and nematodes. We can now use DGGE and TGGE to generate fingerprints for the fungal community of soil because specific primers are available for fungal 18S rRNA.

Phospholipid Fatty Acid (PLFA) Analysis

Another approach used to overcome the problem of selective culturing for assessing the composition of soil microflora is phospholipid fatty acid (PLFA) analysis. This technique is based on the extraction, fractionation, methylation and chromatography of the phospholipid component of soil lipids. Phospholipids are thought to be related to the viable component of soil microflora because they are present as important components of membranes of living cells and break down rapidly when the cells die, and they cannot survive long enough to interact with soil colloids. In addition, they have another prerequisite of a biomarker: they make up a fairly constant proportion of the biomass of organisms. Changes in the phospholipid profiles are generally related to variation in the abundance of microbial groups and can be interpreted by reference to a database of pure cultures and known biosynthetic pathways. Direct extraction of PLFAs or whole fatty acids (WSFAMEs) from soil does not permit detection at the species level and can be used to estimate only gross changes in community structure. However, identification of species by fatty acid analysis is possible with standard cultural-based media and databases (MIDI, Newark, Delaware). Lechevalier and Zelles have listed many fatty acids isolated from specific microbial groups. Bacterial biomass can be calculated by summing up several fatty acids (15:0, a15:0, i15:0, i16:0, i17:0, cy17:0, cy19:0 and 16:1ω7c), whereas fungal biomass can be obtained from the 18:2ω6c fatty acid.

MICROBIAL DIVERSITY IN SOIL

Communities of plants and animals generally compete because the majority of species are rare and a few species are abundant. In these communities the driving factor is the competitive interactions among species. By considering the result of a small subunit rRNA gene-based approach carried out by Zhou, Tiedje suggested that competition in microbial communities of surface soils with prevalence of any microbial species was absent because the various microbial species inhabiting soil are spatially separated for most of the time. They assumed that the contact among microhabitats lasts for a very short time immediately after rain, when water bridges are formed between the various soil particles and aggregates. Rapid drainage maintains the spatial isolation among the various microhabitats of soil. This hypothesis is confirmed by the fact that the saturation of soil with water resulted in a predominance of one or a few species. However, it does not take into account the mixing and transport by soil fauna and the stability of communities in biofilms at the interface between roots and soil which are not so strongly affected by wetting and drying. An alternative hypothesis to explain the large microbial diversity of surface soil is based on the presence of a greater variety and content of organic compounds there than deeper. This presence would be responsible for the diverse heterotroph-dominated microbial community in surface soil. However, microbial diversity of soil from preferential flow paths (cracks, fissures, biopores such as earthworm burrows or root channels) was similar to that of the bulk soil, in spite of the fact that the former sample showed a greater concentration of organic C and organic N and greater microbial biomass values than the latter sample.

Plants exert a strong influence on the composition of microbial communities in soil through rhizodeposition and the decay of litter and roots. The link between plant species and microbial communities in the rhizosphere soil is strict, being the result of co-evolution.

An interesting question about microbial diversity in soil is: are microbial species inhabiting soil ubiquitous or native to particular places only? If micro-organisms are cosmopolitan and no differences exist among different soil types all over the earth then we may assume that micro-organisms can be dispersed everywhere by wind, water currents, birds and human activity. On the other hand, if microbial species have a restricted distribution, that is, are geographically unique, the microbial diversity in soil must be very large globally. To find out Tiedje took samples from the 5–10 cm soil depth (and not from the surface layer so as to avoid recent additions to the surface) from many sites worldwide. They included soil from parks and nature reserves, so as to minimise human effects, from Mediterranean and boreal ecosystems in southwest Australia, southwest South Africa, central Chile and central California, northern Saskatchewan and northern Russia. The DNA of the fluorescent Pseudomonas, isolated from all soil samples, was analysed by ARDRA, intergenic transcribed spacer fragment length polymorphism (ITS-RFLP) and rep-PCR genomic fingerprinting. The third analysis, having the finest resolution, showed that genotypes were peculiar to each sampled site of the same continent region. The same genotypes were found in a 200 m transect, and a positive and significant relationship occurred between genetic distance and geographic distance.

MICROBIAL AND BIOCHEMICAL FUNCTIONS IN SOIL

Microbial and biochemical characteristics are used as potential indicators of soil quality, even if soil quality depends on a complex of physical, chemical and biological properties. The rationale for the use of microbial and biochemical characteristics as soil quality indicators is their central role in cycling of C and N and their sensitivity to change.

Microbial activity is a term used to indicate the vast range of activities carried out by micro-organisms in soil, whereas biological activity reflects not only microbial activities but also the activities of other organisms in the soil, including plant roots. Although the two terms are conceptually different they are confused. Various methods have been used to determine microbiological activity (Table 18.2). Some of them measure the rate of entire metabolic processes; for example, the evolution of CO_2 reflects the catabolic degradation under aerobic conditions; nitrification activity represents the rate of ammonia oxidation to nitrate; thymidine incorporation represents the rate of DNA synthesis in bacteria; and dehydrogenase activity represents the intracellular flux of electrons to O_2 and is due to the activity of several intracellular enzymes catalysing the transfer of hydrogen and electrons from one compound to another. As in other assays in soil biochemistry, dehydrogenase assays are based on the use of synthetic electron acceptors, which are less effective than O_2. Another established method, the hydrolysis of fluorescein diacetate (FDA), a colourless compound, to fluorescein which is coloured, is also a measurement of the contribution of several enzymes such as non-specific esterases, proteases and lipases, all of which are involved in the decomposition of organic matter in soil. Since more than 90 per cent of the energy flow in a soil system passes through microbial decomposers, and since heterotrophic micro-organisms are predominant in soil, FDA hydrolysis is thought to reflect overall soil microbiological activity. However, we must interpret the FDA data cautiously because the measured enzyme activities depend on the contribution of both extracellular and intracellular enzyme activities. As mentioned above, stable extracellular enzyme activities are associated with soil colloids and persist even in harsh environments that would limit intracellular microbiological activity. Thus, only strictly intracellular enzyme activities can truly reflect microbial activity because the contribution of free extracellular enzyme released by active microbial cells is negligible; indeed, these enzymes are short-lived because they are degraded by proteases unless they are adsorbed by clays or immobilised by humic molecules. Unfortunately, the present enzyme assays do not distinguish the contribution of intracellular from extracellular and stabilised enzyme activities, and thus they do not give valid information on the distribution and comparative importance of reactions mediated by microbes. In the case of enzymes maintaining their activity in the extracellular environment, Landi has suggested that calculating the ratio between the measured enzyme activity and the microbial biomass would provide more meaningful information on the location of the measured enzyme activities. However, any change in the ratio does not depend exclusively on variations in the stabilised extracellular enzyme activity because the intracellular enzyme activity can also increase or decrease without any change in microbial biomass.

Table 18.2. Some parameters used to determine microbiological activity.

Basal respiration	Dehydrogenase activity
Substrate induced respiration	Fluorescein diacetate hydrolysis
Nitrogen mineralisation	Heat output
Nitrification rate	Thymidine incorporation
Potential denitrification activity	Leucine incorporation
Nitrogen fixation	Specific enzyme activities
Adenylate energy charge	Arginine ammonification
ATP content	Dimethyl sulphoxide reduction

The need to measure the activities of a large number of enzymes and to combine these measured activities in a single index has been emphasised to provide information on microbial activity in soil. Note that it is conceptually wrong to assume a simple relationship between a single enzyme activity and microbiological activity in soil.

Most of the assays used to determine microbiological activities in soil present the same problem: measuring potential rather than real activities. Indeed, assays are generally made at optimal pH and temperature and at saturating concentration of substrate. Furthermore, synthetic rather than natural substrates are often used, and soil is incubated as a slurry.

Recent measurements, such as the community-level physiological profile (CLPP), otherwise known as the BIOLOG® approach, and the mineralisation kinetics of compounds added to soil have the advantage of combining both functional diversity and degradation rates.

Investigators often measure microbiological activity in soil by determining soil respiration in the absence (basal respiration) or in the presence of specific organic substrates or organic residues. The CO_2 produced by the oxidation of the added compound or residue can be determined only with ^{14}C-labelled material. Indeed, it can be erroneous to subtract the CO_2 evolved from the control soil (basal respiration) from that of the treated soil because of the priming effect (acceleration or inhibition of the mineralisation of native organic matter in the treated soil). Soil respiration can be insensitive to pollution, whereas the lag time prior to mineralisation of amino acid, as determined by evolution of CO_2, can increase in the presence of heavy metals. Hopkins suggested that the L:D respiration ratio of several amino acids can be a sensitive measure of microbial stress in soil because the micro-organisms are less discriminative between the stereoisomeric forms of amino acids. Indeed, acid pH and Cd pollution decreased the L:D glutamic acid respiration ratio of soil. Respiration has also been measured to determine nutrient limitations in soil. In glucose-treated soil, there is an initial increase (2–6 hours) in respiration followed by a second increase at 6–10 hours. The first increase is proportional to soil microbial biomass. The second increase reflects assimilation and microbial growth, and it is larger in the presence than in the absence of the nutrient (nitrogen, phosphorus or sulphur) when microbial growth is limited. Also, the rates of ^{14}C-leucine or ^{3}H-thymidine incorporation increased in the presence of the nutrient limiting bacterial growth in soil. Pennanen and Alden have suggested that by monitoring acetate incorporation in ergosterol, one could determine nutrients that limit fungal growth in soil. Another approach for determining nutrient-limited bacterial growth uses bacteria of a reporter gene. The disadvantage of this method is that one has to use one strain for each limiting nutrient.

BIOLOG® is nowadays much in favour to measure microbial functional diversity in soil because the utilisation of available carbon is the key factor governing microbial growth in soil. In addition, the technique is rapid and simple. However, there are several drawbacks: it is culture dependent, and reproducible results can be obtained only if replicates contain identical community profiles and are of similar inoculation density; changes in the microbial community can occur during the incubation; and the contribution of fungi is not measured because of their slow growth.

Degens and Harris proposed the measurement of the patterns of *in situ* catabolic potential of microbial communities for overcoming the problems with BIOLOG®. They used differences in the individual short-term respiration responses (or substrate induced respiration or SIR) of soils to the addition amino acids, carboxylic acids, carbohydrates and organic polymers to assess patterns of microbial communities. Although the approach of Degens and Harris does not present the problems of the BIOLOG® technique (see above), it has not been widely used since it was first published in 1997. This method permits the determination of catabolic richness, which is a measure of the number of substrates oxidised to CO_2, and catabolic evenness, E, given by

$$E = 1 / \sum_{i=1}^{N} P_i^2$$

where, p_i is the proportion of the total respiration response (i.e. is equal to $r_i/\Sigma r_i$), r_i is the response for substrate i and N is number of substrates.

The application of molecular techniques can improve the accuracy of determining microbial activities in soil. The traditional assays determine the rates of metabolic processes or activities of specific reactions that can be carried out by several microbial species. By using molecular techniques one might be able to determine the actual species involved in the processes being measured. This can be obtained by the direct targeting of 16S rRNA. Although methods for extracting DNA from soil are now well established, only a few extraction methods of RNA frcm soil have been reported. These are less than ideal because they involve multiple steps for purification and are impractical for processing large numbers of samples. In addition, recovery of intact mRNA molecules from soil is difficult because RNA is not stable. Recent protocols have been developed for the co-extraction of DNA and RNA from soil, making possible the characterisation of bacterial diversity with the differentiation of the active bacteria. Glassware and utensils used for RNA extraction are treated (such as by heating at 500°C for 4 hours) to inactivate RNases. In addition, the efficient co-extraction of RNA and DNA makes it possible to calculate the RNA/DNA ratio, which can be an important indicator of the metabolic status of microbial communities in soil.

MICROBIAL DIVERSITY AND SOIL FUNCTIONS

As mentioned above, the links between biodiversity and the functions of terrestrial ecosystems have been studied mainly in above-ground systems. A well-known relation between biodiversity and function is that described by the hump-shaped curve, in which there is an increase in plant production (i.e. the function) concurrent with increasing biodiversity until a certain point is reached; then a further increase in biodiversity results in a decrease in plant production. Other concepts proposed by current macroecological theories include stability defined as the property of an ecosystem to withstand perturbations. Stability includes both resilience (i.e. the property of the system to recover after disturbance) and resistance (i.e. the inherent capacity of the system to withstand disturbance). Tilman found that variations in plant populations in grassland caused by interspecific competition produced compensating effects over the total community and increased stability as measured by primary production. In microbial systems, functional stability is not necessarily related to community stability. Fernandez studied the behaviour of a well-mixed and functionally stable (i.e. constant pH and oxygen demand) metanogenic reactor fed with glucose. They discovered that the structure of the community was dynamic, as revealed by amplified ribosomal DNA restriction analysis (ARDRA).

It is difficult to measure both resistance and resilience in soil. Generally microbe-mediated processes are the most sensitive to perturbations in the soil; for this reason the capacity of soil to recover from perturbations can be assessed by monitoring microbial activities.

The links between microbial diversity and soil functioning, as well those between stability (resilience or resistance) and microbial diversity in soil, are unknown because, as stated above, it is difficult to measure microbial diversity. In addition, we generally measure soil functions by determining the rates of microbial processes, without knowing the microbial species effectively involved in the measured process. According to O'Donnell, the central problem of the link between microbial diversity and soil function is to understand the relations between genetic diversity and community structure and between community structure and function.

The links between microbial diversity and soil functions have been studied by approaches based on the use of (i) soils with the same texture but different microbial composition, (ii) repeated $CHCl_3$ fumigations of soil to decrease microbial diversity, (iii) specific biocides for killing specific soil micro-organisms and (iv) sterile soils inoculated with soil micro-organisms.

The second and third approaches are destructive, whereas the fourth is constructive. Degens investigated the relation between microbial diversity and soil functioning. He measured catabolic evenness in two silty clay loam soils subjected to three different stresses (decline in pH, increase in electrical conductivity and increase in Cu concentration) and two cyclic disturbances (wetting and drying, and freezing and thawing). The two soils had different catabolic evennesses: 21.4 in the soil under permanent pasture for more than 20 years and 19.0 in the soil under arable crops (2 years potatoes, 3 years wheat, 4 years maize and 26 years barley). The arable soil was less resistant to cell stresses and all disturbances than the pasture soil. Since the organic C content, stability of aggregates, cation exchange capacity and microbial biomass were also greater in the pasture soil, these factors might also have increased the resistance of soil micro-organisms to stresses and disturbances. For example, the organic C content and microbial biomass are generally correlated in soil, and this shows that organic matter is a good habitat for micro-organisms. Organic matter can also adsorb compounds toxic to micro-organisms. The catabolic evenness showed the typical hump-shaped pattern, generally observed in plant communities, with an increase with minimal stress and disturbance, followed by a decrease, more marked in the pasture soil at greater stress. Unfortunately, Degens did not measure the composition of the soil microflora, and thus the relation between catabolic evenness and microbial evenness or richness could not be directly assessed.

Griffiths studied stability (resistance and resilience) of soils with different microbial diversities as a result of their different use that were otherwise similar. The soils were clay loam cropped with either a single annual species or six annual species; the B horizons of a petroleum polluted sandy loam soil (formerly a petrol station) under remediation and the relatively unpolluted control soil; and a horticultural loamy sand soil. Only small differences were detected by BIOLOG® analysis, whereas protozoan numbers were significantly different and were smaller in the polluted soil (about one third of the protozoan biomass of the uncontaminated soil). In addition, no fungal growth was recorded in the polluted soil (in contrast to the other) when treated with plant residues. Short-term (25 hours) decomposition of added shoots of Lolium perenne was monitored at 15°C under constant moist conditions after different perturbations: amendment with 500 µg Cu g^{-1} soil; heating at 40°C for 18 hours; freezing at −20°C for 18 hours. Soil resistance was calculated by the following equation:

$$\%\text{Change from the control} = 100\,\frac{(\text{Control CO}_2 - \text{Treated CO}_2)}{\text{Control CO}_2}$$

The grassland soil was more resistant to both Cu and heat than the polluted soil, which was also less resistant to heat than the respective non-contaminated site. However, the polluted soil showed resilience after the Cu stress because soil respiration recovered after 15 days.

Griffiths created different degrees of microbial diversity (determined by DGGE and protozoan identification, respectively) artificially by inoculating a gamma irradiated agricultural clay loam with serially diluted soil suspensions prepared from the parent soil. The number of bacterial, fungal and protozoan taxa decreased with increasing dilution. There was no consistent relationship between microbial diversity and soil functions. Some functions increased with dilution (SIR), others were not affected (thymidine and leucine incorporation, NO$_3^-$ accumulation, respiratory growth response), and some declined only at the greatest dilution (short-term respiration from added grass, nitrification rates, BIOLOG®). In addition, there was no difference in resilience since the decomposition of continuously labelled (^{14}C) barley straw recovered after transient heat stress (40°C for 18 hours) to its pre-stress rate. Using sterile soil inoculated with different dilutions of the non-sterile soil has been criticised because only extractable micro-organisms are used and the resulting community of the inoculated soil has an unrealistically small microbial diversity.

Another destructive approach is to kill certain microbial groups with specific biocides so as to study their influence on soil functioning. The bacteriocide streptomycin and the fungicide cycloheximide have been used to inhibit selectively the respiratory response of bacteria and fungi to glucose with the aim of determining the bacteria:fungi ratio in soil. This method is based on two assumptions: the ratio of bacteria:fungi in the sensitive component of soil microflora is the same as that in the inhibitor-insensitive component; and the fungal and bacterial components of soil microflora respond equally to addition of glucose. Ingham and Coleman used amphotericin (a saprophytic fungicide), diazinon (an insecticide), carbofluran (a carbamate nematicide), cygon (an acaricide), and p-chloronitrobenzene (a fungicide specific for phycomycetous fungi), in addition to streptomycin and cycloheximide. However, biocides have the disadvantages that their effects are not restricted to their target group. In addition, the biocides can be adsorbed by soil colloids or used as an energy and nutrient source by the surviving micro-organisms. Chander found that soil fumigated with chloroform, with a much smaller microbial biomass than the corresponding non-fumigated soil, respired about the same amount of ^{14}C-CO_2 from labelled straw as the non-fumigated soil. They also found that pollution with zinc affected respiration of non-fumigated and fumigated soils in the same way, indicating that the ratio between substrate C-to-microbial biomass C (larger in the fumigated than the non-fumigated soil) was not important. The ratio of CO_2 to ^{14}C-to-microbial biomass ^{14}C was linearly related to Zn pollution. Griffiths found that soil fumigated with chloroform, with greater microbial diversity, was more resistant and resilient than soil with a less diverse community to perturbations such as heating at 40°C for 18 hours or treatment with 500 μg Cu g^{-1} soil. However, they suggested that the observed effects were due to the physiological influence of $CHCl_3$ fumigation on the microbial community rather than differences in microbial diversity.

The studies of Griffiths indicate that the effect of microbial diversity on microbial functions in soil depends on the measured function. According to Dighton, only by functional assays, which integrate both microbial community and species composition, can the functional aspects of microbial community be assessed. In ecology, the negative correlation between variation of functional measurements and diversity has been generally found in fairly uniform systems. Redundancy of functions is believed to be typical in soil, and this could explain the reason for the lack of any observed relation between microbial diversity and soil functions. No relationship has been found between microbial diversity and decomposition of organic matter, and a reduction in any group of species has little effect on overall soil process since other micro-organisms can carry out this function.

In conclusion, despite the recent progress in methods, the relations between genetic diversity and taxonomic diversity are not well understood, and even less is known about the manner in which genetic diversity and taxonomic diversity of micro-organisms affect microbial functional diversity.

HOLISTIC APPROACH AND SOIL FUNCTIONING

The extraction of large amounts of DNA and RNA from soil and the accurate taxonomic and functional characterisation of these acids can increase our knowledge of the composition of soil microflora, of pathogenicity, of trophic interactions between soil micro-organisms and of the production of bioactive molecules by soil micro-organisms. These have important implications for agriculture, the environment, industry and pharmacology. However, the molecular approach does not seem sufficient for a better understanding of soil functioning (Fig. 18.2). In this regard it is also important to understand the link between community structure and functions in microhabitats, that is, the spatial distribution of microbial species and their spatial and temporal links with the various microbemediated reactions *in situ*. For this reason, high-resolution techniques must be applied to soil preparations to be examined by electron microscopy. Ultracytochemical methods have been successfully applied to locate microbial cells in the soil matrix without giving any taxonomic characterisation. It is also difficult to associate microbial activities with the locations of species within the soil matrix; the presence of electron-dense components in the soil has limited the location of active enzymes to microhabitats associated with microbial and root fragments. Thus, acid phosphomonoesterase activity has been detected in roots, soil micro-organisms and small (7 nm × 20 nm) fragments of microbial membranes, but not in clays or humic particles.

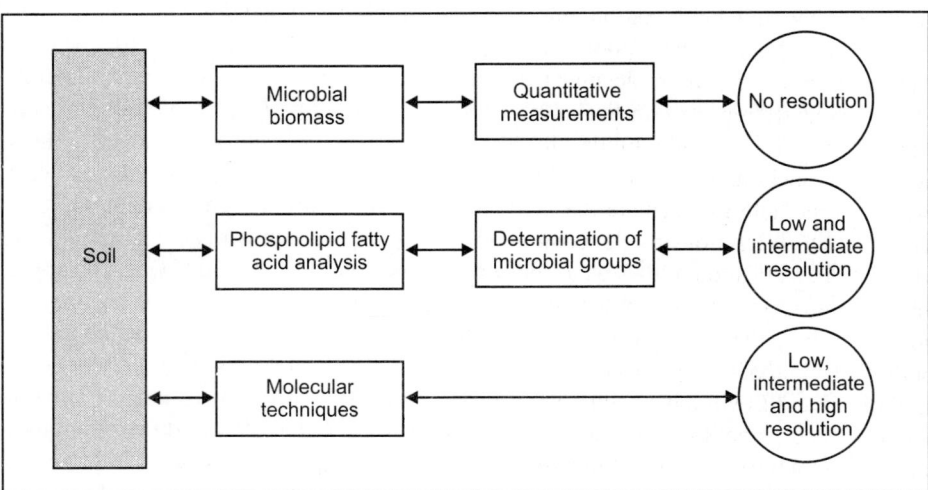

Fig. 18.2. Scientific insights provided by determining the composition of microbial communities or by using the holistic approach in soil.

However, it is not always rewarding to determine the composition of soil microflora and the various microbially mediated activities when assessing the functionality of the soil. Determining the composition of the soil microflora, the concentration of each metabolite, and the rate of each transformation reaction can be not only time-consuming, laborious and expensive, but also unnecessary for quantifying nutrient cycling or assessment of soil quality.

Information on transformation rates of nutrients in soil has been obtained with labelled compounds, such as ^{14}C-, ^{13}C-labelled or ^{15}N-enriched compounds. In the holistic approach, the system is partitioned into pools with a functional meaning, and fluxes between these pools represent physical (such as leaching and volatilisation) or abiotic or biotic transformations. Then, the distribution of the isotope (reflecting the behaviour of the added compound) between the various pools can be followed, and the behaviour of the added compound can be discriminated with respect to that of the native C or N. Particular attention has been given to nutrients, such as N, present in a large organic reservoir. For example, the most popular models for describing N mineralisation and immobilisation consider microbial biomass as an undifferentiated whole, and a significant source and sink of nutrients, and consider microbial decomposition of organic matter and microbial synthesis occurring simultaneously in soil. When determining the mineralisation and immobilisation turnover of N (MIT) conventionally we assume that ammonium is the end product of the decomposition and that the ammonium may be immobilised by a proliferating microbial population or taken up by plant roots. Alternatively we may assume the so-called direct route, which involves the uptake of simple organic molecules (such as amino acids) directly through cell membranes. Once assimilated, amino acids are deaminated, and only the surplus N is released (i.e. mineralised) into the extracellular ammonium pool in the soil. Barraclough demonstrated that both MIT and the direct route can operate concurrently in the soil. The organic pool is also measured as an undifferentiated pool, even though we know that it is heterogeneous in terms of biological activity, because a proportion of it cycles rapidly whilst some components cycle very slowly. A better quantification of N transformations in soil can be obtained if we split the organic N pool into at least the more resistant and less resistant N organic pool, and the microbial biomass into the active and the inactive microbial pool. Current models represent both pools, but their N content is not measurable. Nannipieri proposed to set up methods based on the molecular techniques for determining new microbial pools in soil. Radajewski has recently reported an example of this approach. He separated the ^{13}C-DNA extracted from soil treated with a ^{13}C source by density-gradient centrifugation and characterised the ^{13}C-DNA taxonomically and functionally by gene probing and sequence analysis. This technique is very promising because we might be able to calculate the active microbial C pool by multiplying the ratio between labelled and total DNA fractions by the content of microbial C in the soil. In addition, the taxonomic and functional characterisation allows us to understand more precisely the changes in the composition of microbial communities affected by the added C-substrate.

Instead of measuring the composition of soil microflora, it might be less laborious and time-consuming to monitor the behaviour of key species which can function as indicators of the status of the soil microflora. However, Kimball and Levin have been concerned about the use of single populations to monitor the response of a soil ecosystem to perturbations because this response might not predict what will happen throughout the community and cannot identify all the interactions that may be altered.

CONCLUSION

Our understanding of the links between microbial diversity and soil functions is poor because we cannot measure easily the microbial diversity, even if we can detect unculturable micro-organisms by molecular techniques. In addition, the present assays for measuring microbial functions determine the overall rate of entire metabolic processes, such as respiration or specific enzyme activities, without our identifying the active microbial species involved. According to O' Donnell, the central problem with the link between microbial diversity and soil function is to understand the relations between genetic diversity and community structure and between community structure and function. The recent advances in RNA extraction from soil might permit us to determine active species in soil. Further advances in understanding require us to determine the composition of microbial communities and microbial functions in microhabitats.

Griffiths showed that the effect of microbial diversity on microbial functions depends on the function measured. Some functions increased (substrate induced respiration, SIR) with decreasing microbial diversity in soil, others were not affected (thymidine and leucine incorporation, NO_3^- accumulation, respiratory growth response), and some declined when microbial diversity was small (short-term respiration from added grass, potential nitrification rates, BIOLOG®). No relation exists between microbial diversity and decomposition of organic matter, and a reduction in any group of species has little effect on overall soil process because the surviving micro-organisms can carry out the decomposition of organic matter. The use of molecular techniques has improved the determination of the composition of soil microflora (Fig. 18.2). Recent advances include bacterial artificial chromosomes (BAC) cloning libraries, which can allow the functional and taxonomic analysis of large segments of soil DNA (metagenome) with further insights into pathogenicity, competitiveness, substrate range and bioactive molecule production by soil micro-organisms. However, the determination of the composition of microbial communities in soil can be unnecessary for a better understanding of soil functions. For example, the holistic approach with the use of labelled compounds allows us to determine the distribution of nutrients among the various pools. However, further advances in quantifying nutrient dynamics in soil require the determination of new pools, with new molecular techniques. With regard to this, Radajewski separated ^{13}C- and ^{12}C-DNA, both extracted from soil treated with a ^{13}C source, by density-gradient centrifugation, and then ^{13}C-DNA was taxonomically and functionally characterised by gene probe and sequence analysis. This technique may also allow us to calculate the active microbial C pool if we multiply the ratio between labelled and total DNA by the microbial biomass C content of soil.

Soil microbiologists and biochemists have to consider carefully the meaning of determinations. The measurement of microbial biomass by the fumigation technique was an important step for a better quantification of nutrient cycling in soil because it allowed the determination of the microbial C, N, P and S content. However, it is conceptually wrong to use this method for determining microbial activity. The C/N ratio of microbial biomass can give indications of the relative prevalence of fungi over bacteria and *vice versa*. However, microbial composition can be determined by phospholipid fatty acid (PLFA) analysis or molecular techniques (Fig. 18.3), depending on the degree of resolution required.

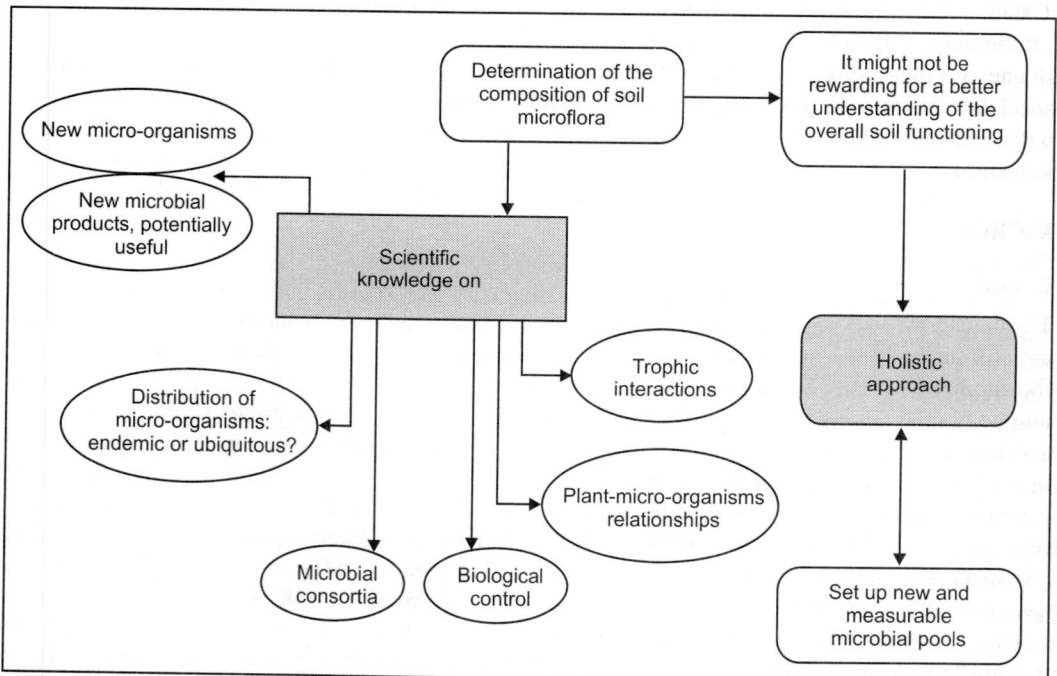

Fig. 18.3. Taxonomic resolution of methods used to determine microbial biomass and microbial composition in soil.

We should use these new techniques with caution. We sometimes tend to accept the new methods without critically considering their limits. With regard to this, we should note the behaviour of some soil microbiologists who use the community level physiological profile (CLPP) as a new technique to determine microbial diversity although it is culture dependent, but criticise the use of plate counts and suggest using molecular techniques to determine microbial diversity.

Chapter 19

Microbial Biodiversity: Strategies for Its Recovery

INTRODUCTION

The microbial world traditionally consists of all organism groups that can be seen only with a microscope, i.e. the fungi, bacteria, archaea, algae, protozoa, and a number of newer and less known lower eucaryotes. If one considers the biodiversity of all these groups, the task is enormous and the subject of many specialised volumes. Several volumes have been devoted to microbial diversity. Hence, this chapter will focus on the bacteria with some comment on the fungi, the two groups of greatest interest in industrial microbiology, and on recovering diversity through retrieving DNA from nature.

MICROBIAL DIVERSITY ON EARTH

Extent of Microbial Diversity on Earth

The extent of bacterial diversity is unknown and without a rationally based extrapolation. This is not the case for fungi. The number of known species is about 72,000. Hawksworth has conservatively estimated the number of fungal species to be about 1.6 million based on an experimentally determined ratio of six unique fungal species per plant species times the 2,70,000 plant species in the world, a reasonably well accepted number. This does not account for species associated with insects and perhaps different ratios of fungal species per plant species in the tropical and polar regions. The case for bacteria is far more primitive. Only about 4200 species have been described. It is widely recognised that this represents, at best, only 0.1 to 1 per cent of the organisms in nature. No rational attempt has been made to extrapolate the global extent of bacterial species as has been done for fungi, because too many coefficients for such an extrapolation are unknown. Several lines of evidence, however, do suggest that the number of bacterial species is much higher than the currently known number of species. First, DNA reannealing studies done by Torsvik on Norwegian forest soil DNA showed that there are 4000 nonhomologous bacterial-sized genomes in a 30 gram sample of soil. If one then factors in the 70 per cent DNA-DNA hybridisation criterion for a species, the unknown relationship of this number to soil sample size as well as a typical species abundance profile, it would not be difficult to reason that a gram of fertile soil could contain several to many thousand species. Second, small subunit ribosomal DNA genes (SSU rDNA) obtained from soil in many studies show extremely high diversity with virtually no resampling of the same clone, even in very large clone libraries. Furthermore, most of the sequences from these clones do not match sequences in the database within 1 to 2 per cent similarity, a rough estimate of species level resolution. This high level of SSU rDNA diversity is consistent with the evidence of high diversity from the DNA reannealing studies. Third, new isolate collections from nature often show that one-third to two-thirds of the strains in the new collection do not match known species with an acceptable level of identity in existing fatty acid methyl ester (FAME), BIOLOG, API, and classical phenotypic databases or descriptions. This indicates that even among the culturable isolates, diversity is much higher than represented by the described species. Fourth, there is no evidence that there is a decline in the reporting of new bacterial taxa, if the annual weight increase in the International Journal of Systematic Bacteriology is any indication.

Furthermore, the new taxa are not simply new species, but often new genera or even families. Some of these taxa would be expected to include species-rich groups. More problematic is the number of very unique organisms that are isolated but die on the shelf because there are few funds to support bacterial systematics, and the effort needed to describe the especially unusual organism can be daunting. We will never know whether these orphans harbour unique physiology, produce valuable pharmaceuticals, or play major roles in biogeochemical cycles.

While the above evidence documents that the extent of bacterial diversity is very high, the absence of information on how different organisms are at different spatial scales and to what degree there are geographic species severely limits our ability to estimate global bacterial diversity. May described a relationship that showed that biodiversity increases as the body length of the organism decreases, at least down to organisms of several millimeters in length. For organisms smaller than this, however, there may not be separate species in different climatic and geographic regions as there are for larger organisms. This is an important issue that remains to be resolved. Furthermore, the above discussion does not consider the variety of different niches known to harbour prokaryotes, e.g. special symbioses, extreme environments (temperature, pH, salt, pressure and their combinations), and novel energy-generating biochemistry. Procaryotes have been on earth perhaps 3.8 billion years, much longer than higher life forms. Evolution should have created enormous diversity during this extensive period of time. Some extinction would of course have occurred, but the planetary changes during this time are not ones thought to have been particularly lethal to most procaryotes. Hence, this long evolutionary period would also argue for high procaryotic diversity. Given all the above arguments, a global bacterial diversity of 10^5 to 10^6 organisms would not be unreasonable. The global biodiversity assessment uses the estimate of 10^6 for bacteria.

Importance of Microbial Diversity

Microbial diversity has several important values to society and to the earth's ecosystem.

1. Micro-organisms are of critical importance to the sustainability of life on our planet, including recycling elements on which primary productivity depends, producing and consuming gases important to maintaining our climate, and destroying the wastes of human civilisation.

2. Discoveries of microbial biodiversity expand the frontiers of knowledge about the strategies and limits of life, including microbes that live at the extreme conditions known for life and ones that have evolved novel redox couples for capturing energy.

3. Microbial diversity represents the largest untapped reservouir of biodiversity for potential discovery of new biotechnology products, including new pharmaceuticals, new enzymes, new speciality chemicals, or new organisms that carry out novel processes.

4. Microbes often play key roles in conservation of higher organisms and in restoration of degraded ecosystems. Hence, microbial diversity goes hand in hand with goals for maintenance of higher organism diversity.

When many think of industrial interest in microbiology, it is often assumed that biotechnology products are the only focus, but all of the above points are important to industry. Microbial treatment of waste streams is critical to acceptable and economic industrial practice in the modern world; microbes and their enzymes that withstand extreme conditions have obvious potential value for new processes; and restoration of mining, production, and harvesting areas are critical to some industries' existence. These topics are also important to the increasing focus by industry on 'green' chemistries and to life cycle analysis of new products.

Beyond the more global values of biodiversity, there are variants in particular organism traits, i.e. particular phenotypes, that are of interest. Classes of traits for which variation (diversity) may be of value include particular kinetic properties, especially K_m and V_{max} values; tolerances to high or low temperatures, high or low pH, and high solute concentrations, e.g. Na^+, high or low pressure; the ability to attach or to be mobile; the use of particular electron donors or acceptors; and the avoidance of predators, i.e. grazing. Other phenotypic traits of biodiversity that can be of interest are the rates, yields, and efficiencies of production of desired products; enzymes with unique properties such as high temperature or alkaline stability, or high turnover at low temperatures; particular regulatory properties; and genetic stability of strains.

The above phenotypic properties are often why particular strains are sought and why particular strains become a focus in research and production. The range of diversity in these traits in nature, or whether there are evolutionary trade-offs, e.g. between K_m and V_{max}, is usually not known. New genetic combinatorial approaches are now used in an attempt to create diversity in the laboratory since it may be easier to recover diversity there than from nature. But, in principle, given the 3-billion-year time span of evolutionary history of prokaryotes, a high degree of natural diversity in many of these traits can be expected to be extant.

The Problem

Given the probable existence of a huge variety of interesting microbial phenotypes, the challenge is the recovery of those traits for study or use. This entails several challenges, including strategies for site selection and sampling, the major problem of nonculturable microbes, DNA recovery and its expression, and efficient screening. Two major strategies are now used for recovery of microbial biodiversity from nature, and they are the subjects of two sections of this chapter—recovery of culturable organisms and recovery of the genetic blueprint, DNA.

Where is New Diversity to be Found

Two factors determine where particular organisms reside: their selection, i.e. growth and colonisation in a particular habitat, and the degree of dispersal. Before evaluating these two forces, however, microbiologists have the problem of establishing whether an organism found in a habitat really successfully competes and lives in that habitat versus being a visitor. The latter is problematic because of the great ability of some microbes to survive long periods in environments outside of where they successfully compete. Such transients can be found in many places and do not factor into using ecological principles to evaluate distribution of diversity. As a practical matter, however, any population that is present in high numbers in a particular habitat must be considered a living resident of that habitat, because only with growth could it achieve such dominance.

A key ecological principle is that the most fit organism is the most successful in its niche. Hence, different niches would be expected to harbour different organisms, i.e. biodiversity. Major selective forces for the resulting populations are the energy source, especially the type and amount of carbon; the electron acceptors, especially oxygen or types of alternative electron acceptors used under anaerobic conditions; and stress factors that might demand tolerance features, e.g. resistance to pH or temperature extremes. Most of these factors are chemical; hence, site chemistry would be expected to be a major determinant of diversity. A sampling strategy that is based on different site chemistries should yield higher diversity. Site chemistry could mean a difference in some populations under oak trees versus maple trees versus a particular grass, or a calcareous soil versus a weathered clay soil. Few studies have been done (indeed few have been feasible) to actually define how the level of microbial diversity changes with a gradient of, for example, soil chemistry. As one example, different serogroups of *Bradyrhizobium japonicum* are known to occupy soils of slightly different pH (1 to 1.5 units), an example of a fine level of phenotypic difference dictated by a small change in soil chemistry.

Since the heterotrophic community in soil and water is usually energy starved, the particular carbon resources should be a dominant driving force of selection. An example of the profound influence of vegetation type on microbial community structure is shown in Fig. 19.1. In this case, soil DNA was extracted from two adjacent soils of identical history and chemistry until the native forest was replaced by pasture on a portion of the site approximately 80 years ago. Approximately one-quarter of the soil DNA had a different guanine + cytosine (G + C) composition as a result of the shift to pasture vegetation. Furthermore, some of the DNA of the same G + C composition under the two vegetation types is likely to be from very different organisms with different traits. This is supported by finding different SSU rRNA sequences in soils of the same G + C content under the two different vegetations. Hence, vegetation is a strong driving force in bacterial community selection and should be a primary consideration in schemes to recover new diversity.

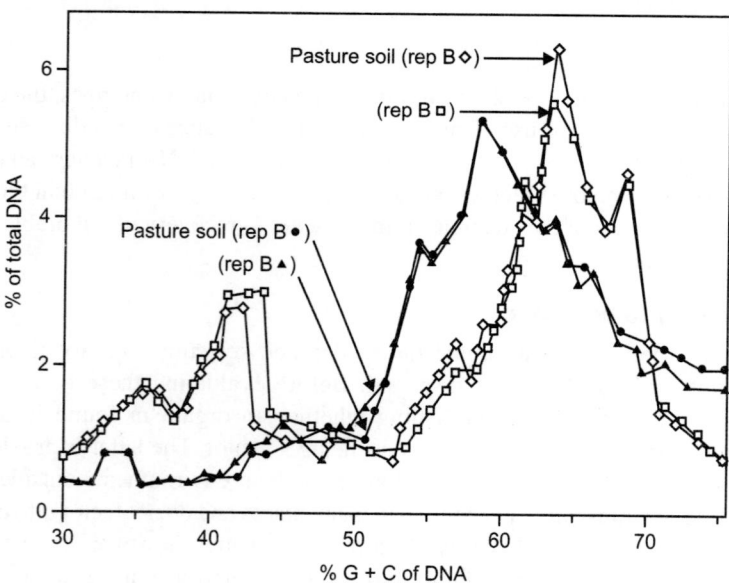

Fig. 19.1. Per cent guanine + cytosine (G + C) profile of DNA extracted from Hawaiian soil of identical parent material, climate, and vegetative cover until one portion was converted to pasture for cattle grazing approximately 80 years ago. Data for two replicate soil samples are shown.

The G + C fractionation method shown in Fig. 19.1 is also a methodology that allows one to recover DNA from more minor members of the community or to target DNA from particular taxonomic groups defined by G + C content. This method is described by Holben and Harris. Briefly, DNA is mixed with *bis*-benzimidazole, which binds to the adenine and thymidine and changes the buoyant density of the DNA in proportion to the G + C content. A gradient of G + C concentrations can then be established by equilibrium density gradient ultracentrifugation. Fractions can be collected with a fraction collector, and the DNA in particular fractions can be amplified by PCR, cloned, hybridised to probes, or analysed as appropriate. The G + C content of each fraction is established by using a standard curve relating G + C content to density measured with a refractometer.

Another type of selective habitat is that provided by a host organism such as a particular part of a plant, insect, animal, or other high organism. Bacterial colonisers of these habitats are often specialists and are not free living. As a result, they may be difficult to cultivate. Nonetheless, such symbionts are thought to be a rich source of microbial diversity. In a few cases where bacterial colonists of invertebrates have been extensively examined, the recovery of novel genera and species has been high, suggesting that the invertebrate hosts are a rich source of novel bacterial diversity. Since there are an estimated 10^8 insect species, this could make insects a very rich source of microbial biodiversity.

The second major factor that determines where new microbial diversity is located is the degree of dispersal of microbial species relative to the local rate of accumulated genetic variation. Another way to evaluate this question is to consider what is the degree of bacterial endemism, i.e. the extent of localised genetic and phenotypic uniqueness. As discussed previously, the answer to the question has a major impact on the extent of microbial diversity on Earth.

The center for microbial ecology has begun to evaluate this question by examining the degree of endemism in organisms that grow on 3-chlorobenzoic acid, a fairly rare trait in nature. Approximately 150 isolates were examined from six continental regions, which included four undisturbed Mediterranean ecosystems (Western Australia, southwest South Africa, central Chile, and central California), and two undisturbed boreal forest ecosystems in north central Canada and in northwest Russia. Using repetitive extragenic palindromic PCR (rep-PCR) to indicate genotype, no globally dispersed genotypes were found but several genotypes were regionally endemic. This study suggested a high degree of endemism in this population. The degree of endemism may, however, vary with the type of species. Some species may disperse, survive, colonise, and be more genetically stable, while others may not.

For host-associated organisms, whose hosts are endemic, one would expect some level of endemism. For the nonhost-associated micro-organisms, the degree of bacterial endemism is far from clear. However, cosmopolitan subspecies or ecovars are likely not the rule. Hence, a sampling strategy that considers distinct geographic regions is probably wise in attempts to recover new diversity. Production of the same antibiotic is a trait that is not often conserved in the same species or subspecies, and hence sampling that uses a geographic as well as site chemistry strategy should be beneficial.

BIODIVERSITY OF CULTURABLE BACTERIA

What Level of Bacterial Diversity Matters

In this sense, level means what level of taxon resolution, e.g. genus, species, subspecies, ecovar (analogous to pathovar), and in a slightly different context, what degree of genetic difference among the organisms. The pragmatic answer is the degree of resolution that is important for the particular problem. If it is a virulent versus a weakly virulent pathogen of the same subspecies, then a fine degree of resolution is critical. If it is a comparison of pine tree and cereal grain rhizosphere populations, a more moderate level of resolution is reasonable, or if it is a comparison of ocean and hydrothermal vent communities, then a coarse level of resolution may be adequate. The pragmatic approach places little emphasis on the species or any other level in a classification hierarchy.

Bacteriologists, however, cannot avoid the need for defining a distinct kind of bacteria, that is, a species, because it is needed to communicate among microbiologists, with other scientists, and in the world of practitioners, e.g. for patents, diagnostics, quality control, quarantine, or international material transport. Hence, we need to continue to evolve the species concept for bacteria. The current recognised bacterial species definition derives from an ad hoc committee on reconciliation of approaches to bacterial systematics. The committee proposed a number of recommendations to combine genotypic and phenotypic data leading to a polyphasic approach, which is the current standard for bacterial taxonomy. This committee proposed 'a bacterial species as a group of strains, including the type strain, sharing 70 per cent or more DNA-DNA relatedness'. Furthermore, phenotypic characteristics should agree with the phylogenetic data, and it was recommended that a genospecies, although distinct, that cannot be differentiated on any known phenotypic grounds cannot be renamed. However, the 70 per cent criterion presents its own technical limitations, including the need for pure cultures, imprecision of the method, lack of suitability for reference databasing, and the virtual impossibility of making all pairwise hybridisation comparisons in a collection. This criterion also lacks a theoretical basis that would explain why two-thirds of the genome should be more highly conserved than some other portion, and it cannot be related to the role of natural selection in determining differential reproductive success that provides the theoretical basis for the species criteria among eucaryotic organisms. It does have the advantage, however, over other species definitions of providing a quantifiable criterion.

Until the relative phylogenetic relevance of any kind of information can be dependably evaluated for bacteria, distinction of one kind of bacterium from all others should be based on every kind of information that can be obtained by the methods available, rather than on one or a few kinds of traits. This has led to some new attempts to define bacterial species. At a biodiversity workshop in 2005, the following definition was proposed. A bacterial species can be defined as 'a group of related organisms that is distinguished from similar groups by a constellation of significant genotypic, phenotypic and ecologic characteristics'. A bacterial species defined in this way is at once a naturally occurring group of like organisms and a taxon sufficiently defined by the collected properties of the group to distinguish the species from closely related groups. With this meaning, a bacterial species should be comparable to eucaryotic species in its adequacy as a unit of diversity in both systematics and ecology. An addition to this definition not present in others is the recognition of ecological characteristics as a determining criterion, which helps bring the microbial criteria in line with reality for higher organisms. Obviously, the particular species definition used is a major factor in any estimate of the global extent of the bacterial species, but it seems reasonable that the definition should have a meaning as similar as possible to that for higher organisms. More recently, DeVoss offered a possible definition of the polyphasic species as 'a group of strains which originated from a common ancestor population in which the steady generation of genetic diversity and recombination after the introduction of foreign DNA, resulted in clones with different degrees of variation but still sharing a significant degree of DNA relatedness and with a common phenotype'. Currently there are a large number of techniques to determine a common phenotype. These include standardised phenotypic tests, such as BIOLOG and API galleries; chemotaxonomic traits such as fatty acid profiles, polyamines, and sodium dodecyl sulphate-polyacrylamide gel electrophoresis patterns of cellular proteins; immunologic data; antibiotic resistance; and morphological characteristics.

For genotypic characterisation there are also a variety of techniques including amplified rDNA restriction analysis; sequencing of the 16S and 23S rRNA gene; ribotyping; rDNA intergeneric spacer region restriction analysis; amplified fragment length polymorphism; rep-PCR using REP, BOX, or ERIC primers; randomly amplified polymorphic DNA analysis; and DNA-DNA hybridisation.

In recent years, SSU rRNA gene sequencing has become an extremely popular method to help identify new isolates as well as nonculturable microbes when their DNA is cloned from microbial communities. The rRNA gene, however, is highly conserved, which makes this methodology alone inadequate for providing insight into the physiology and ecology of the organism. It cannot be relied on to routinely provide a species-level identification according to any of the above definitions. Figure 19.2, illustrates this point. Some strains with greater than 99 per cent 16S rRNA sequence homology do not meet the greater than 70 per cent DNA-DNA reassociation cirterion. It was found that 'the strength of sequence analysis is to recognise the level at which DNA paring studies need to be performed, which certainly applies to (rRNA) similarities of 97 per cent and higher'. Because of the great current interest in and ease of rRNA analysis, there is a danger that the species-level identification and new description will not be adequately done, leading to misrepresentation of microbial diversity information.

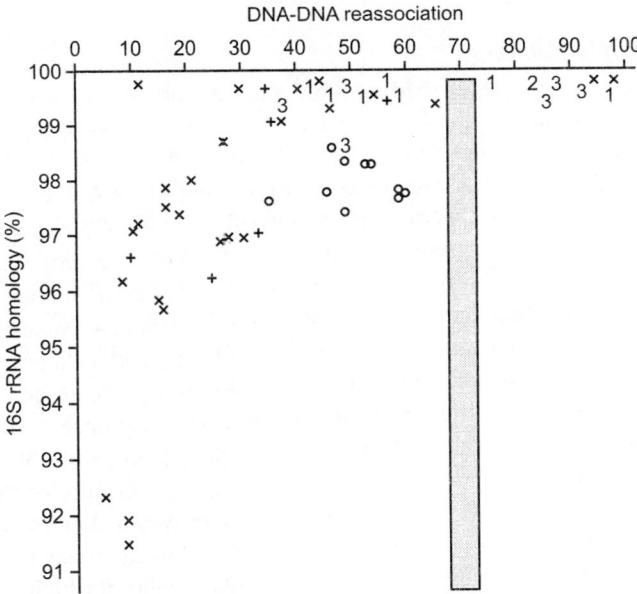

Fig. 19.2. Comparison of 16S rRNA homology and DNA-DNA reassociation values. ×, membrane filter method; #, renaturation rate method; 1, renaturation rate method; 2, renaturation rate method; 3, renaturation rate method; O, S1 nuclease method. The bar indicates the DNA threshold value for species delineation.

Isolation Strategies

Requirements

Skilled microbiologists believe that many nonculturable bacteria will become 'formerly nonculturable' with careful investigation into the physiology and nutrition of the mixed culture, patience, and cleverness. There are many success stories, but all have one thing in common, that it takes considerable time to finally isolate a fastidious nonculturable microbe. Nothing, however, substitutes for the personal skill and insight that some microbiologists have for nurturing difficult-to-grow bacteria, the 'green thumb' of microbiology. The skill of the scientist is the most important strategy.

The ability to track the target population during enrichment is extremely important to gaining insight into what conditions favour or inhibit growth of this organism over the others in the community. Isolation conditions can be constructed from this information. Identification of a means to track the target organism's growth can be challenging without having a pure culture to identify the unique target. Microscopy remains a primary method if there are any morphological clues to distinguish the target organism. Antibodies, oligonucleotide probes, and stains can also aid the recognition in the microscopic field. Non-microscopic molecular methods and chemotaxonomic methods can also be useful if there is a signature to be tracked. Effort spent on identifying a means to track the target population has a good probability of leading to a successful isolation.

One difficulty in isolating nonculturable microbes is that the cell separation step and the cultivation step are usually commingled so that one does not know which is the bottleneck. An ability to resolve those two components to determine which remains the problem helps focus the effort on the real problem preventing isolation.

Mimicking the organism's habitat

The classical enrichment technique was based on mimicking the organism's habitat and using natural selection to enrich the desired population over other members of the community. The technique has been powerful in microbiology, but it can do even more if carried to another level of sophistication. All aspects of mimicking the environment need to be considered. For example, might phosphorus (used as a buffer) be inhibitory, could the atmospheric oxygen concentration be inhibitory, or are the inorganic ions in the medium supplied in inappropriate concentrations or ratios compared to those in the natural habitat? As an example, a very low ionic strength medium was recently found to be necessary to cultivate the formerly nonculturable acidophilic methanotroph; their natural peat bog habitat has an ionic strength much lower than that of conventional mineral media. The carbon content of standard media, such as tryptic soya broth and nutrient broth, has been shown to completely inhibit growth of many important groups of soil bacteria. The RZA medium, originally designed as a relatively low-nutrient medium to recover oligotrophs from waters, has proved much more satisfactory for cultivating a variety of new and diverse types of bacteria from soil and water. A major recommendation in using the mimicry approach is to thoroughly consider all aspects of the targeted population's natural and laboratory environment and to make sure they are as similar as possible.

Rather than trying to construct optimum concentrations of the strain's resources in a defined medium, an alternative is to use gel-stabilised gradients or combinations of gradients. In principle, each of the three spatial dimensions can be used for a gradient to supply different concentrations of resources. The organism is then allowed to find its own optimum for growth within this combination of gradients. Once that combination is identified, those concentrations can be used in a defined medium. Steady-state gradients also more closely mimic the natural condition where the resource pool size is maintained by its natural turnover.

Some bacteria cannot be cultivated because they are members of a tightly interdependent food web. Co-cultures, helper bacteria, and extracts from these bacteria can be used to help cultivate the desired organisms. As an example, isolation of the anaerobic fatty acid-oxidising bacteria was initially successful only because of the use of co-cultures of hydrogen consumers.

Patience

Many of the difficult-to-cultivate organisms from nature seem to be naturally very slow-growing organisms. This is especially true of a number of the obligate anaerobes and some of the aerobic oligotrophs. In the former case, it is not unusual for a skilled microbiologist to take 2 years to isolate a novel anaerobe. Many microbiologists, having grown up with *Escherichia coli* or *Pseudomonas* spp. in laboratory exercises, mistake slow growth for no growth. Hence, patience is needed to determine that growth actually has occurred, and turbidity may never be seen.

Hattori's group has effectively shown how both low-nutrient media and long incubation times (2 months) can be used to recover an order of magnitude higher numbers of bacteria from soil. Furthermore, the physiological and taxonomic features of the bacteria appearing on plates after a 1-month incubation period were very different from those that appeared in the first week of incubation. This cultivation approach shows that the number of CFU can come closer to approximating the number of organisms seen by direct count. The oligotrophic media they use is 1/100-strength nutrient broth. The oligotrophic isolates, which constitute 82 per cent of those colonies found in the second month (Fig. 19.3), do not grow on full-strength complex media. Hence, this method shows that patience for the length of incubation and recognising the appearance of tiny colonies yields a large number of newly culturable strains.

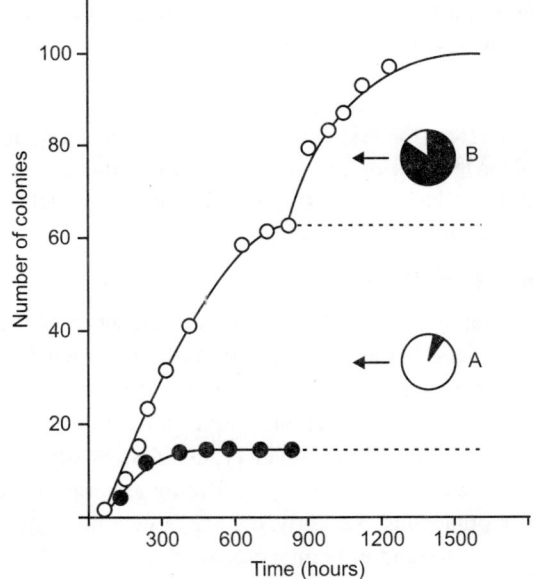

Fig. 19.3. Comparison of the number of colonies on nutrient agar (●) and dilute (1/100) nutrient agar (O). The proportion of oligotrophic isolates at incubation time is shown by the solid portion of the pie chart. (A) 5 per cent of the isolates were oligotrophs; (B) 82 per cent were oligotrophs.

Suppressing weedy bacteria

A major limitation in isolation of new bacterial diversity is fast-growing, nonfastidious bacteria that outcompete slow growers in enrichments or on agar plates. These bacteria, which have the characteristics of weeds, are particularly problematic for isolation of slow-growing bacteria that are present in low numbers. The first strategy to reduce growth of weedy bacteria is to reduce the carbon concentration in the medium, as illustrated in Fig. 19.3. Fast-growing bacteria (high μ_{max}) typically have high K_s values, while slow-growing bacteria have a low K_s. Hence, low substrate concentrations often allow the slow growers to outcompete the fast growers.

Since carbon is the most limiting resource to heterotrophs in nature, it is to be expected that the predominant bacteria in nature are specialised slow growers. A more sophisticated approach to recover low-μ_{max}, low-K_s micro-organisms is by use of a chemostat fed very low concentrations of substrate. This is an effective but labour-intensive isolation strategy.

A second approach to reduce weedy bacteria as well as to impose some selection is to use antibiotics in the enrichment. If one has a means to track the target organism or its activity, appropriate antibiotics can often be identified.

A third approach to reduce weedy bacteria is to omit combined nitrogen from the medium and determine if the desired population fixes N_2. Nitrogen fixation is widespread in bacterial groups, and some slow-growing bacteria can fix nitrogen.

Resuscitation and avoiding the stress of cultivation

Some bacteria are injured in nature, and hence care must be taken for their recovery. Also, some bacteria adapted to natural conditions cannot withstand the conditions of cultivation. Low carbon concentrations in the media, microaerophilic conditions, a supply of critical growth factors, and use of moderate temperatures are means to aid the transition to laboratory cultivation.

Physical separation

Physical separation of target cells from the rest of the community is an alternative to enrichment for the separation phase of isolation. Methods of physical separation include motility to an attractant, micromanipulation more recently by laser tweezers, antibody-linked magnetic beads, and separation by density gradient centrifugation in a nontoxic matrix.

Has the Isolated Strain been Seen Before

Once a new isolate has been obtained, the first question is often whether it is novel and, if so, how different it is from what has been seen before. This question can be directed at two levels: is the strain in the collection and hence already counted or screened, or, at a taxonomic level, has it been identified as belonging to an established taxon? To answer these questions, rapid methods are needed so that resources can be focused on discovery of the novel types. This is often termed dereplication by microbiologists in industry. Methods such as FAME and Biolog are automated rapid systems that can be a first-level screen for certain organisms. Recently, the use of genomic fingerprints obtained by PCR amplification using primers binding to randomly interspersed repetitive DNA sequences (rep-PCR) has gained favour. This method is rapid and reproducible, is suitable for a database, and provides the highest level of taxonomic resolution of any current PCR-based method. There are three primer sets that have been found to work in most bacteria; these are known as REP, BOX and ERIC. Of these, BOX is the most generally useful because it works reliably on new strains and gives the most amplicons and hence a higher degree of resolution among strains. These fingerprints reflect chromosome structure and hence provide resolution at the species-subspecies-ecovar level. Rep-PCR of most bacteria can be done from the cells by using the PCR protocol to lyse enough cells to provide target DNA for the primers. In some cases the DNA must first be extracted, which makes the method more time consuming. The gel images can be digitised, stored in TIFF files, and analysed by software clustering programmes. A complete protocol for use of rep-PCR and GelCompar software for pattern analysis and database construction has been described.

Fungal Biodiversity: Isolation and Identification

Certain fractions of the mycota of temperate soils are reasonably well characterised, and methods for the isolation of these fungi and their identification are available. Other general references that are useful in the identification of soil fungi include von Arx, Farr and Hawksworth provides a synthesis of the biology, identification, and isolation of a unique subset of soil fungi, the nematode-destroying fungi. The saprotrophic basidiomycetes, in all probability the most important group of fungi in the initial delignification of plant debris entering soil, are poorly represented by methods aimed at isolating soil fungi and are greatly under-represented in discussions of the soil mycota.

New methods for the selective isolation of these and other under-represented groups of soil fungi are presented by Thorn and Bills. Since the saprotrophic basidiomycetes in culture lack sufficient morphological characters or literature for their identification, identification is best attempted using DNA sequencing (e.g. nuclear 18S or 25S rDNA) and placement in a phylogenetic framework of known sequences.

RECOVERING BIODIVERSITY USING ENVIRONMENT DNA

Accessing Uncultivated Microbes

The art and science of cultivating microbial isolates has provided the basis for virtually all of our fundamental knowledge of microbial physiology and diversity. Within the past decade, however, there has been a growing appreciation that the microbes amenable to growth in the dense, monocultural state referred to as 'pure culture' may not be representative of most microbes in the environment. This appreciation has been driven largely by results from the application of molecular techniques on nucleic acids extracted directly from environmental samples. With few exceptions, these studies have shown that the organisms readily cultivated from environmental samples are frequently minor components of the resident microbial population. The percentage of (culturable) microbes varies among environments, with a general theme being that nutrient-poor environments tend to harbour microbes that are more recalcitrant to growth in laboratory culture. There have been many explanations given for this recalcitrance, with most centering on the scrupulous nutritional requirements of the microbes, the lack of appropriate surface attachment materials, or on the difficulty in determining the precise conditions required to revive cells from dormant stages. Perhaps the most encompassing explanation for unculturability is that microbes in natural populations have evolved within the context of a microbial community where the cross-feeding and signalling relationships that have developed among individuals are extremely difficult to duplicate in a laboratory setting.

To circumvent the difficulties of cultivation, several general methods are available to directly describe microbial communities, including microscopic examination, flow cytometric methods, immunological approaches, and the analysis of lipids. Perhaps the most widely used approach, however, is to exploit the information content stored in nucleic acids extracted directly from environmental samples to examine the ecology, phylogeny and physiology of microbes from natural populations. The great advantage in analysing nucleic acids lies in the fact that a nucleotide sequence represents a high-resolution map to an organism's evolutionary history. Areas on this map can be read and compared among different organisms to derive information on phylogenetic affiliations and physiological potential. In addition, physiological activity, in the form of gene expression, can be determined by examining mRNA extracted from environmental samples.

By far the most extensively applied nucleic acid analysis method for biodiversity studies is the coupling of the PCR with 16S rRNA gene analysis. A cursory key-phrase search on Medline using the term '16S PCR' revealed 1129 citations, a testimony to the broad application and ease of use of this method. The power of this approach comes from the ability to amplify, clone, and analyse homologous regions of 16S rDNA from vanishingly small amounts of sample DNA. This is particularly important when the sample is from environments where DNA is difficult to extract or separate from contaminants, such as in certain soil environments, or when the resident microbial population is present at very low abundance, such as in oligotrophic waters or in deep subsurface environments. Variations of the 16S approach such as denaturing gradient gel electrophoresis or *in situ* hybridisation have been used to examine the complexity and structure of microbial communities, further extending the utility of this approach. Results from these studies have revolutionised our view of microbial diversity and have provided the information required to organise micro- and macro-organisms into the universal tree of life.

Despite the power of this approach, a limitation to using PCR on the 16S or other gene sequence is that prior knowledge of the sequence is required to construct the oligomers used to prime the reaction. Even the so-called 16S 'universal primers', which contain degenerate sites to accommodate base variations in conserved regions, do not match all 16S sequences and will likely vary considerably from a subset of those yet to be discovered. Another drawback is that 16S sequences allow only limited inference of physiological potential. For example, the 16S-based phylogeny of the recently discovered and apparently ubiquitous Group I archaeoplankton suggests that they are hyperthermophiles when in actuality they are most abundant in cold marine environments and possess enzymes adapted to low temperature. Conversely, protein-coding gene sequences amplified from environmental samples can rarely be definitively tied to the 16S sequence of the source organism, making the sequence difficult to place within a broader phylogenetic context. It would thus clearly be advantageous to couple the power of the 16S probing approach with one that could access other portions of an uncultivated organism's genome. This would allow a coupling of phylotype to phenotype that could rapidly expand our knowledge of the phylogeny, physiology and ecology of uncultivated microbes.

Environmental Genomics

One means to connect phylotype to phenotype in uncultivated organisms is to directly clone DNA from environmental samples, probe the resultant environmental library with labelled 16S oligomers or genes, then sequence the hybridising clones to identify the resident 16S gene along with any protein-coding regions that may provide clues to the organism's physiology. Even though the hybridisation of 16S probes to a library is subject to similar constraints in uncovering novelty as in using PCR, annealing a single probe to a target gene is a less stringent event than PCR priming and may allow the discovery of more 16S variants. The initial report of this approach was by Schmidt, who discovered numerous unique eubacterial phylotypes in an environmental lambda library constructed from marine picoplankton. This study, however, focused on the 16S sequences and did not describe protein-coding regions on the clones carrying the 16S genes. The recent advent of more efficient sequencing methods and robust bioinformatics tools and databases have made practical the rapid analysis of protein-coding regions that are contiguous with the 16S sequence on larger clones. The ends of such large clones can also serve as probes against the library to isolate overlapping clones to identify additional informative sequences. Carrying this approach to its natural end, one can envision that the entire genome of an uncultivated organism could be described given a sufficiently large library. This environmental genomics approach has recently been applied to describe the physiological potential of the group I archaeoplankton that had previously been known only from partial 16S sequences. For this study, an F-factor-based vector, the fosmid, was used to construct a stable 100-million-bp library with an average insert size of approximately 40 kbp per clone. An archaeal-specific probe was used to isolate a clone from this library. Sequence analysis of this clone revealed that it contained the entire 16–23S operon as well as several protein-coding genes, hypothetical proteins and unidentified open reading frames. These sequences were sufficient to provide clues to the physiology of this uncultivated organism. For example, sequence from the end of the rRNA operon showed that the archaeoplankton lack the target sequences for streptomycin and erythromycin, the peptidyl transferase domain of the 23S rRNA, suggesting that they are resistant to these antibiotics. Conversely, a conserved residue in the elongation factor 2 gene indicated a susceptibility to diphtheria toxin protein. This clone also carried a gene encoding a form of glutamate semialdehyde aminotransferase involved in the synthesis of a precursor of chlorophyll, suggesting that the organism may have the ability to harvest energy from light. In a subsequent study the DNA polymerase gene isolated from another group I archaeal clone (*Cenarccheum symbiosum*) was subcloned and expressed. Analysis of the gene product showed that the polymerase was thermolabile, confirming that the archaeoplankton are not inactive refugees from hydrothermal vent environments but rather are likely to be significant members of many cold marine ecosystems. It is clear from these studies that environmental genomics can offer new insights into the physiological potential of uncultivated microbes and their subsequent role in ecological communities. In addition, this approach can help define the metabolic plasticity of microbes in ways that can help predict responses to environmental stresses or the conditions needed to eventually cultivate these organisms in the laboratory.

The constraints involved in cloning environmental DNA differ considerably from those involved in using PCR. Primary among these is the amount of DNA required. Leaving aside for a moment the inherent biases involved in PCR and cloning, consider that a comprehensive library of 16S molecules can be amplified from as little as 1 ng of DNA from a mixed microbial community. This amount of starting DNA would represent the combined genomes of approximately 10^5 bacteria, the amount in approximately 100 μl of coastal seawater or 1 mg of soil. To construct a comprehensive genomic library, by contrast, requires 10 to 1000 times more DNA depending upon the type of vector/host system used and the quality of the DNA preparation.

Screening Environmental Libraries

By overcoming cloning constraints, the complete microbial diversity of an environmental sample can be made accessible in the form of a recombinant library. In effect, one could propagate uncultivated organisms in the form of genome fragments in surrogate hosts. Given the sheer numbers of microbes in natural samples, screening an environmental library for a particular gene fragment or activity can be challenging. For example, Torsvik estimated that a single 30-gram sample of forest soil contained some 4000 different genomes. Converting this sample to a plasmid library of 5-kbp average insert size would require 20×10^6 clones to have the desired five-fold genome coverage needed to ensure complete representation. Screening a library of this size by standard filter lift hybridisation would require 4000 agar plates containing 5000 clones per plate, a daunting task by any measure. Alternative screening approaches are clearly needed to thoroughly screen environmental libraries. One such approach is to multiplex, i.e. create clone pools that can be screened by hybridisation, expression or PCR for an initial signal that can be broken out by subsequent screening. The size of the initial pools is determined by the sensitivity of the assay. For example, an initial PCR screen can be used to reduce the clones from the 4000 plates above to 40 pools. A positive signal from one or more of these pools can be followed by screening successively smaller pools. Another approach, which can be used in concert with multiplexing, is to apply high-throughput robotic screening. Such systems are now in place that can screen tens of thousands of clones per day for multiple enzyme activities.

Barriers and Challenges

Even though the concept of diverse, uncharacterised pools of unculturable microbes populating natural environments appears to have been largely accepted, there are key tasks that remain to be addressed in characterising these organisms. First, we have yet to learn sufficient information about an uncultivated organism from sequence data to grow it under laboratory conditions. Laboratory culture still remains an important goal and is still the only way to accurately measure many physiological traits. Second, the complete genome sequence of an uncultivated organisms remains only an intriguing concept. Funding, refinement of bioinformatic tools, and the consensus on which organism to tackle have yet to be resolved. Likely candidates would be symbiont that can be readily separated from its host or a microbe that represents a large fraction of a natural population. A candidate group that meets both of these criteria is the Group I archaea, which are a major fraction of the Antarctic picoplankton and also occur as a specific symbiont of a temperate-water marine sponge. A final goal that would benefit the field of environmental genomics would be to make a prediction of a physiological trait from sequence data and then later confirm expression of this trait with field or laboratory measurements. Several groups are now poised to accomplish this task and in doing so will further increase the interest in examining members of this largely uncharacterised, yet undoubtedly important group of microbes.

MICROBIAL BIODIVERSITY INVESTIGATION TECHNIQUES

Micro-organisms represent the foundation of life on earth. They impact life on earth through their activities that effect the biogeochemical cycles to their beneficial uses and their pathogenic properties. The survival, propagation and ability of micro-organisms to inhabit a wide variety of environments and habitats demonstrate their evolutionary success. This versatility inspired several scientists to consider how micro-organisms could impact life on other planets. There were several restrictions that limited the ability to study and to explore microbial life on earth. The majority of the traditional microbiology techniques and methods that required growing micro-organisms on lab based media, isolating them, producing pure colonies to study their physiology and morphology. However recent developments such as direct DNA extraction from water and soil samples coupled with the new sciences of Bioinformatics, Proteomics and Genomics enhanced our ability to study and describe new micro-organisms that have not been described previously. The concept that environmental factors influence the functionality of micro-organisms and its correlation with the formation of new proteins gave a new insight into microbial evolution and adaptations. This has enhanced the exploration of a vast number and variety of microbial life with a better understanding of their ability to survive and propagate in new environments. This section describes the techniques used in the isolation and describing of new bacteria from water and soil samples. These activities include periodical water sampling from selected areas of the West Bridgewater section of the Hockomock Swamp. Microbial identification followed two paths through the traditional cultural isolation procedures that included morphology and biochemistry methods to identify pure cultures of isolated micro-organisms. DNA extractions from pure colonies were compared to direct DNA extraction from samples. This added a new dimension of identification of microbial life through non-culturing methods. Other activities include PCR-denaturing gradient gel electrophoresis (PCRDGGE) analysis, 16S rRNA, phylogenic tree, fatty acids analysis and protein sequencing. Once the 16S rRNA base-pairs were obtained basic local alignment search tool (BLAST) searches was carried out to compare the nucleotide sequences were conducted against those in the National Center for Biotechnology Information (NCBI) nucleotide database at: (http://www.ncbi.nlm.nih.gov/index.html) and the European Molecular Biology Laboratories (EMBL) (http://www.ebi.ac.uk/embl/) prokaryote database. The section will include sequences of new bacteria, their fatty acid methyl ester (FAME) analysis results and phylogenies tree alignments.

Biodiversity is the variety of life on earth at all its levels, from genes to ecosystems, and the ecological and evolutionary processes that sustain it. Biodiversity is not limited to describe the various species living in a certain habitat but it also identifies the genetic variation and the functionality of these species within their ecosystems. These parameters would be useful in the identification of the many unknown and undescribed microbial species living on earth. Microbial Biodiversity is gaining more importance not only to understand their evolution but also to further determine their ecological impact. Microbial biodiversity encompasses the variability of micro-organisms. A publication by the Center for Microbial Ecology in 2003 suggests that there may be up to 1 million species of prokaryotes yet only 3100 are known and have been fully described in Bergey's Manual. Micro-organisms are essential for life on earth to function, maintained and continue. They play many roles on land, air and water, including being the first to colonize and transform effects of naturally occurring and man-made disruption to the environment. Nations has committed to protecting the environment, and has included ecological diversity in this goal.

Bacteria were the only free living cellular organisms some 3.5 billion years ago. Their survival, propagation and ability to inhabit a wide variety of environments demonstrate their evolutionary success. This remarkable versatility inspired several scientists to consider the ways in which this may impact other planets. There have been discussions that some form of microbial life does exist on some planets. There have also been discussions on the issue of planetary protection and the affect of possible deposition of bacteria, such as *Bacillus,* on other planets such as Mars reflecting the ability of micro-organisms to survive and propagate in any environment. The fact that there may be microbial life on other planets does signal the need to explore the vast amount of microbial life here on earth. The methods to explore microbial life on earth were limited through the use of traditional microbiology techniques and methods. However, the recent development of new molecular techniques including PCR denaturing gradient gel electrophoresis (PCR-DGGE) analysis, 16S rRNA, phylogenic tree, fatty acids analysis and protein sequencing have increased our knowledge of the depth of microbial life that exist. We now have a greater understanding of the vast diversity of micro-organisms inhabiting the environment from soil to the deep-sea, from fresh-water to volcanic ash.

These advances have assisted in the identification and classification of bacteria through direct DNA extraction from water and soil samples then sequencing the isolated DNA. This section review's both traditional culture dependent techniques that include molecular microbiology methods and the non-cultural that are molecular based in nature. This would include work carried out on the Hockomock Swamp with its associated wetlands and water bodies that comprise the largest vegetated freshwater wetland system in Massachusetts. Some of these procedures were already used in the isolation and identification of a new species of *Bacillus. Bacillus samanii so nov* that was isolated from snow covered soil.

The identification process involved culture based techniques, 16S rRNA sequencing and fatty acid methyl ester analysis (FAME). The phylogenic analysis based on 16S rRNA gene sequences that were deposited with the Genbank (sequence Genbank accession EF036537) indicated that the isolate belonged to the genus *Bacillus*, and the *Bacillus cereus* group. It was closest to *Bacillus cereus* with 0.40 per cent alignments and 0.69 per cent to *Bacillus thuringiensis*. The cellular fatty acid profiles were closest to *Bacillus mycoides* with the iso-C15 making 21.25 per cent of the total cellular membrane fatty acids. On the basis of phenotypic and molecular data, the strain represents a novel species within the *Bacillus cereus* group and it's named *Bacillus samanii*. A culture deposit was made to the Biodefense and Emerging Infections Research Resources Repository (BEI Resources) with the BEI code (NR-4056).

Culture Based Molecular Techniques

Isolation of pure colonies of micro-organisms

Bacteria could be grown on lab based media that would furnish their required growth factors such as sugars, proteins, vitamins and minerals. Sample must be diluted to levels that would reduce the number of growing colonies to a controllable level. Dilution series of 3 test tubes containing sterile 0.9 per cent peptone water in 9 ml tubes with 1 ml of water samples. For soil and river sediment samples 1 gm of the sample should be added to 9 ml of peptone water then the same dilution series will be used for the water samples. Using a mineral solution that would provide sources of nitrogen, potassium, sodium, magnesium, sulphate, calcium and chloride added to basal media providing the resources for the growth of various micro-organisms. The sodium requirement would ensure the media would support the growth of halophilic species which may be present in the collect samples. The same procedure of dilution would be used but this time using a vitamin solution that would support the growth of many types of Methanogens and types of *Clostridium* that would not grow other wise. The vitamin solution would contain Pyridoxine, Thiamine, Riboflavin, Vitamin B12, Biotin and folic acids in addition to others.

Fatty acid analysis

Short chain fatty acid analysis also known as volatile fatty acids (VFA), could be used for the identification of bacteria as they could serve as a second finger print for bacteria. These fatty acids which are found as part of the cell membrane bilipid layer have a length chain of 9–20 carbons. This method was used for genus and species identification that shared most of their fermentative abilities. Gram-positive bacteria had more branched fatty acid chains compared to Gram-negatives that had short chains of fatty acids as this is related to their lipopolysaccharides that make up most of their cell wall. Cellular fatty acids analysis uses gas chromatography of fatty acid methyl esters (FAME) interpret the results. Isolates should be inoculated onto Trypticase soy agar for 24 hours at their optimum growth temperatures. Four reagents are required to cleave the fatty acids from lipids in a five-step process that ends with the GC.

Genomic DNA extraction

The process involves the harvesting of cells then suspending them in TE buffer. Lysis of a microbial pellet using the Lysosomes and RNAase solution for 15–30 min. Lysate solubilisation is to be followed by DNA precipitation. The extracted DNA is washed using—1 ml 75 per cent ethanol then at the last stage DNA solubilisation with TE buffer, water or 8 mM NaOH. The extracted DNA could be amplified through PCR and sequenced to determine relativity to known species.

DGGE-PCR amplification and 16S rRNA Sequencing

Denaturing gradient gel electrophoresis (DGGE) is a molecular fingerprinting method. PCR polymerase chain reaction of the extracted genomic DNA can generate templates of differing nucleotide sequence that once the 16S rRNA fragments sequenced and the results analysed through phylogenies a firm identification and comparison could be made. This technique that has proved to be useful is identification of new species. This method is being used to study the microbial biodiversity of the Hockomock Swamp resulting in the identification of more than 12 new previously undescribed micro-organisms.

Nucleotide sequence accession number

The sequenced obtained will be deposited in the GenBank nucleotide sequence database under accession numbers repeating the steps taken for the *Bacillus samanii* EF036537. There are others that have been deposited at the GenBank such HS 1 *Citrobacter hockomockus*-FJ756447, HS 5 II. *Leucobacter mimi*-FJ756448, HS 2 *Serriatia peterii*-FJ756449, HS 5 I. *Bacillus patriciaus*-FJ756450 and YIL *Rhodosporidium hockomockus*-FJ756452.

Nonculturable Technique

Many of the environmental living bacteria cannot grow on lab based media this had limited the identification of species to about 5 per cent that could grow on lab based media. Nonculture based techniques involves the direct isolation of microbial genomic DNA directly from water and soil samples without the use of any growth media. These methods would enable the identification of micro-organisms from various environment and study their interaction and enzymes functionality. This would allow a better understanding of how such micro-organisms have evolved and their growth requirements.

Direct microbial DNA isolation from water and soil

In the Hockomock Swamp project WaterMaster DNA Purification kit from Epicentre Biotechnologies will be used as one of the methods for direct extraction of DNA from water samples. This Protocol involves filtration of water samples, cell lysis, DNA Precipitation and removal of inhibitors. The other method will again include filtration using multiple Millipore filters that would trap particle-bound micro-organisms. Using proteinase K and lysozyme will be used for purification and cell lysis, SDS and chloroform will be used to enhance DNA extraction and ethanol will be used for precipitation. Bovine serum albumin (BSA) will be added before PCR to remove any humic acid that can attach to the Taq polymerase inhibiting the PCR reaction. DNA isolation from soil by technique involves acid washing, suspension in 0.1 per cent (wt/vol) sodium pyrophosphate, and centrifugation. The pellets are to be resuspended in 0.15 M NaCl-0.1 M EDTA (pH 8.0) and used for DNA isolation by a modified Marmur method. This will be compared with a Soil Master kit for direct DNA extraction from Epicentre Biotechnologies and DNAzol Direct, which is produced by the Molecular Research Centre, Inc for extraction of PCR ready DNA from various sources.

Guanine and cytosine (GC) per cent

One of the most important tests that should be conducted is the determination of the GC per cent in the extracted DNA. The GC ratio is the percentage of guanine and cytosine in the DNA of an organism. This is an important tool that can be used in identification of micro-organisms. There are some guidelines that manifest the importance of this process, such as micro-organisms with a low-GC content which is more likely to be Gram-positive with the exception of *Mycoplasma* which does not have a cell wall and has a low GC content even though it stains as a Gram-negative. Low GC content could indicate either Non-Endospore forming Gram-positive or Endospore forming Gram-positive or a *Mycoplasmas*.

Gram-negative micro-organisms will have a GC content of 65 per cent and higher as will filament bacteria such as Actinomycetes. Related groups would only have a 3–5 per cent difference in their G/C content. DNA denaturation will be at 85C followed then by using a UV-spectrophotometer. GC content will be determined at 260 nm using gradient fraction which is determined by linear regression analysis of data obtained from control gradients containing standard DNA samples of known G + C. This method may be used for DNA of culturable or directly extracted DNA and it will indicate the type of bacteria that is being isolated and will assist in their identification. A library of all isolated bacteria with their GC content need to be established for use as a reference for future studies.

Single-strand-conformation polymorphism analysis and DNA microarrays

Single-strand-conformation polymorphism (SSCP) is a mutation analysis method that will be used in the analysis and differentiation of bacteria. 400 bp fragments of the isolate or sample 16S or 28S rRNA gene is PCR amplified then they will be selectively digested and separated through electrophoresis. Analysis of the bands and comparison with template mixtures of known micro-organisms will be used to distinguish the different types and the presence of any mutations.

DNA-DNA hybridisation is used with DNA microarrays to identify microbial species that could not be grown or cultured on lab based media. This method does not only determine the gentic make up of the identified micro-oragnisms but also could determine the level of relativity and their evolutionary pathway. This method coupled with the widely accepted criteria for delineating species in current bacteriology: these state that strains with DNA relatedness values of less than 70 per cent or with more than 3 per cent difference in their 16S rRNA gene sequences are considered to represent different species.

Construction of Phylogenetic Trees

BLAST searches to compare the nucleotide sequences obtained against those in the national center for biotechnology information (NCBI) nucleotide database (http://www.ncbi.nlm.nih.gov/index.html) and the European Molecular Biology Laboratories (EMBL) (http://www.ebi.ac.uk/embl/) prokaryote database. The use of this practice is very important and is used in the identification of new previously undescribed species such as the identification of *Bacillus samanii*.

Examples of 16S rRNA Sequencing

Isolate HS 7 Gram-positive non-spore forming rod bacteria

TGGAGAGTTTGATCCTGGCTCAGGACGAACGCTGGCGGCGTGCTTAACAC
ATGCAAGTCGAACGCTGAAGCCCCAGCTTGCTGGGGTGGATGAGTGGCGA
ACGGGTGAGTAACACGTGAGTAACCTGCCCATCACTCTGGGATAAGCGCT
GGAAACGGCGTCTAATACTGGATACGAGCAGCGACCGCATGGTCAGCTGC
TGGAAAGACTGGTTCGGTGATGGATGGACTCGCGGCCTATCAGCTTGTTG
GTGAGGTAATGGCTCACCAAGGCGACGACGGGTAGCCGGCCTGAGAGGGT
GACCGGCCACACTGGGACTGAGACACGGCCCAGACTCCTACGGGAGGCAG
CAGTGGGGAATATTGCACAATGGGCGCAAGCCTGATGCAGCAACGCCGCG
TGAGGGATGACGGCCTTCGGGTTGTAAACCTCTTTTAGTAGGGAAGAAGC
CTTCGGGTGACGGTACCTGCAGAAAAAGCACCGGCTAACTACGTGCCAGC
AGCCGCGGTA

Isolate Y1L-yeast

AGACCGATAGCGAACAAGTACCGTGAGGGAAAGATGAAAAGCACTTTGGA
AAGAGAGTTAACAGTACGTGAAATTGTTGGAAGGGAAACGCTTGAAGTCA
GACTTGCTTGCCGGAGCTTGCTTCGGTTTGCAGGCCAGCATCAGTTTTCC
GGGGTGGATAATGGTGGTTTGAAGGTAGCAGCCTCGGCTGTGTTATAGCT
TTCCACTGGATACATCCTGGGGGACTGAGGAACGCAGCGTGCTTTTTGCG
AAGGTTTCGACCTTTTCACGCTTAGGATGCTGGTGTAATGACTTTAAACG
ACCCGTCTTGAAACACGGACCAAG

CONCLUSION

The advancement of molecular microbiology techniques would allow the identification of a good number of microbial species have various habitats to live in. These techniques have proven their worth now as they ended the requirement that micro-organisms must grow on lab based media so they could be isolated and identified. This would also provide further understanding on the interactive functionality of such micro-organisms and their effect on their habitats. This would be a further tool to be used to understand microbial evolution and would provide considerable amount of information on microbial biodiversity. There is however a need to create libraries such as these for GC content and DNA microarrays to enable research a faster more efficient methods of determining if they are dealing with a new or a known species. These molecular techniques would also provide a better understanding to microbial phylogenies and taxonomy.

Chapter 20

Tropical Forest Conservation

INTRODUCTION

The world's tropical forests, which circle the globe, are interestingly diverse. Ranging from the steamy jungles of the rainforests to the dry forests and savannas, they provide habitat for millions of species of plants and animals. Once covering some 15.3 billion acres (6.2 billion ha), these tropical forests have been reduced through cutting and clearing by 210 million acres (85 million ha) between 1985 and 1990. All types of tropical forests are defined and their products and benefits to the environment are presented and discussed. Modern forest practices are shown as a means of halting forest destruction while still providing valuable forest products and protecting and preserving the habitats of many endangered species of plants and wildlife. The Luquillo Experimental Forest is presented as a possible model to exemplify forestry practices and research that could manage and ultimately protect the tropical forests throughout the world. This chapter shows how modern forest practices can help stem the tide of forest destruction while providing valuable forest products for people. The tropical forests of Puerto Rico, which were abused for centuries, were badly depleted by the early 1900's. Widespread abandonment of poor agricultural lands has allowed natural reforestation and planting programs to create a patchwork of private, Commonwealth, and Federal forests across the land. The most frequent example in this publication is the Luquillo Experimental Forest, which could be a model for protecting and managing tropical forests worldwide.

IMPORTANCE OF TROPICAL FORESTS

Environmental Importance

Tropical forests have a special role in the conservation of biodiversity. They are the home to 70 per cent of the world's plants and animals—more than 13 million distinct species. The tropical forests contain 70 per cent of the world's vascular plants, 30 per cent of all bird species, and 90 per cent of invertebrates. Many of the mammals are among the most famous icons of natural history—the great cats, the primates, and the ungulates of the East African woodlands. In tree species alone, tropical rainforests are extremely diverse, often having more than 200 species per hectare. Boreal forests, on the other hand, are biologically much simpler, with as few as one species per hectare for fire-regenerated stands like lodgepole pine in North America.

Forests influence the local and probably global climates. They moderate the diurnal range of air temperatures and maintain atmospheric humidity levels. Forests absorb atmospheric carbon and replenish the oxygen in the air we breathe. The conservation of forest resources in the watersheds that supply water for irrigation, sanitation and human consumption is an important component of water supply strategies. When tropical watersheds have balanced land use, their forests absorb excessive rainfall that is gradually released later. Forests regulate stream flows by intercepting rainfall, absorbing the water into the underlying soil, and gradually releasing it into the streams and rivers of its watershed. This minimises both downstream flooding and drought conditions. Tree cover conserves moisture in the soil by providing shade that reduces the evaporative loss from radiant energy exchange with the atmosphere. Tree roots enhance soil porosity, reduce compaction, and facilitate infiltration. Trees act as windbreaks, reducing the force of desiccating, eroding winds at ground level.

Socioeconomic Importance

Some 500 million people live in or at the edge of the tropical forests. They are some of the least privileged groups in our global society. They depend on the forests for many important products and environmental services. Included in this population of forest-dependent peoples are the world's 150 million native or indigenous peoples who rely on the forests for their way of life. They not only meet their economic needs for food and shelter but also form an integral part of their culture and spiritual traditions.

Forests provide us with a wide range of industrial wood products that we use in daily life—lumber, panels, posts, poles, pulp and paper. Industrial wood products account for US$ 400 billion worth of global production, about two per cent of the global gross domestic product. Tropical forests account for approximately 25 per cent of this production. The international trade of wood products is estimated at over US$ 100 billion, equivalent to three per cent of all goods exchanged. In 1994, the total world production of wood for all end uses is reported to have been 3358 million cubic metres, of which 1318 million cubic metres was produced in developing countries. World production is divided almost equally between industrial uses and fuelwood.

Production of Forest Products in Developing Countries

Table 20.1 gives the production of forest products in developing countries.

Table 20.1. Production of forest products in developing countries.

Forest product	Quantity	World production
Sawnwood (millions m³)	111	26%
Wood-based panels (millions m³)	45	31%
Pulp for paper (millions tons)	41	22%
Paper and paperboard (millions tons)	61	22%
Fuelwood and charcoal (millions m³)	1638	89%

Whereas the developed countries produce most of the world's industrial wood products, developing countries account for the majority of fuelwood consumed. Fuelwood and charcoal makes up 56 per cent of global wood production and developing countries account for almost 90 per cent of it. Wood is by far the most important source of energy for developing countries and the only source of energy for much of the world's rural areas.

In addition to wood products, tropical forests give us a wide range of non-timber forest products, the so-called 'minor' forest products which in many cases are 'major' forest products for the local people.

These include fibres, resins, latexes, fruits and traditional medicines. Forests are often important sources of foodstuffs, particularly in times of drought and famine when conventional agricultural crops have failed. For example, in the state of Madhya Pradesh in India, tribal peoples rely on their forests for up to 25 per cent of their basic food requirements. Most of these 'minor' or nontimber forest products are produced, traded, and consumed outside the cash economy and therefore are not quoted in the national economic statistics.

Tropical forests are also very important economically for plant-improvement breeding. For example, a species of wild maize has been found in Mexican woodlands that is resistant to five of the world's seven most important corn viruses; it is now an important genetic resource for corn-improvement programs. Forests are also important sources of new pharmaceuticals used to fight cancer, AIDS, and other serious human diseases. The periwinkle plant from the Madagascar forests provides a drug that has proven very successful in treating lymphocytic leukemia. The bark of *Prunus africanum* is now an important commodity in world trade as a pharmaceutical for the treatment of prostate disorders. At present, our knowledge about tropical forest plants is limited, but it is improving with ongoing research. Obviously, the great variety of forest products is important by any economic standard.

TYPES OF TROPICAL FORESTS

About half of all the world's forests are in the Tropics, the area between the Tropic of Cancer and the Tropic of Capricorn. This region may be best known for its rainforests—lush, steamy jungles with towering trees, epiphytes, and dense under stories of smaller trees, shrubs and vines. Tropical forests are surprisingly diverse. In addition to rainforests, there are mangroves, moist forests, dry forests and savannas. Such classifications, however, give only a slight indication of the diversity of tropical forests. One study by the Food and Agriculture Organisation (FAO) of the United Nations, which considered 23 countries in tropical America, 37 in tropical Africa, and 16 in tropical Asia, identified dozens of types of tropical forests: open and closed canopy forests, broadleaved trees and conifer forests, closed forests and mixed forest grasslands, and forests where agriculture has made inroads.

Rainforests

The largest remaining areas of tropical rainforests are in Brazil, Congo, Indonesia and Malaysia. Precipitation generally exceeds 60 inches (150 cm) per year and may be as high as 400 inches (1000 cm). Lowland rainforests are among the world's most productive of plant communities. Giant trees may tower 200 feet (60 m) in height and support thousands of other species of plants and animals. Montane (mountain) rainforests grow at higher elevations where the climate is too windy and wet for optimum tree growth.

Mangrove forests grow in the swampy, intertidal margin between sea and shore and are often considered part of the rainforest complex. The roots of mangrove trees help stabilise the shoreline and trap sediment and decaying vegetation that contribute to ecosystem productivity (Fig. 20.1).

Dry Forests

Large areas of tropical dry forests are found in India, Australia, Central and South America, the Caribbean, Mexico, Africa and Madagascar. Dry forests receive low rainfall amounts, as little as 20 inches (50 cm) per year, and are characterised by species well adapted to drought. Trees of dry tropical forests are usually smaller than those in rainforests, and many lose their leaves during the dry season. Although they are still amazingly diverse, dry forests often have fewer species than rainforests.

Fig. 20.1. Rainforests.

Savanna is a transitional type between forest and grassland. Trees are often very scattered and tend to be well adapted to drought and tolerant of fire and grazing. If fire is excluded, trees eventually begin to grow and the savanna is converted to dry forest. With too much fire or grazing, dry forest becomes savanna (Fig. 20.2). This vegetation type has fewer species of trees and shrubs but more grasses and forbs than other forest types in the Tropics.

Fig. 20.2. Dry forests.

VALUE OF TROPICAL FORESTS

All forests have both economic and ecological value, but tropical forests are especially important in global economy. These forests cover less than 6 per cent of the earth's land area, but they contain the vast majority of the world's plant and animal genetic resources. The diversity of life is astonishing. The original forests of Puerto Rico, for example, contain more than 500 species of trees in 70 botanical families. By comparison, temperate forests have relatively few. Such diversity is attributed to variations in elevation, climate and soil, and to the lack of frost.

There is also diversity in other life forms: shrubs, herbs, epiphytes, mammals, birds, reptiles, amphibians, and insects. One study suggests that tropical rainforests may contain as many as 30 million different kinds of plants and animals, most of which are insects (Fig. 20.3).

Fig. 20.3. Tropical rainforests insects.

Wood and Other Products

Tropical forests provide many valuable products including rubber, fruits and nuts, meat, rattan, medicinal herbs, floral greenery, lumber, firewood and charcoal. Such forests are used by local people for subsistence hunting and fishing. They provide income and jobs for hundreds of millions of people in small, medium and large industries.

Tropical forests are noted for their beautiful woods (Fig. 20.4). Four important commercial woods are mahogany, teak, melina and okoume. Honduras mahogany (*Swietenia macrophylla*), grows in the Americas from Mexico to Bolivia. A strong wood of medium density, mahogany is easy to work, is long lasting, and has good colour and grain.

Fig. 20.4. Tropical forests beautiful woods.

It is commonly used for furniture, moulding, panelling and trim. Because of its resistance to decay, it is a popular wood used in boats. Teak (*Tectona grandis*) is native to India and Southeast Asia. Its wood has medium density, is strong, polishes well, and has a warm yellow-brown colour. Also prized for resistance to insects and rot, teak is commonly used in cabinets, trim, flooring, furniture and boats (Fig. 20.5). Melina (*Gmelina arborea*) grows naturally from India through Vietnam. Noted for fast growth, melina has light coloured wood that is used mainly for pulp and particleboard, matches and carpentry. Okoume (*Aucoumea klaineana*) is native to Gabon and the Congo in west Africa. A large fast-growing tree, the wood has moderately low density, good strength-to density ratio, and low shrinkage during drying. It is commonly used for plywood, panelling, interior furniture parts and light construction.

Fig. 20.5. Teak (*Tectona grandis*) trees.

Other Economic Values

Tropical forests are home for tribal hunter-gatherers whose way of life has been relatively unchanged for centuries. These people depend on the forests for their livelihood. More than 2.5 million people also live in areas adjacent to tropical forests. They rely on the forests for their water, fuelwood and other resources and on its shrinking land base for their shifting agriculture. For urban dwellers, tropical forests provide water for domestic use and hydroelectric power. Their scenic beauty, educational value, and opportunities for outdoor recreation support tourist industries.

Many medicines and drugs come from plants found only in tropical rainforests. Some of the best known are quinine, an ancient drug used for malaria; curare, an anesthetic and muscle relaxant used in surgery; and rosy periwinkle, a treatment for Hodgkin's disease and leukemia. Research has identified other potential drugs that may have value as contraceptives or in treating a multitude of maladies such as arthritis, hepatitis, insect bites, fever, coughs and colds. Many more may be found. In all, only a few thousand species have been evaluated for their medicinal value.

In addition, many plants of tropical forests find uses in homes and gardens: ferns and palms, the hardy split-leaf philodendron, marantas, bromeliads and orchids (Fig. 20.6), to name just a few.

Fig. 20.6. Orchids.

Environmental Benefits

Tropical forests do more than respond to local climatic conditions; they actually influence the climate. Through transpiration, the enormous number of plants found in rainforests return huge amounts of water to the atmosphere, increasing humidity and rainfall, and cooling the air for miles around. In addition, tropical forests replenish the air by utilising carbon dioxide and giving off oxygen. By fixing carbon they help maintain the atmospheric carbon dioxide levels low and counteract the global 'greenhouse' effect.

Forests also moderate stream flow. Trees slow the onslaught of tropical downpours, use and store vast quantities of water, and help hold the soil in place. When trees are cleared, rainfall runs off more quickly, contributing to floods and erosion.

DEFORESTATION

Before the dawn of agriculture approximately 10,000 years ago, forests and open woodland covered about 15.3 billion acres (6.2 billion ha) of the globe. Over the centuries, however, about one-third of these natural forests has been destroyed. According to a 1982 study by FAO, about 27.9 million acres (11.3 million ha) of tropical forests are cut each year—an area about the size of the States of Ohio or Virginia. Between 1985 and 1990, an estimated 210 million acres (85 million ha) of tropical forests were cut or cleared. In India, Malaysia and the Philippines, the best commercial forests are gone, and cutting is increasing in South America. If deforestation is not stopped soon, the world will lose most of its tropical forests in the next several decades.

Reasons for Deforestation

Several factors are responsible for deforestation in the Tropics: clearing for agriculture, fuelwood cutting and harvesting of wood products. By far the most important of these is clearing for agriculture. In the Tropics, the age-old practice of shifting, sometimes called 'slash-and-burn', agriculture has been used for centuries. In this primitive system, local people cut a small patch of forest to make way for subsistence farming. After a few years, soil fertility declines and people move on, usually to cut another patch of trees and begin another garden. In the abandoned garden plot, the degraded soil at first supports only weeds and shrubby trees. Later, soil fertility and trees return, but that may take decades. As population pressure increases, the fallow (rest) period between cycles of gardening is shortened, agricultural yields decrease, and the forest region is further degraded to small trees, brush, or eroded savanna.

Conversion to sedentary agriculture is an even greater threat to tropical forests. Vast areas that once supported tropical forests are now permanently occupied by subsistence farmers and ranchers and by commercial farmers who produce sugar, cocoa, palm oil and other products.

In many tropical countries there is a critical shortage of firewood. For millions of rural poor, survival depends on finding enough wood to cook the evening meal. Every year more of the forest is destroyed, and the distance from home to the forest increases. Not only do people suffer by having to spend much of their time in the search for wood, but so does the land. Damage is greatest in dry tropical forests where firewood cutting converts forests to savannas and grasslands.

The global demand for tropical hardwoods, an $8-billion-a-year industry, also contributes to forest loss. Tropical forests are usually selectively logged rather than clear-cut. Selective logging leaves the forest cover intact but usually reduces its commercial value because the biggest and best trees are removed. Selective logging also damages remaining trees and soil, increases the likelihood of fire and degrades the habitat for wildlife species that require large, old trees-the ones usually cut. In addition, logging roads open up the forests to shifting cultivation and permanent settlement.

In the past, logging was done primarily by primitive means-trees were cut with axes and logs were moved with animals such as oxen. Today the use of modern machinery—chain saws, tractors and trucks makes logging easier, faster, and potentially more destructive.

Endangered Wildlife

Forests are biological communities-complex associations of trees with other plants and animals that have evolved together over millions of years. Because of the worldwide loss of tropical forests, thousands of species of birds and animals are threatened with extinction. The list includes many unique and fascinating animals, among them the orangutan, mountain gorilla, manatee, jaguar and Puerto Rican parrot. Although diverse and widely separated around the globe, these species have one important thing in common. They, along with many other endangered species, rely on tropical forests for all or part of their habitat.

Orangutans (*Pongo pygmaeus*) are totally dependent on small and isolated patches of tropical forests remaining in Borneo and Sumatra, Indonesia. Orangutans spend most of their time in the forest canopy where they feed on leaves, figs and other fruit, bark, nuts and insects. Large trees of the old-growth forests support woody vines that serve as aerial ladders, enabling the animals to move about, build their nests, and forage for food. When the old forests are cut, orangutans disappear.

The largest of all primates, the gorilla, is one of man's closest relatives in the animal kingdom. Too large and clumsy to move about in the forest canopy, the gorilla lives on the forest floor where it forages for a variety of plant materials. Loss of tropical forests in central and west Africa is a major reason for the decreasing numbers of mountain gorillas (*Gorilla gorilla*). Some habitat has been secured, but the future of this gentle giant is in grave danger as a result of habitat loss and poaching.

The jaguar (*Leo onca*), a resident of the Southwestern United States and Central and South America, is closely associated with forests. Its endangered status is the result of hunting and habitat loss.

The Puerto Rican parrot (*Amazona vittata*), a medium-sized, green bird with blue wing feathers, once inhabited the entire island of Puerto Rico and the neighbouring islands of Mona and Culebra.

Forest destruction is the principal reason for the decline of this species. Hunting also contributed. Today, only a few Puerto Rican parrots remain in the wild and their survival may depend on the success of a captive breeding program (Fig. 20.7).

Fig. 20.7. Puerto Rican parrot (*Amazona vittata*).

In addition to species that reside in tropical forests year round, others depend on such forests for part of the year. Many species of migrant birds journey 1000 miles or more between their summer breeding grounds in the north and their tropical wintering grounds. These birds are also threatened by tropical forest destruction.

PRACTICE OF FORESTRY

Forestry-loosely defined as the systematic management and use of forests and their natural resources for human benefit has been practiced for centuries. Most often, forestry efforts have been initiated in response to indiscriminate timber cutting that denuded the land and caused erosion, floods or a shortage of wood products.

Ancient Forestry Practices

In ancient Persia (now Iran), forest protection and nature conservation laws were in effect as early as 1700 BC. Two thousand years ago the Chinese practiced what they called 'four sides' forestry-trees were planted on house side, village side, road side and water side. More than 1000 years ago, Javanese maharajahs brought in teak and began to cultivate it. In the African Tropics, agroforestry (growing of food crops in association with trees) has been practiced for hundreds of years.

In the Yucatan Peninsula of southern Mexico, the ancient Mayas cultivated fruit and nut trees along with such staples as corn, beans and squash. Bark, fibres and resin were obtained from plants grown in fields, kitchen gardens and orchards. Early in their civilisation, the Mayas practiced slash-and-burn agriculture. As their population grew, they found more efficient methods of growing crops. They terraced hillsides, learned how to decrease the time between 'rotations' of agricultural land with native forests, dug drainage channels and canals to move water to and from cultivated areas, and filled in swampland to plant crops. The agricultural sophistication of the Mayas enabled their civilisation to grow and flourish. What brought about their decline about AD 820 is not fully known, but some believe that as their society developed, the Mayas made unsustainable demands on their environment.

Relatively little is known about tropical forestry before the mid-1800's in most places. At that time, the European colonial empires—notably the Dutch, English and Spanish—brought modern forest management practices to Indonesia, India, Africa and the Caribbean. Centers for forestry and forestry research were established, and more careful records were kept.

Sustainable Forestry

Modern forestry has its basis in 18th-century Germany. Like the Chinese and the Mayan forest practices, German forestry is essentially agricultural. Trees are managed as a crop. Two concepts are important: renewability and sustainability. Renewability means that trees can be replanted and seeded and harvested over and over again on the same tract of land in what are known as crop 'rotations'. Sustainability means that forest harvest can be sustained over the long-term. How far into the future were foresters expected to plan? As long as there were vast acres of virgin (original) forests remaining, this question was somewhat academic. Today, however, sustainability is a vital issue in forestry. Most of the world's virgin forests are gone, and people must rely more and more on second-growth or managed forests. Perhaps we now face, as never before, the limits to long-term productivity.

In the German forest model, forestry is viewed as a continual process of harvest and regeneration. Harvest of wood products is a goal, but a forester's principal tasks are to assure long-term productivity. That is achieved by cutting the older, mature, and slow-growing timber to make way for a new crop of young, fast-growing trees.

Harvest-Regeneration Methods

Three examples of timber harvest-regeneration methods (silvicultural systems) illustrate how foresters manage stands to produce timber on a sustained basis.

Selection

Individual trees or small groups of trees are harvested as they become mature. Numerous small openings in the forest are created in which saplings or new seedlings can grow. The resulting forest has a continuous forest canopy and trees of all ages. Such systems favour slow-growing species that are shade tolerant.

Clearcutting

In clearcutting, an entire stand of trees is removed in one operation. From the forester's point of view, clearcutting is the easiest way to manage a forest—and the most economical. Regeneration may come from sprouts on stumps, from seedlings that survive the logging operation, or from seeds that germinate after the harvest. If natural regeneration is delayed longer than desired, the area is planted or seeded.

Clearcutting systems are often used to manage fast-growing species that require a lot of light. Resulting stands are even aged because all the trees in an area are cut-and regenerated-at the same time. Clearcutting has become controversial in recent years because it has the potential to damage watersheds and because it tends to eliminate species of wildlife dependent on old growth trees. If clearcuts are kept small and the cutting interval is long enough, however, biological diversity may not be impaired.

Shelterwood

In shelterwood systems, the forest canopy is removed over a period of years, usually in two cuttings. After the first harvest, natural regeneration begins in the understory. By the time the second harvest is made, enough young trees have grown to assure adequate regeneration. Shelterwood systems favour species that are intermediate in tolerance to shade. Such systems are difficult to use successfully and are the least used of the three silvicultural methods described.

Multiple-use forestry

Gifford Pinchot, the first Chief of the US Forest Service, was also this country's first professional forester. Pinchot advocated the use of forest resources—all resources, not just timber—for human benefit. Pinchot was a strong and charismatic leader, and his ideas helped shape the course of forestry in the United States.

Pinchot had a vocal opponent in John Muir, a young naturalist from California who believed that public lands should be preserved rather than used. Eventually Muir and Pinchot became rivals for public approval. Oddly enough, there was no loser in this early conservation battle. Muir's preservation ethic became embodied in the philosophy of the National Parks, and Pinchot's concept of wise use became the guiding principle of the National Forests.

National forests are still managed under the concepts of multiple use and sustained yield. The dominant uses of National forests are considered to be wood, water, wildlife, forage (for domestic cattle and wildlife) and recreation. Extraction of minerals and other valuable products is also considered a legitimate use of National Forests. Because Pinchot's philosophy left room for the 'highest and best use' of a given area, the US National Forests now include a wilderness system of more than 32 million acres (13 million ha) in which timber harvest is not allowed.

Today it is generally recognised that most, if not all, nondestructive uses of forest are valid. Some areas may be set aside as parks; others for wildlife habitat or as wilderness. Still others will be managed for timber harvest or multiple benefits. Today, conflicts arise primarily over where these different uses will be dominant. In the National Forests, such decisions are made through a land-use planning process in which the public has ample opportunities for input and involvement.

FORESTRY RESEARCH

At the turn of the century, very little was known about the world's native forests or how to manage them. In the United States, foresters were quick to recognise the value of information about forests and a branch of research was established in the Forest Service in 1915. Early research was done primarily in support of reforestation efforts, but, as forestry grew in size and complexity, so did the research.

Today, the USDA Forest Service has six regional experiment stations located in important forest regions. Each experiment station has several field laboratories generally with specialised assignments for a geographic region or a specific subject area and numerous sites for field research. In addition, the Forest Products Laboratory in Madison, WI, serves as a nationwide center for research and development of new technology relating to wood, including tropical woods. Two laboratories are dedicated exclusively to tropical forest research: the International Institute of Tropical Forestry in Puerto Rico and the Institute of Pacific Islands Forestry in Hawaii.

Research is vital for modern forest management, which is information intensive. Today's foresters require vast quantities of data and a knowledge of ecology: they must understand not only the parts of ecosystems but how different parts of the environment interact. Scientific investigations are conducted in support of all kinds of forestry activities: silviculture, forest insect and disease control, wildlife habitat management, fire prevention and control, range and watershed management, forest products utilisation, forest survey, reforestation, ecology and economics.

TROPICAL FORESTRY

In the past, timber harvest in the Tropics has seldom been followed by regeneration. Conversion to agriculture is often permanent or results in soil erosion. Timber harvest contracts have usually been short-term and have provided little or no incentive for timber companies to replant. So little reforestation has been done in the Tropics that many people believe these forests cannot be restored. However, there are many examples of successful reforestation in India, Indonesia and the Caribbean.

In the Tropics, as elsewhere, forestry is a mixture of modern innovations and ancient techniques borrowed from local tradition. Plantation forestry is common. Forest reserves have been established for timber harvest, wildlife habitat, scenery, outdoor recreation or watershed protection. And in the Tropics, agroforestry—tree growing combined with agricultural cropping is much more common than elsewhere.

Plantation Forestry

In the Tropics, trees are often planted and grown in plantations for wood production. Often, many species must be tried to determine which will grow best (Fig. 20.8). Plantations must also be supported by major investments in forest management and research. Forest nurseries must be established, and planting techniques and cultural practices (spacing and thinning, pruning, fertilisation, insect and disease control, and genetic improvement) must be developed.

Fig. 20.8. Forest plantation.

Extensive pine plantations have been established in the moist Tropics, mainly in South Africa and Australia. Species most often planted include Caribbean pine (*Pinus caribaea*), ocote pine (*P. oocarpa*), slash pine (*P. elliottii*), and benguet pine (*P. kesiya*). Pines are popular plantation trees because they are generally fast growing, have good survival rates, and are adapted to a wide variety of environments, including degraded forest sites.

Eucalypts, including species such as *Eucalyptus grandis, E. deglupta, E. tereticornis, E. globulus* and *E. camaldulensis* are favoured for the same reasons. Eucalypts are commonly grown for pulp, fuel and lumber. Other species commonly planted include teak (*Tectona grandis*), Honduras mahogany (*Swietenia macrophylla*), melina (*Gmelina arborea*), beefwood (*Casuarina equisetifolia*) and Mexican cypress (*Cupressus lusitanica*).

Forest Reserves

There are many reasons for establishing forest reserves in the Tropics. They can restore watersheds and wildlife habitat, improve scenic beauty and opportunities for outdoor recreation, and produce wood and other products for local use and export. Many forest products contribute to the sustenance and income of local people: wildlife and fish, firewood, rubber, fruits and nuts, rattan, medicinal herbs, floral greenery and charcoal.

Perhaps the most famous of these reserves is the 5600 square mile (14,500 k squared) Serengeti National Park in Tanzania. With its vast herds of grazing ungulates (hoofed animals) and predators, including several endangered species, the Serengeti is a showcase of a savanna ecosystem that has long been protected and managed for wildlife and other natural resources (Fig. 20.9). Although plagued with poachers, the Serengeti promotes the cause of wildlife conservation to the many thousands of 'ecotourists' who pay to experience nature each year. Another type of forest reserve is the 'extractive' reserve, which is dedicated to the production of useful products. Large reserves of this type have been established recently in Brazil. Local residents use them for tapping rubber, for gathering fruits and nuts, for hunting, and for harvesting wood on a sustained yield basis. Such uses provide a sustainable income while maintaining the ecological integrity of the forest.

Fig. 20.9. Forest reserves.

Agroforestry

The practice growing of trees in combination with agricultural crops is fairly common in the Tropics. It is possible to grow food crops year around in many forested areas, and rural poor depend on this source of food as nowhere else on earth.

Taungya System

Various systems have been developed for combining forestry with agriculture. 'Taungya' is a Burmese word meaning cultivated hill plot. This system of agroforestry was developed in Europe during the Middle Ages and probably independently in a number of places in the Tropics. After existing forest or ground cover is removed by burning, trees are planted along with agricultural crops. Both are cultivated until the tree canopy closes. Then the area is left to grow trees, and another site is located for combined forestry agriculture.

Shade Cropping

An overstory of trees is often used to provide shade for agricultural crops. A common practice is to grow tree species such as *guaba Inga vera* over coffee. In Puerto Rico, many forests developed where coffee was once grown in this manner.

Support Crops

Trees can be planted to provide support (and sometimes shade) for vine crops. Vines such as pepper and vanilla need support.

Alley Cropping

Nitrogen-fixing trees are planted in hedges in widely-space parallel rows along the contour of slopes. Food crops are grown in the 'alley' between the rows. The trees add nitrogen and organic matter, protect the soil from erosion, and provide wood and animal forage.

Living Fences

Green fenceposts that will root and sprout often are planted in a closely spaced row. When they sprout, they create a 'living fence' that provides shade and forage for cattle.

Windbreaks

Trees are often planted as windbreaks for agricultural crops, farms or homesites. Such plantings can eventually contribute wood products as well as shelter. Food trees such as citrus, rubber and mango can also provide fuel, lumber and other wood products when they have outlived their original usefulness.

NEW DIRECTIONS IN TROPICAL FORESTRY

The conservation issues of the past seem simple compared with those of today. As we move toward the 21st century, human societies are concerned with global warming, deforestation, species extinction, and rising expectations. Growing populations must be fed, clothed and sheltered, and people everywhere want higher standards of living.

Global Warming

Warming of the earth's atmosphere is a major environmental issue. Air pollution, deforestation, and widespread burning of coal, oil and natural gas have increased atmospheric concentrations of carbon dioxide, methane, nitrous oxide and chlorofluorocarbons. These gases trap heat from the sun and prevent it from radiating harmlessly back into space. Thus, the 64 greenhouse or warming effect is created.

Because of natural variations in climate, it is difficult to measure warming over large areas. Scientists agree, however, that increases in atmospheric concentrations of greenhouse gases will cause higher temperatures worldwide. Even an increase of a few degrees might cause serious melting of the polar icecaps, a gradual rise in sea level, a disruption in normal weather patterns, a possible increase in forest fires, and the extinction of species.

Role of Forests

Trees, the largest of all land plants, act as a kind of environmental 'buffer' for the ecosystem they dominate. They help ameliorate the extremes of climate (heat, cold, and wind) and create an environment where large land mammals, including people, can live comfortably. Trees complement animals in the global environment. Mammals take in oxygen from the air and exhale carbon dioxide. Plants use the carbon dioxide in their growth processes, store the carbon in woody tissues and return oxygen to the atmosphere as a waste product. This process, known as photosynthesis, is essential to life. Carbon captured from the atmosphere by photosynthesis is eventually recycled through the environment in a process known as the carbon cycle. Trees have an especially important role in the carbon cycle. Tree leaves also act as filters to remove atmospheric pollutants from the air. This effect is particularly beneficial in urban areas.

Forestry Issues

Two key issues will dominate forestry in the years ahead: (i) maintaining long-term productivity of managed forests, and (ii) preventing further loss of tropical forests. Both problems will require new approaches to forest management. Traditionally, forestry has focused on growing crops of wood in plantations or in managed natural stands. In this 'agricultural mode', other benefits of forest such as watershed protection, wildlife habitat, climate moderation and outdoor recreation, have received less attention than wood production.

Perhaps more importantly, the sustainability of the full range of forest benefits has not been measured. There is no question that trees can be grown for crops of wood in managed stands. With intensive management—short rotations, species selection, genetic improvement, fertilisation, thinning, and other cultural treatments—more wood can be produced in less time than in natural forests. But for how long? And at what cost in other benefits?

As more and more of the world's original forests have been cut, the ecological value of forests has come to be more appreciated. In recent years, increased emphasis has been put on what some are calling 'ecosystem management'.

In this model, the health and long-term stability of the forest are paramount, and timber production is considered a by-product of good forest management rather than the principal product. In Puerto Rico, for example, wood production is a relatively minor aspect of forestry.

Since the 1930's when timber harvests were curtailed, the forests have been managed primarily for watershed protection, wildlife habitat and outdoor recreation.

There are no easy solutions to the problem of tropical forest destruction, but most experts agree that the problems cannot be solved simply by locking up the forests in reserves. The forests are too important to local people for that to be a workable solution. There is no doubt that tropical forests will be cut. It is better for them to be cut in an ecologically sound manner than to be cleared for poor-quality farmland or wasted by poor harvest practices.

The only real long-term solutions are: (i) more efficient agriculture on suitable farmland, (ii) efficient forestry practice including plantations and (iii) reserves to protect species and ecosystems. Many forestry experts believe that we have only begun to tap the potential for wise use of tropical forests. Many uses have yet to be fully explored. We are only starting to learn the value of tropical forests for medicines, house and garden plants, food and fibre, tourism, and natural resource education.

TROPICAL RAINFORESTS—FOUR WAYS TO STOP DEFORESTATION

It's hard to imagine that we would knowingly destroy something so valuable; could it be that we are destroying them before we realise their worth? Before we truly understand their biodiversity? And even before we fully understand the life and the ecosystems they support?

Massive deforestation brings with it many horrifying consequences—air and water pollution, soil erosion, the release of carbon dioxide into the atmosphere, the eviction and decimation of indigenous Indian tribes, and the extinction of many plants, animals and creatures. Fewer rainforests mean less rain, less oxygen for us to breathe, and an increased threat of global warming.

Confucius said, 'A man who has committed a mistake and doesn't correct it, is committing another mistake.' Clearly deforestation is man's mistake. So how do we correct this mistake? Can we correct this mistake?

If deforestation ceased today, it would help immensely, but unfortunately would not be enough. We have lost complete species, both in plant and animal life; however, all is not lost. What we can hope for in bringing deforestation to an end is a new beginning; new species to evolving and the rebirth of this diminishing treasure. With the rapid loss of earth's rainforests, it's time to correct our mistake. There is no simple solution or quick fix, but there are definitely steps that can be taken to stop the deforestation and restore not only the damaged ecosystems, but the beauty of life that's been lost.

Four invaluable steps to saving our rainforests:

Step 1: Education

In the last 20 years, deforestation has claimed millions of square miles of tropical rainforests, and to protect their future we need to develop sound educational initiatives. Education programs and curricula for each grade level is vital as children of today are our future. Encouraging good global citizenship in school aged children will help them develop a deeper understanding of conservation challenges, as well as a healthy respect for the environment. Education cannot, however, stop with school-aged kids; adults need the same education about deforestation and preventative measures. Educational resources are now becoming widely available to educators. For example, Paradise Earth Scholastic is Paradise earth's academic service and the Internet's premier source for rainforest education, replete with educational curricula for first and secondary education, multimedia educational features, and resources for research and teaching. Paradise Earth Scholastic is available online at www.paradiseearth.com w.e.f January 2009.

Step 2: Conservation Policies

Saving tropical rainforests is a worldwide responsibility, not just the responsibility of the country the forests are home to. Stronger policies prohibiting deforestation need to be written and enforced; our responsibility lies quite a bit deeper. If the international community wants to provide a higher level of protection of these forests, financial resources have to be a major part of the conservation strategy.

Historically, world governments have been willing to grant loans to tropical nations, and in some cases even cancel debts owed by them in exchange for environmental protection. For example, the British government recently assigned $150 million to preservation and sustainable development of tropical forests around the globe. Germany cleared Kenya of its $400 million debt when Kenya agreed to pass environmental legislation.

In 2001, President Clinton proposed $150 million in funds to assist developing countries preserve their tropical forests while strengthening their economies. Under the budget, $100 million would go towards conservation programs (through the US Agency for International Development-USAID), while $37 million would be slated for debt—for-nature swaps under the Tropical Forest Conservation Act.

In addition to financial support, developed nations can also provide their conservation expertise to developing countries and assist in the planning of new protected areas.

Step 3: Restore and Regrow

Though fully restoring our lost rainforests seems impossible, a myriad of studies and rebirth projects have been conducted worldwide.

In September 2008 the announcement came that the first Kihansi spray toadlet was born at the Wildlife Conservation Society's Bronx Zoo. This little creature was last seen in the wild May of 2005. The birth of the Kihansi toadlet has renewed hopes that the species can someday be successfully reintroduced to its natural habitat in a remote gorge in Tanzania.

In other news, researchers from the Boyce Thompson Institute for Plant Sciences (BTI) on the Cornell campus are attempting what many thought was impossible—restoring a tropical rainforest ecosystem. Ten years after the tree plantings, Cornell graduate student Jackeline Salazar counted the species of plants that took up residence in the shade of the new-planted areas. She found remarkably high numbers of species—more than 100 in each plot. And many of the new arrivals were also to be found in nearby remnants of the original forests.

It may take hundreds of years to regain what has been lost, but every year we see evidence that the 'impossible' is actually quite possible.

Step 4: Support Ecotourism

According to United Nations World Tourism Organisation (http://www.unwto.org/sdt/mission/en/mission.php), sustainable tourism is envisaged as leading to management of all resources in such a way that economic, social and aesthetic needs can be fulfilled while maintaining cultural integrity, essential ecological processes, biological diversity and life support systems.

Responsible ecotourism includes programs that minimise the negative aspects of conventional tourism on the environment while enhancing the cultural integrity of local people and their economy. From 1993 to 2007 alone, tourism to 23 countries harbouring biodiversity hot spots grew by 100 per cent.

At first glance, it seems that ecotourism was designed for the traveller, but its intent is much greater. Ecotourism creates jobs in food and beverage service, hotel and resort industry, transportation, and many other industries. Because Ecotourism relies on healthy ecosystems, it provides a powerful incentive to protect our rainforests. People who earn their living from ecotourism are more likely to protect local natural resources and support conservation efforts.

Correcting the 'mistake' of deforestation could still be probable; but not without an overdose of human effort to finally bring an end to the demise of tropical rainforests. No matter how unreachable this goal may seem, our mistake still has a chance of being corrected.

Chapter 21

Forest Fire and Biological Diversity

INTRODUCTION

Fire is a vital and natural part of the functioning of numerous forest ecosystems. Humans have used fire for thousands of years as a land management tool. Fire is one of the natural forces that has influenced plant communities over time and as a natural process it serves an important function in maintaining the health of certain ecosystems. However, in the latter part of the twentieth century, changes in the human-fire dynamic and an increase in El Niño frequency have led to a situation where fires are now a major threat to many forests and the biodiversity therein. Tropical rainforests and cloud forests, which typically do not burn on a large scale, were devastated by wildfires during the 1980s and 1990s.

Although the ecological impact of fires on forest ecosystems has been investigated across boreal, temperate and tropical biomes, comparatively little attention has been paid to the impact of fires on forest biodiversity, especially for the tropics. For example, of the 36 donor-assisted fire projects carried out or ongoing in Indonesia, a megadiversity country, between 1983 and 1998, only one specifically addressed the impact on biodiversity.

Fire ecology is concerned with the processes linking the natural incidence of fire in an ecosystem and the ecological effects of this fire. Many ecosystems, such as the North American prairie and chaparral ecosystems, and the South African savanna, have evolved with fire as a natural and necessary contributor to habitat vitality and renewal. Many plant species in naturally fire-affected environments require fire to germinate. Fire suppression can lead to the build-up of inflammable debris and the creation of less frequent but much larger and destructive wildfires. Fire suppression, in combination with other human-caused environmental changes, has resulted in unforeseen consequences for natural ecosystems. Some uncharacteristically large wildfires in the United States have been caused as a consequence of years of fire suppression and the continuing expansion of people into fire-adapted ecosystems. Land managers are faced with tough questions regarding where to restore a natural fire regime.

A fire regime describes the pattern that fire follows in a particular ecosystem. Its 'severity' is a term that ecologists use to refer to the impact that a fire has on an ecosystem. Ecologists can define this in many ways, but one way is through an estimate of plant mortality. Fire can burn at three levels. Ground fires will burn through soil that is rich in organic matter. Surface fires will burn through dead plant material that is lying on the ground. Crown fires will burn in the tops of shrubs and trees. Ecosystems may experience predominantly one of these fire regimes or a mix of all three. Fires will often break out during a dry season, but in some areas wildfires may also commonly occur during a time of year when lightning is prevalent. The frequency over a span of years at which fire will occur at a particular location is a measure of how common wildfires are in a given ecosystem. It is either defined as the average interval between fires at a given site, or the average interval between fires in an equivalent specified area. Fire has important effects on the abiotic (nonliving) components of an ecosystem, particularly the soil. Fire can affect the soil by direct contact with it and by its effects on the plant community associated with it. By removing overhead vegetation, fire can lead to increased solar radiation on the soil surface by day, resulting in greater warming, and to greater cooling through the loss of radiative heat at night. Fewer leaves left to intercept rain will allow more moisture to reach the soil surface. In addition, plant transpiration (the process by which water travels through plants and evaporates through pores in the leaves) will be reduced following a fire, allowing the soil to retain more moisture. Exposure to sunlight, wind and evaporation, however, will work in the other way, to dry the soil. The fire may have created an impermeable crust at the soil surface, if organic matter on the ground was heated by the fire into a waxy residue, and if this has happened, it may lead to increased soil erosion through surface runoff.

Fire may cause nutrient loss through a variety of mechanisms, including oxidation, volatilisation, and increased erosion and leaching by water. Temperatures must be very high, however, to cause a significant loss of nutrients, which are often replaced by organic matter left behind in the fire. Charcoal is able to counteract some nutrient and water loss because of its absorptive properties.

Overall, soils become more basic (higher pH) following fires because of acid combustion. By driving novel chemical reactions at high temperatures, fire can even alter the texture and structure of soils by affecting the clay content and the soil's porosity.

ECOSYSTEM EFFECTS OF FIRE

Forest fires have many implications for biological diversity. At the global scale, they are a significant source of emitted carbon, contributing to global warming which could lead to biodiversity changes. At the regional and local level, they lead to change in biomass stocks, alter the hydrological cycle with subsequent effects for marine systems such as coral reefs, and impact plant and animal species' functioning. Smoke from fires can significantly reduce photosynthetic activity and can be detrimental to health of humans and animals. One of the most important ecological effects of burning is the increased probability of further burning in subsequent years, as dead trees topple to the ground, opening up the forest to drying by sunlight, and building up the fuel load with an increase in fire-prone species, such as pyrophytic grasses. The consequence of repeated burns is detrimental because it is a key factor in the impoverishment of biodiversity in rainforest ecosystems. Fires can be followed by insect colonisation and infestation which disturb the ecological balance. The replacement of vast areas of forest with pyrophytic grasslands is one of the most negative ecological impacts of fires in tropical rainforests. These processes have already been observed in parts of Indonesia and Amazonia. What was once a dense evergreen forest becomes an impoverished forest populated by a few fire-resistant tree species and a ground cover of weedy grasses. In North Queensland in Australia, it has been observed that where the aboriginal fire practices and fire regimes were controlled, rainforest vegetation started to replace the fire-prone tree-grass savannahs.

EFFECTS OF FIRE ON PLANTS AND ANIMALS: INDIVIDUAL LEVEL

Plants

Obviously plants can't move away when a fire comes through an area, so how are they able to survive and persist?

Most vegetative survival involves the protection of tissue from heat which would otherwise destroy it. Fire resistance and tolerance is exhibited through: bark thickness, other vegetative insulation, above-ground resprouting, underground roots and stems.

Bark thickness: Thick bark insulates and protects the cambium from heat and damage.

Vegetative insulation: Some protection is afforded by leaf sheaths. Grasses have meristems at leaf base so are protected from heat and damage in this way. *Pandanus* also receives some protection from leaf sheaths.

Fire acts as a generalist herbivore removing plant material above the ground surface, thus enabling new herbaceous growth.

Above ground resprouting: While many trees are killed by total defoliation following a fire, some can re-sprout from epicormic buds, which are buds positioned beneath the bark. *Eucalyptus* trees are known for their ability to vegetatively regenerate branches along their trunks from buds. This is because epicormic buds of *Eucalyptus* trees are more protected than on other tree species because they are set much deeper at maximum bark thickness. The ability to survive and re-sprout depends on tree height, scorch and char heights, but also tree species, age, size (height) and the severity of the fire.

Ladder fires running up the trunks of these *Melaleuca* trees might look impressive but they can stimulate regrowth from epicormic buds and do not pose a threat to the trees.

Belowground roots and underground stems: Because soil is a good insulator, buds underground are well protected. Plants can survive fires by resprouting from basal stems, and also from roots and horizontal rhizomes.

Root suckering can result in large clonal populations post-fire. Other plants have woody swellings at the base of their stems called lignotubers. They contain latent buds which are released from dormancy post-fire. Most *Eucalyptus* species have lignotubers. Geophytes survive burning because the storage organs are below ground protected from burning. They persist vegetatively between fires, and are stimulated to flower following a fire.

FIRE ECOLOGY

Fire is a natural component of many ecosystems, which include plants and animals that interact with one another and with their physical environment. Fire ecology examines the role of fire in ecosystems. Fire ecologists study the origins of fire, what influences spread and intensity, fire's relationship with ecosystems, and how controlled fires can be used to maintain ecosystem health.

Physical and Chemical Nature of Fire

For fuel to ignite it must be heated in the presence of oxygen to the ignition point or kindling temperature. Wood must reach about 800°F to burst into flame. As the wood is being heated to this point it dries as water, oils, and resins are boiled away. The chemical structure of the fuel is broken down and flammable gases are produced. The ignition of these flammable gases is known as flaming combustion. Flaming combustion transforms the surface of the wood to charcoal. At cooler temperatures, glowing combustion consumes charcoal, producing ash, water and carbon dioxide. Many factors such as fuel, weather, topography and fire history influence the probability of ignition and combustion.

Fire Behaviour

Fire behaviour is most often described by intensity and spread. Many factors influence this behaviour. Five factors that influence intensity are available fuel, moisture and temperature, fuel composition, wind and topography. Available fuel is quantified by size and arrangement. The more available fuel, the more intense the fire and cool, moist fuels combust more slowly than hot, dry fuels. Fuel composition can make a fire more or less intense. Oils and resins increase the heat yield of the reaction and cause a fire to burn intensely whereas other chemical factors, such as high concentrations of minerals, can reduce flammability. Wind increases oxygen supply, convects heat and can produce 'spot fires' from fragments that blow down-wind. Finally, topography effects intensity. A fire ignited on the top of a slope is likely to spread slowly as it burns downhill, whereas a fire at the bottom of a slope will start rapidly and gain momentum as it burns uphill because warm air rises and preheats uphill fuels. Many of the factors that affect intensity also affect the rate of spread. For example, fires in dry, windy conditions with abundant fuel spread rapidly. Fuel continuity and topography also play a role in spread. Topographic features such as streams and lakes can create firebreaks, thus influencing the distribution of burns across landscapes. Finally, the composition of plant communities affects spread, as some species are more flammable than others.

Effects of Fire on Ecosystems

There is much yet to be learned about how wildland fire affects ecosystems. This is in part because each fire and each ecosystem has unique properties. However, some generalities can be made.

Mosaic patterns

Wildland fires create a mixture of totally burned, partially burned, and unburned sections called a burn mosaic. The varying degrees of burn are a result of many factors including wind shifts, daily temperature changes, moisture levels, and varying chemical composition of the vegetation. The burn mosaic results in varied regrowth rates that creates a vegetation mosaic.

Soil conditions

Wildland fires can be both a detriment and a benefit to soil. The soil can become more nutrient-rich after a fire due to the high mineral content of the ash and charcoal and also due to the warm, moist conditions that increase microbial activity. The intense heat can also cause soil particles to become water-repellant, causing rainwater to run off. As the water runs off it can carry soil particles with it and lead to erosion.

Animal populations

Some animals will perish in wildland fires, especially small animals, insects, and older and weaker individuals. However, fire has a greater affect on habitat than on individuals. While the vast majority of large mammals are able to flee fires, populations often suffer substantial losses in the months following a fire due to a loss of food sources. Food sources are scarce because of the fire itself and also because most natural fires occur shortly before winter. These habitat changes allow other animals to thrive. Scavenging animals find an increased abundance of food sources and predatory animals may benefit from reduced forest cover which makes prey more visible. Nutrient-rich new growth also benefits many animals and animals such as deer will even eat the nutrient-rich charcoal and ashes. Birds also thrive on increased seed availability and nesting sites in snags.

Plant populations

Vegetation composition is one factor that determines how a fire behaves. The fire behaviour in turn determines the extent to which the plant populations are affected. The more intense the fire, the more vegetation is killed. The initial vegetation losses may look harsh, but the reduced number of trees and shrubs minimise competition among the surviving individuals.

The organisms that survive the fire gain more access to nutrients, light and water. Plants may exhibit increased growth, benefiting from the additional minerals in the soil as a result of the fire. Fire may also rid some plants of their parasites, increasing plant health. For example, a high-intensity fire kills dwarf mistletoe, a parasitic plant of the lodgepole. Some plant species have adaptations that allow them to survive, thrive, and even require fire for survival. The giant sequoia can produce bark that is 2-feet thick as protection from fire. Other plants such as the chaparral snowbush require the heat of wildland fires to crack their seed coats.

Fire regimes

Fire regimes are the patterns of wildland fires that include factors such as frequency, extent, intensity, type and season. Regimes vary by ecosystem because each ecosystem has a different composition and structure determined by climate conditions. vegetation types, and ignition sources. Humans have altered many aspects of natural fire regimes over time. Currently, ecologists are studying evidence to try to determine historical fire records or natural fire regimes. Techniques include sampling fire scars on trees for evidence of a sequence of fires in the growth rings, sampling lake and reservoir sediments for extreme or unusual runoff events, using written and oral histories, and extrapolating from current patterns of weather, fuel build-up and lightning fires. Understanding natural fire regimes should lead to the most appropriate resource management policies. The variety of ecosystems and regimes dictates that there should be a variety of techniques and practices in any comprehensive management policy. One solution does not fit all.

Lodgepole pine

The Lodgepole Pine (*Pinus contorta*) is a dominant tree of the northern United States and Canada. A lodgepole stand may live 250 to 400 years. During the first century of the stand's existence surface fires are unable to climb into the forest canopy because the lower branches of a lodgepole die and drop-off as the tree grows taller. Slowly shade-tolerant spruces and firs start to grow on the dark floor of the stand, blocking out the sun and preventing lodgepoles from sprouting. Eventually spruces and firs would dominate if there were no forest fires. However lightning usually ignites a fire that destroys the stand. After a fire, lodgepole pines are often the first trees to reappear because the bare, sunlit soil that remains after a fire is ideal for lodgepole seedling growth. Lodgepoles also have special serotinous cones that are coated with a hard waxy substance. Fire melts this coating, allowing thousands of seeds to be released. The Lodgepole Pine is just one of many species that is dependent upon fire.

Human influence on wildland fire

The way a fire operates is largely determined by a region's wildlife. Altering these biotic components impacts the fire's effects. Humans have had one of the greatest influences on the biota of ecosystems. Native Americans and early settlers used fire extensively in their land management practices. Today, we clear vegetation for farming, homes, commercial buildings, and roads. We introduce nonnative species. We use forests to harvest timber. Our influences are countless. As a result, it is impossible to fully understand the extent to which humans have altered natural fire regimes. This makes fire management a complex and often controversial topic.

IMPACTS OF HUMAN-INDUCED OR SEVERE NATURAL WILDFIRE ON PLANT DIVERSITY

Wildfire is unusual in most undisturbed, tall, closed-canopy, tropical rainforests because of the moist microclimate, moist fuels, low wind speeds and high rainfall. However, rainforests may become more susceptible to fire during severe droughts, as experienced during El Niño years. In these forests which are not adapted to fire, fire can kill virtually all seedlings, sprouts, lianas and young trees because they are not protected by thick bark. Damage to the seed bank, seedlings and saplings hinders recovery of the original species. The degree of recovery and need for rehabilitation interventions depends on the intensity of burning. Tropical forests are also subject to fires started by humans for agricultural clearing. Deforestation fires, which are more common in disturbed forests, can vary in intensity and burn standing trees, at the worst completely burning the forest leaving nothing but bare soil.

There is some concern that salvage logging (removal of dead timber from severely burned logged-over forest or burned primary forest), used as a management and financing tool after fires in Indonesia in 1997–1998, may adversely affect the course of vegetation succession.

Although fire is a frequent natural disturbance in boreal forests and they usually regenerate easily after fire, frequent high-intensity fires can offset this balance. As a result of extremely severe fires in the Russian Federation in 1998, more than 2 million hectares of forest have lost most of their major ecological functions for a period of 50 to 100 years. Severe fires have had a significant negative impact on plant diversity. Southern species that are at the northern edge of their geographic range are particularly vulnerable. For example, in Primorsky Kray, human-induced fires have contributed to drastic reductions in the populations of 60 species of vascular plants, ten fungi, eight lichens and six species of mosses during the past two or three decades.

Wildfire

A wildfire is any uncontrolled fire in combustible vegetation that occurs in the countryside or a wilderness area. Other names such as brush fire, bushfire, forest fire, grass fire, hill fire, peat fire, vegetation fire, veldfire, and wildland fire may be used to describe the same phenomenon depending on the type of vegetation being burned. A wildfire differs from other fires by its extensive size, the speed at which it can spread out from its original source, its potential to change direction unexpectedly, and its ability to jump gaps such as roads, rivers and fire breaks. Wildfires are characterised in terms of the cause of ignition, their physical properties such as speed of propagation, the combustible material present, and the effect of weather on the fire.

Wildfires occur on every continent except Antarctica. Fossil records and human history contain accounts of wildfires, as wildfires can occur in periodic intervals. Wildfires can cause extensive damage, both to property and human life, but they also have various beneficial effects on wilderness areas. Some plant species depend on the effects of fire for growth and reproduction, although large wildfires may also have negative ecological effects.

Strategies of wildfire prevention, detection and suppression have varied over the years, and international wildfire management experts encourage further development of technology and research. One of the more controversial techniques is controlled burning: permitting or even igniting smaller fires to minimise the amount of flammable material available for a potential wildfire. While some wildfires burn in remote forested regions, they can cause extensive destruction of homes and other property located in the wildland-urban interface: a zone of transition between developed areas and undeveloped wilderness.

Characteristics

The name wildfire was once a synonym for Greek fire but now refers to any large or destructive conflagration. Wildfires differ from other fires in that they take place outdoors in areas of grassland, woodlands, bushland, scrubland, peatland, and other wooded areas that act as a source of fuel or combustible material. Buildings may become involved if a wildfire spreads to adjacent communities. While the causes of wildfires vary and the outcomes are always unique, all wildfires can be characterised in terms of their physical properties, their fuel type, and the effect that weather has on the fire.

Wildfire behaviour and severity result from the combination of factors such as available fuels, physical setting and weather. While wildfires can be large, uncontrolled disasters that burn through 0.4 to 400 square kilometres (100 to 1,00,000 acres) or more, they can also be as small as 0.0010 square kilometres (0.25 acre) or less. Although smaller events may be included in wildfire modelling, most do not earn press attention. This can be problematic because public fire policies, which relate to fires of all sizes, are influenced more by the way the media portrays catastrophic wildfires than by small fires.

Causes

The four major natural causes of wildfire ignitions are lightning, volcanic eruption, sparks from rockfalls, and spontaneous combustion. The thousands of coal seam fires that are burning around the world, such as those in Centralia, Burning Mountain, and several coal-sustained fires in China, can also flare up and ignite nearby flammable material. However, many wildfires are attributed to human sources such as arson, discarded cigarettes, sparks from equipment, and power line arcs (as detected by arc mapping). In societies experiencing shifting cultivation where land is cleared quickly and farmed until the soil loses fertility, slash and burn clearing is often considered the least expensive way to prepare land for future use. Forested areas cleared by logging encourage the dominance of flammable grasses, and abandoned logging roads overgrown by vegetation may act as fire corridors. Annual grassland fires in Southern Vietnam can be attributed in part to the destruction of forested areas by herbicides, explosives, and mechanical land clearing and burning operations during the Vietnam War.

The most common cause of wildfires varies throughout the world. In the United States, Canada and Northwest China, for example, lightning is the major source of ignition. In other parts of the world, human involvement is a major contributor. In Mexico, Central America, South America, Africa, Southeast Asia, Fiji and New Zealand, wildfires can be attributed to human activities such as animal husbandry, agriculture and land-conversion burning. Human carelessness is a major cause of wildfires in China and in the Mediterranean Basin. In Australia, the source of wildfires can be traced to both lightning strikes and human activities such as machinery sparks and castaway cigarette butts.

Fuel type

The spread of wildfires varies based on the flammable material present and its vertical arrangement. For example, fuels uphill from a fire are more readily dried and warmed by the fire than those downhill, yet burning logs can roll downhill from the fire to ignite other fuels. Fuel arrangement and density is governed in part by topography, as land shape determines factors such as available sunlight and water for plant growth. Overall, fire types can be generally characterised by their fuels as follows:

1. Ground fires are fed by subterranean roots, duff and other buried organic matter. This fuel type is especially susceptible to ignition due to spotting. Ground fires typically burn by smoldering, and can burn slowly for days to months, such as peat fires in Kalimantan and Eastern Sumatra, Indonesia, which resulted from a riceland creation project that unintentionally drained and dried the peat.

2. Crawling or surface fires are fuelled by low-lying vegetation such as leaf and timber litter, debris, grass and low-lying shrubbery.
3. Ladder fires consume material between low-level vegetation and tree canopies, such as small trees, downed logs and vines. Kudzu, Old World climbing fern, and other invasive plants that scale trees may also encourage ladder fires.
4. Crown, canopy or aerial fires burn suspended material at the canopy level, such as tall trees, vines and mosses. The ignition of a crown fire, termed crowning, is dependent on the density of the suspended material, canopy height, canopy continuity, and sufficient surface and ladder fires in order to reach the tree crowns. For example, ground-clearing fires lit by humans can spread into the Amazon rainforest, damaging ecosystems not particularly suited for heat or arid conditions.

Physical properties

Wildfires occur when all of the necessary elements of a fire triangle come together in a susceptible area: an ignition source is brought into contact with a combustible material such as vegetation, that is subjected to sufficient heat and has an adequate supply of oxygen from the ambient air. A high moisture content usually prevents ignition and slows propagation, because higher temperatures are required to evaporate any water within the material and heat the material to its fire point. Dense forests usually provide more shade, resulting in lower ambient temperatures and greater humidity, and are therefore less susceptible to wildfires. Less dense material such as grasses and leaves are easier to ignite because they contain less water than denser material such as branches and trunks. Plants continuously lose water by evapotranspiration, but water loss is usually balanced by water absorbed from the soil, humidity or rain. When this balance is not maintained, plants dry out and are therefore more flammable, often a consequence of droughts.

A wildfire front is the portion sustaining continuous flaming combustion, where unburned material meets active flames, or the smoldering transition between unburned and burned material. As the front approaches, the fire heats both the surrounding air and woody material through convection and thermal radiation. First, wood is dried as water is vapourised at a temperature of 100°C (212°F). Next, the pyrolysis of wood at 230°C (450°F) releases flammable gases. Finally, wood can smolder at 380°C (720°F) or, when heated sufficiently, ignite at 590°C (1000°F). Even before the flames of a wildfire arrive at a particular location, heat transfer from the wildfire front warms the air to 800°C (1470°F), which preheats and dries flammable materials, causing materials to ignite faster and allowing the fire to spread faster. High-temperature and long-duration surface wildfires may encourage flashover or torching: the drying of tree canopies and their subsequent ignition from below.

Wildfires have a rapid forward rate of spread (FROS) when burning through dense, uninterrupted fuels. They can move as fast as 10.8 kilometres per hour (6.7 mph) in forests and 22 km per hour (14 mph) in grasslands. Wildfires can advance tangential to the main front to form a flanking front, or burn in the opposite direction of the main front by backing. They may also spread by jumping or spotting as winds and vertical convection columns carry firebrands (hot wood embers) and other burning materials through the air over roads, rivers, and other barriers that may otherwise act as firebreaks. Torching and fires in tree canopies encourage spotting, and dry ground fuels that surround a wildfire are especially vulnerable to ignition from firebrands. Spotting can create spot fires as hot embers and firebrands ignite fuels downwind from the fire. In Australian bushfires, spot fires are known to occur as far as 10 km from the fire front.

Especially large wildfires may affect air currents in their immediate vicinities by the stack effect: air rises as it is heated, and large wildfires create powerful updrafts that will draw in new, cooler air from surrounding areas in thermal columns. Great vertical differences in temperature and humidity encourage pyrocumulus clouds, strong winds, and fire whirls with the force of tornadoes at speeds of more than 80 km per hour (50 mph). Rapid rates of spread, prolific crowning or spotting, the presence of fire whirls, and strong convection columns signify extreme conditions.

Effect of weather

Heat waves, droughts, cyclical climate changes such as El Niño, and regional weather patterns such as high-pressure ridges can increase the risk and alter the behaviour of wildfires dramatically. Years of precipitation followed by warm periods can encourage more widespread fires and longer fire seasons. Since the mid-1980s, earlier snowmelt and associated warming has also been associated with an increase in length and severity of the wildfire season in the Western United States. However, one individual element does not always cause an increase in wildfire activity. For example, wildfires will not occur during a drought unless accompanied by other factors, such as lightning (ignition source) and strong winds (mechanism for rapid spread). Fire intensity also increases during day time hours. Burn rates of smoldering logs are up to five times greater during the day due to lower humidity, increased temperatures, and increased wind speeds. Sunlight warms the ground during the day which creates air currents that travel uphill. At night the land cools, creating air currents that travel downhill. Wildfires are fanned by these winds and often follow the air currents over hills and through valleys. Fires in Europe occur frequently during the hours of 12:00 pm and 2:00 pm. Wildfire suppression operations in the United States revolve around a 24 hour fire day that begins at 10:00 am due to the predictable increase in intensity resulting from the daytime warmth.

Ecology

Wildfires are common in climates that are sufficiently moist to allow the growth of vegetation but feature extended dry, hot periods. Such places include the vegetated areas of Australia and Southeast Asia, the veld in southern Africa, the fynbos in the Western Cape of South Africa, the forested areas of the United States and Canada, and the Mediterranean Basin. Fires can be particularly intense during days of strong winds, periods of drought, and during warm summer months. Global warming may increase the intensity and frequency of droughts in many areas, creating more intense and frequent wildfires.

Although some ecosystems rely on naturally occurring fires to regulate growth, many ecosystems suffer from too much fire, such as the chaparral in southern California and lower elevation deserts in the American southwest. The increased fire frequency in these ordinarily fire-dependent areas has upset natural cycles, destroyed native plant communities, and encouraged the growth of fire-intolerant vegetation and non-native weeds. Invasive species, such as *Lygodium microphyllum* and *Bromus tectorum*, can grow rapidly in areas that were damaged by fires. Because they are highly flammable, they can increase the future risk of fire, creating a positive feedback loop that increases fire frequency and further destroys native growth.

In the Amazon Rainforest, drought, logging, cattle ranching practices, and slash-and-burn agriculture damage fire-resistant forests and promote the growth of flammable brush, creating a cycle that encourages more burning. Fires in the rainforest threaten its collection of diverse species and produce large amounts of CO_2. Also, fires in the rainforest, along with drought and human involvement, could damage or destroy more than half of the Amazon rainforest by the year 2030. Wildfires generate ash, destroy available organic nutrients, and cause an increase in water runoff, eroding away other nutrients and creating flash flood conditions. Wildfires can also have an effect on climate change, increasing the amount of carbon released into the atmosphere and inhibiting vegetation growth, which affects overall carbon uptake by plants.

Plant adaptation

Plants in wildfire-prone ecosystems often survive through adaptations to their local fire regime. Such adaptations include physical protection against heat, increased growth after a fire event, and flammable materials that encourage fire and may eliminate competition. For example, plants of the genus *Eucalyptus* contain flammable oils that encourage fire and hard sclerophyll leaves to resist heat and drought, ensuring their dominance over less fire-tolerant species. Dense bark, shedding lower branches, and high water content in external structures may also protect trees from rising temperatures. Fire-resistant seeds and reserve shoots that sprout after a fire encourage species preservation, as embodied by pioneer species. Smoke, charred wood, and heat can stimulate the germination of seeds in a process called serotiny. Exposure to smoke from burning plants promotes germination in other types of plants by inducing the production of the orange butenolide.

Atmospheric effects

Most of the earth's weather and air pollution reside in the troposphere, the part of the atmosphere that extends from the surface of the planet to a height of about 10 km. The vertical lift of a severe thunderstorm or pyrocumulonimbus can be enhanced in the area of a large wildfire, which can propel smoke, soot, and other particulate matter as high as the lower stratosphere. Previously, prevailing scientific theory held that most particles in the stratosphere came from volcanoes, but smoke and other wildfire emissions have been detected from the lower stratosphere. Pyrocumulus clouds can reach 6100 metres (20,000 ft) over wildfires. Increased fire by-products in the stratosphere can increase ozone concentration beyond safe levels.

Human involvement

The human use of fire for agricultural and hunting purposes during the Paleolithic and Mesolithic ages altered the preexisting landscapes and fire regimes. Woodlands were gradually replaced by smaller vegetation that facilitated travel, hunting, seed-gathering and planting.

Prevention

Wildfire prevention refers to the pre-emptive methods of reducing the risk of fires as well as lessening its severity and spread. Effective prevention techniques allow supervising agencies to manage air quality, maintain ecological balances, protect resources, and to limit the effects of future uncontrolled fires. North American firefighting policies may permit naturally caused fires to burn to maintain their ecological role, so long as the risks of escape into high-value areas are mitigated. However, prevention policies must consider the role that humans play in wildfires, since, for example, 95 per cent of forest fires in Europe are related to human involvement. Sources of human-caused fire may include arson, accidental ignition, or the uncontrolled use of fire in land-clearing and agriculture such as the slash-and-burn farming in Southeast Asia. Wildfires are caused by a combination of natural factors such as topography, fuels and weather. Other than reducing human infractions, only fuels may be altered to affect future fire risk and behaviour. Wildfire prevention programs around the world may employ techniques such as wildland fire use and prescribed or controlled burns. Wildland fire use refers to any fire of natural causes that is monitored but allowed to burn. Controlled burns are fires ignited by government agencies under less dangerous weather conditions.

Vegetation may be burned periodically to maintain high species diversity, and frequent burning of surface fuels limits fuel accumulation, thereby reducing the risk of crown fires. Using strategic cuts of trees, fuels may also be removed by handcrews in order to clean and clear the forest, prevent fuel build-up, and create access into forested areas. Chain saws and large equipment can be used to thin out ladder fuels and shred trees and vegetation to a mulch. Multiple fuel treatments are often needed to influence future fire risks, and wildfire models may be used to predict and compare the benefits of different fuel treatments on future wildfire spread. However, controlled burns are reportedly 'the most effective treatment for reducing a fire's rate of spread, fireline intensity, flame length, and heat per unit of area' according to Jan Van Wagtendonk, a biologist at the Yellowstone Field Station. Additionally, while fuel treatments are typically limited to smaller areas, effective fire management requires the administration of fuels across large landscapes in order to reduce future fire size and severity.

Building codes in fire-prone areas typically require that structures be built of flame-resistant materials and a defensible space be maintained by clearing flammable materials within a prescribed distance from the structure. Communities in the Philippines also maintain fire lines 5 to 10 metres (16 to 33 ft) wide between the forest and their village, and patrol these lines during summer months or seasons of dry weather. Fuel buildup can result in costly, devastating fires as new homes, ranches and other development are built adjacent to wilderness areas. Continued growth in fire-prone areas and rebuilding structures destroyed by fires has been met with criticism.

However, the population growth along the wildland-urban interface discourages the use of current fuel management techniques. Smoke is an irritant and attempts to thin out the fuel load is met with opposition due to desirability of forested areas, in addition to other wilderness goals such as endangered species protection and habitat preservation. The ecological benefits of fire are often overridden by the economic and safety benefits of protecting structures and human life. For example, while fuel treatments decrease the risk of crown fires, these techniques destroy the habitats of various plant and animal species. Additionally, government policies that cover the wilderness usually differ from local and state policies that govern urban lands. A new and ecologically evolutionary practice, termed 'Hydro-Pyrogeography', promises to bound wildfire from passing through any such wildland-urban interface anywhere on earth that the practice is put into place, and thereby diminishing, even eliminating the above-referred oppositions and concerns to traditional fuel management techniques.

EFFECTS OF FIRE ON FOREST FAUNA

In forests where fire is not a natural disturbance, it can have devastating impacts on forest vertebrates and invertebrates—not only killing them directly, but also leading to longer-term indirect effects such as stress and loss of habitat, territories, shelter and food. The loss of key organisms in forest ecosystems, such as invertebrates, pollinators and decomposers, can significantly slow the recovery rate of the forest.

Estimates from the 1998 fires in the Russian Federation suggest that mammals and fish were badly affected. Mortality of squirrels and weasels, estimated immediately after the fires, reached 70 to 80 per cent; boar 15 to 25 per cent; and rodents 90 per cent.

Loss of Habitat, Territories and Shelter

The destruction of standing cavity trees as well as dead logs on the ground has negative effects on most small mammal species (e.g. tarsiers, bats and lemurs) and cavity-nesting birds. Fires can cause the displacement of territorial birds and mammals, which may upset the local balance and ultimately result in the loss of wildlife, since displaced individuals have nowhere to go. The severe fires of 1998 in the Russian Federation led to increased water temperatures and high carbon dioxide levels in lakes and waterways, which adversely affected salmon spawning. In areas where frequent burning occurs on a broad scale, preserving a range of microhabitats can make a substantial contribution to conserving biodiversity.

Loss of Food

Loss of fruit-trees results in overall decline in bird and animal species that rely on fruits for food; this effect is particularly pronounced in tropical forests. A few months after the 1982–1983 fires in Kutai National Park, East Kalimantan, fruit-eating birds such as hornbills declined dramatically, and only insectivorous birds such as woodpeckers were common because of the abundance of wood-eating insects.

Burned forests become impoverished of small mammals, birds and reptiles, and carnivores tend to avoid burned over areas. The reduction in densities of small mammals such as rodents can adversely affect the food supply for small carnivores. Fires also destroy leaf litter and its associated arthropod community, further reducing food availability for omnivores and carnivores.

Fire-adapted Fauna

Not all species suffer from fire. For instance, grass-layer beetle species in Australia's savannahs show remarkable resilience to fire, although fires affect abundance, species and family richness. In the fire-prone Mediterranean region, the current fire regime has probably contributed to maintaining the bird diversity at the landscape level in Portugal. In Israel, richness of fauna species in certain areas was the highest two to four years after a fire followed by a decrease over time.

Fire can have positive effects on wildlife populations in boreal forests, where fire is a major natural disturbance mechanism. In North America, although moose are occasionally trapped and killed by fire, fire generally enhances moose habitat by creating and maintaining seral communities, and is considered beneficial to moose populations. The beneficial effects of fire on its habitat is estimated to last less than 50 years, with moose density peaking 20 to 25 years following fire.

Fire has contributed to the reduction in populations of grey wolf (*Canis lupus*) in Minnesota, United States, by limiting its prey—including beaver (*Castor canadiensis*), moose and deer, fire-dependent species that require the plant communities that persist following frequent fires.

EFFECTS OF SUPPRESSION OF THE NATURAL FIRE REGIME

Temperate forests in the United States and Australia in which fire was deliberately suppressed are now experiencing devastating wildfires because of an unnatural accumulation of fuel. Deliberate human suppression of fire can also have direct negative impacts on species. In forests where fire is a natural part of the system, plant and animal species are adapted to a natural fire regime and benefit from the aftermath of a fire. In North America, fire suppression in some areas has contributed to decline in the numbers of grizzly bear, *Ursus arctos horribilis*. Fires promote and maintain many important berry-producing shrubs, which are an important food source for bears, as well as providing habitat for insects and in some cases carrion. The 1998 fires in Yellowstone National Park increased availability of some food items for grizzly bears, especially carcasses of *elk*.

In boreal forests, exclusion of fire induces the build-up of organic layers that prevents melting of the upper soil during spring and summer and rise of the permafrost layer, resulting in impoverishment of forests, decrease in productivity and conversion of forests to marshes.

ECONOMICS OF FIRE MANAGEMENT POLICY

Similar to that of military operations, fire management is often very expensive in the US. Today, it is not uncommon for suppression operations for a single wildfire to exceed costs of $1 million in just a few days. Although fire suppression offers many benefits to society, other options for fire management exist. While these options can't completely replace fire suppression as a fire management tool, other options can play an important role in overall fire management and can therefore affect the costs of fire suppression.

The application of fire management tools requires making certain trade-offs. Below is a sample of some costs and benefits associated with the tools currently used in fire management. Current approaches to fire management are an almost complete turnaround compared to historic approaches. In fact, it is commonly accepted that past fire suppression, along with other factors, has resulted in larger, more intense wildfire events which are seen today. In economic terms, expenditures used for wildfire suppression in the early 20th century have contributed to increased suppression costs which are being realised today. As is the case with many public policy issues, costs and benefits associated with particular fire management tools are difficult to accurately quantify. Ultimately, costs and benefits should be weighed against one another on a case-by-case basis in planning wildland fire management operations.

Depending on the trade-offs that a land manager is willing to make, a combination of the following fire management tools could be used. For instance, prescribed fire and/or mechanical fuels reduction could be used to help prevent or lessen the intensity of a wildfire thereby reducing or eliminating suppression costs. In addition, prescribed fire and/or mechanical fuels reduction could be used to improve soil conditions in fields or in forests to the benefit of wildlife or natural resources. On the other hand, the use of prescribed fire requires much advanced planning and can have negative impacts on human health in nearby communities (Table 21.1).

Table 21.1. Cost and benefits of wildland fire management tools.

	Costs	Benefits
Suppression	Labour intensive	Can reduce human health impacts
	Requires high level of planning	Can protect forest resources and agricultural resources
	Can be very expensive	
	Particular strategies can be very inefficient (i.e. aerial retardant drops)	Can save private dwellings and commercial buildings
	Can increase intensity and likelihood of future wildfires	
	Inhibits natural ecological processes in many cases	
Prescribed fire	Can be expensive to implement	Can provide habitate for wildlife
	Requires skilled workforce to implement Requires high level of planning	Can improve forest resources and agricultural resources
	Can impact human health (e.g. smoke and its effect on those with asthma or allergies)	Can reduce hazardous fuel loading
		Mimics natural processes but under more controlled circumstances
Mechanical fuels reduction	Requires use of heavy machinery (resulting in fossil fuel consumption, soil compaction, etc.)	Can provide habitat for wildlife
		Can improve forest resources and agricultural resources
	Can be expensive to implement	Can reduce hazardous fuel loading
	Does not mimic natural processes	Does not produce large amounts of smoke

Detection

Fast and effective detection is a key factor in wildfire fighting. Early detection efforts were focused on early response, accurate results in both daytime and night time, and the ability to prioritise fire danger. Currently, public hotlines, fire lookouts in towers, and ground and aerial patrols can be used as a means of early detection of forest fires. However, accurate human observation may be limited by operator fatigue, time of day, time of year, and geographic location. Electronic systems have gained popularity in recent years as a possible resolution to human operator error. These systems may be semi- or fully-automated and employ systems based on the risk area and degree of human presence, as suggested by GIS data analyses.

An integrated approach of multiple systems can be used to merge satellite data, aerial imagery and personnel position via global positioning system (GPS) into a collective whole for near-realtime use by wireless Incident Command Centers. A small, high risk area that features thick vegetation, a strong human presence or is close to a critical urban area can be monitored using a local sensor network. Detection systems may include wireless sensor networks that act as automated weather systems: detecting temperature, humidity and smoke. These may be battery-powered, solar-powered or tree-rechargeable: able to recharge their battery systems using the small electrical currents in plant material. Larger, medium-risk areas can be monitored by scanning towers that incorporate fixed cameras and sensors to detect smoke or additional factors such as the infrared signature of carbon dioxide produced by fires. Additional capabilities such as night vision, brightness detection, and colour change detection may also be incorporated into sensor arrays.

Satellite and aerial monitoring can provide a wider view and may be sufficient to monitor very large, low risk areas. These more sophisticated systems employ GPS and aircraft-mounted infrared or high-resolution visible cameras to identify and target wildfires. Satellite-mounted sensors such as Envisat's Advanced Along Track Scanning Radiometer and European Remote-Sensing Satellite's Along-Track Scanning Radiometer can measure infrared radiation emitted by fires, identifying hot spots greater than 39°C (102°F).

The National Oceanic and Atmospheric Administration's Hazard Mapping System combines remote-sensing data from satellite sources such as geostationary operational environmental satellite (GOES), moderate-resolution imaging spectroradiometer (MODIS), and advanced very high resolution radiometer (AVHRR) for detection of fire and smoke plume location.

Wildfire Modelling

Wildfire modelling is concerned with numerical simulation of wildfires in order to comprehend and predict fire behaviour. Wildfire modelling can ultimately aid wildfire suppression, increase the safety of firefighters and the public, and minimise damage. Using computational science, wildfire modelling involves the statistical analysis of past fire events to predict spotting risks and front behaviour. Various wildfire propagation models have been proposed in the past, including simple ellipses and egg- and fan-shaped models. Early attempts to determine wildfire behaviour assumed terrain and vegetation uniformity. However, the exact behaviour of a wildfire's front is dependent on a variety of factors, including windspeed and slope steepness.

Modern growth models utilise a combination of past ellipsoidal descriptions and Huygens' Principle to simulate fire growth as a continuously expanding polygon. Extreme value theory may also be used to predict the size of large wildfires. However, large fires that exceed suppression capabilities are often regarded as statistical outliers in standard analyses, even though fire policies are more influenced by catastrophic wildfires than by small fires.

VOLCANOES

A volcano is an opening or rupture, in a planet's surface or crust, which allows hot magma, volcanic ash and gases to escape from below the surface. Volcanoes are generally found where tectonic plates are diverging or converging. A mid-oceanic ridge, for example the Mid-Atlantic Ridge, has examples of volcanoes caused by divergent tectonic plates pulling apart; the Pacific Ring of Fire has examples of volcanoes caused by convergent tectonic plates coming together. By contrast, volcanoes are usually not created where two tectonic plates slide past one another. Volcanoes can also form where there is stretching and thinning of the earth's crust (called 'non-hotspot intraplate volcanism'), such as in the East African Rift, the Wells Gray-Clearwater volcanic field and the Rio Grande Rift in North America. Volcanoes can be caused by mantle plumes. These so-called hotspots, for example at Hawaii, can occur far from plate boundaries. Hotspot volcanoes are also found elsewhere in the solar system, especially on rocky planets and moons.

Plate Tectonics and Hotspots

Divergent plate boundaries

Divergent plate boundaries are locations where plates are moving away from one another. This occurs above rising convection currents. The rising current pushes up on the bottom of the lithosphere, lifting it and flowing laterally beneath it. This lateral flow causes the plate material above to be dragged along in the direction of flow. At the crest of the uplift, the overlying plate is stretched thin, breaks and pulls apart (Fig. 21.1).

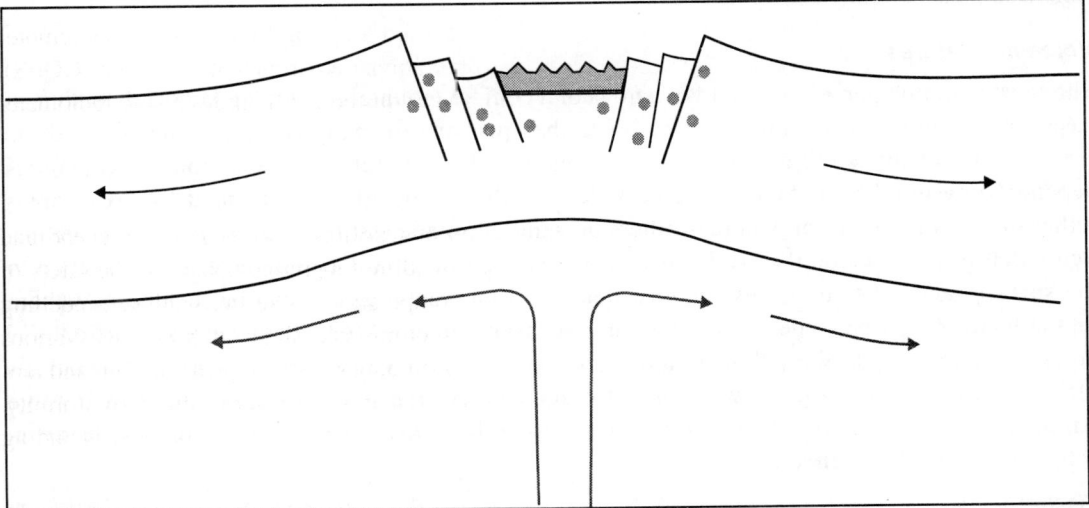

Fig. 21.1. Divergent plate boundaries.

At the mid-oceanic ridges, two tectonic plates diverge from one another. New oceanic crust is being formed by hot molten rock slowly cooling and solidifying. The crust is very thin at mid-oceanic ridges due to the pull of the tectonic plates. The release of pressure due to the thinning of the crust leads to adiabatic expansion, and the partial melting of the mantle causing volcanism and creating new oceanic crust. Most divergent plate boundaries are at the bottom of the oceans, therefore most volcanic activity is submarine, forming new seafloor. Black smokers or deep-sea vents are an example of this kind of volcanic activity. Where the mid-oceanic ridge is above sea-level, volcanic islands are formed, for example, Iceland.

Convergent plate boundaries

Subduction zones are places where two plates, usually an oceanic plate and a continental plate, collide. In this case, the oceanic plate subducts, or submerges under the continental plate forming a deep ocean trench just offshore. Water released from the subducting plate lowers the melting temperature of the overlying mantle wedge, creating magma. This magma tends to be very viscous due to its high silica content, so often does not reach the surface and cools at depth. When it does reach the surface, a volcano is formed. Typical examples for this kind of volcano are Mount Etna and the volcanoes in the Pacific Ring of Fire.

Hotspots

Hotspots are not usually located on the ridges of tectonic plates, but above mantle plumes, where the convection of the earth's mantle creates a column of hot material that rises until it reaches the crust, which tends to be thinner than in other areas of the earth. The temperature of the plume causes the crust to melt and form pipes, which can vent magma. Because the tectonic plates move whereas the mantle plume remains in the same place, each volcano becomes dormant after a while and a new volcano is then formed as the plate shifts over the hotspot. The Hawaiian Islands are thought to be formed in such a manner, as well as the Snake River Plain, with the Yellowstone Caldera being the part of the North American plate currently above the hotspot.

Volcanic features

The most common perception of a volcano is of a conical mountain, spewing lava and poisonous gases from a crater at its summit. This describes just one of many types of volcano, and the features of volcanoes are much more complicated. The structure and behaviour of volcanoes depends on a number of factors. Some volcanoes have rugged peaks formed by lava domes rather than a summit crater, whereas others present landscape features such as massive plateaus. Vents that issue volcanic material (lava, which is what magma is called once it has escaped to the surface, and ash) and gases (mainly steam and magmatic gases) can be located anywhere on the landform. Other types of volcano include cryovolcanoes (or ice volcanoes), particularly on some moons of Jupiter, Saturn and Neptune; and mud volcanoes, which are formations often not associated with known magmatic activity. Active mud volcanoes tend to involve temperatures much lower than those of igneous volcanoes, except when a mud volcano is actually a vent of an igneous volcano.

Fissure vents

Volcanic fissure vents are flat, linear cracks through which lava emerges.

Shield volcanoes

Shield volcanoes, so named for their broad, shield-like profiles, are formed by the eruption of low-viscosity lava that can flow a great distance from a vent, but not generally explode catastrophically. Since low-viscosity magma is typically low in silica, shield volcanoes are more common in oceanic than continental settings. The Hawaiian volcanic chain is a series of shield cones, and they are common in Iceland, as well.

Lava domes

Lava domes are built by slow eruptions of highly viscous lavas. They are sometimes formed within the crater of a previous volcanic eruption (as in Mount Saint Helens), but can also form independently, as in the case of Lassen Peak. Like stratovolcanoes, they can produce violent, explosive eruptions, but their lavas generally do not flow far from the originating vent.

Cryptodomes

Cryptodomes are formed when viscous lava forces its way up and causes a bulge. The 1980 eruption of Mount St. Helens was an example. Lava was under great pressure and forced a bulge in the mountain, which was unstable and slid down the north side.

Volcanic cones

Volcanic cones or cinder cones are the result from eruptions that erupt mostly small pieces of scoria and pyroclastics (both resemble cinders, hence the name of this volcano type) that build up around the vent. These can be relatively short-lived eruptions that produce a cone-shaped hill perhaps 30 to 400 metres high. Most cinder cones erupt only once. Cinder cones may form as flank vents on larger volcanoes, or occur on their own. Parícutin in Mexico and Sunset Crater in Arizona are examples of cinder cones. In New Mexico, Caja del Rio is a volcanic field of over 60 cinder cones.

Stratovolcanoes

Stratovolcanoes or composite volcanoes are tall conical mountains composed of lava flows and other ejecta in alternate layers, the strata that give rise to the name. Stratovolcanoes are also known as composite volcanoes, created from several structures during different kinds of eruptions. Strato/composite volcanoes are made of cinders, ash and lava.

Cinders and ash pile on top of each other, lava flows on top of the ash, where it cools and hardens, and then the process begins again. Classic examples include Mt. Fuji in Japan, Mayon Volcano in the Philippines, and Mount Vesuvius and Stromboli in Italy.

In recorded history, explosive eruptions by stratovolcanoes have posed the greatest hazard to civilisations, as ash is produced by an explosive eruption. No supervolcano erupted in recorded history. Shield volcanoes have not an enormous pressure build up from the lava flow. Fissure vents and monogenetic volcanic fields (volcanic cones) have not powerful explosive eruptions, as they are many times under extension. Stratovolcanoes (30°–35°) are steeper than shield volcanoes (generally 5°–10°), their lose tephra are material for dangerous lahars.

Supervolcanoes

A supervolcano is a large volcano that usually has a large caldera and can potentially produce devastation on an enormous, sometimes continental, scale. Such eruptions would be able to cause severe cooling of global temperatures for many years afterwards because of the huge volumes of sulphur and ash erupted.

They are the most dangerous type of volcano. Examples include Yellowstone Caldera in Yellowstone National Park and Valles Caldera in New Mexico (both western United States), Lake Taupo in New Zealand, Lake Toba in Sumatra, Indonesia and Ngorogoro Crater in Tanzania, Krakatoa near Java and Sumatra, Indonesia.

Supervolcanoes are hard to identify centuries later, given the enormous areas they cover. Large igneous provinces are also considered supervolcanoes because of the vast amount of basalt lava erupted, but are non-explosive.

Submarine volcanoes

Submarine volcanoes are common features on the ocean floor. Some are active and, in shallow water, disclose their presence by blasting steam and rocky debris high above the surface of the sea. Many others lie at such great depths that the tremendous weight of the water above them prevents the explosive release of steam and gases, although they can be detected by hydrophones and discolouration of water because of volcanic gases. Pumice rafts may also appear. Even large submarine eruptions may not disturb the ocean surface.

Because of the rapid cooling effect of water as compared to air, and increased buoyancy, submarine volcanoes often form rather steep pillars over their volcanic vents as compared to above-surface volcanoes. They may become so large that they break the ocean surface as new islands. Pillow lava is a common eruptive product of submarine volcanoes. Hydrothermal vents are common near these volcanoes, and some support peculiar ecosystems based on dissolved minerals.

Subglacial volcanoes

Subglacial volcanoes develop underneath icecaps. They are made up of flat lava which flows at the top of extensive pillow lavas and palagonite. When the icecap melts, the lavas on the top collapse, leaving a flat-topped mountain.

These volcanoes are also called table mountains, tuyas or (uncommonly) mobergs. Very good examples of this type of volcano can be seen in Iceland, however, there are also tuyas in British Columbia. The origin of the term comes from Tuya Butte, which is one of the several tuyas in the area of the Tuya River and Tuya Range in northern British Columbia.

Tuya Butte was the first such landform analysed and so its name has entered the geological literature for this kind of volcanic formation. The Tuya Mountains Provincial Park was recently established to protect this unusual landscape, which lies north of Tuya Lake and south of the Jennings River near the boundary with the Yukon Territory.

Mud volcanoes

Mud volcanoes or mud domes are formations created by geo-excreted liquids and gases, although there are several different processes which may cause such activity. The largest structures are 10 kilometres in diameter and reach 700 metres high.

Effects of Volcanoes

There are many different types of volcanic eruptions and associated activity: phreatic eruptions (steam-generated eruptions), explosive eruption of high-silica lava (e.g. rhyolite), effusive eruption of low-silica lava (e.g. basalt), pyroclastic flows, lahars (debris flow) and carbon dioxide emission. All of these activities can pose a hazard to humans. Earthquakes, hot springs, fumaroles, mud pots and geysers often accompany volcanic activity (Fig. 21.2).

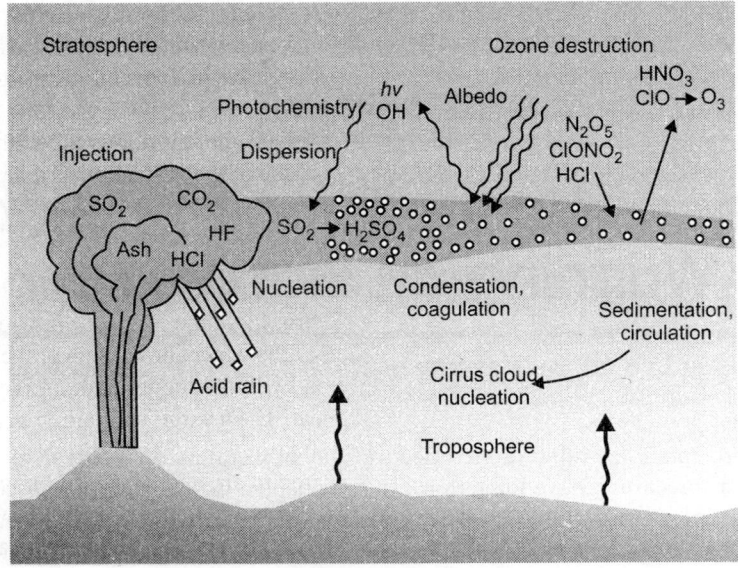

Fig. 21.2. Volcanic injection.

The concentrations of different volcanic gases can vary considerably from one volcano to the next. Water vapour is typically the most abundant volcanic gas, followed by carbon dioxide and sulphur dioxide.

Chapter 22

Forecasting the Effects of Global Warming on Biodiversity

INTRODUCTION

The demand for accurate forecasting of the effects of global warming on biodiversity is growing, but current methods for forecasting have limitations. In this chapter, we compare and discuss the different uses of four forecasting methods: (i) models that consider species individually, (ii) niche-theory models that group species by habitat (more specifically, by environmental conditions under which a species can persist or does persist), (iii) general circulation models and coupled ocean–atmosphere–biosphere models and (iv) species–area curve models that consider all species or large aggregates of species. After outlining the different uses and limitations of these methods, we make eight primary suggestions for improving forecasts. We find that greater use of the fossil record and of modern genetic studies would improve forecasting methods. We note a Quaternary conundrum: While current empirical and theoretical ecological results suggest that many species could be at risk from global warming, during the recent ice ages surprisingly few species became extinct. The potential resolution of this conundrum gives insights into the requirements for more accurate and reliable forecasting. Our eight suggestions also point to constructive synergies in the solution to the different problems.

Now that it is widely accepted that global warming is happening, there is a growing demand for accurate forecasts of its effects, and much concern about its effects on biological diversity. Specialists know that theoretical models of these effects are limited—although useful in certain contexts when all the provisions, preconditions, and limitations of a given model are understood—and should not be taken literally. Often, however, the media do not convey these caveats. It is no wonder that policy-makers and the general public are confused.

The purpose of an environmental forecast is either to support a decision process or to test a scientific hypothesis. To support a decision process, it must be clear which decisions the forecast expects to improve. To mitigate the effects of global warming on biodiversity, two distinct kinds of actions are needed: long-term actions, such as reducing emissions of greenhouse gases, and short-term ones, such as designing an appropriate nature reserve.

Fossil evidence and recent ecological and genetic research, along with specific problems with present forecasting methods, lead us to believe that current projections of extinction rates are overestimates. Previous work has failed to adequately take into account mechanisms of persistence. We note a Quaternary conundrum: While current empirical and theoretical ecological forecasts suggest that many species could be at risk from global warming, during the recent ice ages few extinctions are documented. The potential resolution of this conundrum gives insights into the requirements for more accurate and reliable forecasting.

EIGHT WAYS TO IMPROVE BIODIVERSITY FORECASTING

Reliable ways to forecast rates of extinction, both in relation to global warming and in general, still elude us. In the face of growing demand for accurate, timely forecasts, we consider how these forecasts can be improved, and make eight primary suggestions. One suggestion concerns the fundamentals of formal models; two concern the use of the concept of biodiversity; two, the better use of available data; and three, specific modelling types.

Select a specific definition of biodiversity. Among the many meanings of the term 'biodiversity', it is important to select one as the focus for each specific forecast. Most of the existing literature on forecasting the effects of global warming on biodiversity seems to assume that 'biodiversity' has some universally accepted meaning, and that readers already know what this is. However, biodiversity is a complex concept, and its meanings are becoming both more complex and more quantitative as greater emphasis is placed on DNA analysis as a determinant of genetically distinct units. As if distinctions among different levels of organisation of biodiversity (genes, species, ecosystems, etc.) were not already complex, there are even more fundamental distinctions between different ways of valuing biodiversity: intrinsic value (species' value independent of human use and needs) and use value (the human use of diversity, ranging from the desire to harvest one species to the ability to see and appreciate complex ecosystems).

Evaluate models before using them: Models that forecast the impacts of climate change on biodiversity are difficult to validate, and it may be many years before anyone can conclude whether a given forecast of the effects of global warming on biodiversity was nearly right or not. However, scientists can and should evaluate a prospective forecasting method before using it to generate forecasts, and there are well-known methods, applied widely in other disciplines, for doing so. The evaluation should include the accuracy of the method (e.g. its ability to reproduce past situations) and sensitivity analyses. For example, if small changes in one parameter in a model lead to large changes in results, one must ask whether the model is sufficiently robust to be used. This principle is widely acknowledged in many applications of forecasting, though not often acknowledged or used in forecasts of climate change and extinctions.

Of the modelling papers we have reviewed, only a few were validated. Commonly, these papers simply correlate present distribution of species with climate variables, then replot the climate for the future from a climate model and, finally, use one-to-one mapping to replot the future distribution of the species, without any validation using independent data. Although some are clear about some of their assumptions (mainly equilibrium assumptions), readers who are not experts in modelling can easily misinterpret the results as valid and validated. For example, Hitz and Smith discuss many possible effects of global warming on the basis of a review of modelling papers, and in this kind of analysis the unvalidated assumptions of models would most likely be ignored. Furthermore, those who have attempted to validate their models have commonly used 'resubstitution', whereby data used to calibrate the models are also used to evaluate them or have left a portion of the original data set apart for model evaluation. A problem with these approaches is that the test data are not independent or are only partially so, from those used to fit the models, and models evaluated thus are bound to provide overoptimistic assessments of predictive accuracy. More appropriate tests of models for forecasting include using a model fitted to current data to reproduce the past, using a model fitted to past data to reproduce the present, and using data from one geographic region to develop and calibrate a model and data from another distinct geographic area to test that model.

Account for multiple causes of changes in biodiversity. Climate change is only one way in which the environment and human activities are affecting biodiversity. Forecasts must integrate other human impacts as well, but generally they do not. Biodiversity is also under pressure from humans' conversion of natural and seminatural habitats, wildlife trade, war, pollution, physical infrastructure (e.g. roads), and introduction of invasive exotic species (including disease organisms), as well as from natural environmental change, including catastrophes. Forecasts should disentangle the effects of climate change from these other sources of change or at least account for the climate component and its interaction with the other components (e.g. how current landscape fragmentation could affect species migrations, compared with past migrations in more continuous landscapes).

Obtain good information and make better use of it: The data that scientists and policy-makers need most are usually inadequately available — sometimes no data exist or, more commonly, the available data are sparse, poorly collected, statistically insufficient, and biased. These include basic information on the abundance and geographic patterns of most species, as well as the data necessary to estimate the probability of extinction for a species. For example, scientists have no knowledge at all about the status of more than 40 per cent of marine fauna (every taxonomically identified species ever recorded) within the Swedish parts of Skagerak, Kattegat and the Baltic Sea, even though these areas are among the most intensively studied marine areas in the world (Table 22.1).

Table 22.1. Summary of preliminary results from work with the Swedish Red List on every recorded species within selected marine taxa.

Taxon	Red-listed species	Species of least concern	Species for which knowledge is lacking	Total
Porifera	3	28	114	145
Anthozoa (Cnidaria)	9	22	22	53
Priapulida	0	1	1	2
Sipuncula	0	4	7	11
Phoronida	0	2	2	4
Echiura	0	2	1	3
Decapoda (Crustacea)	5	45	28	78
Mollusca	52	218	145	415
Chaetognatha	0	5	3	8
Brachiopoda	1	2	1	4
Echinodermata	16	35	20	71
Hemichordata	0	1	2	3
Total (percentage)	86	365	346	797

Note: Red-listed species fall into four categories: critically endangered, endangered, near threatened, and vulnerable. Species for which knowledge is lacking fall into two categories: not evaluated (because of lack of knowledge) and data deficient.

Furthermore, although Table 22.1 gives the impression that we have enough knowledge about more than half of the fauna to assign a risk of extinction 11 per cent of the species have been red listed, indicating some level of threat, and 46 per cent classified as of 'least concern', suggesting that their populations are vigorous and robust — in fact the information is so meager and poor that the evaluations in many cases are merely intelligent guesses. Current initiatives to assemble large data sets from natural history collections, such as the Global Biodiversity Information Facility or to assemble knowledge about ecosystems and biodiversity, such as the Millennium Ecosystem Assessment, are among present efforts to obtain the necessary data. Forecasting methods must not only target key information gaps but also make the best possible use of existing data. For example, models of species distribution may combine available environmental layers with data from museum collections, compensating to some extent for the weaknesses of either form of data on its own.

Use the Quaternary fossil record to understand mechanisms that preserve biodiversity, and use these in forecasting models: Current forecasting methods suggest that global warming will cause many extinctions, but the fossil record indicates that, in most regions, surprisingly few species went extinct during the Quaternary (from approximately 2.5 million years BCE to the present) — in North America, for example, only one tree species is known to have gone extinct. Large extinctions were reported mainly for tree species in northern Europe (68 per cent loss of tree genera) and for large mammals (> 44 kg) in the Northern Hemisphere. We refer to this contrast between the implications of modern forecasts and the observed fossil record as the 'Quaternary conundrum'. The resolution of this conundrum is key to improving forecasts of climate-change effects on biodiversity. Among the possible explanations are that climate change during the Quaternary was greatly different from climate change forecasted for the future; that genetic and ecological mechanisms, not accounted for in formal forecasting methods, allow the persistence of many species even under rapid climate change; and that factors in addition to climate change could decrease rates of extinction.

Some recent ecological genetics research further deepens the puzzle. For example, the risk of extinction for a species in response to climate change depends on the demography and evolution of genetically differentiated populations across their geographic ranges. If populations are locally adapted, climate change will cause conditions to deteriorate across the species' range, rather than just at the margins of the range. Modern reciprocal transplant experiments, in which spatial gradients in climate serve as proxies for temporal climate change in the future, show that these fitness losses can be large. For example, a reciprocal transplant experiment on lodgepole pine in Canada indicated that global warming would slow tree growth and increase mortality, resulting in a 20 per cent loss of productivity. Likewise, a study of a prairie annual in the Great Plains of the United States showed a 30 per cent reduction in seed production in climates similar to those predicted for future decades. Ecological genetic data, in each of these cases, predicted different rates of adaptive evolution in different parts of the species' range but generally suggested that evolutionary rates would be slower than the anticipated rate of climate change.

Until recently, it was thought that past temperature changes were no more rapid than 1°C per millennium, but recent information from both Greenland and Antarctica, which goes back approximately 4,00,000 years, indicates that there have been many intervals of very rapid temperature change, as judged by shifts in oxygen isotope ratios. Some of the most dramatic changes (e.g. 7° to 12°C within approximately 50 years) are actually of greater amplitude than anything projected for the immediate future. Although these changes were probably not equally severe everywhere on the globe, a well-documented rapid warming did occur around the shores of the North Atlantic at the end of the last glaciation, when melting of the ice cover on the ocean suddenly allowed the Gulf Stream to reach the shores of northern Europe. There, temperatures rose rapidly, perhaps as rapidly as anticipated today for the next several decades.

What could explain the Quaternary conundrum? One possibility is that migrations were faster than has been thought possible. A large literature examines late-Quaternary range shifts deduced from the pollen record, and recent papers consider models and seed-dispersal mechanisms that may account both for migration across geographic barriers and for rapid invasion of new territory. Sparse populations of several tree species are now known (from genetic and macrofossil evidence, supplemented by detailed analysis of mapped pollen data) to have persisted during the last glacial maximum in regions where very few, if any, pollen grains have been observed—regions that for this reason would be judged well outside the climate envelope for these species. These populations serve as advance colonists, allowing rapid population growth in newly available habitat.

A second explanation is that low extinction rates during Quaternary climate change may be partially attributable to ongoing adaptive evolution. Theoretical models suggest that adaptive evolution can enhance the persistence of populations in a changing environment even when migration is possible. And rapid genetic adaptation to climate has already been documented for a few wild organisms for which long-term studies of field populations have been conducted. Invasive species have also evolved since their arrival in a new habitat in the 20th century, at surprisingly rapid rates of evolution.

A long-standing controversy regarding the role of people in Quaternary extinctions of large mammals speaks to the difficulty of quantifying impacts of multiple factors on species loss. The high extinction rate of large mammals has been widely recognised since the 19th century, and extinctions of large mammals and island birds over the past 1,00,000 years have been the subject of much conjecture. Paul Martin has made the now well-known case that the timing of extinctions followed human dispersal from Afro-Asia to other parts of the globe and that these extinctions resulted from human 'blitzkrieg' overkill. But careful analysis of well-documented extinctions in Beringia suggests that human hunting was superimposed on a preexisting trend of diminishing animal population density. These data suggest that the interaction of environmental change and human resource use can have a larger negative impact on biodiversity than either factor alone.

Improve widely used modelling methods: Theoretical models are essential to quantitative forecasts of effects of global warming on biodiversity. Four methods are in use today for these forecasts: (i) modelling individuals, (ii) using groups of species as the units of interest in a model, (iii) integrating biodiversity within general circulation models (GCMs) and (iv) using species–area models, that is, modelling based on theories of total biodiversity.

Clearly, since there are 1.5 million named species and many more that are as yet undocumented but believed to exist, methods that require specific information on all species for forecasting overall biodiversity are not practical. At the heart of the choice about which method to use is the question of appropriate spatial and temporal scales, a problem common to many disciplines. The larger the scale of the primary units of a model, the simpler it is to estimate effects over large areas and times, but also the cruder the approximation is and the more likely that undesirable assumptions will prevail. The smaller the scale, the greater the detail that can be considered, but the more detailed the information, the greater the number of calculations that must be made. The question boils down to whether it is better to know a lot of detail about fewer points or much less information about much greater areas. At present, it is not clear which approach is more useful for forecasting the effects of global warming on biodiversity, and one of the results is that research is being carried out at multiple scales.

Models of individuals: Researchers have considerable experience with models that use the individual as the primary unit with fixed characteristics, forecasting population and species responses from the sum of the responses of individuals. Computer simulations of vegetation responses to climate and habitat have been available since 1970, beginning with the JABOWA forest model, which forecasts the growth and mortality of individual trees and the regeneration of species in small forested areas. Such methods have been applied to reconstruct Holocene forest changes and to forecast possible effects of global warming on forests. These models have been widely used to forecast changes in tree species composition under expected climate change. They have also been used to make forecasts about how global warming will affect ecological communities and ecosystems. For example, the JABOWA model forecasts that increases in carbon dioxide (CO_2) concentrations in the atmosphere will speed up forest succession. Individual-based models have also been used to forecast the effects of climate change on individual endangered species: The JABOWA model forecasts that global warming will further threaten the Kirtland's warbler, an endangered bird that nests only in jack pine stands in southern Michigan. With global warming, this region will no longer be suitable for jack pine, and jack pine is unlikely to grow in the sandy soils farther north. Curiously, no attempt has been made to validate either the ecosystem forecast or the endangered species forecast.

An advantage of individual-based models is that they do not impose steady-state assumptions about species, the environment or the relationships between the two. Nonsteady-state conditions can easily be integrated, and subspecies can be treated separately. However, the required species-specific empirical data are not always available, and the spatial scale is small, necessitating many separate simulations to form a general picture.

Considerable work has been done to add to the details of plant physiology incorporated in these models. These models might be improved further in several ways. One way is to increase model realism in the areas of dispersal, temperature response, life history trade-offs, and disturbance response. Another is to use shorter time steps (rather than the usual annual time steps), but this would greatly increase the models' computational loads. A third way is to improve the models' empirical basis. Yet another way is to use better methods to represent the effects of changing CO_2, ozone, and other environmental factors. But arguably the most important need is to attempt validation of this class of models against actual temporal changes in forests, because this kind of forecasting has been available for several decades.

Niche-theory models: A large group of models considers a species or a group of species with similar niches, as a single unit with fixed characteristics, which can be viewed as an integrated measure of a set of environmental variables. Thus the set of environmental conditions alone forecasts the distribution of a species or group of species. These models, based on ecological niche theory, are known by various names — niche-theory models, habitat distribution models, and bioclimatic-envelope models — and have been applied to a variety of taxa, most commonly to plants, but also to animals, including tropical rainforest vertebrates of many kinds, amphibians and reptiles and butterflies. This approach traces back to G. Evelyn Hutchinson's definition of the realised niche — the set of all environmental conditions within which a species can persist — and to a generalised theory by Smith for predicting biomes or vegetation types. Two advantages of niche-theory models are that only a few variables are necessary to predict ranges for many species and that even small-scale, patchy museum collections data can be used.

However, niche-theory models have a number of limitations. First, they are primarily correlative, using observed statistical relationships between occurrences of a species and its environment. Second, they assume that observed distributions are in equilibrium (or quasi-equilibrium) with their current environment, and that therefore species become extinct outside the region where the environment, including the climate, meets their present or assumed requirements — contradicting the data reviewed earlier, as well as many natural history observations of transplanted species, that show species have survived in small areas of unusual habitat or in habitats that are outside the well-established geographic range but actually meet their requirements. Thus niche-theory models are likely to overestimate extinctions, even when they realistically suggest changes in ranges of many species.

Another problem with niche-theory models, as with most models, is that they are difficult to validate, and few have been adequately validated. For example, Lawler and colleagues compare six approaches to modelling the effects of global warming on fauna, but do not attempt to validate any of the models independently. Indeed, bioclimatic models vary greatly in their projections of extinction.

An additional complication is that the relationship between the occurrence of a species and climatic variables is not always correlated with the mean. For example, amphibian declines due to outbreaks of a pathogenic chytrid fungus (*Batrachochytrium dendrobatidis*) are related to the annual range of temperatures, not to the mean temperature.

Unknown and untested biases are also problems for these models. More computationally intensive approaches, using the 'consensus' of many models, have been shown to improve the accuracy of models at predicting observed shifts in distribution. On the other hand, differences in forecasts from different niche models can be as large as the change in predicted distributions due to climate change. Thus, global estimates of extinctions due to climate change may have greatly overestimated the probability of extinction as a result of the inherent variability in niche modelling. It is a problem when a paper reports on minor uncertainties and does not describe major uncertainties.

Most applications of bioclimatic-envelope models do not consider dispersal and migration rates, nor do they consider the biotic interactions of symbiosis, competition and predation or other dynamic processes, such as fire all of which could change the future distribution of the species. Some species may be more constrained locally by biotic interactions than by climate *per se*, with climate operating at broader landscape scales. If some ecosystems display a high degree of inertia and their responses lag behind changes in climate, then at least some of the component species (e.g. understory plants in a forest) could be buffered from climate change, at least in the short-term. For example, some clones of the alpine sedge *Carex curvula* in the European Alps could be as old as 2000 years, and possibly older. If so, the clone would have passed through a series of climatic variations (e.g. medieval optimum, little ice age) without shifting to lower or higher elevations. Unfortunately, disentangling direct climatic constraints from other habitat constraints requires quite detailed ecological knowledge, which is usually lacking.

Most niche models assume fixed genetic and phenotypic characteristics for the species and thus do not discriminate among various genotypes and phenotypes across a species' range. This is in contrast to what geneticists and experimental ecologists believe to be the case. If these attributes were available, they could be integrated in niche models by fitting a different model for each genotype or phenotype.

One promising use of niche models is to identify gaps in nature reserve systems that will occur as the climate changes. For flora of the Cape of Good Hope, South Africa, Hannah and colleagues applied a niche model to forecast whether the climate where existing nature reserves are located would be adequate, under global warming, for the species that each reserve is intended to conserve.

Improve ecological principles embedded in general atmosphere-ocean-biotic coupled circulation models: Niche-theory modelling has been used both by ecologists and by those involved with coupling the biosphere to ocean–atmosphere GCMs. Curiously, these two activities have gone on relatively independently, with the GCMs historically using a simpler and more static kind of niche-theory than the models applied directly by ecologists. In part this independence is the result of different goals. Climate modellers have primarily been interested in the influence of vegetation on climate through albedo, surface roughness, water evaporation, and exchange of greenhouse gases and aerosols.

Recently, GCMs have been extended in dynamic global vegetation models or DGVMs, to consider vegetation and ecosystem dynamics, and therefore the effects of land-use change in a dynamic climate. These models have led to further refinements in forecasts of atmospheric greenhouse gas concentrations and climate change. But again the interest has been primarily in the feedbacks of changes in vegetation, and in biogeochemical and biophysical processes, on CO_2 concentrations and climate dynamics rather than on biodiversity. Thus biodiversity has not been the primary interest of researchers employing DGVMs; their focus has been less on what happens to vegetation *per se* than on the effects on climate dynamics. Meanwhile, those concerned primarily with the effects of global warming on biodiversity have developed the other kinds of models discussed in this section. These two modelling communities need to communicate better in order to improve modelling and understanding on both sides.

In favour of the simplest Box modelling approach is that only a few climate variables seem necessary to predict ranges for many plant species. Also in favour of this approach is that some applications are mechanistic, based on first principles. For example, some simulate photosynthetic processes, albeit in a conceptual or quasi-biochemical way. However, the rate coefficients for many of the processes, while empirically based, are often poorly quantified. Therefore, although these dynamic vegetation models can simulate the response of plant functional types to climate and other environmental drivers, and show how biomes may shift, they cannot predict biodiversity changes within or between biomes or the genetic variations discussed earlier. Also, there are large uncertainties at these scales about the data, so these models may have significant biases, particularly in simulating the margins of current biome distributions.

There is a need for continued evaluation of the appropriate level of diversity that must be represented. Ways of incorporating the non-steady-state characteristics of vegetation and the relationship between climate and vegetation need to be developed as the models mature so they can be used to forecast changes in vegetation. Comparison of model output with Quaternary vegetation data is validating the models and providing insight into drivers of vegetation change.

Develop better models for forecasting total biodiversity: One of the simplest and most straightforward methods of forecasting the effects of global warming on total biodiversity is based on the species–area curve. In simplest terms, the number of species is correlated with area. There are many species–area curves, but most commonly such curves are given by the two-parameter curve with the formula $S = cA^z$, where S is the number of species, A is the area, c is the number of species on one unit of area, and z is the rate at which log S increases with log A.

This empirical relationship works quite well, provided some care is taken in its application. Most often, it is an equilibrium relationship applying to areas that have existed in their present circumstances for a long time. Rosenzweig points out that the power law is only approximate, with the power (exponent z) changing in accordance with the circumstances (e.g. type of area, such as island, subarea in a continent or whole continent) and scale. Rosenzweig and others argue that this relationship has a fundamental ecological basis in the number of species actually present on a contiguous area of size A. Another kind of relationship, sometimes called a species–area relationship, is used in sampling to extrapolate to the total number of species on a defined piece of land as the sample size (area) increases. It is commonly applied to diversity of broad taxa, such as birds, vascular plants and mammals.

The species–area method of forecasting changes in biodiversity under global warming has six limitations. First, it assumes an equilibrium (or very slowly changing) relationship between species number and area. Second, the future climate probably will not be an exact analogue of the current one, so 'moving' a bioclimatic zone for an ecological type may not be accurate. Third, topographic variation, which affects the species–area curve shape, may be greater or less in the future zone. Fourth, factors relating to the shape of areas and the amount of their fragmentation suggest that an alternative 'endemics–area curve' may enable more accurate predictions. Fifth, the correct z value must be chosen: It must apply to the entire area under consideration, and it must also consider the type of area and timescale applicable. Sixth, many species are not confined to a particular vegetation zone or type. For the species–area relationship to predict species extinctions, the area must be for closed communities. Thomas and colleagues used individual species distributions as the basis for their analysis. They examined changes in realised niches without taking into account the likelihood of changed interactions and adaptation, and thus the new areas that they predicted were probably too small. How these area changes relate to changes in area of closed communities is unclear.

These problems need to be overcome and the method tested, perhaps with historical data, before too much weight can be placed on this method. But even if the species–area curve is a correct causal relationship, the required data are limited, so the quantitative estimations may have little value. For example, Thomas and colleagues used the species–area curve to predict diversity changes by estimating the change in area that species would experience after climate change.

In one case they concluded that the area of the boreal forest would decline from 13 'undisturbed' units to 12.5 units globally — a loss of about 4 per cent. But a study of the biomass stored in the boreal forest of North America showed that botanical maps of the North American forests led to the possibility that the area defined as boreal forest differed by a factor of two 200 per cent eclipsing the forecast loss of 4 per cent. If scientists cannot agree within a factor of two on the size of the boreal forest, then the forecast loss of 4 per cent means little.

Although in theory one can separate the use of the species–area curve for estimating total species present from an attempted use of this relationship as some kind of causal basis for biodiversity, it is not always clear which is behind an application. If applying this curve to changes in biodiversity simply reverses the sampling operation, so that a decrease in area is assumed to lead to a decline in species number that follows back down the asymptotic curve, then this reverse operation is illogical. Because removing area over an ecological timescale is not a direct reversal of the processes that have added species over evolutionary time, there is no *a priori* reason to expect that the change in species number corresponding to a change in area will track a species–area curve, although there are indications that a power law with a lower z value would apply to predictions for short time horizons.

A solution to these problems is suggested by tackling the reason for the observed correlation between species and area. Rosenzweig, and Connor and McCoy summarise a variety of evidence suggesting that the primary, though not the only, cause of increases in species with increased area is that larger areas tend to have a greater variety of habitats, and thus a greater variety of niche opportunities. If the relative habitat diversity of different-sized areas can be measured directly, this information can be used as a causative predictor of diversity, whereas area is mostly correlative, which brings us back to the basis of the niche-theory models discussed earlier. If changes in habitat diversity can be predicted, they should predict changes in species diversity more accurately than would changes in area.

The limitations of the species–area curve may be corrected by using different subsets of areas to generate a species–area curve that is linked to the habitat heterogeneity captured by a given subset. Scenarios of climate change are then interpreted as shifts in patterns of habitat heterogeneity, rather than simply as changes in habitat area.

This approach allows the use of a range of available biotic and abiotic information, and also may help address shape and fragmentation issues, providing a link between endemics–area and species–area curves so that these can be used in concert. Further, such an approach may allow for assessment not just of climate-change 'losers' but also of 'winners'. The species–area curve offers the simplest theoretical model to link biodiversity to climate change and perhaps, with the modifications suggested here, might be useful in predicting the effects of climate change on biodiversity.

RECOMMENDATIONS

In this chapter we pointed eight ways to improve forecasts of the effects of global warming on biodiversity. We have considered four kinds of models used to forecast the effects of global warming on biodiversity. Three,share a foundation in ecological niche-theory. The first group of models, represented by JABOWA-type vegetation models, makes forecasts for individuals and species and has the flexibility to involve non-steady-state relationships between a species and its environment, but requires species-specific data that are not always available.

The second group, ecological niche-theory models, makes forecasts based on the environmental conditions that are possible for a set of species or a single species. Traditionally, these have been applied primarily with climate change taken into account as a driver, but some recent studies expand this approach to include other habitat characteristics. Also traditionally, these models assume equilibrium relationships between species and the environment, and among environmental variables, which limits their utility. Recent advances are making these models more flexible, with the result that this kind of model has broad appeal among scientists and is likely to remain an important and useful approach.

The third group, used by climatologists, consists of bioclimatic-envelope models in even simpler forms than those in use by ecologists. These models have necessarily been simple because of the complexity of all climate models. Although advances have been made, they remain crude from a biological point of view, and static (especially in the relationship between climate and vegetation). They are used to model the feedback between climate and vegetation. The fourth group comprises models based on species–area theory. These models are appealing in the abstract because they do not require detailed knowledge of species or habitats, but they have often been applied inappropriately for forecasts of the effects of global warming on biodiversity. One promising approach is to focus on links between the species–area curve and models of habitat heterogeneity. Such models build on a combination of available habitat and species data as surrogate information for overall biodiversity patterns.

The effectiveness of these four useful classes of models will depend on the extent to which our recommendations are adopted, especially with respect to obtaining necessary data. Curiously, although three of the approaches — JABOWA based models, niche-theory models and models used by climatologists — make use of similar niche-theory ideas about the relationship between a species and its environment, scientists using each of those approaches tend not to communicate with each other or read each other's literature.

We suggest that there is now much scope for an integrated framework for forecasting the impacts of global change on biodiversity. Such a framework could integrate models for species persistence and consider multiple causes of biodiversity change. This emerging framework awaits more of the important evaluation steps, including case studies.

What, then, is the answer to the Quaternary conundrum? The answer appears to lie in part with the ability of species to survive in local 'cryptic' refugia, that is, to exist in a patchy, disturbed environment whose complexity allows faster migration than forecast for a continuous landscape, within which species move only at a single rate. The answer also lies in part with greater genetic heterogeneity within species, including local adaptations, which allows rapid evolution. For example, populations close to latitudinal borders are likely to be better adapted to some environmental changes than the average genotype. However, the conundrum is not completely solved, and some important genetic research suggests that species are more vulnerable than the fossil record indicates. A fuller solution to the conundrum will be important for improving forecasts of climate change effects on biodiversity, suggestions.

Chapter 23

Effect of Global Warming on Biodiversity in India

INTRODUCTION

It is now well established that the world is heating up. The average temperature of the earth's surface increased by an estimated 0.6°C in the 20th century, and according to the most recent projections of the Intergovernmental Panel on Climate Change, could rise 1.4° to 5.8°C above the 1990 average by 2100. Much of this predicted increase is attributed by scientists to increasing concentrations of greenhouse gases such as carbon dioxide (CO_2) in the atmosphere.

The effects of such a temperature increase might include:

1. More frequent extreme high maximum temperatures and less frequent extreme low minimum temperatures.
2. A decrease in snow cover: satellite observations suggest that the area of the planet covered by snow has already declined by 10 per cent since the 1960s.
3. Rising sea levels.
4. An increase in the variability of climate, with changes in both the frequency and severity of extreme weather events.
5. And alterations to the distribution of certain infectious diseases.

According to India's National Action Plan on Climate Change (IPCC), multi-model averages show that the temperature increases during 2090–2099 relative to 1980–1999 may range from 1.1 to 6.4°C and sea level rise from 0.18 to 0.59 meters. These could lead to impacts on freshwater availability, oceanic acidification, food production, flooding of coastal areas and increased burden of vector borne and water borne diseases associated with extreme weather events.

At the national level, increase of –0.4°C has been observed in surface air temperatures over the past century. A warming trend has been observed along the west coast, in central India, the interior peninsula, and northeastern India. However, cooling trends have been observed in northwest India and parts of south India. A trend of increasing monsoon seasonal rainfall has been found along the west coast, northern Andhra Pradesh, and northwestern India (+10 to +12 per cent of the normal over the last 100 years) while a trend of decreasing monsoon seasonal rainfall has been observed over eastern Madhya Pradesh, northeastern India, and some parts of Gujarat and Kerala (– 6 to – 8 per cent of the normal over the last 100 years). There has been an overall increasing trend in severe storm incidence along the coast at the rate of 0.011 events per year. While the states of West Bengal and Gujarat have reported increasing trends, a decline has been observed in Odisha. The available monitoring data on Himalayan glaciers indicates that while recession of some glaciers has occurred in some Himalayan regions in recent years, the trend is not consistent across the entire mountain chain. It is accordingly, too early to establish long-term trends or their causation, in respect of which there are several hypotheses.

Now we should expect annual mean surface temperature rise by the end of century, ranging from 3° to 5°C under A2 scenario and 2.5° to 4°C under B2 scenario of IPCC, with warming more pronounced in the northern parts of India, from simulations by Indian Institute of Tropical Meteorology (IITM), Pune. Some simulations by IITM, Pune, have indicated that summer monsoon intensity may increase beginning from 2040 and by 10 per cent by 2100 under A2 scenario of IPCC.

BIODIVERSITY: WHAT IS IT, WHERE IS IT AND WHY IS IT IMPORTANT

Biodiversity reflects the number, variety and variability of living organisms. It includes diversity within species, between species and among ecosystems. The concept also covers how this diversity changes from one location to another and over time. Indicators such as the number of species in a given area can help in monitoring certain aspects of biodiversity.

There has been a decline in biodiversity in recent years. The Living Planet Index, compiled by the WWF, provides an indication of the declines in the overall abundance of wild species.

Global warming is having impact on biodiversity in various ways. Some impacts are discussed here.

GLOBAL WARMING EFFECTS ON BIODIVERSITY AND ANIMALS

In its most recent assessment, the IPCC reiterates that 20–30 per cent of species assessed so far are likely to be at increased risk of extinction if increases in global average warming exceed 1.5°–2.5°C (relative to 1980–1999) and as global average temperature increase exceeds about 3.5°C, model projections suggest significant extinctions (40–70 per cent of species assessed) around the globe. Global warming does not only make vegetation 'gasp for air' but also leads to animal habitat loss. This is an especially big problem for sensitive species.

The loss of these habitats leads to extinction of the amphibians dependent on these forests for their survival. Many species may be seriously affected by the spread of viruses and bacteria which normally thrive in warmer conditions. This, among many other things, may push these animals even closer to the brink of extinction.

It is not only the habitat loss and spread of diseases that may cause animal extinction. It is also the availability of food and water for animals that will likely be made more scarce as a result of global warming. Thus are just some examples of animals affected. The water level has risen considerably resulting in frequent floods, eroding of riverbanks and mingling of freshwater and seawater sources thus leading to the extinction of several marine species. The shift in climatic conditions has an adverse effect on sea levels, availability of food, amount of rainfall, the composition of an ecosystem and temperature levels. In fact, early instances of life extinction have been attributed to climate change.

In the past, many species have managed to thwart the risk of extinction by migrating to greener pastures. But given the current scenario, it is extremely difficult to tackle the consequences of global warming, since human beings have made it all the more difficult by splitting up, transforming and at times obliterating the existing habitats and thereby leaving no scope for migration.

As far as different species are concerned, the effect of global warming is clearly visible with some of them shifting their habitats. Moreover, it is also becoming more difficult to preserve huge land tracts, which is affecting the chances of preserving biodiversity of a particular region. It is believed that if the situation is not taken stock of immediately, then around 2050 species will disappear from the surface of the earth as a result of global warming.

Global warming has already threatened the existence of the alpine meadows located in the Rocky Mountains. Mangroves and tropical montane are also exposed to the threat of extinction in future owing to global warming. The melting of polar ice has already taken a toll on the population of penguins and polar bears. Global warming has also threatened the existence of coral reefs.

The list of animals at risk of climate change will, of course, be longer and longer as the planet gets hotter and hotter.

GLOBAL WARMING EFFECTS ON VARIOUS SEASONAL PROCESSES OF PLANTS AND ANIMALS

Many seasonal processes are also affected by global warming. We are starting to witness:
1. Earlier leaf production by trees.
2. Earlier greening of vegetation.
3. Changed timing of egg-laying and hatching.
4. Changes in migration patterns of birds, fish and other animals.

Reductions and redistributions in populations of algae and plankton; this threatens the existence of fish and other animals that rely on algae and plankton for food.

IMPACTS ON FORESTS AND WILDLIFE

Based on future climate projections of Regional Climate Model of the Hadley Centre (HadRM3) using A2 and B2 scenarios and the BIOME 4 vegetation response model, it is predicted that 77 and 68 per cent of the forest areas in the country are likely to experience shift in forest types, respectively under the two scenarios, by the end of the century, with consequent changes in forests produce. Correspondingly, the associated biodiversity is likely to be adversely impacted. India's NATCOM I projects an increase in the area under xeric scrublands and xeric woodlands in central India at the cost of dry savannah in these regions.

A mean Sea Level Rise (SLR) of 15–38 cm is projected along India's coast by the mid 21st century and of 46–59 cm by 2100. India's NATCOM I assessed the vulnerability of coastal districts based on physical exposure to SLR, social exposure based on population affected, and economic impacts. In addition, a projected increase in the intensity of tropical cyclones poses a threat to the heavily populated coastal zones in the country.

Thus, climate change could have dramatic effects on a wide range of India's plants and animals.

SHIFTS IN CLIMATIC ENVELOPES

To estimate the effect of climate change on species, scientists use what they call a climatic envelope (sometimes also referred to as a bioclimatic envelope), which is the range of temperatures, rainfall and other climate-related parameters in which a species currently exists.

As the climate warms, the geographic location of climatic envelopes will shift significantly, possibly even to the extent that species can no longer survive in their current locations. Such species will need to follow their climatic envelopes by migrating to cooler and moister environments, usually uphill or northwards in the northern hemisphere. There is some evidence that plants and animals are already responding to warmer temperatures.

The tree line (above which there are no trees) may move up in altitude in coming years. In many cases, however, such migration might not be possible because of unsuitable soils and other unfavourable environmental parameters, geographical or human-made barriers and competition from species already in an area.

As human activities, particularly agriculture but also settlement and industrial development, have expanded over the last few centuries, natural vegetation—such as forests, grasslands and heathlands— have been cleared in large patches. Once-extensive plant communities have been reduced in size and broken into smaller patches. This habitat reduction and fragmentation poses a problem because it limits the ability of many species to migrate to favourable conditions. Species on mountain-tops, islands and peninsulas will have a similar problem.

In general, those species with restricted climatic envelopes, small populations and limited ability to migrate are most likely to suffer in the face of rapid climate change. A number of species will be affected physiologically by global warming. There is evidence that some species are physiologically vulnerable to temperature spikes.

CORAL BLEACHING

Warmer sea surface temperatures are blamed for an increase in a phenomenon called coral bleaching, which a whitening of coral is caused when the coral expels a single-celled, symbiotic alga called zooxanthellae. This alga usually lives within the tissues of the corals and, among other things, gives them its spectacular range of colours. Zooxanthellae are expelled when the coral is under stress from environmental factors such as abnormally high water temperatures or pollution. Since the zooxanthellae help coral in nutrient production, their loss can affect coral growth and make coral more vulnerable to disease. Major bleaching events took place on the Great Barrier Reef in 1998 and 2002, causing a significant die-off of corals in some locations.

INCREASES IN EXTREME EVENTS

Predicted changes in the intensity, frequency and extent of disturbances such as fire, cyclone, drought and flood will place existing vegetation under stress and favour species able to rapidly colonise denuded areas. In many cases this will mean the spread of 'weed' species and major changes in the distribution and abundance of many indigenous species.

RISES IN CONCENTRATIONS OF CARBON DIOXIDE

The basic ingredients of photosynthesis are carbon dioxide and water. Increased carbon dioxide in the atmosphere causes increased growth rates in many plant species. This is good news for farmers, but only if this carbon dioxide 'fertilisation' effect is matched by adequate soil moisture and other nutrients. Leaf-eating animals may not be so lucky: increased concentrations of carbon dioxide could diminish the nutritional value of foliage. Rising levels of atmospheric carbon dioxide could also decrease the calcification rates of corals, meaning that reefs dam aged by bleaching or other agents would recover more slowly.

Sea-level Rise

In most climate-change models, sea levels are predicted to rise by 9 to 88 centimetres by 2100, due to the thermal expansion of the oceans and the melting of polar ice-caps, coupled with the effects of storm surges, which are expected to be of a greater magnitude in a warmer world. Coastal ecosystems, such as mangrove forests and low-lying freshwater wetlands could be severely affected.

The First Victims

The early victims of all these factors will be the endangered species. The following list describes the list of endangered species in India including all animals and birds which occur in India and are rated as Critically Endangered (CR), Endangered (EN) or Vulnerable (VU) in the 2008 International Union for Conservation of Nature and Natural Resources (IUCN) and Wildlife Institute of India (WII).

Critically Endangered

1. Jenkin's Shrew (*Crocidura jenkensii*) (Endemic to India).
2. Ganges Shark (*Glyphis gangeticus*) (Endemic to India).
3. Himalayan Wolf (*Canis himalayensis*) (Endemic to India and Nepal).
4. Indian Vulture (*Gyps indicus*).
5. Malabar Large-spotted Civet (*Viverra civettina*).
6. Nam dapha Flying Squirrel (*Biswamayopterus biswasi*) (Endemic to India).
7. Pygmy Frog (*Sus salvanius*).
8. Salim Ali's Fruit Bat (*Latidens salimalii*) (Endemic to India).
9. Wroughton's Free-tailed Bat (*Otomops wroughtoni*) (Endemic to India).
10. Jerdon's Courser (*Crows bitorquatus*) (Endemic to India).

Endangered

1. Andaman Shrew (*Crocidura andamanensis*) (Endemic to India).
2. Andaman Spiny Shrew (*Crocidura hispida*) (Endemic to India).
3. Asian Arowana (*Scleropages formosus*).
4. Asiatic Black Bear (*Selenarctos thibetanus*).
5. Asiatic Lion (*Panthera leo persica*).
6. Asiatic Wild Dog/Dhole (*Cuon alpinus*).
7. Banteng (*Bos javanicus*).
8. Blue Whale (*Balaenoptera musculus*).
9. Capped Leaf Monkey (*Trachypithecus pileatus*).
10. Chiru (Tibetan Antelope) (*Pantholops hodgsonii*).
11. Wild Cat (*Felis silvestris ornata*).
12. Fin Whale (*Balaenoptera physalus*).
13. Ganges River Dolphin (*Platanista gangetica*).
14. Golden Leaf Monkey (*Trachypithecus geei*).
15. Great Indian Rhinoceros (*Rhinoceros unicornis*).
16. Hispid Hare (*Caprolagus hispidus*).
17. Hoolock Gibbon (*Bunipithecus hoolock*) (Previously Hylobates hoolock).
18. Indian Elephant or Asian Elephant (*Elephas maximus*).
19. Indus River Dolphin (*Platanista minor*).
20. Kashmir Stag/Hangul (*Cervus elaphus hanglu*).
21. Kondana Soft-furred Rat (*Millardia kondana*) (Endemic to India).
22. Lion-tailed Macaque (*Macaca silenus*) (Endemic to India).
23. Loggerhead Sea Turtle (*Caretta caretta*).
24. Malabar Civet (*Viverra civettina*).
25. Markhor (*Capra falconeri*).

26. Marsh Mongoose (*Herpestes palustris*) (Endemic to India) (Previously considered to be a subspecies of *Herpestes javanicus*).
27. Narcondam Hornbill (*Rhyticeros narcondami*).
28. Nicobar Shrew (*Crocidura nicobarica*) (Endemic to India).
29. Nicobar Tree Shrew (*Tupaia nicobarica*) (Endemic to India).
30. Nilgiri Leaf Monkey (*Presbytis johni*).
31. Nilgiri Tahr (*Hemitragus hylocrius*) (Endemic to India).
32. Olive Ridley Turtle (Endemic to Odisha, Andhra Pradesh, India).
33. Particoloured Flying Squirrel (*Hylopetes alboniger*).
34. Peter's Tube-nosed Bat (*Murina grisea*) (Endemic to India).
35. Pygmy Hog (*Sus salvanius*).
36. Red Panda (Lesser Panda) (*Ailurus fulgens*).
37. Royal Bengal Tiger (*Panthera tigris tigris*).
38. Sei Whale (*Balaenoptera borealis*).
39. Servant Mouse (*Mus famulus*). (Endemic to India).
40. Snow Leopard (*Uncia uncia*).
41. Wild Water Buffalo (*Bubalus bubalis*) (Previously Bubalus arnee).
42. Woolly Flying Squirrel (*Eupetaurus cinereus*).

Vulnerable

'Endangered Mammal List'. Wildlife Institute of India (WII).

1. Andaman Horseshoe Bat (*Rhinolophus cognatus*) (Endemic to India).
2. Andaman Rat (*Rattus stoicus*) (Endemic to India).
3. Argali (*Ovis ammon*).
4. Himalayan W-toothed Shrew (*Crocidura attenuate*).
5. Sri Lankan Highland Shrew (*Suncus montanus*).
6. Asiatic Black Bear (*Ursus thibetanus*).
7. Asiatic Golden Cat (*Catopuma temminckii*).
8. Assamese Macaque (*Macaca assamensis*).
9. Back-striped Weasel (*Mustela strigidorsa*).
10. Barasingha (*Cervus duvauceli*).
11. Bare-bellied Hedgehog (*Hemiechinus nudiventris*) (Endemic to India).
12. Blackbuck (*Antilope cervicapra*).
13. Brow-antlered Deer (*Cervus eldi eldi*).
14. Brown Bear (*Ursus arctos*).
15. Brown fish owl (*Ketupa zeylonensis*) (Endemic to India).
16. Brown Palm Civet (*Paradoxurus jerdoni*).
17. Central Kashmir Vole (*Alticola montosa*) (Endemic to India).
18. Clouded Leopard (*Neofelis nebulosa*).
19. Day's Shrew (*Suncus dayi*) (Endemic to India).
20. Dhole (*Cuon alpines*).
21. Dugong (*Dugong dugon*).
22. Eld's Deer (*Cervus eldi*).
23. Elvira Rat (*Cremnomys Elvira*) (Endemic to India).

24. European Otter (also known as Eurasian Otter) (*Lutra lutra*).
25. Fishing Cat (*Prionailurus viverrinus*).
26. Four-horned Antelope (*Tetracerus quadricornis*).
27. Ganges River Dolphin (*Platanista gangetica*).
28. Gaur (*Bos gaurus*).
29. Golden Jackal (*Canis aureus*).
30. Goral (*Nemorhaedus goral*).
31. Himalayan Musk Deer (*Moschus chrysogaster*).
32. Himalayan Shrew (*Soriculus nigrescens*).
33. Himalayan Tahr (*Hemitragus jemlahicus*).
34. Humpback Whale (*Megaptera novaeangliae*).
35. Indian Fox (*Vulpes bengalensis*).
36. Indian Giant Squirrel (*Ratufa indica*) (Endemic to India).
37. Indian Wolf (*Canis lupus indica*).
38. Irrawaddy Squirrel (*Callosciurus pygerythrus*).
39. Jerdon's Palm Civet (*Paradoxurus jerdoni*) (Endemic to India).
40. Kashmir Cave Bat (*Myotis longipes*).
41. Kerala Rat (*Rattus ranjiniae*) (Endemic to India).
42. Khajuria's Leaf-nosed Bat (*Hipposideros durgadasi*) (Endemic to India).
43. Kolar Leaf-nosed Bat (*Hipposideros hypophyllus*) (Endemic to India).
44. Lesser Horseshoe Bat (*Rhinolophus hipposideros*).
45. Lesser Panda (*Ailurus fulgens*).
46. Mainland Serow (*Capricornis sumatraensis*).
47. Malayan Porcupine (*Hystrix brachyuran*).
48. Mandelli's Mouse-eared Bat (*Myotis sicarius*).
49. Marbled Cat (*Pardofelis marmorata*).
50. Mouflon (or Urial) (*Ovis orientalis*).
51. Nicobar Flying Fox (*Pteropus faunulus*) (Endemic to India).
52. Nilgiri Leaf Monkey (*Trachypithecus johnii*) (Endemic to India).
53. Nilgiri Marten (*Martes gwatkinsii*) (Endemic to India).
54. Nonsense Rat (*Rattus burrus*) (Endemic to India).
55. Asiatic Wild Ass (*Equus hemionus*).
56. Pale Grey Shrew (*Crocidura pergrisea*) (Endemic to India).
57. Palm Rat (*Rattus palmarum*) (Endemic to India).
58. Red Goral (*Naemorhedus baileyi*).
59. Rock Eagle-owl (*Bubo bengalensis*) (Endemic to India).
60. Rusty-spotted Cat (*Prionailurus rubiginosus*).
61. Sikkim Rat (*Rattus sikkimensis*).
62. Sloth Bear (*Melursus ursinus*).
63. Slow Loris (*Loris tardigradus*).
64. Smooth-coated Otter (*Lutrogale perspicillata*) (Previously *Lutra perspicillata*).
65. Sperm Whale (*Physeter macrocephalus*).
66. Sri Lankan Giant Squirrel (*Ratufa macroura*).

67. Stumptail Macaque (*Macaca arctoides*).
68. Takin (*Budorcas taxicolor*).
69. Wild Goat (*Capra aegagrus*).
70. Wild Yak (*Bos grunniens*).
71. Tiger.

Threatened

1. Indian Wild Ass (*Equus hemionus khur*).
2. Leopard (*Panthera pardus*).
3. Red Fox (*Vulpes vulpes montana*).
4. Kashmir Stag (*Praygnaa*).

WHAT WOULD RAPID SPECIES EXTINCTION MEAN FOR INDIA

Global warming is predicted to take place faster in the next century than at any time for at least the last 10,000 years. Coupled with other factors, such as continued land-clearing, this could mean the extinction of species at a rate even greater than when the dinosaurs disappeared about 65 million years ago. Some species not under immediate threat of extinction might nonetheless suffer decreases in population size, diminishing intra-species' genetic diversity (and therefore face increased vulnerability).

Does it really matter if many species go extinct? The world would certainly be a less interesting place with less biodiversity, but would it affect us?

A diversity of species increases the ability of ecosystems to do things like hold soils together, maintain soil fertility, deliver clean water to streams and rivers, cycle nutrients, pollinate plants (including crops), and buffer against pests and diseases—these are sometimes called 'ecosystem functions' or 'ecosystem services'. A loss of species could reduce this ability, particularly if environmental conditions are changing rapidly at the same time. It is therefore possible that as the climate changes and as species are eliminated from an area we will see a change in some ecosystem functions; this could mean more land degradation, changes in agricultural productivity and a reduction in the quality of water delivered to human populations.

ADAPTING TO CHANGE

Scientists agree that human-induced global warming is happening, and that the world will continue to warm for some time even if greenhouse gas emissions are somehow curbed. Some species, particularly insects, might be able to adapt to changing conditions or evolve in response to global warming. But for many, especially those that are already rare and have limited climatic envelopes, global warming could pose an insurmountable challenge.

The plan, which was developed in consultation with scientists, conservationists and national, state and local governments, contains seven objectives, along with actions that should be taken to achieve the objectives. At this early stage of development many of these actions are aimed at improving our understanding of the impacts of global warming on biodiversity, while others are general or strategic in nature.

Some of the impacts of global warming may be sudden, but in many cases societies will have some years to adapt their management of biodiversity as conditions change. Increasing our understanding of the effects of climate change on biodiversity, and developing practical ways of mitigating such effects, are critical to limit the damage. Even so, the dangers are great for humans as well as our native plants and animals.

NATIONAL MISSION FOR SUSTAINING THE HIMALAYAN ECOSYSTEM

The Himalayan ecosystem is vital to the ecological security of the Indian landmass through providing forest cover, feeding perennial rivers that are the source of drinking water, irrigation, and hydropower, conserving biodiversity, providing a rich base for high value agriculture, and spectacular landscapes for sustainable tourism. At the same time, climate change may adversely impact the Himalayan ecosystem through increased temperature, altered precipitation patterns, and episodes of drought.

Concern has also been expressed that the Himalayan glaciers, in common with other entities in the global cryosphere, may lose significant ice-mass, and thereby endanger river flows, especially in the lean season, when the North Indian rivers are largely fed by melting snow and ice. Studies by several scientific institutions in India have been inconclusive on the extent of change in glacier mass, and whether climate change is a significant causative factor.

It is accordingly necessary to continue and enhance monitoring of the Himalayan ecosystem, in particular the state of its glaciers, and the impacts of change in glacial mass on river flows. Since several other countries in the South Asian region share the Himalayan ecosystem, appropriate forms of scientific collaboration and exchange of information may be considered with them to enhance understanding of ecosystem changes and their effects.

It is also necessary, with a view to enhancing conservation of Himalayan ecosystems, to empower local communities, in particular through the Panchayats, to assume greater responsibility for management of ecological resources.

The National Environment Policy, 2006, *inter alia* provides the following relevant measures for conservation of mountain ecosystems:

1. Adopt appropriate land use planning and watershed management practices for sustainable development of mountain ecosystems.
2. Adopt 'best practice' norms for infrastructure construction in mountain regions to avoid or minimise damage to sensitive ecosystems and despoiling of landscapes.
3. Encourage cultivation of traditional varieties of crops and horticulture by promotion of organic farming enabling farmers to realise a price premium.
4. Promote sustainable tourism through adoption of 'best practice' norms for tourism facilities and access to ecological resources, and multistake-holder partnerships to enable local communities to gain better livelihoods, while leveraging financial, technical and managerial capacities of investors.
5. Take measures to regulate tourist inflows into mountain regions to ensure that these remain within the carrying capacity of the mountain ecology.
6. Consider particular unique mountain scapes as entities with 'Incomparable Values', in developing strategies for their protection.

NATIONAL MISSION FOR A GREEN INDIA

Forests are repositories of genetic diversity and supply a wide range of ecosystem services thus helping maintain ecological balance. Forests meet nearly 40 per cent of the energy needs of the country overall, and over 80 per cent of those in rural areas, and are the backbone of forest-based communities in terms of livelihood and sustenance. Forests sequester billions of tons of carbon dioxide in the form of biomass and soil carbon. The proposed national program will focus on two objectives, namely increasing the forest cover and density as a whole of the country and conserving biodiversity.

INCREASE IN FOREST COVER AND DENSITY

The report of the Working Group on Forests for the 11th Five-Year Plan puts the annual rate of planting during 2001/02 to 2005/06 at 1.6 million hectares and proposes to increase it to 3.3 million hectares during the 11th Plan. The final target is to bring one-third of the geographic area of India under forest cover.

The Greening India Program has already been announced. Under the program, 6 million hectares of degraded forest land would be afforested with the participation of Joint Forest Management Committees (JFMCs), with funds to the extent of Rs 6000 crores provided from the accumulated additional funds for compensatory afforestation under a decision of the Supreme Court in respect of forest lands diverted to nonforest use. The elements of this program may include the following:

1. Training on silvicultural practices for fast-growing and climate-hardy tree species.
2. Reducing fragmentation of forests by provision of corridors for species migration, both fauna and flora.
3. Enhancing public and private investments for raising plantations for enhancing the cover and the density of forests.
4. Revitalising and upscaling community-based initiatives such as Joint Forest Management (JFM) and Van Panchayat Committees for forest management.
5. Implementation of the Greening India Plan.
6. Formulation of forest fire management strategies.

CONSERVING BIODIVERSITY

Conservation of wildlife and biodiversity in natural heritage sites including sacred groves, protected areas and other biodiversity 'hotspots' is crucial for maintaining the resilience of ecosystems. Specific actions in this program will include:

1. *In situ* and *ex situ* conservation of genetic resources, especially of threatened flora and fauna.
2. Creation of biodiversity registers (at national, district and local levels) for documenting genetic diversity and the associated traditional knowledge.
3. Effective implementation of the Protected Area System under the Wildlife Conservation Act.
4. Effective implementation of the National Biodiversity Conservation Act, 2001.

Chapter 24

Biodiversity in Asia and The Pacific

INTRODUCTION

The year 2010 marks the International Year of Biodiversity, which is a celebration of life on earth and of the value of biodiversity for our lives. The International Year of Biodiversity also represents what may be viewed as a global deadline for halting the loss of biodiversity, set in April 2002 at the sixth meeting of the Conference of the Parties to the Convention on Biological Diversity. At that meeting parties adopted a strategic plan, including the target of achieving by 2010 a significant reduction in the current rate of biodiversity loss at the global, regional and national levels as a contribution to poverty alleviation and for the benefit of all life on earth. Thus the 2010 biodiversity target was subsequently endorsed by the World Summit on Sustainable Development, held in Johannesburg, South Africa, in 2002, and by the United Nations General Assembly, and was incorporated as a new target under the Millennium Development Goals.

The rapidly changing state of biodiversity has been chronicled in the third edition of the Global Biodiversity Outlook (GBO-3), in order to assess progress towards the attainment of the 2010 target and to consider future action. On the basis of the information available to date and analysed for GBO-3, a common message emerges: biodiversity is in decline globally, in most regions, and in most of its forms. Most governments missed their 2010 target and stated their view that assessing each country's progress towards achieving the biodiversity target had posed a challenging task in the absence of nationally agreed baselines, targets and indicators.

INTRODUCTION TO ASIA AND THE PACIFIC

Asia and the Pacific encompasses some of the world's greatest biological, cultural and economic diversity. It covers 8.6 per cent of the earth's total surface area and nearly 30 per cent of its land area. It is also host to certain wildlife species unique to the region such as the giant panda, the tiger, the Asian elephant, the Javan rhinoceros and the orang-utan. The region's wealth in biological diversity and associated traditional knowledge is evidenced by the fact that 5 of the 17 members of the group known as the Like-minded Megadiverse Countries are from this region: China, India, Indonesia, Malaysia and the Philippines.

This region is also home to approximately 4 billion people, representing some 60 per cent of the world's population. They live in communities ranging from large urban centres to remote rural communities, and together speak more than 2000 languages. The collective economic activity of the region — significantly boosted by current rapid growth in China and India — accounts for about a quarter of the global domestic product.

The vast scale of human activities in this region means that it poses a direct challenge to the resilience of the regional ecosystem. The serious effects of these activities are already evident today: rapid economic development in this region has led to massive changes in lifestyle and increases in correlated indirect drivers of biodiversity loss. As a result, nature has come under great pressure and much valuable biodiversity has been lost or continues to be degraded.

KEY BIODIVERSITY CHALLENGES IN ASIA AND THE PACIFIC

In 2008 Asia and the Pacific recorded the world's highest number of threatened species. Many of the most serious problems are to be found in South-East Asia, where 6 of the 10 countries in the region with the highest numbers of threatened animal and plant species (Fig. 24.1) are to be found. Over the period 2002–2009, nearly 2500 species in Asia and the Pacific were recorded in the Red List of the International Union for Conservation of Nature and Natural Resources (IUCN) as 'critically endangered', 'endangered' or 'vulnerable'. In all, 13 of the 34 biodiversity hotspots designated by Conservation International are also to be found in this area (East Melanesian Islands, Himalayas, Indo-Burma, Japan, mountains of south-west China, New Caledonia, New Zealand, Philippines, Polynesia-Micronesia, south-west Australia, Sundaland, Wallacea, Western Ghats and Sri Lanka). In particular, bird species have faced an especially steep increase in extinction risk in South-East Asia and on the Pacific islands, while mammals have suffered the steepest increase in risk in South and South-East Asia compared to the global average.

Where plant species are concerned, medicinal plants face a high risk of extinction in Asia and the Pacific, where there is continuing dependence on wild collection. The region has seen a net overall gain of forests over the period 2000–2009, but high rates of fragmentation and net loss of forests have continued in many countries in South and South-East Asia. Over the period 2000–2005, the rates of loss of primary forests were fastest in Cambodia, the Democratic People's Republic of Korea, Indonesia, Mongolia, Papua New Guinea and Viet Nam, accounting for a quarter of the world's total losses over that period (Statistics Division, United Nations Economic and Social Commission for Asia and the Pacific). In particular, the rapid increase in the large-scale plantation of oil palms for biofuel processing in some countries, particularly in areas previously covered by primary tropical forest, is a major factor in biodiversity loss, causing both land degradation and habitat loss for many bird species.

Turning to coastal ecosystems, Asia and the Pacific contains a large proportion of the world's remaining mangrove forests and coral reefs; both systems continue, however, to suffer from various direct and indirect pressures. Mangroves are notably, although not exclusively, affected by shrimp farming and other forms of mariculture: this is a matter of particular concern since this is the only region in the world in which the rate of loss of mangrove forests has not slowed in recent years. Meanwhile, the region's coral reef system, which includes the world's two largest coral formations (the Great Barrier Reef and the New Caledonia Barrier Reef) and has the highest level of coral diversity in the world, has seen its extent of coral cover decline from 40 per cent in the early 1980s to approximately 20 per cent by 2003, partly owing to global-scale stressors such as climate change. While an increasing proportion of the world's surface has been designated as protected areas (Fig. 24.2), progress in this regard in Asia and the Pacific appears relatively modest, particularly considering the large number of threatened species in this region. The terrestrial area designated as legally protected constitutes less than 9 per cent of the total surface area, below the global average (Fig. 24.2). Within the region, East and North-East Asia has the highest proportion of protected areas, while North and Central Asia has the lowest. Where marine protected areas are concerned, the area designated as legally protected in 2007 constituted less that 5 per cent of the region's territorial waters (Fig. 24.3).

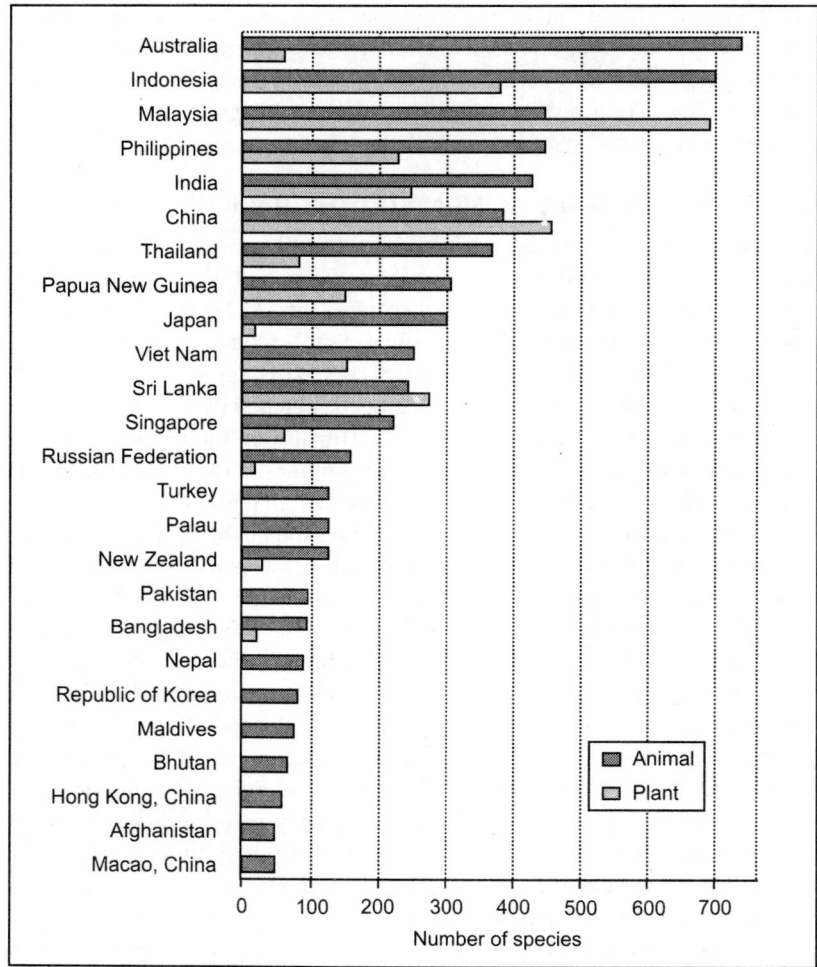

Fig. 24.1. Threatened plant and animal species, Asia and the Pacific, 2009 (Statistics Division, United Nations Economic and Social Commission for Asia and the Pacific).

REGIONAL EFFORTS TO ACHIEVE THE BIODIVERSITY TARGETS SET FOR 2010 AND BEYOND

While the countries failed to achieve the 2010 biodiversity targets and the prospects for biodiversity in Asia and the Pacific remain shaky, some encouraging signs of progress have also been observed. Thus, 87 per cent of the parties to the Convention on Biological Diversity have developed national biodiversity strategies and action plans, and such strategies and action plans are currently under development in most of the remaining countries. Almost half of these have been developed or updated since 2002, demonstrating a significant increase in the willingness manifested by these countries to protect their biodiversity since the 2010 targets were established.

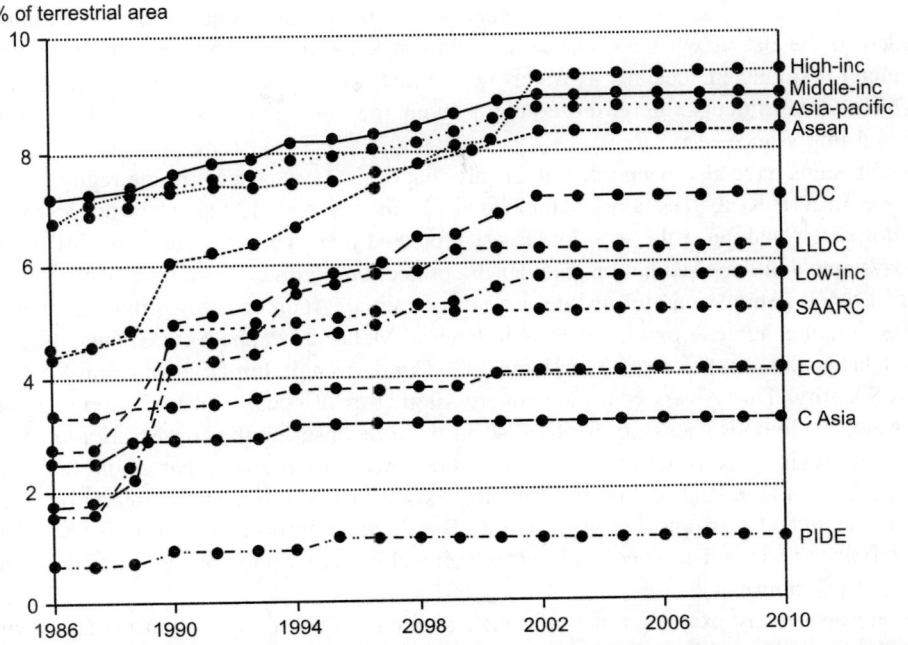

Fig. 24.2. Ratio of protected terrestrial areas to surface area, 1986–2010 (Statistics Division, United Nations Economic and Social Commission for Asia and the Pacific).

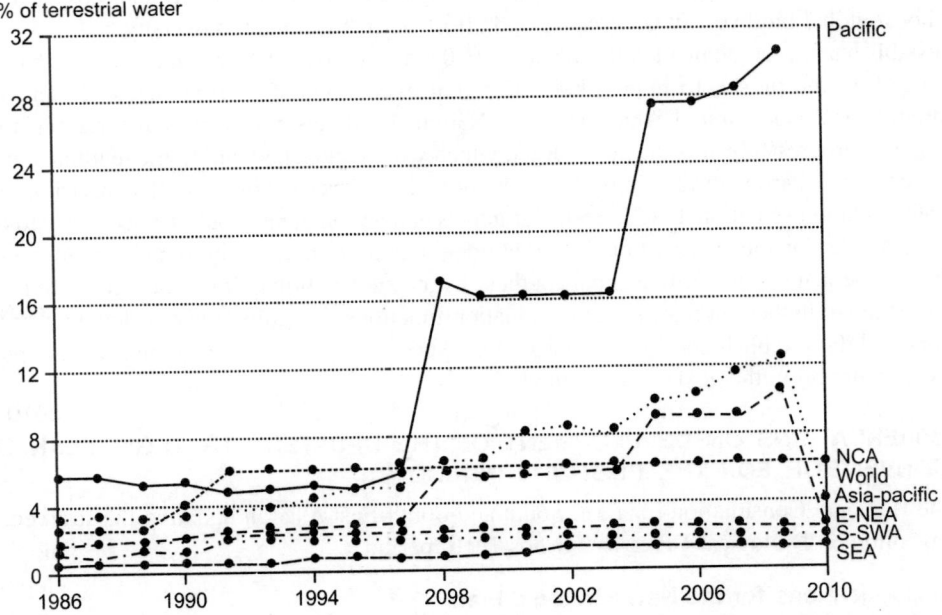

Fig. 24.3. Ratio of protected marine areas to territorial water, 1986–2010 (Statistics Division, United Nations Economic and Social Commission for Asia and the Pacific).

The region's commitment is evident in other indicators as well. Since 2002, 739 additional sites were added to the list of what are known as 'Ramsar sites' under the Convention on Wetlands of International Importance, Especially as Waterfowl Habitat, 143 of which were from Asia and the Pacific. This reflects Governments' increased concern about the ecological condition of wetland sites of international importance.

Important steps have also been taken to ensure that critical ecosystems in the region are protected. For example, in 2008 Kiribati established the Phoenix Islands Protected Area, which covers over 4,00,000 square kilometres, making it the world's largest protected area. The area contains globally important coral ecosystems and deep-sea habitats; under the protection regime commercial fishing is prohibited, while subsistence fishing and sustainable use are permitted. Another noteworthy achievement is the 10 years action plan agreed upon in 2009 by Indonesia, Malaysia, Papua New Guinea, the Philippines, Solomon Islands and Timor-Leste, to implement the Coral Triangle Initiative on Coral Reefs, Fisheries and Food Security. This covers 6 million square kilometres of ocean and will support marine-based industries and livelihoods necessary for food security while ensuring the conservation of the area.

At a recent regional consultation, governments reported on initiatives that would help steer a way forward for Asia and the Pacific. For example, the Asia-Pacific Biodiversity Observation Network and the recently launched East and South-East Asia Biodiversity Information Initiative are designed to establish biodiversity information networks at the regional level to support planning and decision-making in the area of taxonomy.

There are also many examples of traditional, community-based ecosystem management systems that are enjoying increased attention in the region, including Japan's Satoyama landscape concept and the locally managed marine areas in the South Pacific, which can be developed to demonstrate good practice for the planned intergovernmental science-policy platform on biodiversity and ecosystem services and to boost global biodiversity governance and efforts to mainstream biodiversity.

To establish a new strategic plan for the post-2010 era, countries in Asia and the Pacific have agreed on the need, at the outset, to undertake a thorough evaluation of current successes and failures. Accordingly, the Association of Southeast Asian Nations Biodiversity Centre is undertaking a regional assessment of progress towards the 2010 target, which will be launched during the tenth meeting of the Conference of the Parties to the Convention on Biological Diversity in 2010. The revision, updating and implementation of national biodiversity strategies and action plans would be necessary to translate any new strategic plan into clear national commitments. Countries must set minimum quantitative targets and allocate adequate domestic funds to meet these targets, in addition to international funding. There is also a need to strengthen monitoring and evaluation functions, including the development of national indicators and the establishment of baselines. Other specific inputs from the region for the post-2010 targets can be found in the next section below.

RECOMMENDATIONS ON DEVELOPMENT OF THE UPDATED STRATEGIC PLAN OF THE CONVENTION FOR THE POST-2010 PERIOD

From the Regional Consultation for East, South and South-East Asia on updating the strategic plan of the Convention on Biological Diversity for post-2010 period.

Overall Suggestions for the New Strategic Plan

1. It is important to analyse why countries failed to achieve the 2010 target while elaborating the possible vision, mission, strategic goals, targets and support mechanisms for the post-2010 period.

2. The global strategy should be broad and provide a framework to allow regions and countries to develop goals and targets to address priority issues they face, taking into consideration national situation of countries in different development stages.

3. Strategic goals and headline targets for the post-2010 period should be well defined for implementation, monitoring, reporting and evaluation.

4. Goals should be really strategic and targets should be really SMART.

5. Interim targets and milestones should be developed to ensure that momentum will not be lost for achieving strategic goals.

6. Goals and targets should embrace all components of biodiversity and address drivers, both direct and indirect, of biodiversity loss, and obstacles to implementation.

7. It is important to recognise the value of biodiversity and associated traditional knowledge and benefits of conservation for human well-being and link biodiversity conservation with poverty reduction, sustainable development and ecosystem management.

8. The Strategic Plan should address the three objectives of the Convention in a balanced manner, with adequate focus given to ABS-related issues.

9. Inputs should be drawn from the development of strategies of other related conventions such as UNCCD and the Strategic Plans for the Convention and the Biosafety Protocol should be complementary.

Specific Suggestions for the Strategic Plan

1. Capacity constraints for implementation should be identified as one major issue that should be addressed by the new Strategic Plan.

2. Mission should be simple and focused.

3. Strategic goals should be simple and direct and overlaps among them should be avoided.

4. Headline targets need further elaboration to ensure that they are clear enough to achieve strategic goals and easy for monitoring and evaluation.

5. Support mechanisms proposed are weak and need further elaboration. Some mechanisms such as adequate funding, technology transfer, regional cooperation and support to monitoring should be added.

Suggestions for the Implementation of the New Strategic Plan

1. Communication and outreach to all relevant stakeholders should be a crucial priority for the implementation of the new Strategic Plan.

2. Mechanisms to support implementation should be an integral part of the updated Strategic Plan.

3. Ownership of post-2010 target(s) by stakeholders is crucial for future implementation.

4. It is important to develop indicators and baselines for monitoring and reporting.

5. Global targets need to be translated into national commitments including through updating national biodiversity strategies and action plans in light of the new Strategic Plan.

6. A program of capacity-building will be needed to support implementation of the new Strategic Plan, including regional and sub-regional workshops to facilitate the translation of the new Strategic Plan into updated national biodiversity strategies and action plans, and a strengthened clearing house mechanism to facilitate the exchange of relevant information, expertise and experience among and within countries.

BIODIVERSITY: WEST ASIA

Resources

The region has wide variations in terrestrial and aquatic ecosystems. Main terrestrial habitats include Mediterranean forests, rangelands and deserts. Marine ecosystems include mudflats, mangrove swamps, sea grass and coral reefs. Rivers in the Mashriq and springs in the whole region represent freshwater ecosystems.

The estimated number of endemic vascular species in the region is 800, and in some hot spots such as the Socotra Islands of Yemen, 34 per cent of the total number of vascular plants are endemic. There are seven endemic mammal species and ten endemic birds.

The seas are rich in species diversity with 200 species of crabs, 20 species of marine mammals and more than 1200 species of fish and more than 330 species of corals in the Red Sea and the Gulf. More than 11 per cent of the corals are endemic to the Arabian Peninsula sub-region. There are up to 12,000 marine species in the Mediterranean, representing 8–9 per cent of the world sea species richness. Substantial numbers of vertebrates are threatened with extinction in the region (Fig. 24.4).

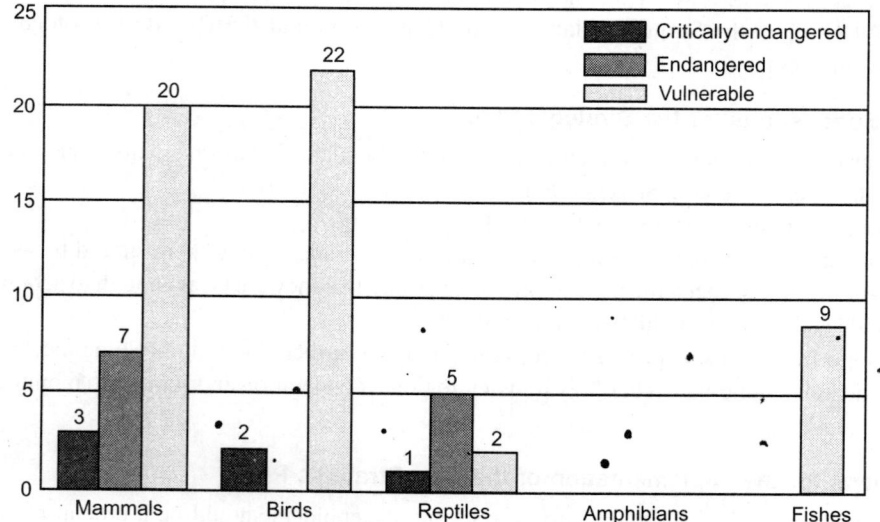

Fig. 24.4. Numbers of threatened vertebrates: West Asia.

Habitat destruction and fragmentation have increased dramatically in most countries over the past three decades due to human population and resource consumption growth. Degradation of unique terrestrial and aquatic ecosystems and loss of genetic resources are the main biodiversity issues in West Asia. Water resource management and the maintenance of inland water biodiversity, as well as overhunting of large mammals and birds, are therefore among the most important issues affecting biodiversity in the region.

Habitat Degradation and Loss

Rapid population increases and changes in lifestyle have contributed to the degradation of wetland ecosystems due to increased exploitation of surface and groundwater. In Jordan, groundwater extraction for urban needs increased from around 2 million m^3 in 1989 to around 25 million m^3 in 2003 while an additional 25 million m^3 per year was used for irrigated agriculture. As well as water extraction, pollution and impacts from refugee camps in the area have led to the deterioration and drying up of the Azraq wetlands natural reserve. As a consequence tourism in Azraq has declined. In the eastern part of the Arabian Peninsula, many of the date palm oases and natural freshwater springs have been lost in the past two decades.

By far the most serious wetland change in West Asia over the past three decades has occurred in the lower Mesopotamian marshlands, where serial satellite images confirm a loss of around 90 per cent of the area of lake and marshlands. This loss may be attributable in part to the large number of dams now present on upstream parts of the Tigris-Euphrates system, but appears to be primarily a result of major hydrological engineering works in southern Iraq, notably the completion of the Major Outfall Drain (or 'Third River') which diverts water to the head of the Gulf. However, despite some negative impacts of damming on indigenous biodiversity, the loss of some habitats such as wetlands has been offset by the creation of large artificial habitats elsewhere in the region. For example, the 630 km^2 Assad Lake in Syria on the Euphrates River is considered an important site for migratory and wintering birds in West Asia.

The rapid decline of the lower Mesopotamian marshlands represents one of the most significant environmental events to have occurred globally during the past 30 years. Loss of such an important habitat illustrates the pressures on wetlands in the region, which are likely to intensify in future as demand for water continues to increase.

Food self-sufficiency policies in the region have resulted in the cultivation of marginal lands for irrigated intensive agriculture. This has strained water resources and caused salinisation, with negative effects on freshwater biodiversity. The breakdown of traditional systems of resource management has also had a major impact on biodiversity. For example, the traditional Al-Hema system, which facilitated the sustainable use of rangelands and other natural resources by setting aside large reserves during times of stress was abandoned in the 1960s in the Arabian Peninsula and Mashriq countries. While about 3000 hema reserves existed in Saudi Arabia in 1969, only 71 were still in existence under various degrees of protection in 1984 and only nine were on the 1997 Protected Areas list (Fig. 24.5).

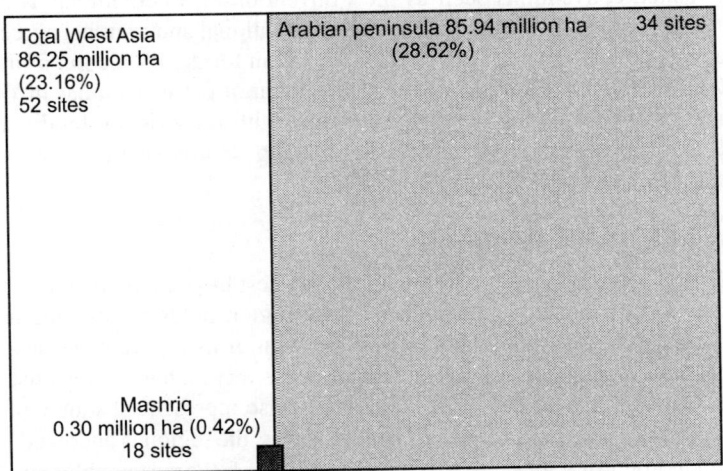

Fig. 24.5. Protected areas: West Asia

Coastal and marine biodiversity is threatened by several human activities including pollution (oil spills, industrial and domestic discharges into the sea), physical alteration of habitats (sand dredging and landfills), climate variability and alien species introduced by ballast water. The extent of mangroves has been decreasing along the shores of the Gulf over the past 30 years due to unplanned coastal development to the extent that only 125–130 km² of mangrove patches remain. In Saudi Arabia, more than 40 per cent of the Gulf coastline has been reclaimed and almost 50 per cent of the mangroves lost. In the Arabian Peninsula seas, about 20,000 km² of coral reefs or 7.9 per cent of the total area of world corals have been exposed to bleaching due to increases in seawater temperature caused by El Niño. It is feared that global warming will intensify this phenomenon. In the Mashriq sub-region many marine species, including Mediterranean monk seals, marine turtles and marine sponges, are threatened by the continuous deterioration of coastal water quality due to sedimentation, nutrient discharge and eutrophication.

Loss of Terrestrial Species

A comprehensive decline in the larger terrestrial species has been recorded. This is primarily because of excess hunting resulting from the decline of traditional resource management practices, and the increased availability of four-wheel drive vehicles and automatic weapons. While wild goat (*Capra ibex*), and gazelles (*Gazella gazella, G. dorcas* and *G. subgutturosa*) are still present in the region they have been much reduced in range and numbers. The leopard, which was formerly widespread, persists in a few isolated areas. The cheetah is on the verge of extinction, if not already extinct, the last confirmed specimen being taken in 1977. The Arabian oryx (*Oryx leucoryx*) was extinct in the wild but has been successfully reintroduced using captive stock. The ostrich is believed extinct, the Arabian bustard (*Ardeotis arabs*) has been reduced in numbers and is possibly extinct in Saudi Arabia, and the Houbara bustard (*Chlamydotis undulata*) now winters in much reduced numbers. Programmes for captive breeding of threatened species have operated since the 1980s, with re-introduction programmes for the Arabian oryx, Houbara bustard and some gazelle species in Jordan, Oman, Saudi Arabia and Syria.

Addressing Biodiversity Loss

Most countries have ratified the Convention on Biological Diversity. In addition, some have ratified other biodiversity-related conventions such as the Convention on International Trade in Endangered Species (CITES). Countries are also adhering to other international and regional agreements such as the Mediterranean Action Plan (MAP) and Regional Organisation for the Protection of the Environment of the Red Sea and Gulf of Aden (PERSGA). The establishment of protected areas in West Asia has been gaining momentum. Local people are generally unhappy with the existing biodiversity conservation programs because they are not involved in decision-making. However, the situation is improving in some countries such as Lebanon and Jordan.

LATIN AMERICA AND THE CARIBBEAN

Latin America and the Caribbean is the region with the greatest biological diversity on the planet and it hosts several of the world's megadiverse countries. The region holds almost one half of the world's tropical forests, 33 per cent of its total mammals, 35 per cent of its reptilian species, 41 per cent of its birds and 50 per cent of its amphibians. Levels of endemism are very high in the region: thus, 50 per cent of the plant life of the Caribbean is unique. This biodiversity also represents a source of abundant genetic resources for Latin America and the Caribbean. In other words, the region is endowed with exceptionally rich biodiversity and its countries are keen to harvest benefits from the sustainable use of that biodiversity and their ecosystems, in order to promote social and economic growth and equality. There is a growing recognition of the value of this biodiversity and of its associated ecosystem services, a number of protected areas have been created and many countries have adopted regulatory frameworks and policies designed to ensure the protection of their biodiversity and the sustainable use of its components.

Biodiversity in 2010

The report entitled 'Millennium Development Goals: advances in environmentally sustainable development in Latin America and the Caribbean', launched in February 2010, highlights considerable advances achieved in the region in some environmental areas, such as the increase in protected areas that are rich in biodiversity: thus, between 1990 and 2008 the number of gazetted marine and terrestrial protected areas in Latin America and the Caribbean more than doubled. The region, however, continues to face major challenges such as halting deforestation. Between 1990 and 2005, it lost nearly 69 million hectares of forest, equivalent to 7 per cent of the region's entire forest cover: the region now has the highest deforestation rate in the world. Notwithstanding, the lack of official historical data evidence points to a concomitant loss of biodiversity. If that loss is to be reversed, strengthened arrangements are required to internalise the benefits of conserving biodiversity.

Terrestrial Ecosystems

The biological diversity of Latin America and the Caribbean constitutes one of its essential features. As noted above, the high rates of deforestation place a large number of species at varying levels of risk, both regionally and nationally. Deforestation has a considerable impact on the region's environment and economies. At the same time, in recent years the pace of deforestation has slowed in some parts of the region, including the Brazilian Amazon forest and Mexico. Thus, recent satellite data suggest that annual deforestation of the Brazilian portion of the Amazon has slowed significantly, from a peak of more than 27,000 square kilometres in 2003–2004 to just over 7000 square kilometres in 2008–2009, a decrease of over 74 per cent. Cumulative deforestation of the Brazilian Amazon is nevertheless substantial, representing more than 17 per cent of the original forest area. In Mexico, deforestation rates have declined: thus, the annual rate of change of the country's forest area fell by 20 per cent over the period 2000–2005, by comparison with the period 1990–2000. In recent years, the Mexican Government has substantially increased the allocation of resources to the forest sector, with a positive impact on sustainable forest management (Fig. 24.6).

Fig. 24.6. *Dendrobates Auratus* (also known as the Green and Black Poison Dart Frog) at Panama Canal forest, Panama.

In 2005, the most recent year for which data are available, Latin America 23 per cent of the world's forested areas, equivalent to some 915 million hectares. The region has both the largest unfragmented tropical forests (Amazon region) and some of the most fragmented and endangered ones. Highly fragmented terrestrial habitats threaten the viability of species and their ability to adapt to climate change. For example, the remaining South American Atlantic forest, largely comprises fragments of less than one square kilometre in area and more than 50 per cent of this forest lies within 100 metres of the forest edge. When ecosystems become fragmented they may be too small for some animals to establish a breeding territory or force plants and animals to breed with close relatives

Inland Waters Ecosystems

Latin America and the Caribbean holds more than 30 per cent of all the planet's available freshwater and some 40 per cent of its total renewable water resources. Although it is one of the world's most water-abundant regions, the resource is distributed very unequally and subject to multiple pressures, including increasing pollution, degradation of watersheds and the depletion and unsustainable use of aquifers as a result of population growth, climate change, social and economic development and societies' increasing interference in the hydrological cycle.

Water quantity and quality are a cause for particular concern in some parts of the region, especially the Caribbean. The subregion faces serious water quantity challenges with a per capita water availability level that measures less than one-third of the global average; this situation has been exacerbated by water pollution. Almost one quarter of the world's inland water fish species are found in Latin America and the Caribbean region: however, human pressures on freshwater resources and water pollution have adversely affected catches, with negative effects for local populations who depend on this source of protein for their nutrition.

The Andes hold 90 per cent of the world's tropical's glaciers, producing 10 per cent of the planet's freshwater. The glaciers of the Andes are a vital source of water for this subregion and, according to the Intergovernmental Panel on Climate Change, most Andean glaciers will melt over the coming 10–20 years. It is expected that many vulnerable communities throughout the subregion will suffer from water shortages. Glacier melting and a drop in the availablility of water are currently two of the major concerns facing Andean countries.

Marine and Coastal Ecosystems

A large percentage of the region's population and its development activities are concentrated in the coastal area. This concentration of population and activities has resulted in increased pressure on these ecosystems, which are being severely degraded. This degradation poses a threat to the very resources that directly or indirectly attracted people to the coastal regions in the first place. The eastern Atlantic coast of South America, the western coast of Central America, and the Caribbean are the worst affected shorelines in the region.

Some of the most degraded ecosystems in Latin America and the Caribbean are mangroves, wetlands and coral reefs, resulting in the loss of valuable ecosystem services, such as sewage treatment by mangrove wetlands systems and the eco-tourism essential for many Caribbean economies. These coastal habitats also play an important protection and stabilisation role. Almost two-thirds of the Caribbean coral reefs are threatened by coastal urbanisation, sedimentation, contamination by toxic substances, water acidification and overfishing.

In the Caribbean region 30 per cent of coral reefs have either been wiped out or are at serious risk. In a business-as-usual scenario it is expected that 20 per cent more will be lost over the coming 10–30 years. Over the past 10 years the marine and coastal ecosystems of the region have provided between 15 and 30 per cent of the world's total fish supply. GEO Uruguay reports that in 30 years' time 90 per cent of fisheries will be overexploited or at maximum capacity. GEO Barbados reports that all fisheries whose status is known are being overexploited.

Species Populations and Extinction Risks

Recent assessments suggest that the immense biodiversity of Latin America and the Caribbean is being lost or seriously threatened by human activities at all levels and practically throughout the region's territory. The region includes 5 of the 20 countries with the highest numbers of species of fauna endangered or threatened, and 7 of the 20 countries whose plant varieties are the most threatened.

The most severe recent increase in extinction risk has been observed among coral species, probably due in large part to the widespread bleaching of tropical reef systems in 1998, a year of exceptionally high sea temperatures. The region is among those with the greatest numbers of tree species in danger of extinction, threatened or vulnerable. Amphibians are suffering the ravages of the chytrid fungus owing to changes in macroclimatic and microclimatic conditions, and similar situations obtain in a variety of groups of organisms. According to the estimations, over the past 100 years Latin America and the Caribbean has lost 75 per cent of the genetic diversity of its agricultural crops.

Genetic Diversity

Latin America and the Caribbean region is the world's richest genetic reservoir, for which ecosystems are vital as sources of new useful traits in food crops, active components for pharmaceutical products, potential industrial (chemical) applications, and useful genes and their corresponding functions. For example, many of the currently most important and widely cultivated crops, such as potato, tomato, cocoa and maize, are native to the region, where they were domesticated by native Americans. Notwithstanding this, genetic diversity in the region is being lost in natural ecosystems and in systems of crop and livestock production. Countries of the region aim to make major strides in conserving genetic diversity, in particular by using *ex situ* seed banks, and the regions is already endowed with several species diversification centres (namely, in Colombia, Brazil, Mexico and Peru).

Access and benefit-sharing and traditional knowledge, associated with genetic resources, are issues of crucial importance for the region and Latin American and Caribbean countries have been very active in the design of the international regime on access and benefit-sharing. The countries have also reaffirmed the importance of respecting the rights of indigenous and local communities over their traditional knowledge associated with genetic resources, and ensuring their participation in the benefits resulting from its use. The future regime will pose many implementation challenges at the domestic level, which will require regional and subregional cooperation. Important incentives for the conservation of biodiversity can emerge from systems that ensure fair and equitable sharing of the benefits arising out of the use of genetic resources.

Current Pressures on Biodiversity

The region presents a generally rising trend in all the five main pressures on biodiversity (land degradation, climate change, pollution from nutrients, unsustainable use and invasive alien species). The third Global Environment Outlook report for Latin America and the Caribbean underlines, however, that the greatest risks to biodiversity stem from the change of land use, with the consequent reduction, fragmentation and even loss of habitats. Land use change has often been unregulated and not been based on environmental criteria. Large tracts of tropical forests (wet and dry), temperate forest, drylands and coastal areas have been — and continue to be — converted to such extent that in Latin America and the Caribbean many species are threatened or endangered.

In Latin America and the Caribbean, the most important driving force of change in the area of land use and habitat loss has been the significant expansion, in recent years, of commercial agriculture for exportation (e.g. soya, biofuels, livestock, fruits, vegetables and flowers) that is responsible for close to one half of the deforestation in the region. Infrastructure, particularly roads, represents an important factor in increasing deforestation rates, mainly in Central and South America. Roads open the way for agriculture to expand, but also for illegal logging—as a rule, deforestation takes place less than 30 kilometres away from an official road. Along the coast and in marine areas, the main pressures come from tourism and unplanned urban expansion, inland contamination and aquaculture. Attempts to safeguard the coastal and marine zone by gazetting protected areas remain very modest. Only 0.1 per cent of the exclusive economic zone of Latin America and the Caribbean countries is under some sort of protection and most of the 255 marine reserves are not properly managed.

The reduction in the pressures on biodiversity is one of the goals pursued by Latin American and Caribbean countries, which aim to integrate biodiversity issues into broader policies, strategies and programs. Governments and local stakeholders in Latin America and the Caribbean region are promoting the application of best practices in agriculture, sustainable forest management and sustainable fisheries. Measures to address the underlying driving forces of biodiversity loss, including demographic, economic, technological, social, political and cultural pressures, in meaningful ways, have to be handled with long-term strategies and short-term and medium-term specific actions.

REGIONAL BIODIVERSITY TRENDS FOR THE TWENTY-FIRST CENTURY

In view of the high vulnerability of Latin America and the Caribbean to climate change, ecosystem-based adaptation approaches and mechanisms (such as reducing emissions from deforestation and forest degradation—known as 'REDD') could support efforts by countries of the region to build essential links and to harness greater synergies between the needs to safeguard biodiversity and to mitigate and adapt to the impacts of climate change. Such efforts will help meet the need for functional approaches designed to improve the flow of resources to those people whose lives are most closely affected by both climate change and biodiversity loss and whose futures stand to be most significantly improved through such efforts.

Because of their valuable biodiversity assets, Latin American and Caribbean countries have a comparative advantage that could be harnessed to stimulate much needed economic growth and social development. There are some clear examples in the region of how biodiversity-related sources, including tourism, timber and nontimber forest products, can produce important revenues. It has been estimated that Guatemala receives over $50 million annually from these sources and Ecuador $100 million from nature-based tourism alone. Indeed, tourism accounts for around 12 per cent of gross domestic product (GDP) in Latin America and the Caribbean, employing approximately 10 million people.

Payment for ecosystem services (for example, the maintenance of a forest to provide water to the supply of a city, reforesting degraded areas to capture the atmospheric carbon dioxide, etc.) is a mechanism that contributes to the creation of green jobs and provides an income to the rural population that preserves and takes care of these services. Countries such as Colombia, Costa Rica and Nicaragua that promoted agro-forestry practices have reported increases of between 10 and 15 per cent in farmers' income levels. Examples such as these suggest that a global shift towards a new economic model could generate large numbers of jobs and help build social equality.

Terrestrial Ecosystems

Protected areas have increased in the region, together with regulatory tools for their integrated management. A number of challenges relating to representativeness, management and resource availability, however, must be overcome for these safeguards to be genuinely effective. Even in conjunction with other strategies to contain biodiversity losses, such as afforestation and community forest management, payment for environmental services and land management and certification, it has not been possible to stem the loss of biodiversity. Biodiversity is endangered by strong pressure exerted on natural habitats. The number of protected areas, however, does not provide a complete picture of the issue. To reduce biodiversity loss, better management of protected areas and more resources are needed. For protected areas to be an effective mechanism for biodiversity conservation, they must be representative of biomes and ecosystems. In addition to protected areas, other conservation techniques must be used, and national and international regulatory and financial structures must be reoriented to internalise the environmental and social cost of the loss of biodiversity or of the benefits of its conservation.

Inland Water Ecosystems

Latin America and the Caribbean will continue to face challenges related to water availability and quality, as the region's growing water demands are exacerbated by a combination of climate change, the introduction of alien species, pollution and unsustainable dam construction, putting further pressure on freshwater biodiversity and the services which it provides. The state of the water ecosystem is directly related to the capacity to promote adequate management of the watercourse and its related terrestrial ecosystem. Accordingly, the countries of the region are strengthening the governance of the integrated water resource management process. To achieve this aim, their national legislation must take due account of three essential elements: participation, transparency and accountability. To that end, effective measures are being implemented to ensure participation by all stakeholders, including economic resources, help in capacity-building among participants and the creation of effective opportunities for interaction. At the regional level, it is also important to promote the integrated management of transboundary water resources, formalising the relevant mechanisms by means of international instruments. The Trifinio plan is an example of an effort to implement transboundary integrated water management in Central America (El Salvador, Guatemala and Honduras), taking into consideration not only the environmental aspect but also the economic, social and political dimensions related to the main shared water basin of the region — the Lempa river.

Coastal and Marine Ecosystems

Combined pressures threaten many coastal ecosystems in the region. The reduction of some forms of stress on coral systems (i.e. land-based contamination and invasive species) may render them less vulnerable to the impacts of acidification and warmer waters. In the case of other coastal ecosystems, planning policies that allow marshes, mangroves and lagoons to migrate inland will make them more resilient to the impact of sea-level rise, and thus help to protect the vital services that they provide.

The sounder management of coastal fisheries can follow a number of paths, including the stricter enforcement of existing rules to prevent illegal, unreported and unregulated fishing. The development of low impact aquaculture, dealing with sustainability issues, would also help to meet the rising demand for fish without adding pressure on wild stocks.

Strategy and Vision for Reducing Biodiversity Loss

Effectively integrating the principles of biodiversity conservation and sustainable use into policies and program is a long and complex process in which the specific features of individual countries and Governments and each type of policy should be taken into account. Some relevant applicable guidelines include the following:

1. Need to improve decision-makers' understanding of the economic and social importance of the biodiversity and ecosystems as part of a country's wealth.
2. Need to lay the foundations for a development model that incorporates the external costs of biodiversity loss, alongside the external benefits of activities that do not damage the integrity of ecosystems.
3. Need to improve the coordination and consistency of public action to guarantee the conservation and sustainable use of biodiversity.
4. Need to strengthen the national and regional systems for data collection, analysis and monitoring related to biodiversity and ecosystems in order to ensure the availability of adequate, structured and comparable data, statistics and information.
5. Need to promote the science-policy interface on the links between biodiversity, ecosystem services and human well-being.

Although Latin American and Caribbean countries have adopted policies and strategies for biodiversity conservation, their biological heritage is threatened by the loss of natural habitats. This is occurring mostly in high mountain areas, drylands, cloud and tropical moist forests and coastal and marine ecosystems.

Some trade-offs between conservation and development are inevitable, and it is important that decisions are informed by the best available information and that the trade-offs are clearly recognised up front. It is more evident than ever that decision-makers should expand their knowledge about the social and economic importance of biodiversity and ecosystems and consider it as a part of the countries' strategic assets. It is also important to expand this knowledge beyond the environment sector and environment ministries to include relevant and related economic sectors such as planning, finance, development, agriculture, fishing, health, transportation, infrastructure and mining. Mainstreaming therefore needs to be seen as the genuine understanding by the government machinery as a whole that the future well-being of society depends on defending natural infrastructure.

The debate at the regional level about the post-2010 targets on biodiversity is also providing impetus to countries to redouble their efforts to conserve biodiversity. It will be crucial for baselines and well-articulated targets to be clearly defined using agreed metrics. An agreed suite of indicators will then enable the monitoring of progress and the early adjustment of key policies and actions based on the assessment of achievements along the way. This work could be greatly assisted through improved access to and sharing of existing biodiversity-relevant data and information, making it more readily and openly available to a wider community of users, including policy-makers. These targets must also be translated into action at the national level though national biodiversity strategies and action plans, and treated as a mainstream issue throughout all sectors of the government.

The implementation of the intergovernmental science-policy platform on biodiversity and ecosystem services is an initiative that could help to reduce biodiversity loss. The need to strengthen the generation of knowledge at the national, regional and global levels and the general agreement on the importance of capacity-building for the generation, assessment and use of knowledge at various levels has been recognised and represents a significant issue for the region. Capacity-building for scientists, policy-makers and members of civil society, including local communities, should be catalysed to enable them to participate more effectively in the science-policy interface and to increase the participation of scientists from the least developing countries.

Changes in the global climate and biodiversity are closely linked. While climate change is a significant driving force in biodiversity loss, healthy biodiversity and resilient ecosystems provide a natural means of adapting to climate change. The concept of REDD is swiftly developing and could contribute to preserve our forests and maintain their ability to regulate carbon. In addition, reducing emissions from deforestation and forest degradation, together with the sustainable management of forests and enhancing forest cover—known as 'REDD-plus'—is undoubtedly of great importance to Latin America and the Caribbean region, and could become an important mechanism for the integration of the sustainable use of biodiversity and climate change adaptation and mitigation measures.

In summary, of all the world's regions, Latin America and the Caribbean has the greatest biological diversity. To reduce the rate of biodiversity degradation, we need a better understanding of the value asset that we are losing. In Latin America and the Caribbean region, habitat alteration and transformation are generating the current biodiversity crisis. Protected areas and conventional conservancy strategies alone will not suffice to mitigate the consequences of this massive change. Biodiversity needs to be properly recognised and valued by political and market systems in order to reverse this trend. At the same time, existing and new sustainable practices need to be applied when producing goods and services, especially to address the needs of the rural poor, who are particularly reliant on the sound functioning of local and regional ecosystems.

Chapter 25

Thailand's Biodiversity

INTRODUCTION

Biodiversity means the variety and variability of living organisms and their ecosystems. Diversity between varieties within a single species is called genetic diversity, which can be easily observed by the existence of hundreds of rice varieties with different pest resistance in Thailand. At a higher level, species diversity can be seen in the difference between the number of tree species contained in a hectare of tropical rainforest (>100 species), mixed-deciduous forest (approximately 31 species) and dry evergreen forest (approximately 54 species). Finally, ecosystem diversity can be indicated in the various types of forest ranging from tropical rainforest to mangrove forest in Thailand.

Thailand is situated in a hot and humid climatic zone which supports a variety of tropical ecosystems. Unlike those in temperate zone, tropical ecosystems provide wider niches for organism's survival and hences, are able to support a much larger variety of plant, animal and microbe species. Thailand has approximately 15,000 species of plant which account for 8 per cent of estimated total number of plant species found globally. These numbers clearly exceed the numbers of plant species in temperate countries such as Norway and Sweden which have approximately 1800 plant species each. Thailand also has approximately 1721 species of terrestrial vertebrate (mammals, birds, reptiles and amphibians) in comparison to 299 and 328 species found in Norway and Sweden, respectively.

Since the Indo–Malaysian region is a centre of distribution of marine organisms, Thai waters have served as habitats for enormously diversified marine organisms. Thai waters support more than 2000 marine fish species, accounting for 10 per cent of total fish species estimated worldwide. Thailand also has approximately 2000 marine mollusk species and 11,900 species of marine invertebrate.

BIOGEOGRAPHY

Thailand is situated within two major biogeographical regions, the Indochinese region in the North and the Sundaic region in the South. Apart from the effect of these two regions within the Indomalayan Realm, some elements of Thailand's flora and fauna are also influenced by biogeographical characteristics of the Indian and Palearctic region. Thailand can be further divided into six biogeographical units which include the restricted ranges of many local and endemic species as follow:

1. The Northern Highland is surrounded by mountain ridges and wide valleys that extend southward from the border with Myanmar and Laos to about 18° North. Prior to anthropogenic disturbance, a number of evergreen montane forests were supported in the area above 1000 metres with mixed deciduous and dry dipterocarp forests on the lower slopes. The valley and upland areas

have now been extensively cultivated especially by hill tribes, resulting in widespread deforestation.

2. The Korat Plateau includes the Northeastern area of Thailand between the Petchabun range in the west and the Donglak range in the South along the Cambodian border. The plateau is now widely deforested with some tropical rainforests and dry evergreen forests persisting in the hills.

3. The central plain of the chao phraya river is now almost entirely cultivated as paddy fields which has completely wiped out the previously existing freshwater swamps and monsoon forests.

4. The southeast upland extends from the Cardamom mountain in Cambodia. Semi-evergreen forests mostly cover the upland's area.

5. The Tenasserim hills extend southward along the border of Myanmar and rise steeply to about 1000 metres above sea level. Even though the hills are situated in the rain shadow of the higher Myanmar side of the range, the hills have supported semi-evergreen forests on the higher elevations. Previously existing deciduous forests on the sides of the hill are now heavily encroached or deforested and have been replaced with grasslands or bamboo forests.

6. The southern peninsula includes the area of Thailand South of Kra isthmus to the Malaysian border. There is a distinct boundary at the Kra isthmus where a considerable number of Indochinese and Malaysian flora and fauna species reach their Southern and Northern limits, respectively. High precipitation in the peninsula was able to support dense rainforests in the area. At present, however, most of these forests in the lowland have been almost entirely cleared for agriculture and those forests on the hills are now threatened by ever-extending encroachment for rubber plantations and other cultivation.

These six biogeographical units have distinct floral and fauna associations. For example, many bird and mammal species of the Northern Highland have Chinese affinities, which cannot be found in other biogeographical units. Similarly, a number of mammal and bird species found in the Southern Peninsula have characteristics related to those of the Sundaic regions.

PLANT DIVERSITY

The majority of plant species in the country are closely related to the species in neighbouring countries. Thus, Thailand could be considered as a collective centre of botanical species from 3 major regional elements: Indo–Burmese element, Indo–Chinese element and Malaysian element.

Approximately 2819 species of Thailand's plants have been recorded in botanical references (Flora of Thailand) and are accounted for estimated 23 per cent of total vascular plant species in Thailand.

It is estimated that there are approximately 12000 vascular plant species in Thailand including 658 species of pterophyte and over 1000 species of orchids. More than 1000 species of vascular plants possess active herbal ingredients used for traditional medicines, not including over 3000 species of mushroom and fungi.

At present, the forest herbarium at the National Park, Wildlife and Plant Conservation Department has collected approximately 2,00,000 plant specimens and approximately 255 specimens are 'type specimens', and all specimens collected account over 80 per cent of vascular plants in the country.

It is expected that, with continuous site-specific sample collections and surveys, more new species can be found. For example, the surveys of plant species in a 50,000 rai area of Pru To Dang peat swamp forest, Narathiwat province, have resulted in the finding of 316 species of vascular plant from 101 families in which 48 species were found for the new record in Thailand. The continuous discovery of new plant species is also demonstrated by the finding of 1100 species of flowering plant in Doi Suthep–

Pui National Park up from 679 species found in a survey in 1964. At present, the continuous researches have now found up to 2247 species of vascular plants.

It is estimated that approximately 80 per cent of 1000 well-known plant species, such as fruits, flowers and vegetables, are introduced (the remaining 20 per cent are indigenous). This clearly shows the lack of knowledge of plant resources among Thai researchers which, sometimes, results in the introduction of ecologically harmful weed species such as *Mimosa pigra*, *Eichhornia crassipes*, *Eupatoriam odoratum*, *Peninstum polystachyon*, etc.

Excluding thallophytes and bryophytes, Thailand's plant species comprise of approximately 303 families, 1363 genera and 10,234 species which can be divided as follow:

1. Ferns comprise of 658 species from 132 genera of 34 families. These species have been completely revised.
2. Gymnosperms comprise of 25 species from 7 genera of 6 families. These species have been completely revised.
3. Angiosperms comprise of approximately 9551 species from 1224 genera of 263 families. There are 2136 species from 705 genera of 109 families have been completely revised.

As mentioned earlier, Thailand does not have significant unique floristic elements of its own which results in the relatively low number of endemic species. From the study conducted under the Flora of Thailand Project, there are 248 species from 94 genera of 43 families of endemic plant species including 24 species (from 22 genera of 13 families) of fern, 224 species (from 72 genera of 30 families) of angiosperm which can be divided into 25 species (from 8 genera of 6 families) of monocotyledon and 199 species (from 64 genera of 24 families) of dicotyledon.

ANIMAL DIVERSITY

From the records, there are 294 species of mammal in Thailand of which 42 per cent originated from the southern part of the region, 34 per cent from Indochinese or Indo–chinese and Indian sub-region and the remaining 24 per cent are species that distributed throughout the Asian continent. Five of these mammal species are endemic to Thailand. Many of the mammal species found in Thailand are bats which account for 38 per cent (108 species) of the total number of mammals found. Eighteen species of these bats are frugivores while 89 species are insectivores with 1 carnivorous species. The second most common mammals are in Order Rodentia which account for 25 per cent.

Regarding avifauna, a Guide to the Birds of Thailand has indicated the finding of 915 bird species in 1998 of which 2 species, White eyed-river Martin (*Pseudochelidon sirintarae*) and Deignan's Martin (*Stachyris rodolphei*) are endemic to Thailand. The publication has also suggested that an additional 62 bird species may be found in the future since these species have already been recorded in neighbouring countries. The prediction made in the book is proving correct with the total finding of 923 species. At present, the continuous researches have now found up to 942 species of birds.

It is confirmed that there are 325 species of reptile in Thailand. The majority of these species are snakes, which account for 54.15 per cent. Second most common species include the organisms belong to gecko, lizard and skink groups and account for 36.6 per cent of total species found while crocodiles account for the least numerous species (3 species). There are 27 turtle species found in Thailand (of 257 species found globally), which comprise of 3 species of land turtle, 1 species of big-headed turtle, 13 species of freshwater turtle, 5 species of trionyx and 5 species of sea turtle. Thirty one species of reptile are endemic to Thailand.

One hundred and forty-one species of amphibian are found in Thailand of which 134 species or 95.05 per cent are frogs. Only one species of salamander found in Thailand.10 amphibian species are endemic to Thailand.

Presently, at least 570 species of freshwater fish have been reported in which 56 species are endemic to Thailand. 1160 species of fish have been found existed in estuarine and seawater. Additional 30 species are deep-sea fishes. Of all marine fish species, 78 species are cartilaginous fishes and 1664 species are bony fishes.

Of other marine organisms, there are 2 species of horse-shoe crab and Giant-king crab. One hundred and eighty-three species of marine shrimp are found in the Gulf of Thailand. There are 1538 species of marine shell-fish in Thai waters (634 gastropods and 382 bivalves). Twenty-eight species of cephalopod are found in the Gulf of Thailand including 11 species of squid, 7 species of cuttle-fish, 5 species of octopus and 5 species of dwaft cuttle-fish.

There are enormous numbers of insect species in Thailand especially hard-shell (wing) and Hawk moth species. However, the extensiveness of knowledge on the species is very limited in comparison to the amount of the species in Thailand. From data of the Department of Agriculture, 7000 species of insect are known which account for only 10 per cent of insect samples collected by the department. It's certain that the collections contain many fewer species than those existing in nature and 90 per cent of the collections have not yet been or cannot be identified. Currently, list of endangered species in Thailand comprises of 2 species of amphibia, 15 species of reptile, 69 species of bird and 34 species of mammal.

Many animal species have been bred and domesticated for various purposes. For example, 19 species of mammal have been utilised for meat production or as labour in traditional agriculture. Some species have also been used for industrial, scientific and medical purposes.

Birds account for the majority of domesticated animals in Thailand. 96 species of these bird species, such as parrots, and red-breasted parakeets, are imported from overseas. There are 11 true domesticated birds that have been utilised for a long time for local consumption and as pet animals.

LOST OF BIODIVERSITY

Many biologists have concluded that the earth is losing at least 21,000 animal and plant species in tropical forest per year. They have also found that there is also the reduction in biological diversity in other natural ecosystems such as coral reefs, wetlands, islands and mountainous areas, even though the loss in these ecosystems are still collectively less than the loss in tropical forests.

When evaluating the loss of species from every natural ecosystem, it appears that, at present, the overall extinction rate is more than 30,000 species per year. From available evidence, it is found that, prior to the appearance of Homo sapien, the average extinction rate was 1 species per 4 years. Thus, the present extinction rate may be 1,20,000 times higher than the rate in prehistoric times and clearly indicates humans as a major cause of the increase in the extinction rate.

The biologists expect that, without significant conservation effort, earth will loose 20 per cent of species within the next 30 years and 50 per cent by the end of the next century.

Even though extinction is a natural process, the high rate of extinction can be considered as unnatural phenomena and indication of serious crisis facing every living organism on earth. For humans, the conservation of biodiversity has greater meaning than preservation of specific species or ecosystems. Conservation is also the means to ensure the existence of the human race itself since the conservation of biodiversity is a direct measure in preserving food, medicine, and other necessary resources required for human survival.

Schombergk's Deer (*Curvus schomburgki*) has been extincted from Thailand (and the surface of the earth) since 1942. In addition, Giant Ibis (*Pseudibis gigantea*) and Large gress Warbler (*Graminicola bengalensis*), have now been confirmed extinct. Furthermore, Javan Rhinoceros (*Rhinoceros soundaicus*), Black Ibis (*Pseudibis papillosa*), Kouprey (*Bos saureli*), Milky Stork (*Mycteria cinerea*), Sarus Crane (*Grus antigone*), and False Gharial (*Tomistoma schlegelii*) can no longer be found in natural environment.

5 species of freshwater fish have now been extincted from Thailand. These species are *Balantiocheilus melanopterus*, *Platytropius siamensis*, *Cyclocheilichthys lagleri*, *Longiculture caihi* and *Oxygaster williaminae*.

There are approximately 30 endangered species of freshwater fish in Thailand. Most of these species have been captured for food or to supply aquarium business. *Catlacarpio siamensis*, *Hilsa toli*, *Cirrhinus microlepsis*, *Ceratoglanis scleronema* and *Pangasius sanitnougeei* are species captured for consumption, while *Botia sidthimunlei*, *Tetrodon baileyi*, *Dadnoides microlepis*, *Notopterus blanci* and *Scleropages formosus* are popular species for aquarium.

The last dugong group in Thailand is found in Choa Mai beach area and Talibong island, Trang province. The population of wild elephants in the natural environment is now greatly reduced and is causing alarming concern among most conservationists. There are only 1975 elephants left in the wild and they can only be found in 47 protected areas.

Similar to the faunal organism, many plant species in Thailand are now placed on the endangered or rare list. Thailand was once world famous for the presence of over 1000 species of orchid. At present, however, a number of local orchid species such as *Paphiopedilum niveum*, *Paphiopedilum sukhakuluii*, *Rhynchostylis coelestis*, *Rhynchostylis gigentea*, *Vanda coerulea*, *Vanda denisoniana*, *Dendrobium scabrilingue*, *Dendrobium tortile*, etc. are now endangered. *Vatida diospyroides* which is a large tree specie with pleasant odour flowers is now regarded as a rare species. In Thailand, there are currently, about 107 endemic plant species, 400 endangered species and 600 rare species.

The loss of crop, and pastured domesticated animal varieties may seem to be insignificant when considering the overall status of biodiversity. However, the loss of these varieties, which are responsible for food production, is indeed a great concern for enhancing performance of varieties through selective breeding, especially when the demand for better varieties is increased by the continued reduction of cultivated land.

The earliest rice cultivation is believed to take place in where is now Thailand. However, many of wild rice species in Thailand are now extinct. At present, a limited number of the 2,00,000 rice varieties existing worldwide are cultivated in Thailand. Some of the wild fruit species such as wild durian and Wild mangosteen now appear to be extinct. There is also a report suggested that the remaining population of local pig species, i.e. Raad, Hailum and Kwai, are now very small and may soon be extinct.

CAUSES OF THE BIODIVERSITY LOSS

The causes of the reduction of biodiversity are usually over-exploitation, illegal trading of animal and plant species, disturbance to natural habitat and the loss of habitat. Biological resources are regarded as renewable resources since the resources or organisms are able to reproduce and, hence, continue supplying the demands of humans. In the past, the over-hunting of wildlife had resulted in the reduction in both populations and variety of wildlife.

Illegal trading of wildlife is another direct threat to biodiversity. The demand for rare animal and plant species has pushed the price of these species through the ceiling and resulted in extensive hunting and gathering of these species for export as well as to satisfy local consumption. These practices have caused the rapid decline in a number of wild animal and plant populations and, have wiped out some species. The most serious threat to the biodiversity is human disturbance. The disturbance to natural habitat and ecosystem such as forestry land reform of both evergreen and mangrove forest, construction of water reservoirs and hydro-electric dams, urbanisation, tourism and pollution have all threatened and contributed to reduction of wildlife populations.

Over-logging is one example of over-exploitation of biological resources. In the past, the logging concession process did not provide incentives for concessionaires to sustainably utilise the forest resources and thus resulted in the logging at a rate that exceeded the growth of the forest through reforestation. This is mainly because the reforestation is not efficiently operated due largely to rapid change in forest land caused by the clearing of forest land by local population following the logging of large and medium trees by concessionaires. The cleared lands are usually converted into agriculture lands which phase out any further reforestation programs in the areas.

FUTURE PROSPECT

The loss of biodiversity in Thailand will continue over a certain period of time in the future. There is also an indication suggesting that rate of the loss will not slow down even though, a significant amount of mitigation actitivites have been implemented by private and public agencies/organisations. Institutional based activities to conserve forests, marine ecosystems and freshwater environments will continue to require efforts to conserve biodiversity.

The continuous loss of biodiversity in Thailand at such an alarming rate is mainly caused by the lack of social awareness and consciousness in preserving natural resources for the coming generations.

The lack of awareness may largely due to the fact that Thai society does not have enough information on the loss of biodiversity and, hence, is unaware of the problem or does not fully provide obvious signs of the problem to the youth.

The efforts by biodiversity related institutions have yet to successfully yield beneficial results since these institutions have not been sufficiently supported either financially or institutionally. At present, there are only about 30 taxonomists in Thailand even though taxonomic researches are an extremely crucial element in better understanding of biodiversity. The research have also not been carried out under initiative of the institutions but rather by individual interests which clearly indicates shameful insufficient and unorganised support to biodiversity research in Thailand. There is also the lack of proper amount of training programs for local authorities assigned for protected area. Institutionally, plans and policies on the conservation of biodiversity by biodiversity-related institutions have never, at any significant extent, been integrated into the policies and plans for utilisation of natural resources. For example, the planting of identical species in both reforestation and agriculture projects by governmental agencies has increased pressure on the diversity of natural forests. Many other governmental agencies also consider the issue of biodiversity as mere 'theory' and continue to undertake development activities that wastefully destroy biological resources, such as transportation, energy production and irrigation, only to satisfy short-term economic return. Some other agencies, in response to accelerated economic development, even promote and support the harvesting of natural resources with no regard to the long-term existence of such resources.

The prospect of biodiversity in Thailand is very much a crisis: The present conservation efforts have not been able to hold back the alarming rate of biodiversity loss. There are still many factors supporting the wasteful use of biodiversity while supporting factors for the conservation and sustainable utilisation of the biodiversity are still greatly insufficient.

Chapter 26

Biodiversity in China

INTRODUCTION

Biodiversity or biological diversity is the total sum of life's variety in a region or the world. It deals with three levels species diversity, genetic diversity and ecosystem diversity. Species diversity is the variety and abundance of all living organisms existing on earth. The total number of species has been estimated to range from 30 to 50 million. Genetic diversity relates to the sum total of genetic information contained in genes of individual organisms inhabiting the earth. Ecosystem diversity refers to the variety and abundance of different ecosystems, including the variety of habitats, biotic communities and ecological processes in the biosphere.

Biodiversity builds our life-support system and forms the fundamental basis for human life, development and prosperity. In no case can human being exists in isolation from the rest of the natural world. It provides us with food, medicines, timber, ornamental plants, oils, gums, fibres and industrial materials. The wild species is the genetic base from which new varieties are bred. The basic food of human beings, such as crops, fruits, poultries, livestock, etc. derived from wild species. Although genetic engineering or the biotechnology of gene recombination can be used to create new genetic materials for human needs, still it depends greatly on those natural products from native plants and animals.

Modern scientific research proves that the variety of species are indispensable parts of ecological system. Extinction of certain species may result in imbalance of systematic functions and even the disintegration of whole ecosystem.

Plants play a very important role in soil and water conservation, prevention of winds and sand erosion, protection of sand dunes, regulation of climate, purification of air and afforestation of environment.

Since we depend entirely on biological diversity, the only source of sustainable productivity, we must have a clear and direct interest in understanding it using it wisely, and preserving it. Other forms of life deserve human respect and should be allowed to proceed in harmony and balance with man. It is unwise and wrong to over-exploit biological resources for the benefits of present generation.

At present, the precise role that each species plays within the life-support system is unknown. It is unwise to interrupt its natural ecological balance by reducing its diversity. Biodiversity has immense potential value to human being. Some species might be used as our future food supplies or medicines. With the destruction of nature ecosystems, we are loss forever the ability to gather knowledge about large proportion of living organisms, because they decline so rapidly. Due to deforestation, habitat change, over-exploitation, indiscriminate use of pesticides and other human impacts, we are now experiencing the most rapid extinction rate since the end of the Cretaceous Period. It is estimated that current rate of extinction is over 1000 times greater than the natural rate of extinction. Conserving biodiversity means leaving more options for future generations to adapt to the possible climatic and environmental changes (Figs. 26.1–26.4).

Fig. 26.1. Golden Pheasant (*Chrysolophus pictus*) threatened due to uncontrolled harvesting.

Fig. 26.2. *Parakmeria omeiensis*, a rare and critically endangered species with only three trees surviving.

HABITAT AND ECOSYSTEM DIVERSITY

China has a vast territory with complex climate, varied geomorphic types, a large river network, many lakes and long coastline. Such complicated natural conditions inevitably form diversified habitats and ecosystems. The terrestrial ecosystem can be divided into several types such as forest, tundra, marsh, etc. The aquatic ecosystem can be classified as marine, rivers and lakes (Fig. 26.5).

Fig. 26.3. *Bos frontalis*, 'Mithun', a semi-domesticated animal in N.W. Yunnan and considered to be an important genetic resource.

Fig. 26.4. Bearded Vulture (*Gypaetus barbatus*) a rare and endangered species inhabiting the plateau in W. China.

Fig. 26.5. Tropical seasonal rainforest dominated by *Tetrameles nudiflora* on limestone in Xishuangbanna.

Forest

The forested area of China is small with unbalanced distribution and variety of types. In total, it is approximately 11.5 million ha with an average coverage about 12 per cent. The proportion of forest coverage ranges from 55 per cent to 4$ in different provinces.

Forest mainly has 212 formations indicated by the dominant species, codominant species or characteristic species in arbor layer. The bamboo forests of China are equally rich, with 36 formations. The shrublands are fairly complicated with 113 formations.

The forest of China can be roughly classified as coniferous forest, broad-leaved deciduous forest, and coniferous and deciduous mixed forest, etc.

The coniferous forest in China consists of taiga (44 types), warm temperate coniferous forest (5 types) and subtropical and tropical coniferous forest (27 types).

The temperate coniferous and deciduous broad-leaved mixed forest is mainly distributed in Northeast China.

There are 42 main types of broad-leaved deciduous forest which are widely distributed over hilly areas and mid or lower mountainous areas in the temperate zone, the warm temperate zone and the subtropical zone of China. The broad-leaved deciduous forest is the zonal vegetation of the warm temperate zone.

The broad-leaved evergreen forest is composed of many broad-leaved evergreen tree species in the subtropical zone. Animals inhabiting this forest are in imminent danger of extinction due to habitat destruction.

Tropical seasonal rainforest and tropical rainforest covers a small area of Southern China. The 24 main types are dominated by more than one characteristic species. Tropical forests have been seriously destroyed especially on Hainan Island. To protect the remaining tropical forest is therefore a priority of conservation of ecosystem biodiversity in China (Fig. 26.6).

Fig. 26.6. Evergreen broad-leaved forest in Fanjingshan Mountain.

Meadow

The meadows are those communities which developed under appropriate moisture conditions. The dominant species are mesophytes and perennial plants. They can be divided into several types: typical meadows (27 formations), saline meadows (20 formations), marsh meadows (9 formations) and high cold meadows (21 formations). In total, there are 77 meadow formations (Fig. 26.7).

Fig. 26.7. Meadow-steppe in the Inner Mongolia.

Steppe

Steppe consists of perennial xeric herbs, occurring from temperate to tropical zone. It is an ecosystem developed under certain hydrothermal conditions of semi-humid and semi-arid region. There are 45 formations of steppe, roughly classified as meadow steppe, typical steppe, desert steppe and high cold steppe. The steppe can be found in temperate semi-arid zones, such as the Qinghai-Xizang (Tibet) Plateau and the mountainous areas of arid regions with dominant species of *Stipa*, *Festuca*, *Aneurolepidium*, *Cleistogenes* and *Artemisia* respectively. The total area of temperate steppe in China is 315 million ha. Due to over-exploitation and over-grazing, the steppe have deteriorated and the degenerative area has been estimated at 30 per cent of the total.

Savanna

Influenced by the warm air mass from the South China Sea and Indian Ocean, the tropical regions are covered with tropical forests, while the arid savanna can only be found in the xerothermic valley in the southern Yunnan and some parts of Hainan Island. In addition, some tropical forests felled repeatedly become secondary savanna.

Desert

The desert covers a total of 20 per cent of landmass and is mainly found in northwestern region of the country. There are 52 formations of desert (small wood desert, shrub desert, small semi-shrub desert and cushion-like small semi-shrub desert).

Marsh

It is a kind of wetland composed of helophytes birds, fishes and waterfowl. There are approximately 19 marsh formations which can be recognised as follows: herbaceous marsh 14, woody marsh 4 and peatbog one. Chinese mangrove (a tropical marsh forest), has 18 formations. The total area of marsh is about

11.5 million ha. It occurs in the mountainous area of Northeast, Sanjiang Plain and Qinghai-Xizang (Tibet) Plateau. The largest marsh is the Sanjiang Plain, Heilongjiang Province, now mostly reclaimed and turned into farmland. The remaining 2.27 million ha. are protected to some extent through establishment of some reserves. In addition, there are 17 formations of tundra, alpine cushion-like and alpine mobile sand vegetation, with small distribution area.

Freshwater and Marine Ecosystems

Freshwater and marine ecosystem are homes to a tremendous diversity of fish, amphibians, invertebrates, aquatic plants and micro-organisms. The richness of biological diversity in coral reef could be sometimes compared with that of tropical forest. Marine ecosystems are far more diverse than terrestrial ones at higher taxonomic level, and marine organisms are highly diverse at the genetic level. Freshwater habitats are relatively discrete units causing hydro-biodiversity highly localised. Accordingly, freshwater ecosystems contains very high degree of endemism. Even though biodiversity is very high in freshwater and marine ecosystems, they are relatively little known. Scientists believe that the deep-sea floor may contain as many as a million undescribed species. Among numerous rivers, streams and lakes in China, there are 22 rivers longer than 1000 km and 2848 lakes larger than 1 km^2. China is one of the largest producers of freshwater fish in the world. The silver carp (*Hypophthalmichthys molitrix*), bighead (*Aristichthys nobilis*) and grass carp (*Ctenopharyngodon idellus*) are well-known aquacultures. China is also rich in marine fish. There are 1694 species recorded from China seas, consisting of 175 chondrichthyes and 1519 teleosts. From the total of 2804 fish species recorded, 440 are endemic.

SPECIES DIVERSITY

Megadiversity

China is a vast country with complex topogeography and climate. It crosses frigid, temperate and tropical zones from north to south. The plateau and high mountains occupy over 50 per cent of land. Biogeographically, China is situated in both the Palaearctic and Oriental Realms. Meanwhile, during the late Tertiary, most regions had not been effected by glaciation, thus the fauna and flora is characterised by having many endemic and relic species. Therefore, it is considered internationally that China is one of the megadiversity countries in the world, where the number of species, as a whole, make-up about one tenth of the total number of species of the world (Figs. 26.8 and 26.9).

Fig. 26.8. Wild Yak (*Bos grunniens*) threatened in the wild in Qinghai-Xizang (Tibet) Plateau.

Fig. 26.9. Tibetan Antelope (*Pantholops hodgsoni*), an endemic species of the Qinghai-Xizang (Tibet) Plateau but threatened by poaching and habitat destruction.

Surveys of animals, plants and cryptogam revealed that some differences still exist in the studies of different populations. New taxa and new records are gradually being published. Through 1980–1986, the number of new species of insects and angiosperms have increased by 500 respectively. The invertebrates and cryptogams are only a little known which accounts for the difference between known and actually existing number of species. For example, insects account for 80 per cent of the species known in the animal kingdom worldwide. However, insects in China account only for one-tenth of the total number of the world. The number of species which had been named does not exceed 40,000, which is about one fourth of the total. Approximately 70 per cent of insect species remain to be described. On the other hand, the number of species of mammals, birds, reptiles, amphibians and fishes, as well as mosses, ferns, gymnosperms and angiosperms are more or less better documented (Table 26.1).

Table 26.1. Numbers of Species in China and the World.

Taxa	Spp. of China (SC)	Spp. of world (SW)	SC/SW(%)	Estimated nos. in the world
Mammals	499	4000	12.5	5000
Birds	1186	9040	13.1	11,000
Reptiles	376	6300	6.0	
Amphibians	279	4184	7.0	
Fishes	2804	19,056	12.1	28,000
Insects	40,000	7,51,000	5.3	15,00,000
Bryophytes	2200	16,600	13.3	
Pteridophytes	2600	10,000	26.0	
Gymnosperms	200	520	37.8	
Angiosperms	25,000	2,20,000	11.4	
Fungi	8000	46,983	17.0	15,00,000
Bacteria	500	3060	16.3	30,000
Algae	5000	26,900	18.6	60,000

Endemism

A species distribution at certain limitation is known as specific phenomenon. For example the monotypic genus of *Cathaya argyrophylla* only occurs in southern parts of China, the Yangtze River Dolphin (*Lipotes vexillifer*) only lives in lower reaches of the Yangtze River, the Giant Panda (*Ailuropoda melanoleuca*) is confined to the S.W. mountainous region, Sichuan, Gansu and Shaanxi. These are the endemic genera and species in China (Fig. 26.10).

Fig. 26.10. *Cathaya argyrophylla*, monotypis genus, endemic to China.

In addition, some species, e.g. Cryptomeria found in China as well as in Japan, are regarded endemic taxa to east Asia. Therefore, the term endemism means a taxon with its distribution only in a specified region as compared with those taxa occurring throughout the world. Study of endemism is important in understanding the characteristics and formation of fauna and flora, and also in determining the priorities of conservation and sustainable use of biodiversity. A number of endemic taxa including higher ones are recognised and the main endemic taxa in China are shown in following Table 26.2.

Table 26.2. Number of endemic genera or species in China.

Taxa	Known gen. or spp.	Endemic gen. or spp.	%
Mammals	499 spp.	73 spp.	14.6
Birds	1186 spp.	99 spp.	8.3
Reptiles	376 spp.	26 spp.	6.9
Amphibians	279 spp.	30 spp.	108
Fishes	2804 spp.	440 spp.	15.7
Bryophytes	494 gen.	8 gen.	1.6
Pteridophytes	224 gen.	5 gen.	2.2
Gymnosperms	32 gen.	8 gen.	2.5
Angiosperms	3116 gen.	232 gen.	7.4

Status

In the process of evolution, the speciation and extinction of species were more or less kept on par. Today, however the extinction is greater than evolution of new species. Due to human interference as well as loss of natural habitat, biological resources are being exhausted at an alarming speed. It is reported that two species of birds become extinct every three years and, by the year 2015, this could reach the level of one species every year. It is estimated that by the end of this century, there will be 50 or 60 thousand plant species becoming threatened in various degrees, and at present the extinction of plant species goes at the rate of one species every day worldwide. In that case, half or one million species of animals and plants may become extinct within next two decades. The present few million species are the modern-day survivors of several billion species that have ever existed. All past extinction occurred by natural processes but today human interference is responsible for rapid extinction of species. Scientists have conducted a series of surveys on biotic and natural resources, accumulating valuable materials. A rough estimation shows that in China about 398 vertebrate species are endangered amounting to 7.7 per cent of the total vertebrate. In plants, the rare and endangered species are as follows: Bryophytes 28, Pteridophytes 80, Gymnospermae 75, Angiospermae 836, in total 1019 species, amounting to 3.5 per cent of the higher plants (Table 26.3).

Table 26.3. Estimated number of Endangered species in China

	Taxa	*No. of End. spp.*
Vertebrate	Mammals	94
	Birds	183
	Reptiles	17
	Amphibians	7
	Fishes	97
	Total	398
Higher plant	Bryophytes	28
	Pteridophytes	80
	Gymnosperms	75
	Angiosperms	836
	Total	1019

It is estimated, however, that there are 10 per cent of plant species endangered or vulnerable in temperate zone, while in tropical and subtropical zones it is much higher. Owing to heavy deforestation, the estimated proportion of endangered or vulnerable plant species in China is about 15–20 per cent, i.e. about 4000–5000 species of higher plants (Fig. 26.11).

Based on the records of the extinction of plants and animals in Chinese history, animals species of *Rhinoceros* sp., *Elaphurus davidianus, Saiga tatarica, Panthera tigris lecoqi, Pygathrix nemaeus*, as well as plant species of *Thuja sutchuensis, Ombrocharis dulcis, Machilus minutiloba*, etc. disappeared for decades or even centuries. It is believed that the Saiga (*Saiga tatarica*) extirpated in Xinjiang just after 1950's (Fig. 26.12).

The species threatened at the brink of extinction are: *Nipponia nippon, Panthera tigris allaica* and *P.t. amoyensis, Neofelis nebulosa brachyurus, Ailuropoda melanoleuca, Presbytis* spp., *Hylobates* spp., *Dugong dugung, Cervus eldi, Lipotes vexillifer, Archinecttia gaudissartii, Diplandrorchia sinica, Cycas hainanensis, Cepha.lotaxus mannii, Panax zingiberensis, P. ginseng, Gastrodia elata, Cistanche deserticola, Boschniakia rossica, Paeonia suffruticosa* var. *papaveracea,* etc. Among these species, the Crested Ibis (*Nipponia nippon*) was still abundant in a rather wide range in the 1950's but only a small population was found in the late 1970's.

Fig. 26.11. Dioecism of *Cycas panzhihuaensis* (Cycadaceae), found in 1970's in Guizhou, northern recorded limit of the family.

Fig. 26.12. Crested Ibis (*Nipponia nippon*) a critically endangered bird of the world with its distribution confined only in a small area of the Qingling Mountain and its population size estimated no more than few dozens.

There are difficulties in estimating the endangered species of many groups such as insects, snails and other invertebrates, fungi and mosses, etc. Many of them are not described. There might be many more endangered species in lower plants, but at present even an estimate numbers are unavailable (Figs. 26.13–26.19).

Fig. 26.13. *Paphiopedilum armeriacum*, an extremely rare and precious ornamental plant, found in 1980's.

Fig. 26.14. *Bretschneidera sinensis*, monotypic genus, confined to China.

Fig. 26.15. *Camellia chrysantha*, 'Golden Hue', a rare species of the genus.

Fig. 26.16. A *diantum reniforme* var. sinensis, discovered recently for the first time on the continent of Asia.

Fig. 26.17. Golden Monkey (*Rhinopithecus r. roxellanae*) an unique monkey confined to China but also threatened by poaching and by habitat loss.

Fig. 26.18. Siberian Tiger (*Panthera tigris altaica*) critically endangered in China.

Fig. 26.19. Chinese Monal (*Lophophorus lhuysii*), an endangered species and one of the 1st Category Protected Species in China.

Threats

Habitat destruction

The main cause of extinction is destruction or loss of habitat, whether it is through land reclamation, an excessive forest felling or draining of fresh and salt marsh. For example, Hainan Island had 25 per cent of natural forest in 1956, however it amounted to 7.2 per cent in 1983. Large numbers of animals and plants become endangered and some extinct, such as *Cephalotaxus mannii, Firmiana hainanensis, Paranephelium hainanensis, Sonneratia hainanensis*. The coral reef in the South China Sea has been harvested excessively, resulting in disappearance of the coral reef fish through destruction of their habitat (Fig. 26.20).

Fig. 26.20. Destroyed tropical rainforest in S. Yunnan.

Overexploitation

In 1950's, the macaque was captured in large numbers, and combined with habitat loss, the population declined on a large scale and have not yet recovered. Animal resources such as gazelles, deer, fur-bearing animals as well as freshwater fish have declined because of over-harvesting of fishing.

Pollution

The freshwater in China is seriously polluted by industrial waste water causing major decline of aquatic fauna and flora. Serious pollution occurs in the sea and near the seashore from oil spillage. Some aquatic species are endangered while others became extinct such as Grass Carp, Silver Carp, Variegated Carp, Crucian Carp, Catfish, Pike, Black Carp and Triangular Bream, etc. in the Songhua River. Air pollution, acid rain and pesticides should be considered as important factors of species decline.

Besides, introduction of exotic species, the construction of new cities, dams and reservoirs, mining industries, as well as natural disasters such as earthquake and volcano eruptions, floods, forest fire, snowstorms, droughts, animal and plant diseases, insect pests, etc. are also threatening causes of plant and animal species reduction.

In conclusion, biological resources are renewable and will serve mankind forever only if they are properly conserved and rationally utilised.

GENETIC DIVERSITY

Due to human encroachment the habitat for animals and plants is continuously diminishing. The number of animal populations and plant species is decreasing. This creates such problems as inbreeding, genetic drift, bortleneck effect, gene flow break, i.e. reduction of genetic diversity or genetic exhaustion. Thus the conservation of genetic diversity is a pressing issue.

Macaca mulatta, a common primate in China, possesses rich differentiation of subspecies. Different geographical populations and subspecies show different mtDNA polymorphism. The subspecies in North China has a unique type of mtDNA, i.e. unique genetic diversity and it should be considered as one of the major primates for protection. In 1988, a new species of deer (*Muntiacus gongshanensis*) was discovered in north-western Yunnan. Its chromosomes ($2n = 8$, female; 9, male) are large and easy to distinguish.

China with its ancient civilisation and a history of several millenia has a wide variety of different domestic animal breeds and crops which are an important component of global bank of genetic resources. In addition, China is one of the original centers of crop plants in the world. Many important crops originated here, such as soybean, rice, barley, tea, apples, etc. Besides these, there are many wild-related species and ancestral forms of domestic animals and cultivated plants, such as Red Junglefowl, Wild Ox, Wild Quail, wild soyabean, wild barley and wild rice. Only in Xishuangbanna area are more than one hundred wild species of cultivated crops. Gayal (*Bos frontalis*) is a semi-domesticated large cattle occurring in the Dulong River area. It is well adapted to local harsh conditions and can be easily reared in tropical and subtropical mountainous regions as a potential and valuable animal resource (Fig. 26.21).

Fig. 26.21. *Oryza sativa*, 'Wild Rice', Yunnan.

The wild rice discovered in Southern China are abundant (*Oriza sativa, O. meyeriana, O. officinalis,* etc.). These plant germplasm resources are valuable in modern plant cultivation and gene transfer biotechnology.

Conservation of genetic diversity of animals and plants, has been initiated recently mainly in collection, preservation and utilisation of genetic resources as well by using modern biotechnology. The Institute of Crop Germplasm Resources of the Chinese Academy of Agricultural Sciences has collected and preserved a large number of seeds of cultivated crops and their allied wild forms. The Institute of Cell Biology and the Kunming Institute of Zoology of Chinese Academy of Sciences (KIZ) have established cell and gene banks. This 'frozen zoo' of KIZ contains cells and tissues of more than 200 species of animals including sperms and ova of some rare and endangered animals.

RECENT PROGRESS

Survey and Bioinventory

Since 1950's, large-scaled surveys on Chinese flora and fauna were sponsored by the Chinese Academy of Sciences (CAS) and then followed by some colleges, universities and other institutions. The surveys covered nearly the whole country including islands and surrounding seas. About 20 million specimens of animals and in the institutions of CAS. Presently, computer inventories of systematic collections are being prepared. Many publications have been published, such as volumes of fauna and flora of China or regions, and taxonomical papers including descriptions on new taxa. However, due to limited funds and manpower, still a large number of species or higher taxa need to be investigated and inventoried. For example, of the 1,50,000 species of insects estimated in China, less than 40,000 have been recorded, only about 27 per cent of the total. At the same time, much work has been carried out on the investigations of Chinese vegetation and plant or animal geography.

Species Conservation

Wild plants and animals are increasingly threatened by population explosion and economic development. As a result, many animals and plants, especially the rare and endemic species, as well as those of great economic value, become endangered and threatened. Since the 1950's, regulations for protection of plants and animals have been enacted but they are not sufficient to stop the decline of the species. The Environmental Law, Forest Law, The Wildlife Protection Law and other related legislations have been issued for preserving species and their habitats, but their implementation and enforcement need to be strengthened (Fig. 26.22).

Fig. 26.22. Chinese Alligator (*Alligator sinensis*) an endangered species but successfully bored in captivity in a breeding center located in Anhui Province; wild population has increased in reserves too.

Meanwhile, the compilation and publication of *China Red Data Books* are in progress. The recently published *China Plant Red Data Book* Vol. 1 contains in the first version 388 species of endangered and rare plants.

Endangered species conservation and research work started in the 1950's. Research on Chinese Alligator was among the earliest projects. The program resulted in establishment of reserves and breeding centers based on ecological and biological research. The project was internationally recognised and acclaimed.

Surveys and research work on the flagship species, the Giant Panda, was initiated in the late 1960's and expanded and intensified in the past twenty years. These programs lead to establishment of a number of reserves as well as breeding centers. Currently, research initiatives on ecology, behaviour, reproductive biology and population genetics are still in progress. Conservation projects on other critically endangered species such as the Yangtze River Dolphin, Guizhou Snub-nosed Monkey, Crested Ibis, etc. have also been conducted and some of them were jointly sponsored with NGOs of the world.

In Situ Conservation-Reserves

Since the first reserve (Guangdong Dinghushan Mountain Nature Reserve) was established in 1956 by Chinese Academy of Sciences, a network of reserves at national and local levels was developed. Among them, the majority concentrates on the conservation of forests and wildlife, while others preserve different types of wetland, high plateau, grassland, desert, small island, and coral reefs. For instance, there are altogether 13 reserves established mainly for protecting the Giant Panda, Golden Monkey, Takin, etc. in Sichuan, Gansu and Shaanxi provinces (Table 26.4).

Table 26.4. Panda reserves.

Name	County	Province	Year established	Area (sq.km)
Foping	Foping	Shaanxi	1978	350
Baishuijiang	Wenxian	Gansu	1978	953
Baihe	Nanping	Sichuan	1963	200
Jiuzhaigou	Nanping	Sichuan	1978	600
Wanglang	Pingwu	Sichuan	1965	277
Tabgjiahe	Qingchuan	Sichuan	1978	300
Xiaozhaizigou	Beichuan	Sichuan	1979	167
Fengtongzhai	Baoxing	Sichuan	1975	400
Wolong	Wenchuan	Sichuan	1975	2000
Labahe	Tianquan	Sichuan	1963	120
Dafengding	Mabian	Sichuan	1978	300
Dafengding	Meigu	Sichuan	1978	160
Huanglongshi	Songpan	Sichuan	1983	400

Since 1980's, six reserves including Changbaishan Mountain Natural Reserve were listed in international network of Man Biosphere (MAB), providing the bases for joint research in different areas by scientists from home and abroad. The total coverage of different types of reserves accounts for about 2.1 per cent of state's land area. However, most of them are still on a primary stage, characterised by the lack of systematic scientific management although training courses for different purposes had been conducted. In most reserves, bioinventory has not yet been developed. In spite of the problems existed and to be solved, the reserve network will play its important role in biodiversity conservation in the future.

Ex Situ Conservation

Botanical gardens

Botanical gardens provide important bases for *ex situ* conservation of plants, especially endangered and rare plant species. Approximately 104 botanical gardens contribute substantially to plant introduction, domestication, collection and cultivation. So far some of wild forms and relatives of cultivated plants as well as around 300 species listed in *China plant red data book* vol. I have been introduced (Table 26.5).

Table 26.5. Botanical gardens in different Floristic regions.

Floristic Region (Subregion)	Number
Holarctic Region	89
Eurasia Forest Subregion	1
Asia Desert Subregion	7
Eastern Asia Steppe Subregion	3
Qinghai-Xizang (Tibet) Plateau Subregion	1
Sino-Japan Forest Subregion	72
Sino-Himalaya Forest Subregion	5
Palaetropical Region	15
Malay Subregion	15

Many botanical gardens also conduct surveys and experiments of certain introduced rare and endangered species regarding their ecology, reproduction, cultivation and biological characters. Some gardens were established aiming mainly at research and conservation of particular groups, such as Magnoliaceae, Cycadaceae and Zingiberaceae in South China Botanical Garden, Theaceae and Ericaceae in Kunming Botanical Garden, Dipterocartaceae and Myristicaceae in Xishuangbanna Tropical Botanical Garden, and Rosaceae in Beijing Botanical Garden. While some others are engaged in resource sustainable utilisation including research and experiments of introduction, domestication and production of wild economic plants, such as Rhododendron, camelia, orchid, Chinese ginseng, Chinese yam, elevated gastrodia and yangtao, etc. In addition, the establishment and improvement of nationwide network of plant *ex situ* conservation are now in progress.

Zoos and other breeding bases

Although zoos in China are administered by Urban Construction Department with long-standing efforts in planning and education, more attention has been given to research on breeding and reproduction of rare and endangered animals, and studbooks of certain species are now being established. In fact, reproduction and artificial insemination of Giant Panda succeeded in the 1950's and 1970's respectively. Moreover, captive breeding of certain species has also proved promising, such as Manchurian Tiger, South China Tiger, Lesser Panda, Clouded Leopard, Golden Monkey, cranes, eared-pheasants, Chinese Alligator, etc. Meanwhile, a number of breeding farms for wild animals are under construction. These animals include Chinese Alligator, Manchurian Tiger and some primates (Fig. 26.23).

Fig. 26.23. Milu (*Elaphurus davidianus*), an indigenous deer, extinct in the wild for at least 1500 years. Reintroduced first into China in 1985 at Nanhaizi, Beijing and 1986 in Dafeng, Jiangsu.

Reintroduction efforts of some species native to China but already extinct in the wild or recently disappeared as Milu, Przewalski's Horse and Saiga have also been initiated. Two reintroduction projects of Milu appear successful and total population approaches nearly 250. A second stage project-releasing to the wild — is now under preparation. Reintroduction of the extirpated species Saiga is also targeted to the development and sustainable utilisation of Saiga horns as an important and traditional medicinal resource.

Germplasm and gene banks

To preserve genetic material by incorporating latest achievements in cell and molecular biology is another important supplementary measure in conservation of species and genetic diversity. This is especially necessary for some highly endangered species or important domesticated and cultivated species.

Currently, a national germplasm bank has been established to preserve crops and economic forest in China. To preserve germplasm and microbes under low temperature also has a history in the past decades. In animals, 118 kinds of cells belonging to 92 species of higher or lower animals are being preserved.

RECOMMENDATIONS

The conservation and sustainable use of biological diversity are closely related to different aspects of social development. The loss of biodiversity is irreversible. Once a species or natural habitat such as tropical forest is lost it is impossible to recover them. Thus biodiversity conservation is the most urgent issue of environmental conservation in the world as well as in China.

It is essential to have a well designed strategy and action plan in order to effectively carry out biodiversity conservation. The following are some general recommendations.

General Recommendations for Strategy and Actions

Better understanding and evaluating the significance and value of biological diversity. To develop and integrated approach and more effective way of conservation and sustainable use of biological diversity; correct handling of relationship between the conservation and sustainable use of biodiversity, especially in those areas rich in biodiversity where conservation is closely combined with economic development.

Appropriate coordination between different governmental and non-governmental agencies, institutions and organisations which are relevant to biodiversity issues; different agencies and organisations should work closely together and it is desirable that a commission on environment be established under the National People's Congress.

In long-term view it is necessary to work out a general strategy and an action plan for conservation of biodiversity at national level where species and ecosystems of global and national significance, as well as priorities for conservation actions should be determined. Following the strategy and action plan the same should be done at the local or regional.

Long-term plan of rational and sustainable use of wildlife based on status survey of wildlife resource should be determined in order to monitor and regulate the uncontrolled and unsustainable use of wildlife resources still existing in China. Similar plan should also be worked out at the local or regional level. Strengthening the planning and management of protected areas in existence so as to further conserve the habitat and ecosystems and to improve the management of the reserve system throughout the country.

Restoration or enhancement of ecosystem diversity in areas which have been used for agriculture, forestry, husbandry, fishery and logging, etc. so as to improve the ecological process to ensure the long-term and sustainable development.

Assessment of the *ex situ* conservation facilities already in existence including zoos, botanical gardens, breeding facilities, gene or germplasm banks, etc. and working out a strategy and action plan for its improvement to meet the needs of biodiversity conservation.

Assessment and evaluation of the existing education and training facilities and meet the real needs for biodiversity conservation in the coming decade; to work out a long-term plan for education and training of professionals. To work out a long-term basic and applied research in the field of biodiversity conservation. Planning of long-term basic and applied research in the field of biodiversity conservation and sustainable use.

Evaluation and assessment of the implementation and enforcement of the existing law and regulations in this field to improve and strengthen law enforcement procedures and to identify area for further development of national legislation.

Expanding and broadening international cooperation in biodiversity conservation and sustainable use. Evaluating the existing programs—effectiveness, channels, appropriateness, etc. and improvement and further needs. Evaluation of the financial resources so far available domestically and internationally, and estimation of costs of implement the actions in the forthcoming decade.

Scientific Research

Bioinventory-based on the detail information available at present it is fundamental to establish a bioinventory information system at the national and local levels. This information system should be a network well organised throughout the country with headquarters in Beijing and regional offices throughout China.

1. Species diversity database.
2. Systematic collection database.

3. Ecosystem diversity, habitat and reserve database.
4. Conservation database or information system.
5. *Ex situ* conservation database.
6. Continuation and strengthening fauna and flora survey, especially of remote areas and those of less-known taxa.

Status survey of groups with national and global significance and endangerness:

1. Status survey-distribution, population size, utilisation and trade if any; threats and necessary protection measures taken and to be taken, etc.
2. Red list and red data books.
3. Evaluation of law implementation and enforcement.
4. Strategy and action plan for species or species groups.
5. Category of endangerment of species-methodology and ecological basis.

In situ and *ex situ* conservation studies on endangered species—conservation biology, population dynamics, reproductive biology, population genetics, viability analysis, etc.

Biological study on sustainable use of wildlife—wildlife management demonstration projects. Genetic resources-survey and research on their potential usage and development.

Preservation of genetic materials from endangered species, ancestral forms of cultivated plants and domesticated animals. Technology of preservation of cells, embryos, semen, ova, germplasms, seeds, etc. Study on biodiversity conservation and sustainable use combined with integrated development in rural areas. Restoration of ecosystem diversity and integrated development.

Glossary

Adaptations	:	Responses that decrease the negative effects and capitalise on positive opportunities associated with impacts.
Alien species	:	A species occurring in an area outside of its historically known natural range as a result of intentional or accidental dispersal by human activities (also known as an exotic or introduced species).
Benthic	:	Associated with the bottom of a waterbody or the sea.
Bioclimatic range	:	Areas of land that are likely to provide a climate appropriate for a species.
Biodiversity	:	The variety of life forms: the different plants, animals and micro-organisms, the genes they contain, and the ecosystems they form. It is usually considered at three levels: genetic diversity, species diversity and ecosystem diversity.
Biome	:	A major portion of the living environment of a particular region (such as a fir forest or grassland), characterised by its distinctive vegetation and maintained largely by local climatic conditions.
Biotic	:	Associated with living organisms.
Breakaways	:	Land features in lateritic landscape, usually consisting of small escarpments.
Buffer zone	:	The region adjacent to the border of a protected area; a transition zone between areas managed for different objectives.
Carrying capacity	:	The maximum number of people or individuals of a particular species, that a given part of the environment can maintain indefinitely.
Climate envelope	:	The range of climate conditions within which a species or ecological community can be found.
Cohesion	:	The sum total of forces or systems that hold a species together. The term is used especially in the interbreeding and cohesion species concepts. Cohesion mechanisms include isolating mechanisms in sexual species as well as 'stabilising' ecological selection, which may cause cohesion even within asexual lineages.
Conservation	:	The management of human use of nature so that it may yield the greatest sustainable benefit to current generations while maintaining its potential to meet the needs and aspirations of future generations.
Conservation of biodiversity	:	The management of human interactions with genes, species, and ecosystems so as to provide the maximum benefit to the present generation while maintaining their potential to meet the needs and aspirations of future generations; encompasses elements of saving, studying and using biodiversity.

481

Cultural diversity	:	Variety or multiformity of human social structures, belief systems and strategies for adapting to situations in different parts of the world. Language is a good indicator of cultural diversity, with over 6000 languages currently being spoken.
Detectors	:	Species occurring naturally in an area of interest that may show measurable responses to environmental change, such as changes in distribution or behaviour.
Detritivores	:	Consumers of detritus or dead biological materials.
Dinoflagellate algae	:	Mobile planktonic algae with more than one flagellum (for propulsion), very important in marine and freshwater ecosystems.
Disruptive selection	:	Selection acting to preserve extreme phenotypes in a population. Speciation usually involves disruptive selection, because intermediates (hybrids between incipient species) are disfavoured.
DNA bar coding	:	A means of delimiting species via DNA sequence clustering, usually from mitochondrial DNA.
Ecology	:	A branch of science concerned with the interrelationship of organisms and their environment; the study of ecosystems.
Ecosystem	:	Ecosystems are self-regulating communities of plants and animals interacting with each other and with their non-living environment—forests, wetlands, mountains, lakes, rivers, deserts and agricultural landscapes. Ecosystems are vulnerable to interference as pressure on one component can upset the whole balance. They are also very vulnerable to pollution. Many ecosystems have already been lost, and many others are at risk. The world's forests house about half of global biodiversity. But they are disappearing at a rate of 0.8 per cent per year. Tropical forests are vanishing at an annual rate of 4 per cent.
Ecosystem approach	:	The ecosystem approach is a strategy for the integrated management of land, water and living resources that promotes conservation and sustainable use in an equitable way. The Ecosystem Approach places human needs at the centre of biodiversity management. It aims to manage the ecosystem, based on the multiple functions that ecosystems perform and the multiple uses that are made of these functions. The ecosystem approach does not aim for short-term economic gains, but aims to optimise the use of an ecosystem without damaging it.
Ecosystem diversity	:	The variety of ecosystems that occurs within a larger landscape, ranging from biome (the largest ecological unit) to microhabitat.
Ecosystem services	:	Ecosystem services are processes by which the environment produces benefits useful to people, akin to economic services. They include: (i) provision of clean water and air, (ii) pollination of crops, (iii) mitigation of environmental hazards, (iv) pest and disease control and (v) carbon sequestration.
Ecotone	:	Transition zone between two or more vegetation communities, such as between grassland and forest.
Ecotourism	:	Travel undertaken to witness sites or regions of unique natural or ecologic quality, or the provision of services to facilitate such travel that have the least impact on biological diversity and the natural environment.
ElNiño	:	A combined ocean and atmosphere pattern which results in an increase in sea-surface temperatures in the eastern Pacific off the east coast of North America but reduced sea-surface temperatures around Australia, resulting in lower than average rainfall and a smaller likelihood of cyclones in Australia.

Endangered species	:	A technical definition used for classification referring to a species that is in danger of extinction throughout all or a significant portion of its range. IUCN The World Conservation Union defines species as endangered if the factors causing their vulnerability or decline continue to operate.
Endemic species	:	Species restricted to a particular geographic region or locality.
Endemism	:	Occurrence of endemic species.
Endothermic	:	Capable physiologically of maintaining a constant body temperature, as humans do; *ecto*thermic species have body temperatures that change with the temperature of their environment.
ENSO	:	El Niño Southern Oscillation.
Evolution	:	Any gradual change. Organic evolution is any genetic change in organisms from generation to generation.
Exclosure	:	A fence around a patch of ground, meant to exclude animals.
Ex situ conservation	:	A conservation method that entails the removal of germplasm resources (seed, pollen, sperm, individual organisms, from their original habitat or natural environment. Keeping components of biodiversity alive outside of their original habitat or natural environment.
Extinction	:	The evolutionary termination of a species caused by the failure to reproduce and the death of all remaining members of the species; the natural failure to adapt to environmental change.
Fauna	:	All of the animals found in a given area.
Flora	:	All of the plants found in a given area.
Gene	:	The functional unit of heredity; the part of the DNA molecule that encodes a single enzyme or structural protein unit.
Genetic	:	Determined by or relating to genes.
Gene flow	:	Movement of genes between populations, usually via immigration and mating of whole genotypes, but sometimes single genes may undergo horizontal gene transfer via transfection by micro-organisms.
Gene bank	:	A facility established for the *ex situ* conservation of individuals (seeds), tissues, or reproductive cells of plants or animals.
Gene pool	:	The sum total of the genetic variation within a reproductively isolated species population; this term is mostly used by supporters of the interbreeding species concept.
Genomic cluster	:	A synonym for genotypic cluster.
Genotypic cluster	:	In a local area, a single genotypic cluster (or species) is recognised if there is a single group of individuals recognisable on the basis of multiple, unlinked inherited characters or genetic markers. A pair of such genotypic clusters (or species) is recognisable if the frequency distribution of genotypes is bimodal. Within each genotypic cluster in a local region, allele frequencies will conform to Hardy–Weinberg equilibrium, and the different unlinked loci will be in approximate linkage equilibrium. The presence of more than one species or genotypic cluster can then be inferred if the distribution of genotypes is bimodal or multimodal, and strong heterozygote deficits and linkage disequilibria are evident between the clusters.
Genetic diversity	:	The variety of genes within a particular populaton, species, variety, or breed.

Genetic paucity	:	Very little genetic variation.
Global warming	:	Anticipated increase in global temperatures resulting from human-induced emissions of greenhouse gases.
Greenhouse climate change	:	Changes in global and regional temperatures, rainfall and other climate-related factors as a result of human-induced net emissions of greenhouse gases.
Greenhouse gases	:	Gases such as carbon dioxide, methane and nitrous oxide, which act as a 'blanket' in the atmosphere, keeping in more outgoing radiation than there would otherwise have been, maintaining higher temperatures in the lower atmosphere and at the surface of the land and oceans.
Habitat	:	A place or type of site where an organism or population naturally occurs.
Habitat degradation	:	The diminishment of habitat quality, which results in a reduced ability to support flora and fauna species. Human activities leading to habitat degradation include polluting activities and the introduction of invasive species. Adverse effects can become immediately noticeable, but can also have a cumulative nature. Biodiversity will eventually be lost if habitats become degraded to an extent that species can no longer survive.
Habitat generalists	:	Species not confined to their range by a specific habitat.
Habitat loss	:	The outcome of a process of land use change in which a 'natural; habitat-type is removed and replaced by another habitat-type, such as converting natural areas to production sites. In such process, flora and fauna species that previously used the site are displaced or destroyed. Generally this results in a reduction of biodiversity.
Habitat specialists	:	Species specialised to suit a particular habitat or niche.
Hotspot	:	An area on earth with an unusual concentration of species, many of which are endemic to the area, and which is under serious threat by people.
Hypo-osmotic	:	A situation in which sea-water is less saline than usual, so that relatively fresh-water diffuses into the cells of sea-living organisms by osmosis, damaging them.
Impact	:	An effect of increased greenhouse gas emissions and atmospheric CO_2 concentration (and other global changes) on physical and biological processes in both natural and managed systems.
Indicator	:	An environmental indicator is a species or group of species that responds predictably and sensitively, in ways that are readily observed and quantified, to an environmental disturbance.
Indicator species	:	A species whose status provides information on the overall condition of the ecosystem and of other species in that ecosystem.
Indigenous people	:	People whose ancestors inhabited a place or country when persons from another culture or ethnic background arrived on the scene and dominated them through conquest, settlement or other means and who today live more in conformity with their own social, economic, and cultural customs and traditions than with those of the country of which they now form a part.
In situ conservation	:	A conservation method that attempts to preserve the genetic integrity of gene resources by conserving them within the evolutionary dynamic ecosystems of the original habitat or natural environment.
IPCC	:	Intergovernmental Panel on Climate Change.
Intellectual property rights	:	Rights enabling an inventor to exclude imitators from the market for a certain period of time.

Invasive species	:	Invasive species are those that are introduced intentionally or unintentionally to an ecosystem in which they do not naturally appear and which threaten habitats, ecosystems or native species. These species become invasive due to their high reproduction rates and by competing with and displacing native species, that naturally appear in that ecosystem.
Inventory	:	On-site collection of data on natural resources and their properties.
La Niña	:	A combined ocean and atmosphere pattern which leads to an increase in seasurface temperatures in the western Pacific and around Australia, resulting in higher than average rainfall and a greater likelihood of cyclones in Australia.
Land use	:	Land use refers to how a specific piece of land is allocated: its purpose, need or use (e.g. agriculture, industry, residential or nature).
Land use requirements	:	The requirements are related to growth and yield of crops and trees, animal husbandry, land management and conservation.
Mining of data	:	Searching existing databases for patterns and relationships not part of the original data-collection scheme.
Millennium Ecosystem Assessment (MA)	:	An international work program designed to meet the needs of decision-makers and the public for scientific information concerning the consequences of ecosystem change for human well-being and options for responding to those changes.
Monophyletic	:	A grouping that contains all, of the descendants of a particular node in a phylogeny. Monophyly is the state of such groupings. Compare paraphyletic polyphyletic. Butterflies (*Rhopalocera*) and birds (*Aves*) are examples of two groups thought to be monophyletic.
Monotonically	:	Steadily, in one direction.
Mitigating measures	:	Measures that allow an activity with a negative impact on biodiversity, but reduce the impact on site by considering changes to the scale, design, location, process, sequencing, management and/or monitoring of the proposed activity. It requires a joint effort of planners, engineers, ecologists, other experts and often local stakeholders to arrive at the best practical environmental option. An example is the unacceptable impact on biodiversity of the construction of a certain road, that is mitigated by the construction of a wildlife viaduct.
Mutualistic symbiosis	:	Members of two different species living together, so that each benefits from the presence and metabolic activities of the other.
Native species	:	Flora and fauna species that occur naturally in a given area or region. Also referred to as indigenous species.
Natural environment	:	The natural environment comprises all living and non-living things that occur naturally on earth. In its purest sense, it is thus an environment that is not the result of human activity or intervention. The natural environment may be contrasted to 'the built environment', and is also in contrast to the concept of cultural landscape.
Net radiative forcing	:	Increase in energy exchange at the earth's surface caused by an imbalance between incoming and outgoing radiation.
Overexploitation	:	Overexploitation occurs when harvesting of specimens of flora and fauna species from the wild is out of balance with reproduction patterns and, as a consequence species may become extinct.
Palaeoecology	:	Reconstruction of the relationships between past organisms and their environments, usually studied via fossils, pollen and charcoal in layers of soil.

Parapatric	:	Having geographic or spatial distributions that abut but do not overlap.
Paraphyletic	:	A grouping that contains some, but not all, of the descendants of a particular node in a phylogeny. Paraphyly is the state of such groupings. Compare monophyletic, polyphyletic. Moths (*Lepidoptera*, excluding butterflies) and reptiles (amniotes, excluding birds and mammals) are examples of two groups thought to be paraphyletic.
Phenetic	:	A classification or grouping based purely on overall similarity. Pheneticists use matrices of overall similarity rather than parsimony to construct a 'phenogram' as an estimate of the phylogeny. Examples of phenetic methods of estimation include unweighted pair group analysis (UPGMA) and neighbour joining. 'Cladists' reject phenetic classifications on the grounds that they may result in paraphyletic or polyphyletic groupings.
Phylogenetic	:	Pertaining to the true (i.e. evolutionary) pattern of relationship, usually expressed in the form of a binary branching tree or phylogeny. If hybridisation produces new lineages, as is common in many plants and some animals, the phylogeny is said to be 'reticulate'. Phylogenies may be estimated using phenetics, parsimony (cladistics) or methods based on statistical likelihood.
Propagule	:	Seed, spore, tuber, egg — something from which new life germinates or hatches.
Patent	:	A government grant of temporary monopoly rights on innovative processes or products.
Protected areas	:	An area of land and/or sea especially dedicated to the protection and maintenance of biological diversity, and of natural and associated cultural resources, and managed through legal or other effective means. A protected area can be under either public or private ownership.
Red list	:	The IUCN Red List of Threatened Species provides taxonomic, conservation status and distribution information on taxa that have been globally evaluated using the IUCN Red List Categories and Criteria. This system is designed to determine the relative risk of extinction, and the main purpose of the IUCN Red List is to catalogue and highlight those taxa that are facing a higher risk of global extinction (i.e. those listed as Critically Endangered, Endangered and Vulnerable). The IUCN Red List also includes information on taxa that are categorised as Extinct or Extinct in the Wild; on taxa that cannot be evaluated because of insufficient information (i.e. are Data Deficient); and on taxa that are either close to meeting the threatened thresholds or that would be threatened were it not for an ongoing taxon-specific conservation program (i.e. are Near Threatened).
Rehabilitation	:	The recovery of specific ecosystem services in a degraded ecosystem or habitat.
Restoration	:	The return of an ecosystem or habitat to its original community structure, natural complement of species, and natural functions.
Riparian zone	:	Riverside zone of varying width, depending on context.
Seedbank	:	A facility designed for the *ex situ* conservation of individual plant varieties through seed preservation and storage.
Sentinels	:	Sensitive organisms introduced into the environment as an early-warning device.
Sibling species	:	A pair of closely related, morphologically similar species (usually sister species).
Signal-to-noise ratio	:	Ratio between the value that is being measured and interference from background occurrences.
Speciation	:	The evolutionary process of the origin of a new species.

Specific mate recognition systems (SMRSs)	:	Fertilisation and mate recognition systems in the recognition concept of species, the factors leading to premating compatibility within a species.
Species	:	A group of organisms capable of interbreeding freely with each other but not with members of other species.
Species diversity	:	The number and variety of species found in a given area in a region.
Succession	:	The more or less predictable changes in the composition of communities following a natural or human disturbance.
Stabilising selection	:	Selection which favours intermediate phenotypes.
Sustainable development	:	Development that meets the needs and aspirations of the current generation without compromising the ability to meet those of future generations.
Sustainable use	:	The use of components of biological diversity in a way and at a rate that does not lead to the long-term decline of biological diversity, thereby maintaining its potential to meet the needs and aspirations of present and future generations.
Symbionts	:	Species that live together closely.
Symbiosis	:	The situation of species living together closely.
Taxonomy	:	The classification of animals and plants based upon natural relationships.
Taxonomic inflation	:	The process whereby the numbers of species in the checklist of a group increases due to a change in species concept rather than due to new discoveries of previously unknown taxa.
Threatened species	:	A technical classification referring to a species that is likely to become endangered within the foreseeable future, throughout all or a significant portion of its range.
Threatening processes	:	Situations, usually caused by human activities, that reduce the quality of a habitat for a species, e.g. land-clearing, presence of feral animals.
Translocation	:	Moving a species into a new area of suitable habitat which the species could not have reached without human assistance.
Wild species	:	Organisms captive or living in the wild that have not been subject to breeding to alter them from their native state.
Wildlife	:	Living, nondomesticated animals. Some experts consider plants also as part of wildlife.

References

Allen, W. and Lloyd, T., *Basic Concepts of Ecology*, Butterworths, London.

Andrew, D., *Global Biodiversity*, Marcel Dekker Inc., New York.

Baker, J.M., *Population Ecology*, Applied Science Publishers, London.

Benn, R.L., *Principles of Conservation Biology*, Springfield, New York.

Blackman, G.W., *Creative Conservation*, Academic Press, London.

Brill, G., *Grassland Ecosystems*, Cambridge University Press, Cambridge.

Cavallaro, S., *Biodiversity*, Pergamon Press, New York.

Connell, D.W., *Growing Threats of Species Invasions*, John Wiley & Sons, New York.

Coolingwood, R.T., *Essentials of Conservation Biology*, John Wiley & Sons, New York.

Curtis, C. and Hawkes, H.A., *Ecological Aspects of Air Pollution*, Academic Press, London.

Dix, N.M., *Biodiversity and Ecosystem Function*, John Wiley & Sons, New York.

Dugan, P.P., *Animal Ecology*, Plenum Publishing Corporation, London.

Fruh, E. and Eckenfelder, W.W., *Biodiversity and Conservation*, University of Texas Press, Austin.

Gascon, Tyler, E.G. and Wayne, A.P., *Trends in Ecology and Evolution*, McGraw-Hill, New York.

Golterman, E.D., *Threatened Birds of the World*, Gordon and Breach, Science Publishers, New York.

Hardin, C.M., *Community Ecology in the Changing World*, Castle Housing Publications Ltd., London.

Jim, A. and Evison L., *Biodiveristy: An Introduction*, John Wiley & Sons, New York.

Kenn, R.L., *Population Ecology, Species*, Harwood Academic Publishers, New York.

Lenihan, F.G., *Species Extinction*, National Academy of Sciences, Washington DC.

Levin, S., *Encyclopedia of Biodiversity*, Academic Press, London.

Mark, M., *Species Diversity in Ecological Community*, John Wiley & Sons, New York.

McCaull, J.J., *Species Diversity in Space and Time*, Harcourt Brace Jovanovich, New York.

Michael, J., *Biodiversity and Conservation*, Routledge, Canada.

Neil, J., *Biodiversity Under Threat*, Applied Science Publishers, London.

Palmer, E.P., *Fundamentals of Ecology*, Saunders, Philadelphia.

Smith, T.E., *Ecological Methods*, Chapman and Hall, London.

Stumm, C., *Tropical Moist Forests*, Applied Science Publishers, London.

Tayler, H.Y., *Conservation Biology*, Van Nostrand Reinhold, New York.

Vinogradov, N.L., *Functional Roles of Biodiversity*, Addison-Wesley Publishing Company, Philippines.

Watts, B.G., *Ecology and Tropical Biology*, Chilton Book Company, Radnor, Pennsylvania.

William, M., *The Diversity of Life*, Chapman and Hall, London.

Index

A

Abiotic and biotic factors, 183
Accessing uncultivated microbes, 373
Achieving a consensus classification of organisms, 258
Adapting to climate change, 222
Addressing biodiversity loss, 446
Aesthetic value, 13
Aggregate effects of diversity, 92
Agricultural biodiversity, 191
Agricultural biodiversity and biotechnology, 200
Agricultural innovations and their impact on agrobiodiversity, 224
Agricultural relevance, 25
Agriculture and Mitigation, 223
Agriculture and the conservation of biodiversity, 49
Agroforestry, 394
Alley cropping, 395
Allopatric speciation, 124
Alpha diversity, 161
Alternative methods for eliminating pesticides, 299
Alternative species concepts, 74
Alternatives to timber management for biodiversity conservation, 61
Altruism in animals, 136
Ancient forestry practices, 390
Animal dispersal mechanisms, 109
Animal diversity, 456
Animal natality, 103
Animal populations, 402
Applications of biotechnology and its effect on biodiversity, 196
Aquaria, 40
Archaebacteria, 325
Artificial speciation, 123
Assessment of biodiversity, 6

B

Bacterial flora in mangrove ecosystems, 329
Bacterial life in extreme environments, 327
Barriers and challenges, 376

Basic information on salt marsh, 311
Berger-Parker index, 180
Beta diversity, 163
Beta diversity in the strict sense, 163
Biodiversity, 1, 333
Biodiversity and ethics: do we have a responsibility to preserve 13
Biodiversity as a hierarchy, 279
Biodiversity conservation and traditional agroecosystems, 48
Biodiversity contribution to Indian economy, 150
Biodiversity database initiatives in India, 260
Biodiversity hotspots, 149
Biodiversity in a life-zone context, 57
Biodiversity in space and time, 84
Biodiversity informatics standards and protocols, 259
Biodiversity informatics, 257
Biodiversity loss and the use of pesticides, 299
Biodiversity of culturable bacteria, 367
Biodiversity, desertification and climate interactions in drylands, 273
Biodiversity: West Asia, 444
Biofuels and agrobiodiversity, 226
Biogeographic classification of India, 146
Biogeography, 454
Biological dispersal, 107
Biotic exchange, 286
Biotic hypotheses, 176
Bird poisonings caused by pesticides, 302
Bird species decline owing to pesticides, 301
Black pepper, 263
Botanical gardens, 477

C

Calculating alpha diversity, 162
Calculating gamma diversity, 164
Calculating population growth, 101
Calculation of diversity, 158
Calculations where natality is a factor, 104

Cancún initiative and declaration of like-minded megadiverse countries, 189
Cartagena protocol on biosafety, 193
Causes of the biodiversity loss, 458
Characteristic patterns of marine biodiversity, 338
Chemoautotrophic eubacteria, 326
Chemoheterotrophic eubacteria, 326
Climate change, 32
Clumped distribution, 181
Coastal and marine ecosystems, 451
Coextinction, 97
Cohesion concept of species, 77
Collection techniques, 332
Combined species concepts, 77
Commercial forest management, 62
Community and ecosystem diversity, 336
Conservation of biodiversity, 3
Conservation strategies, 196
Conserving biodiversity, 437
Construction of phylogenetic trees, 380
Consumptive use value, 12
Continental drift, 137
Convention on biological diversity (CBD), 193
Convergent plate boundaries, 414
Coping with poor genetic diversity, 25
Coral bleaching, 431
Coral reefs, 148
Coral-reef parks, tourists and fishermen, 60
Critically endangered, 432
Critiques of hotspots, 188
Crop diversity, 230, 234
Culture based molecular techniques, 378
Curcuma species, 263
Current biodiversity informatics activities, 259
Current biodiversity informatics issues, 258
Current pressures on biodiversity, 449

D

Darwin's morphological species criterion, 72
Darwinian species criteria, 72
Darwinism, 131
Definition and measurement of microbial diversity, 350
Deforestation, 388
Density and dispersion, 104
Desert, 464
Development planning and biodiversity conservation, 70
DGGE-PCR amplification and 16S rRNA sequencing, 379
Diagnostic species concept, 76
Different alpha diversity concepts, 162
Different gamma diversity concepts, 164
Direct gene transfer to crops and farm animals, 197

Direct microbial DNA isolation from water and soil, 379
Dispersal mechanisms, 109
Divergent plate boundaries, 413
Diversified planting, 299
Diversity indices, 159
Driving forces of biodiversification in drylands, 267
Dry forests, 384
Dryland biodiversity status and trends, 267

E

Earth summit, 5
Ecological effects of biodiversity, 20
Ecological species concept, 74
Ecology, 407
Economics of fire management policy, 411
Ecosystem approaches, 66
Ecosystem effects of fire, 400
Ecosystem engineering, 88
Ecosystem science, 65
Ecosystem uses and forest products, 8
Ecotourism, 61
Effect of pesticides on plant communities, 306
Effects of fire on ecosystems, 402
Effects of fire on forest fauna, 409
Effects of fire on plants and animals: individual level, 400
Effects of fishing and other forms of over exploitation, 342
Effects of suppression of the natural fire regime, 410
Effects of volcanoes, 416
Effects on biota, 296
Effects on community productivity, 21
Effects on community stability, 23
Eight ways to improve biodiversity forecasting, 419
Endangered wildlife, 389
Endemism, 467
Environmental benefits, 388
Environmental genomics, 374
Essence of biodiversity, 190
Eubacteria, 325
Evolutionary and lineage concepts, 77
Evolutionary diversification, 27
Ex situ conservation, 275, 477
Ex situ conservation (outside natural habitats), 155
Examples of 16S rRNA sequencing, 380
Extent of microbial diversity on earth, 363
Extinction, 93
Extraction of non-timber products, 61

F

Farmers and the future of agrobiodiversity, 221
Farm-scale diversity, 230, 235
Fatty acid analysis, 378

Fire behaviour, 401
Fire ecology, 401
Fire regimes, 403
Fire-adapted fauna, 410
Fish and shellfish habitat, 316
Fish habitat in restored marsh, 316
Flagship and keystone species, 4
Floral diversity: bacteria and fungi, 322
Fluctuation in population size, 104
Forest dynamics, 8
Forest reserves, 394
Forestry issues, 396
Forestry research, 392
Freshwater and marine ecosystems, 465
Functional diversity, 336
Fungal biodiversity: isolation and identification, 373
Fungal flora in mangrove ecosystems, 332

G

Gaia hypothesis, 138
Gamma diversity, 164
Genealogy, 75
General recommendations for strategy and actions, 479
Generality of the latitudinal diversity gradient, 177
Genetic divergence, 124
Genetic diversity, 24, 334, 449
Genetic modification and the sourcing of genes, 206
Genetic pollution, 31, 96
Genetically engineered (GE) crops, 230
Genetically modified (GM) crops, 233
Genetics of reproductive isolation barriers, 116
Genomic DNA extraction, 378
Genotypic cluster or genomic cluster criterion, 80
Germplasm and gene banks, 478
Global action for agriculture, 228
Global change and biodiversity scenarios in grasslands, 280
Global climate change, 341
Global list of all species, 258
Global system on plant genetic resources, 210
Global warming, 395
Global warming effects on biodiversity and animals, 429
Global warming effects on various seasonal processes of plants and animals, 430
Governance and agriculture, 227
Gram-negative bacteria, 327
Gram-positive bacteria, 326
Grasslands, 147
Grazing by domestic herbivores, 281
Guanine and cytosine (GC) per cent, 379

H

Habitat and ecosystem diversity, 461
Habitat degradation and loss, 444
Habitat degradation, 96
Habitat degradation, fragmentation and loss, 340
Habitat destruction, 30, 473
Habitat diversity, 337
Harvest-regeneration methods, 391
Holistic approach and soil functioning, 359
Holocene extinction, 32
Hotspot conservation initiatives, 186
Hotspots of biodiversity, 3
Human influence on wildland fire, 403
Human perceptions of the oceans, 345
Human population expansion, 192
Hunting and fishing reserves, 58
Hybrid gender: Haldane's rule, 116
Hybridisation, genetic pollution/erosion and food security, 31

I

Identification method, 329
Identification using diagnostic kits, 332
Identification using microbial identification system (MIS), 332
Impact of pesticides on butterflies, bees and natural enemies, 304
Impact of pesticides on wildlife populations and species diversity, 301
Impacts of biofuels on agrobiodiversify, 226
Impacts of human-induced or severe natural wildfire on plant diversity, 404
Impacts on forests and wildlife, 430
Importance of biodiversity, 256
Importance of genetic diversity, 24
Importance of microbial diversity, 364
Importance of species concepts for biodiversity and conservation, 83
Importance of tropical forests, 382
In situ and *ex situ* conservation, 3
In situ conservation, 275
In situ conservation (within natural habitat), 41, 152
In situ conservation-reserves, 476
Increase in forest cover and density, 437
Increases in extreme events, 431
Increasing pressure of agriculture on habitats and biodiversity, 300
Indicator traits, 133
Indices that measure diversity, 178
Indirect indicators, 232

Information and biodiversity conservation, 70
Inland water ecosystems, 451
Innovations and impacts, 225
Integrating conservation and development, 4
Integrating park and regional planning through an ecosystem approach, 64
Introduced and invasive species, 30
Introduction to Asia and the Pacific, 438
Isolate Y1L-yeast, 381
Isolation mechanisms that occur after breeding or copulation (postzygotic isolation), 114
Isolation mechanisms that occur before breeding or copulation (pre-zygotic isolation), 112
Isolation method, 329
Isolation of population, 124
Isolation of pure colonies of micro-organisms, 378
Isolation strategies, 369

K

Key biodiversity challenges in Asia and the Pacific, 439
Keystone species, 185

L

Land reclamation, 313
Land use and land cover changes, 7
Land use effects on biodiversity, 281
Land-capability assessment, 57
Landscape-scale diversity, 231, 239
Larus gulls, 130
Latin America and the Caribbean, 446
Latitudinal gradients in species diversity, 173
Latitudinal pattern of diversity, 338
Legal framework of biodiversity conservation, 345
Leisure, cultural and aesthetic value, 29
Less common species as keystone species, 91
Life histories and the structure of populations, 99
Life tables and the rate of population growth, 100
Linkages between policies, institutions and processes within the framework, 220
Livelihood outcomes, 217
Livelihood strategies, 217
Living fences, 395
Lodgepole pine, 403
Longitudinal pattern of tropical diversity, 338
Loss of agrobiodiversity: plants and animals for food and agriculture, 208
Loss of biodiversity and conservation, 194
Loss of food, 410
Loss of habitat, territories and shelter, 409
Loss of terrestrial species, 446

Losses of biodiversity, 36
Lost of biodiversity, 457
Low toxicity pesticides, 299

M

Macro-micro linkages, 213
Major biodiversity threats, 144
Major problems with biodiversity conservation, 144
Management of nature reserves, 42
Mangroves, 148
Marine and coastal ecosystems, 448
Marine bacterial diversity, 324
Marine fungal diversity, 328
Marine microbial diversity, 323
Marsh, 464
Mass extinctions, 97
Meadow, 464
Measures of genetic diversity, 26
Megadiverse countries, 189
Megadiversity, 465
Microbial and biochemical functions in soil, 354
Microbial biodiversity investigation techniques, 376
Microbial diversity and soil functions, 357
Microbial diversity in soil, 353
Microbial diversity on earth, 363
Mimicking the organism's habitat, 370
Mobilising primary biodiversity information, 258
Modern evolutionary synthesis, 136
Modern extinctions, 98
Modes of population growth, 106
Molecular techniques, 351
Monitoring the dynamics of biodiversity, 7
Monophyly, 75
Mosaic patterns, 402
Mosquito control, 314
Multiple faces of biodiversity, 55
Multiple mechanisms, 115
Multiple-use forestry, 392
Mutualists, 185

N

Natality in humans, 104
Natality in population ecology, 103
National mission for a green India, 436
National mission for sustaining the himalayan ecosystem, 436
Native biodiversity and biotechnology, 198
Native biodiversity, 190
Natural speciation, 121
Need and potential for private biodiversity conservation, 54

Need for a biodiversity rescue plan, 309

Negative impacts of pesticides on food sources of birds, 302

New directions in tropical forestry, 395

Nitrogen inputs and atmospheric CO_2 increases, 285

Nitrogen loading, 313

Nonculturable technique, 379

Nucleotide sequence accession number, 379

Nutmeg, 264

Nutrient cycling, 86

O

Operation of reproductive isolation, 128

Option value, 13

Origin of species, 110

Other marine biodiversity patterns, 339

Other meanings of beta diversity, 163

Other measures of diversity, 26

Overall suggestions for the new strategic plan, 442

Overexploitation, 31, 473

Over-hunting, 9

Overpopulation, 32

P

Parapatric speciation, 128

Pesticides affecting amphibians and aquatic species, 305

Phenetic species concept, 79

Philosophisation of species, the 'interbreeding' concept, 72

Phospholipid fatty acid (PLFA) analysis, 353

Phyletic diversity, 335

Physical and chemical nature of fire, 401

Physical separation, 372

Phytophthora spp., 264

Pillars for a plan of action, 276

Placement and design of nature reserves, 10

Plant collections, 41

Plant diversity, 455

Plant natality, 103

Plant populations, 402

Plantation forestry, 393

Plate and direct counts, 350

Plate tectonics and hotspots, 413

Policies and methods for biodiversity conservation, 308

Policies, institutions and processes (PIPs), 216

Policy and research needs, 227

Policy innovations, 225

Political will and biodiversity conservation, 69

Pollution and marine litter, 343

Polytypic species, 72

Population bioenergetics, 107

Population density and growth, 99

Postcopulation or fertilisation isolation mechanisms in plants, 119

Postzygotic isolating mechanisms, 111

Practice of forestry, 390

Predation, competition and disease, 97

Predators, 185

Prezygotic mechanisms, 111

Private conservation, 59

Problems with genus and species scientific names as unique and persistent identifiers, 258

Prodigality of production, 131

Production of forest products in developing countries, 383

Productive use value, 12

Productivity and stability as indicators of ecosystem health, 21

Protected sites or reserves, 41

Protecting and restoring biodiversity, 10

Pseudoextinction, 94

R

Rainforests, 384

Random distribution, 183

Rare species and ecosystem functioning, 89

Reasons for deforestation, 388

Recognition concept of species, 74

Recommendations on development of the updated strategic plan of the convention for the post-2010 period, 442

Recovering biodiversity using environment DNA, 373

Recovery of threatened species, 43

Reduction of biodiversity, 194

Regional biodiversity trends for the twenty-first century, 450

Regular or uniform distribution, 182

Reintroduction of species to the wild poses several different problems, 40

Relative species abundance, 165

Rényi entropy, 180

Replacement of grassland by other land cover types, 283

Reproductive isolation, 111

Restoration of the marsh, 315

Resuscitation and avoiding the stress of cultivation, 372

Ring species, 129

Rises in concentrations of carbon dioxide, 431

Risk to mammals of hazardous pesticides, 303

S

Savanna, 464

Screening environmental libraries, 375

Sea-level rise, 431

Sexual selection, 132

Sexual selection in primates, 135
Shade cropping, 395
Shannon's diversity index, 179
Shifts in climatic envelopes, 430
Simpson's diversity index, 178
Single-strand-conformation polymorphism analysis and DNA microarrays, 380
Social concerns and biodiversity conservation, 67
Social values, 12
Socioeconomic importance, 383
Soil as a microhabitat, 348
Soil conditions, 402
Special features of dryland biodiversity, 268
Speciation, 120
Species concepts based on history, 75
Species conservation, 475
Species distribution, 180
Species distribution model, 183
Species diversity and biodiversity, 19
Species diversity, 158, 334, 465
Species evenness, 161, 178
Species introductions/invasions, 344
Species loss rates, 30
Species populations and extinction risks, 449
Species richness, 160, 177
Species traits, 85
Species, 2
Species-area curve, 171
Specific suggestions for the strategic plan, 443
Sperm competition, 134
Status of biodiversity in India, 146
Status of dryland biodiversity, 268
Strategy and vision for reducing biodiversity loss, 452
Support crops, 395
Suppressing weedy bacteria, 371
Survivorship curves, 100
Sustainable forestry, 391
Sympatric speciation, 127

T

Taungya system, 395
Taxonomic and size relationships, 34
Technical and technological innovations, 225
Technical aspects of agricultural biotechnology at the interface with agrobiodiversity, 205
Temporal variability in species abundance, 92
Terrestrial ecosystems, 447, 451

Theory and preliminary effects from examining food webs, 24
Threatened habitats, 4
Threatened species, 3
Threatening reports on hazardous effects of pesticides, 298
Threats to biodiversity, 9
Threats to dryland ecosystems and species diversity, 272
Threats to marine biodiversity, 339
Threats to traditional agroecosystems, 50
Tidal flooding and vegetation zonation, 312
Trends in species diversity, 160
Trends in species richness, 160
Tropical forestry, 393
Tropical rainforests—four ways to stop deforestation, 397
Types of diversity, 161
Types of tropical forests, 384

U

Understanding relative species abundance patterns, 167
Unreality of species in space and time, 82
UV-B radiation, 342

V

Value of biodiversity, 11
Value of tropical forests, 385
Visual quality/aesthetics in restored marsh, 321
Visual quality/aesthetics under present conditions, 321
Visual quality/aesthetics, 321
Volcanic features, 414
Volcanoes, 413
Vulnerability context, 215

W

Watershed alteration and physical alterations of coasts, 344
Wetlands, 148
Wildfire modelling, 412
Wildfire, 404
Wildlife habitat, 318
Wildlife habitat in a restored marsh, 318
Wildlife habitat under existing conditions, 318
Windbreaks, 395
Wood and other products, 386

Z

Zoos and other breeding bases, 477
Zoos, 39